Alan Aylward

D1823579

Non-LTE Radiative Transfer in the Atmosphere

SERIES ON ATMOSPHERIC, OCEANIC AND PLANETARY PHYSICS

Editor: F. W. Taylor (University of Oxford)

Published:

Series on Atmospheric, Oceanic and Planetary Physics — Vol. 3

Non-LTE Radiative Transfer in the Atmosphere

M. López-Puertas

Instituto de Astrofísica de Andalucía,
CSIC, Spain

F. W. Taylor

University of Oxford, UK

World Scientific
Singapore • New Jersey • London • Hong Kong

Published by

World Scientific Publishing Co. Pte. Ltd.

P O Box 128, Farrer Road, Singapore 912805

USA office: Suite 1B, 1060 Main Street, River Edge, NJ 07661

UK office: 57 Shelton Street, Covent Garden, London WC2H 9HE

British Library Cataloguing-in-Publication Data
A catalogue record for this book is available from the British Library.

ISBN 981-02-4566-1

Printed in Singapore by Mainland Press

Preface

Atmospheric science involves three main sub-disciplines: radiation, chemistry, and dynamics. Many books have been written about each of these, and non-LTE (we shall come in a moment to what it means) plays a part in most of them. However, no book exists which deals primarily with non-LTE, and this has become a serious omission. It is now realised that the upper part of the atmosphere is an important part of the whole system, as well as the scene of some complex and fascinating physical processes. It is also clear that we cannot make measurements of the middle and upper atmosphere by remote sensing from satellites, nor produce meaningful computerised models of the region, without taking detailed account of non-LTE processes. The need for a basic text which deals with the phenomenon and its related processes is therefore quite obvious.

Non-LTE is shorthand for non local thermodynamic equilibrium. It is often abbreviated NLTE, but this can be misleading; it refers to an absence of local thermodynamic equilibrium, not at all to thermodynamic equilibrium which is not local. Pedants sometimes write it as non-(local thermodynamic equilibrium)! It arises when the 'internal' temperature of a representative molecule, as determined by the statistics of the relative populations of the vibrational and rotational energy levels, becomes different from the 'external' temperature, as determined by the statistics of the velocity distribution of the many molecules making up a parcel of gas. Whether or not any difference exists between these two concepts of the 'temperature' of the gas depends primarily on the mean rate at which collisions take place between the individual molecules, which in turn depends mainly on the number density, and so on the pressure and temperature of the sample. Since the pressure dependence usually dominates, non-LTE effects in atmospheres are mainly of importance only at the higher altitudes, for example in the stratosphere and above on the Earth.

The basic concept of LTE and its breakdown is often quite difficult for someone new to atmospheric research to grasp. Since we intend this book to be useful to such a beginner, as well as to the experienced scientist wanting to be brought right up-to-date, we devote quite a lot of space in the first chapter to discussing exactly what it means. In summary, it deals with those situations where the internal energy of a molecule, represented by its excited energy levels, cannot be treated as being in equilibrium with its external or kinetic energy. The latter is of course a function of the translational velocity of the molecule, which in turn is part of a Maxwellian distribution of velocities determined by the temperature of the gas, in the usual meaning of that term. Situations in which the internal and external energies diverge can occur for many reasons, but the most common one in the atmosphere is when there are not enough inter-molecular collisions to keep the two pools of energy related to each other. In other words, non-LTE situations are common in the middle and upper atmosphere, where the pressure is relatively low. Once non-LTE sets in, even the concept of temperature becomes complicated as several definitions are possible and they no longer have the same values. It also becomes much more difficult to compute the absorption and emission of radiation by atmospheric gases in this state. Quite difficult and complex numerical models, programmed on a substantial computer, are required. It is one of the goals of this book to explain how such a model is constructed and used.

Much of the study of non-LTE involves the use of computation-intensive numerical models, because many processes are going on simultaneously and the objective is to determine how the balance between all of these determines the interaction between the molecules and the photons in the atmosphere. The former are of many different species, including of course the critical 'greenhouse' gases of such current notoriety, such as CO_2. The photons span several orders of magnitude in wavelength, from the ultraviolet, the visible, and the near infrared (originating mostly in the Sun) to the middle and far infrared frequencies which the atmosphere and surface of the Earth emits continuously. Mostly, we are concerned with the internal energy of molecules which is stored as rotations and vibrations. When it is realised that, not only are many complicated physical processes involved, but also that we are generally interested in solving the sets of equations which result in at least two dimensions (usually height and latitude or solar zenith angle), it is easy to appreciate why non-LTE models have been neglected in spite of their importance.

Once the problem has been tackled, we gain not only insight into the physics of the atmosphere, but an essential tool for modelling its behaviour. Dynamical models of the upper atmosphere are very limited if they do not include realistic calculations of the heating and cooling which drive the motions in the first place, and these are strongly effected by non-LTE. Most of our measurements of the atmosphere, except near the surface, come these days from satellites, using the technique known as remote sensing. Instruments on the satellites measure emission or absorption by atmospheric gases, in order to determine their temperature or abundance. These can only be interpreted if the rate of emission per molecule is known from a non-LTE model calculation. We shall look at these applications in the later chapters of the book.

Basic non-LTE models deal with the transfer of energy as (primarily infrared) radiation within an atmosphere with a fixed composition and no fluid dynamics. In places, we digress into basic photochemical modelling, i.e. the situation where photon-molecule encounters lead to the break-up of the molecule and the subsequent formation of a different stable or quasi-stable species. This is, of course, another example of a non-equilibrium process, and since it depends on the nature of the photon field (and often the internal state of the molecule also) it forms a natural part of the book. Also, a knowledge of the composition of the atmosphere, including key photochemical products such as ozone or carbon monoxide, is crucial to the success of any non-LTE radiative transfer model. Although, as already noted, any dynamical model of the atmosphere requires a good radiation scheme, the converse is not usually true and so we disregard atmospheric motions, referring the reader to the many good texts on that subject which already exist.

Overall, then, our objective is three-fold: to provide an introduction to the subject at post-graduate level, to describe in detail how non-LTE calculations may be performed, with a 'hand-book' approach to setting up a suitable model, and finally to describe, with the aid of such a model, what is actually happening in the atmosphere. This last goal is mainly concerned with the computation of heating and cooling rates, i.e. the energy balance of the upper atmosphere, and with the interpretation of sophisticated satellite measurements and what they reveal. Finally, remembering that physics is universal in spite of our parochial interest in the behaviour and evolution of our own planet, we look at what is known or can be expected about non-LTE behaviour in the atmospheres of our sister worlds elsewhere in the Solar System.

References can be a problem in a book of this scope and complexity. To avoid clutter in the main text, and to provide a convenient source for those who wish to refer to original work or to delve deeper into the subject, we have generally collected them all into a substantial 'References and Further Reading' section at the end of each chapter. A complete bibliography appears at the end of the book.

We thank our many colleagues at Granada, Oxford and elsewhere who have worked with us on non-LTE problems over the years and contributed directly or indirectly to the creation of this book. In particular we would like to thank Gail Anderson and Anu Dudhia, for transmission calculations in the UV/visible and in the infrared respectively; David Edwards for material used in Chapter 8; Jean-Marie Flaud for unpublished spectroscopic data on O_3 and NO_2; Bernd Funke for non-LTE modelling of NO; Francisco J. Martín Torres for non-LTE models of O_3, CH_4, and N_2O; and Guillermo Zaragoza for the calculation of the H_2O cooling rates. For reading all or part of the manuscript in detail and offering very useful comments and corrections we thank the above-named colleagues, plus Maia García Comas, who also compiled appendices B, D, and E; Miguel Á. López Valverde for long discussions on a whole range of topics; Chris Mertens, who pointed out very subtle errors in the maths; and Richard Picard, Jeremy Winick, Peter Wintersteiner and Clive Rodgers for important corrections and constructive comments. Finally, we thank Gustav Shved for a very careful and detailed review of the whole book.

M. López Puertas
F.W. Taylor
September 2001

Contents

Chapter 1

Introduction and Overview

1.1 General Introduction

The atmosphere is never in true thermodynamic equilibrium. A given parcel of air is always in the process of changing its properties through the exchange of particles and photons with its surroundings, through collisions between molecules, and through chemical changes. It can, however, be in a state known as *local thermodynamic equilibrium* or LTE, a concept which permits calculations of the state of an atmosphere, and its energy exchanges (both internally and with its upper and lower boundaries), to be made relatively simply. In general, LTE applies if the populations of the energy levels within a molecule are the same, or nearly the same, as they would be under true thermodynamic equilibrium conditions. LTE occurs when collisions are so frequent that the energy level populations depend predominantly on the local kinetic temperature, as defined by the Maxwellian statistics of molecular motion, rendering negligible the various other processes which may be going on at the same time.

Obviously, it is necessary for the atmospheric scientist to understand when LTE applies and when it does not. As we shall see throughout much of this book, there are many important situations where LTE does not apply, especially in the upper atmosphere. Then it is necessary to have a formalism for calculating energy transfer in which the physical processes involved are dealt with properly. Without this, the whole field of numerical modelling of atmospheres and the important technique of remote sensing from satellites, mostly in the infrared, would both be invalid, except in the higher pressure regimes where experience has shown that LTE is generally a safe assumption. On the Earth, this would correspond to the altitude range below about 30 km, where pressures are high and collisions between molecules are frequent. In the middle and upper atmosphere, few research studies can proceed without at least considering the possibility that non-LTE may be important. The structure, dynamics and to some extent the composition of these regions would be quite different from that observed if it were entirely in LTE throughout.

The purpose of this book is to describe the various non-LTE processes which

occur in planetary atmospheres, and to provide a detailed mathematical and computational prescription for the determination of temperature profiles, heating/cooling
rates, and outgoing radiance fields in atmospheres in non-LTE. All of this will be illustrated with practical examples, and the discussion, while centring on the Earth's
atmosphere, will be extended to other planetary atmospheres and comets as well.

The non-LTE situations which we deal with in this book are mainly concerned
with the vibrationally excited states of atmospheric molecules, which interact with
the atmospheric infrared radiation field. In Chapter 2 we discuss the allowed energy
levels of atmospheric molecules under the types of conditions which arise in the
middle and upper atmospheric regions, the allowed transitions between states, and
the populations which result if the molecule is part of a gas in LTE. Departures from
LTE are most commonly, although not exclusively, caused by radiative excitation
or de-excitation of molecules into levels which causes them to exceed or fall below
their thermally (i.e. collisionally) induced populations. Because the source of the
radiation responsible may lie at a great distance from the molecule being excited,
and because in general a range of sources is involved, the solution of the non-LTE
problem is particularly complicated when radiative processes are a factor. For these
reasons a significant part of this book deals with the theory of radiative transfer,
beginning with the basics in Chapter 3 and proceeding to a more detailed treatment,
including LTE conditions, in Chapters 4 and 5.

The theory of radiative transfer under non-LTE conditions has not changed
in essence since it was first developed (see Secs. 1.6 and 1.10), but it required the
advent of fast computers before it became realistic to compute sophisticated models
of the populations of vibrationally excited states and the cooling rates resulting from
the transitions between them. In the past few years, several practical formulations
have been employed. The most significant formulations of non-LTE for practical
cases are reviewed in Chapter 5, with a detailed description of the Curtis matrix
method. Its application to the important case of carbon dioxide is described in
Chapter 6. Chapter 7 describes the results of calculations for a further range of
specific molecules and transitions of atmospheric interest.

In Chapters 8 and 9, the use of non-LTE models is illustrated. First we discuss
the effects of non-LTE upon the radiance emission and transmission by atmospheric
gases and its subsequent effects on the retrieval of atmospheric parameters (kinetic
temperature, species abundances and excitation/relaxation rates) from measurements of atmospheric infrared emissions by satellite instruments. Then we present
the effects of non-LTE on the radiative balance through changes in the cooling and
heating rates in infrared bands. Attention is focused primarily on the Earth's upper
stratosphere, mesosphere and lower thermosphere (40 to 120 km), except for some
special cases such as nitric oxide which have interesting behaviour up to 200 km
or more. In the final chapter we will consider the extension of the problem to the
atmospheres of other planets, where the same physical processes are at work under
different conditions of pressure, temperature, composition, and solar illumination,

and to the satellite Titan and the coma regions of comets.

1.2 Basic Properties of the Earth's Atmosphere

1.2.1 *Thermal structure*

The various regions of the atmosphere are usually named for their temperature structure (Fig. 1.1). Many investigators, however, frequently refer to the different regions as the *lower atmosphere* (the troposphere), the *middle atmosphere* (stratosphere and mesosphere) and the *upper atmosphere* (above about 100 km).

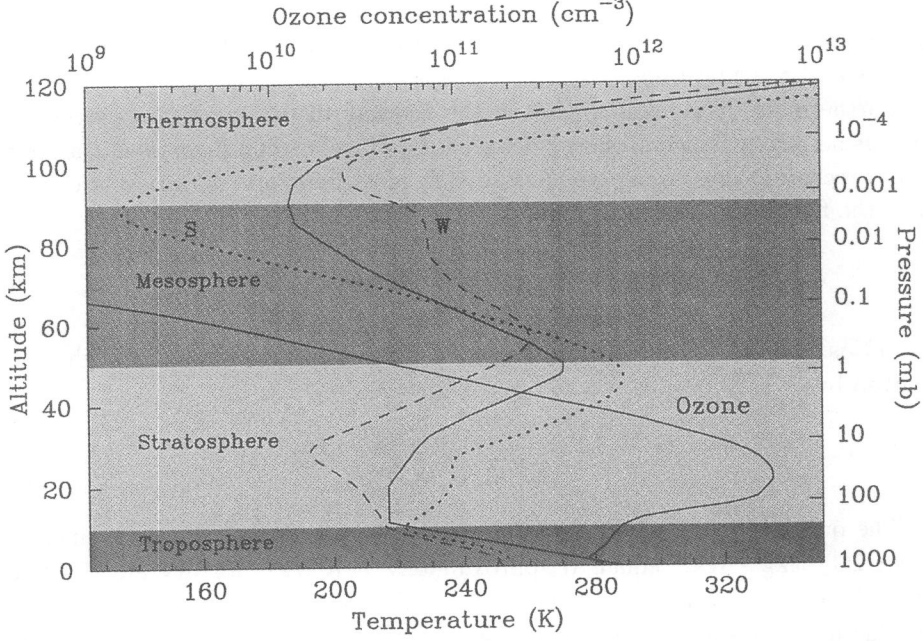

Fig. 1.1 Typical atmospheric temperature profiles for mid-latitudes (solid line), polar summer (S, dotted), and polar winter (W, dashed) conditions. The figure also shows the names given to regions, and a global mean ozone profile.

In simplified terms, these regions exist for the following reasons. Much of the electromagnetic radiation emitted by the Sun and reaching the Earth is at wavelengths in or near the visible part of the spectrum (see Fig. 1.5a). These are not strongly absorbed by the atmosphere and so, unless there are clouds to scatter the photons and reflect a portion of the radiation back to space, most of the energy reaches the ground where it is absorbed (see Fig. 1.5c). The lower atmosphere is then heated by the ground, and will become unstable against convection if a layer nearer to the ground is warmer than that above it by a sufficient amount to become less dense in spite of the vertical pressure gradient. Convective instability exists in the lowest 10 km or so of the atmosphere, which is known as the *tropo-*

sphere ('turning–region'). The upper boundary is the level where the overlying atmosphere is of such a low density that a substantial amount of radiative cooling to space can occur in the thermal infrared region of the spectrum. At this level, called the *tropopause*, radiation cools rising air so efficiently that the temperature tends to become constant with height and convection ceases. Figs. 1.2 and 1.3 show that the tropopause varies in height over about 6 km with latitude, being highest (around 16 km) in the tropics, where solar heating is greatest, but it is generally quite a sharp feature in the temperature profile everywhere.

Assuming that hydrostatic equilibrium applies, then

$$\mathrm{d}p = -g\,\rho\,\mathrm{d}z, \tag{1.1}$$

where p is the pressure, g is the Earth gravitational acceleration constant, ρ is the density and z the altitude.

Furthermore, if we assume that in the vertical displacement of a parcel of air, there is no net exchange of energy between it and its surroundings then, the specific heat at constant pressure c_p, temperature T, pressure p, and density ρ, are related, after the first law of thermodynamics, by

$$c_p\,\mathrm{d}T + \frac{\mathrm{d}p}{\rho} = 0. \tag{1.2}$$

From Eqs. (1.1) and (1.2), we find that the temperature gradient with height $\mathrm{d}T/\mathrm{d}z$ is given by

$$-\frac{\mathrm{d}T}{\mathrm{d}z} = \frac{g}{c_p}. \tag{1.3}$$

The quantity g/c_p, called the *adiabatic lapse rate*, is constant and, since c_p is about $1000\,\mathrm{J\,kg^{-1}\,K^{-1}}$, equal to approximately $10\,\mathrm{K\,km^{-1}}$ for dry air. For moist air, $-\mathrm{d}T/\mathrm{d}z$ is less and can be as small as $3\,\mathrm{K\,km^{-1}}$. A useful average value for the Earth's troposphere is $8\,\mathrm{K\,km^{-1}}$. Also, as shown in Fig. 1.1, the temperature gradient is smaller in the polar regions, where the surface is cooler because less energy per unit area is absorbed from the Sun.

The lapse rate above the tropopause, where convection stops, tends to zero (i.e. constant temperature with height) because there is no longer enough absorption above the layer at most wavelengths to stop emitted photons from reaching space. Each layer is then heated by radiation from the optically thick atmosphere below, and cooled by radiating to space; to first order height is no longer important. This region is called the stratosphere; it is stratified in the sense that density decreases monotonically with height, and therefore the layers do not try to move up or down through each other as in the troposphere. If the stratosphere is modelled as an optically thin slab of (vanishingly small) emissivity ε, its temperature T_{strat} can be related to the effective radiative, or equivalent blackbody, temperature of the Earth, T_{Earth}. The stratosphere is heated from below by an infrared flux proportional to

$\varepsilon T_{\mathrm{Earth}}^4$, and cooled by identical fluxes in the upward and downward directions, each proportional to $\varepsilon T_{\mathrm{strat}}^4$. Thus we must have $\varepsilon T_{\mathrm{Earth}}^4 = 2\varepsilon T_{\mathrm{strat}}^4$ whence

$$T_{\mathrm{strat}} = \frac{T_{\mathrm{Earth}}}{\sqrt[4]{2}} \approx \frac{250}{\sqrt[4]{2}} \approx 210\,\mathrm{K}. \tag{1.4}$$

Although these simple arguments suggest that the stratospheric temperature remains constant to a great height, in fact it is observed to increase above about 20 km as a consequence of the absorption of solar ultraviolet radiation by photochemically-produced ozone in the stratosphere. Ozone absorbs in the Hartley band, which forms a continuum from 0.2 to 0.3 μm (see Fig. 1.5c). Below 70 km, virtually all of the energy absorbed is converted to heat.

The ozone concentration in the stratosphere peaks near 25 km, but a calculation of the heating rate at height z:

$$c_p\, \rho(z)\, \frac{\mathrm{d}T(z)}{\mathrm{d}t} = \int_{0.2}^{0.3} \frac{\mathrm{d}F_\lambda(z,\chi)}{\mathrm{d}z}\, \mathrm{d}\lambda, \tag{1.5}$$

where $F_\lambda(z,\chi) = F_\lambda(\infty)\exp\left[-\int_z^\infty k_\lambda n_{\mathrm{O}_3}(z)\mathrm{d}z\right]$ is the spectral flux of radiant energy, $n_{\mathrm{O}_3}(z)$ the O_3 number density, χ is the solar zenith angle, and k_λ the O_3 absorption coefficient at wavelength λ, finds that maximum heating occurs at a height of about 50 km. The temperature also peaks at this level, which is known as the *stratopause*. Above the stratopause, the temperature declines again, reaching an absolute minimum at the *mesopause* near 90 km. The pressure here is only a few microbars. With such low densities of gas above, very energetic solar photons in the extreme ultraviolet, and particles too, penetrate into the region causing ionization and dissociation and releasing kinetic energy. The heating thus produced causes the temperature to increase rapidly with height, leading to the name *thermosphere*.

By using the perfect gas law in Eq. (1.1) and integrating we have the equivalent relationship

$$p(z) = p_0 \exp\left(-\int_0^z \frac{\mathrm{d}z}{H}\right), \tag{1.6}$$

where p_0 is the pressure at zero altitude, and we have introduced the *scale height* $H = kT/mg$, with m being the mean molecular mass of air. This expression is frequently used in the studies of the upper mesosphere and thermosphere. The quantity H is the increase in altitude necessary to reduce the pressure by a factor e. Hence it has the useful meaning of the thickness that the atmosphere above z would have if it were at a constant pressure and temperature equal to the values at z. Sometimes, mostly in studies of the middle atmosphere (stratosphere and mesosphere), the term *scale height* is used to mean $-\ln(p/p_0)$, a dimensionless quantity that, according to Eq. (1.6), is equivalent to $\int_0^z (\mathrm{d}z/H)$.

Between the tropopause, where large-scale vertical convection ceases, and the mesopause, the atmosphere remains fairly well mixed by turbulence produced by a

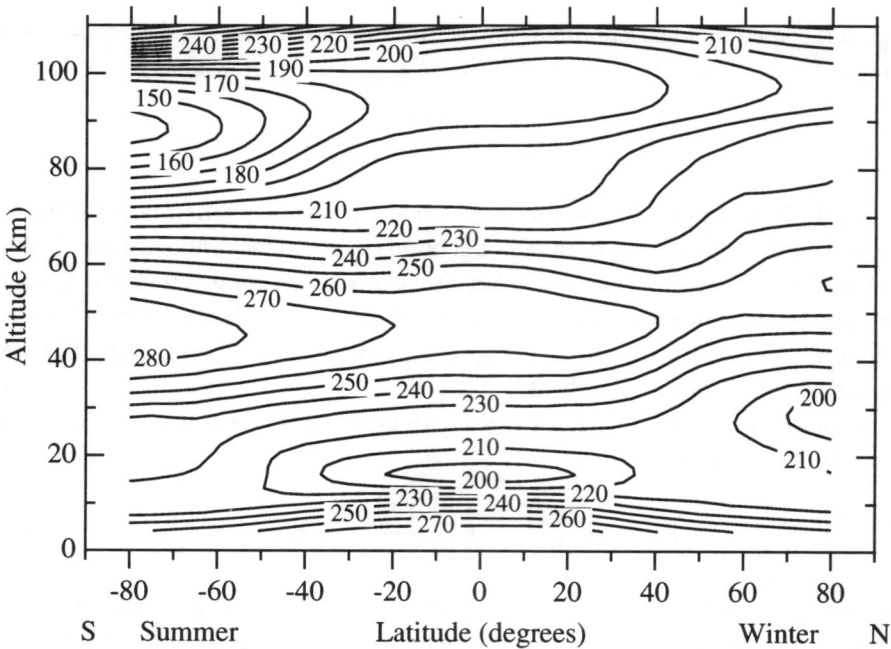

Fig. 1.2   Latitude/altitude temperature structure of the atmosphere for solstice (December) conditions. From the CIRA 1986 reference atmosphere.

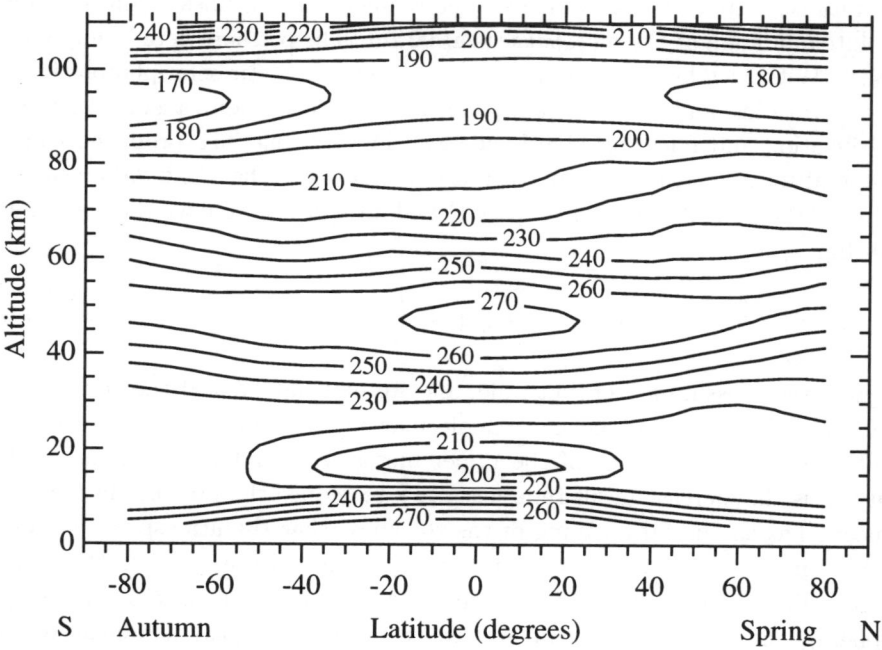

Fig. 1.3   Latitude/altitude temperature structure of the atmosphere for equinox (March) conditions. From the CIRA 1986 reference atmosphere.

variety of instabilities in wave motions and the mean flow. Near the base of the thermosphere, diffusion takes over as the dominant process and the atmosphere starts to separate into its lighter and heavier components. This is clear from Eq. (1.6), which, in these regions, has meaning only when applied to each individual species separately. Each compound is then distributed according to its molecular or atomic mass m_i, with its own scale height $H_i = kT/m_i g$, and the heavier compounds start decreasing with altitude more rapidly than the lighter. Carbon dioxide, the heaviest compound in the upper atmosphere, starts to deviate from mixing at lower altitudes, around 80–90 km. Some 500 km higher, the base of the *exosphere* is reached. At this level only hydrogen and helium are still present in significant numbers and the mean free path between collisions becomes less than the scale height, allowing the possibility of escape to space.

Figures 1.2 and 1.3 show zonally averaged temperatures for December and March as compiled in the CIRA 1986 reference atmosphere. It can be seen that the tendency for the tropopause to be situated at lower altitudes in the polar regions and higher in the tropics occurs in all seasons. The stratopause is significantly warmer at the summer than at the winter pole, mainly caused by the longer solar illumination and absorption of solar radiation by ozone. In the mesopause region, however, the situation reverses, and the summer polar mesopause is the coldest place in the atmosphere. There, the temperature can be so low (below 135 K) that water vapour condenses into crystals to form the so-called noctilucent clouds (NLC), not to be confused with the polar stratospheric clouds (PSC) occurring in the polar winter stratosphere. The reason for the summer mesopause cold temperatures is not *in situ* radiation, since the absorption of solar radiation is certainly larger than in the winter, but dynamical processes, in particular the meridional circulation. This progresses from the summer to the winter pole, with upwelling motion suffering nearly adiabatic expansion and hence net cooling over the summer pole while the opposite occurs in the polar winter. Dynamics thus plays an essential role in explaining the observed atmospheric temperature structure. Another example of dynamical control of cooling is the tropical tropopause, which is much colder than its counterpart at middle and high latitudes (even at winter latitudes), as a result of upwelling air in the tropical troposphere.

1.2.2 *Composition*

The atmosphere is composed of a mixture of chemical constituents, including some which are important for radiative transfer and photochemistry despite being present only in extremely small amounts. Table 1.1 shows the composition of dry air near the surface. The amount of water which is present in addition to these non-condensable species varies, but can be as high as a few percent. Some of the other species, particularly ozone, are also very variable, especially in polluted environments. Fig. 1.4 shows typical vertical profiles of the mixing ratios of atmospheric chemical constituents for mid-latitude and daytime conditions. Many of them vary with latitude and season, and those affected by photochemical and photoabsorption processes also show dramatic diurnal changes, for example $O(^3P)$ below about 80 km, NO below 60 km and $O(^1D)$.

Table 1.1   Composition of clean dry air near the surface.

Species	Mixing ratio (by volume)
Nitrogen, N_2	0.78084
Oxygen, O_2	0.20948
Argon, Ar	0.00934
Carbon dioxide, CO_2	0.00037^*
Neon, Ne	18.2×10^{-6}
Helium, He	5.2×10^{-6}
Krypton, Kr	1.1×10^{-6}
Xenon, Xe	0.1×10^{-6}
Ozone, O_3	0.01–0.5×10^{-6}
Hydrogen, H_2	0.55×10^{-6}
Methane, CH_4	1.7×10^{-6}
Nitrous oxide, N_2O	0.3×10^{-6}
Carbon monoxide, CO	0.05–0.2×10^{-6}

*In 2000 (see the 'References and Further Reading' section).

Molecular nitrogen and oxygen are the most abundant species. These compounds, like the noble gases, have very long chemical lifetimes and hence are well mixed by winds, waves and turbulence in the lower and middle atmosphere. Molecular oxygen is partially photo-dissociated into atoms and ions above approximately 90 km, as is molecular nitrogen at rather higher altitudes. Water vapour is very variable in the troposphere, and falls off rapidly near the tropopause, reaching a stratospheric value of only a few parts per million. It, too, is photo-dissociated above the mesopause, falling to negligible amounts in the lower thermosphere.

Ozone is the most important of the minor species which, despite their low abundances, play key roles in determining the atmospheric thermal structure, energy balance and the conditions at Earth's surface. Ozone is produced after photodissociation of molecular oxygen and subsequent recombination of molecular and atomic oxygen. It is mainly controlled by dynamical processes below about 30 km and is highly variable. The maximum abundance normally occurs at about 25 km (see Fig. 1.1). In the upper stratosphere and lower mesosphere ozone is short-lived because of rapid photo-dissociation, so it is approximately in photochemical equilibrium. In the upper mesosphere and lower thermosphere it is also short lived but significantly affected by the $O(^3P)$ concentration. The large abundance of $O(^3P)$ in the upper mesosphere, maintained by downwards transport of $O(^3P)$ produced at higher altitudes from molecular oxygen photodissociation, gives rise to the sec-

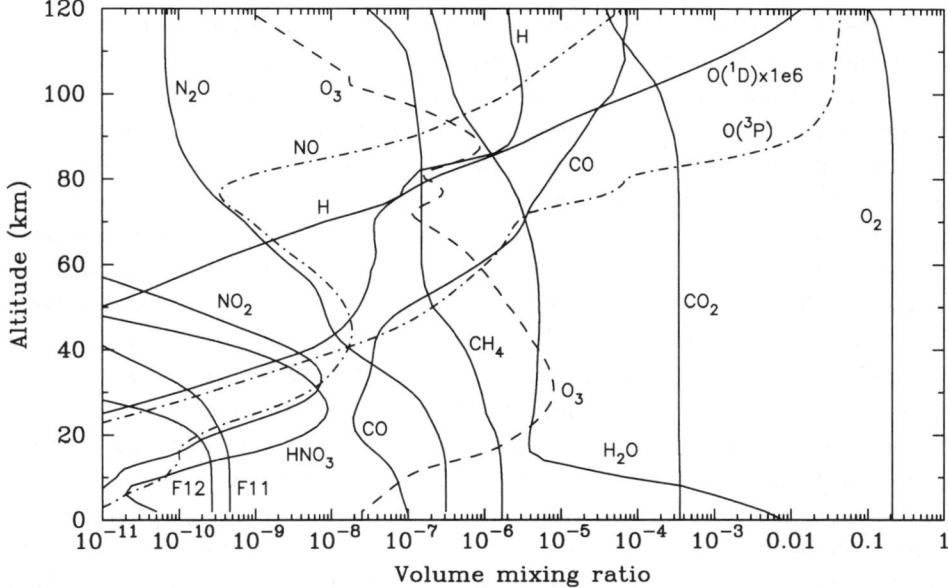

Fig. 1.4 Typical mixing ratio profiles for selected chemical constituents at mid-latitudes and daytime conditions. The mixing ratio of the excited $O(^1D)$ is multiplied by 10^6. F11 = $CFCl_3$ and F12 = CF_2Cl_2. Based on the U.S. Standard Atmosphere (1976), Clarmann *et al.* (1998) and Garcia and Solomon (1983, 1994).

ondary maximum of ozone in this region.

Carbon dioxide plays a major role in the infrared radiation field at practically all altitudes below 130 km. It is not chemically reactive. There are very large amounts of CO_2 in the atmosphere, in the oceans and on the Earth's surface and fluxes between them are bidirectional and nearly in balance. Thus, the net exchange of CO_2 between the atmosphere and the oceans and biosphere is small compare to the size of the reservoirs and difficult to measure. Nevertheless, the total anthropogenic release of CO_2 into the atmosphere represents a significant perturbation of the natural carbon cycle. Because of fossil fuel combustion and biomass burning its mixing ratio in the atmosphere has been slowly increasing at a rate of approximately 1.5–2 ppmv year^{-1} in recent years. This increase propagates upwards (also to the tropospheric polar regions where it is not produced locally) and takes about 5–6 years to reach the stratosphere and mesosphere. It is well mixed up to the upper mesosphere/lower thermosphere, where it begins to decrease due to molecular diffusion and destruction by UV solar radiation.

Methane and nitrous oxide are, after water vapour and CO_2, the most abundant greenhouse gases in the troposphere. They are produced at the surface by natural and anthropogenic sources and removed from the atmosphere by photochemical processes in the stratosphere. Observations show that their tropospheric concentrations are slowly increasing with time.

Carbon monoxide, although not a major greenhouse gas, is important in tropospheric chemistry through its reaction with the OH radical, which in turn plays a central role in the chemistry of the troposphere and stratosphere. CO is produced at the surface from biomass and fossil fuel burning, which results in high spatial and temporal variability; it is about twice as abundant in the Northern Hemisphere troposphere compared to the Southern. Its concentration may have doubled in the last 50 to 80 years in the Northern Hemisphere, although there is some evidence for a reduction over the last decade. CO is produced in the stratosphere and lower mesosphere from methane oxidation and destroyed by chemical reaction with OH. CO_2 photolysis in the lower thermosphere is a major source of upper-atmospheric CO, and this production and its subsequent downward transport into the middle atmosphere means that carbon monoxide generally increases with height throughout the middle atmosphere. With important sources in all three major atmospheric regions —lower, middle and upper— and an intermediate chemical lifetime of a few months, the detailed distribution of CO in all four dimensions can be quite complex.

Atomic oxygen plays an important role in the radiative cooling of the thermosphere by transferring kinetic energy to the internal energy of infrared-active atmospheric molecules like CO_2 and NO. Its major sources are the photo-dissociation of molecular oxygen and ozone, but above the mesopause its distribution depends mainly on the dynamical conditions and does not exhibit much diurnal variation. Below that height, it disappears rapidly after sunset, recombining with molecular oxygen to form ozone. Atomic oxygen in its excited metastable electronic state, $O(^1D)$, has a very low abundance but is still a significant source of vibrational excitation for the more abundant species in the middle atmosphere and lower thermosphere.

Electronically excited $O(^1D)$ is also highly reactive and is the main source for the OH radical throughout the atmosphere (particularly in the troposphere and stratosphere) and of chemically-active nitrogen compounds in the stratosphere. The two most important of these, nitric oxide (NO) and nitrogen dioxide (NO_2) are closely tied through photolytic and chemical exchange in the lower troposphere and stratosphere. Because of that the two constituents are often grouped together as NO_x. These 'active nitrogen' species are primarily responsible for the photochemical production of ozone and smog in the troposphere. They also play a crucial role in the catalytic destruction of ozone in the stratosphere. NO_x is released into the troposphere as nitric oxide at the surface as a result of human activity (fossil fuel combustion and biomass burning) and natural processes in the soil. It is also produced in lightning discharges and to a lesser extent by subsonic aircraft. In the stratosphere, NO is produced by the oxidation of nitrous oxide (N_2O) and lost by reacting with ozone to form NO_2. During the day, NO_2 undergoes rapid reaction with atomic oxygen and is also photochemically destroyed, both processes restoring NO. This catalytic cycle is the main sink for odd oxygen (O_3 and O) in the stratosphere. NO is also the main source of infrared cooling in the thermosphere, where it is formed by the reaction of nitrogen atoms with molecular oxygen and by auroral processes, and is highly variable in abundance.

The chlorofluorocarbons (CFCs), mainly $CFCl_3$ (F11) and CF_2Cl_2 (F12), are of industrial origin and reside initially in the troposphere. Because of their very long chemical lifetimes and low solubility, they mostly survive to reach the stratosphere eventually, where they are fragmented by short-wave ultraviolet radiation and liberate ozone-destroying chlorine atoms. Their abundance has steadily increased with time in the last few decades but, because of international regulations limiting their production, we are beginning to witness their levelling–off and eventual depletion.

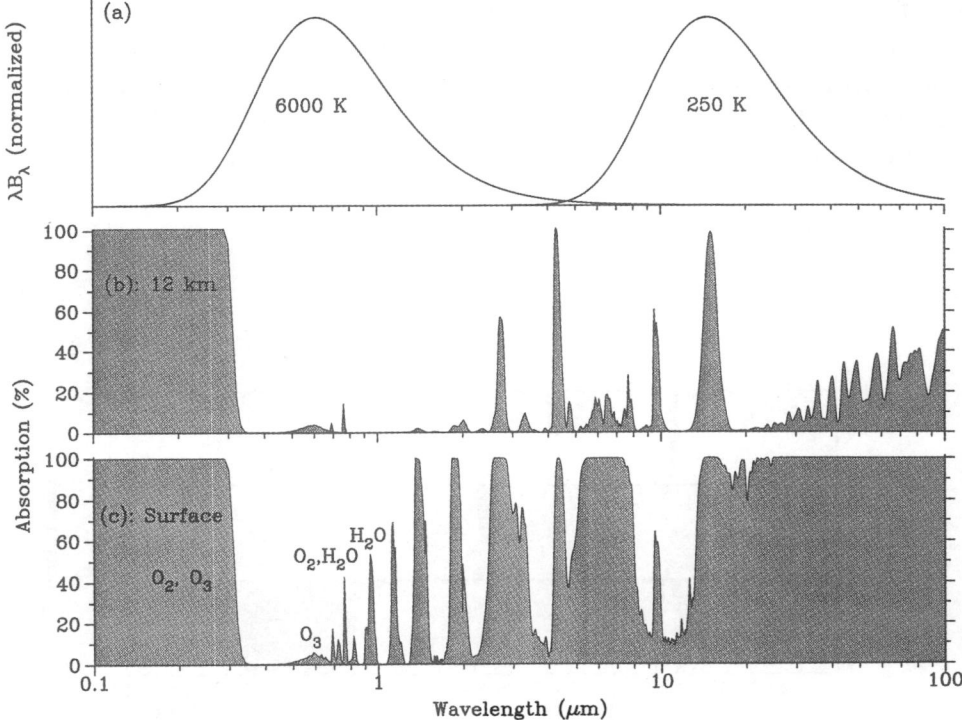

Fig. 1.5 Atmospheric absorption. (a) Blackbody curves for 6000 K and 250 K normalized to the energy emitted. (b) Absorption spectra for 12 km to the top of the atmosphere. (c) Atmospheric absorption spectra at the Earth's surface. The absorption features of the atmospheric compounds at $\lambda > 1\,\mu$m are identified in Fig. 1.6. The U.S. Standard atmosphere (1976) was used. Spectra up to $1\,\mu$m were computed by G. Anderson. Those above $1\,\mu$m were calculated by A. Dudhia.

1.2.3 *Energy balance*

Virtually all of the energy received by the Earth comes from the Sun in the form of solar radiation and, since the mean temperature of the Earth changes little with time, it is essentially all returned to space, again in the form of radiation. The Sun emits radiation approximately as a blackbody, with the maximum of this emission falling in the visible region (see Fig. 1.5a). At shorter wavelengths the solar radiation better approximates blackbodies at temperatures of 5100 K for the 200–250 nm range and 4600 K for the 130–170 nm range. In the infrared part of the spectrum, solar radiation is actually approximated better by a 5770 K blackbody.

The most energetic solar photons at UV and shorter wavelengths are removed at various, mostly high, levels in the atmosphere, where they participate in the dissociation and ionization of oxygen, ozone and nitrogen molecules. O_2, $O(^3P)$, and N_2 absorb at wavelengths shorter than 100 nm, O_2 in the Schumann-Runge continuum (100–175 nm) and in the Schumann-Runge bands (175–205 nm) and more weakly in the Herzberg continuum (200–245 nm). Ozone is the dominant attenuator of solar radiation in the Hartley band (210–300 nm). At wavelengths larger than approximately 290 nm solar radiation penetrates down to the Earth's surface, attenuated partially by the ozone Huggins (310–400 nm) and Chappuis (400–850 nm) bands (see Fig. 1.5).

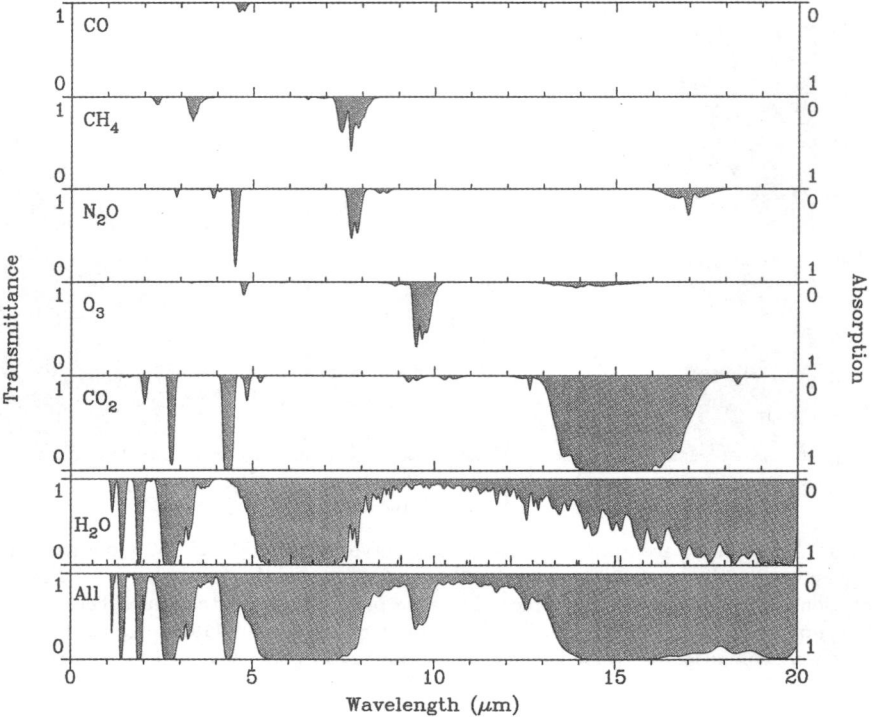

Fig. 1.6   Spectral transmittance/absorption from the top of the atmosphere down to the Earth's surface by the major infrared-active atmospheric constituents and for all species together. The U.S. Standard atmosphere (1976) was used. Computed by A. Dudhia.

The longer infrared wavelengths from the Sun (1 to 5 μm) are absorbed in different regions of the near infrared spectrum by molecular vibration-rotation bands, particularly those of water vapour and carbon dioxide (Fig. 1.6). The emissions from these bands often take place under non-LTE conditions in the middle and upper atmosphere, and will be the main subject of Chapters 6 through 9. At wavelengths shorter than about $4\,\mu$m, they absorb solar radiation (see Fig. 1.5), part of which is redistributed to lower energy levels by collisions with other atmospheric molecules and emitted back to space in the longer wavelengths bands.

The solar radiation at wavelengths in or near the visible part of the spectrum mainly propagates to the ground unless obstructed by clouds or aerosols. On a global average, about 30% of the energy from the Sun is reflected back to space (most of it, 21%, by clouds), 21% is absorbed in the atmosphere before reaching the ground, and 49% is absorbed at the ground. The surface transfers the energy back to the atmosphere, mostly by infrared radiation and latent heat transfer which is absorbed in the lower atmosphere and eventually emitted to space, together with that directly reflected by backscattering in the atmosphere.

Because of its lower temperature relative to that of the Sun, the Earth's surface re-emits the absorbed energy at a much longer wavelength than that at which it was received. Some of this thermal infrared radiation is transmitted directly to space, but at most wavelengths longer than the visible the atmosphere is very opaque. Interestingly, the source of this opacity is not, for the most part, the main constituents nitrogen and oxygen, but the minor and trace constituents, especially water vapour, carbon dioxide, ozone, methane and nitrous oxide (see Fig. 1.6). The reason for this is the fact that the two homopolar molecules N_2 and O_2 have no permanent dipole moment and hence no strong vibration-rotation bands in the infrared. The polar molecules, on the other hand, have very rich infrared spectra and collectively absorb photons at all wavelengths except in a few small 'window' regions, such as those near $3.7\,\mu$m and $12\,\mu$m.

1.3 What is LTE?

The concept of LTE was introduced by Schwarzschild in 1906 in his study of stellar atmospheres. He realised that, in a star just as in a planetary atmosphere, any individual parcel of gas is not isolated and, in principle, no equilibrium can be defined. However, the concept of equilibrium is so valuable in practice that it is worth looking for situations where it can be assumed as a close approximation to reality.

Consider an atmospheric gas parcel composed of atoms and molecules with excited electronic, vibrational and rotational energy levels which interact with radiation. If the gas parcel is completely enclosed and isolated from the rest of the atmosphere, it will eventually come to thermodynamic equilibrium and a single temperature T could be defined. Under these conditions almost all properties of the gas parcel are described by this temperature, i.e., the distribution of molecular velocities is given by the Maxwellian distribution at temperature T, and the excited states are populated accordingly to Boltzmann's law at the same temperature. Also, the radiating properties of matter depend only on temperature, and the radiative field inside the parcel is characterised by the well-known Planck blackbody function.

In any gas under atmospheric conditions, the kinetic energy of molecules respond to any external change more quickly than the internal energy states, so that exchanges of translational energy to and from other energy forms are not sufficient to cause the atmosphere to depart from the Maxwellian distribution. Changes in solar illumination, for example, might vary the temperature from a region to another, but the local equilibrium distribution is always reached. Thus, a Maxwellian distribution of molecular velocities for a local kinetic temperature can be assumed at each atmospheric height up to the exosphere. In other words, we can assume a local kinetic temperature $T(z)$ at each atmospheric altitude z. A particular excited (electronic, vibrational or rotational) state at a given atmospheric altitude z is then said to be in LTE if the distribution of atoms or molecules in the excited state in question is given by Boltzmann's law at the local kinetic temperature. One of the most important consequences, sometimes itself referred to as a definition of LTE, is that the radiating properties of matter for the transition between that state and the fundamental state (or another state in LTE) can be described solely in terms of temperature, in particular, the source function of the transition becomes equal to the Planck function at that temperature.

Although this definition of LTE is rather simple, there are some underlying consequences which are not so obvious. For example, unlike full thermodynamic equilibrium, the LTE condition does not impose restrictions upon the radiation field of the emission involved. That is, while in thermodynamic equilibrium the gas parcel radiates according with the blackbody function, in LTE the local radiation field is not necessarily described by Planck's formula, even though the source function is given by the Planck function. Hence, LTE can be compatible with a net gain or loss of radiative energy (heating or cooling) by the gas in the transition in question, provided that collisions are quick enough to supplement the energy sink and to keep the population of the excited state coupled with the translational reservoir. This is the case for the most important atmospheric emissions, CO_2 at $15\,\mu m$, H_2O at $6.3\,\mu m$ and O_3 at $9.6\,\mu m$ in the stratosphere, although not necessarily at greater heights, where the energy levels of these molecules deviate significantly from LTE.

Another important point is that, in a given atmospheric parcel containing different kinds of molecules, some can be in LTE and others not. In fact, it is possible (and indeed quite usual) for one internal energy state to be in LTE, and another not, in the same individual molecule. For example, carbon dioxide, which is one of the most important constituents of the atmosphere where radiative transfer is concerned, has two fundamental modes of vibration, ν_2 and ν_3 (this terminology will be explained in Chapter 2), which interact with photons. Both of these are in LTE in the lower atmosphere, but above about $40\,km$ the ν_3 mode is no longer, and above about $100\,km$ neither is. The reason for the dependence on height is the dependence on pressure of the collision frequency, and the different efficiencies

for transferring energy to vibrational modes during collisions, as further discussed
below.

1.4   Non-LTE Situations

Our main concern in this book is to emphasise the physical basis for non-LTE,
distinguishing between the different processes involved, and the circumstances which
cause them to occur in planetary atmospheres. To see under which conditions LTE
holds we have to consider all the microscopic processes that affect the population
of the excited levels. This is treated in a general way in Chapter 3 and particular
cases are described in Chapters 6 and 7. We consider here some general ideas about
the different non-LTE situations and when we expect a given transition to depart
from LTE.

There are basically two ways in which the internal states of any given molecule
can 'know' what the ensemble of molecules in the gas is doing. One is through
collisions, and the other through exchanging photons. In simple terms, a given
internal mode of the molecule is in LTE if the population distribution among its
energy levels is determined predominantly by collisions. On the other hand, if
dominated by radiative processes (absorption and emission of photons) it will be, in
general, in non-LTE since the radiative field is non-local in nature, with an absolute
intensity, a directional distribution, and a frequency spectrum that usually have
little or no relation to the Planck function at the local kinetic temperature. In
the atmosphere, pressure and therefore frequency of collisions falls off with height.
Thus we expect a given state to be more likely in non-LTE in the upper atmosphere.
Another important parameter in a crude estimation of the height at which non-LTE
occurs is the energy of the transition. For those corresponding to small energy jumps
the average number of collisions required to keep the levels in equilibrium is smaller,
so they can be in LTE up to higher altitudes in the atmosphere. For example, the
CO_2 $4.3\,\mu m$ band starts to depart from LTE at approximately $40\,km$; while the
$O(^3P)$ states which emit at $63\,\mu m$ and the purely rotational levels of virtually all
atmospheric molecules are in LTE up to the high thermosphere. Another way of
thinking about maintaining LTE is to compare the mean free path of photons to
a typical distance over which temperature changes. For example, if the mean free
path of photons is much smaller than this distance, or if $dT/dz \simeq 0$, few collisions
are needed to maintain LTE because the practically entire local photon flux derives
from a region within which the source function is uniform and characteristic of the
same local temperature.

Less commonly, collisions by molecules with internal non-LTE excited energy
states are sometimes responsible for bringing other states of the same or different
molecules out of LTE. Some examples are enumerated in Chapter 3 and described in
detail in Chapters 6 and 7. There are also some situations, under optically very thick
conditions, where the radiative field is so strong that a very few thermal collisions are

enough to *maintain* the population of an excited state in LTE. For these situations some authors say that the LTE state is maintained by radiation. More precisely, what happens is that a few thermal collisions are enough to maintain LTE, and it is these that are responsible for the LTE state since, if no collisions occurred, LTE would not normally be present. For example, this happens in the CO_2 15 μm fundamental band in parts of the Venusian and Martian middle atmospheres (see Chapter 10).

Thus, depending on the processes at work, we can generally distinguish between the following non-LTE situations:

(i) The classical non-LTE situation when, in the absence of a strong radiative source, thermal collisions are not fast enough to supply the energy lost by spontaneous emission. Under these conditions the population of the excited state is smaller than that corresponding to LTE. The altitude where a transition departs from LTE can be estimated in such cases as that where the collisional relaxation time is of the same order as the spontaneous radiative lifetime. The height of the LTE breakdown rise with the optical depth of the transitions. For optically thick conditions a better approximation is the altitude at which the deactivation by collisions is comparable to the spontaneous emission multiplied by the probability of photon escape to space. Some examples of this will be found in Chapter 3.

(ii) The internal atmospheric radiation field (earthshine) can cause a breakdown of LTE when it is responsible for overpopulating the excited level with respect to the Boltzmann distribution. This situation is common for molecular states which give rise to weak infrared bands in the cold upper mesosphere, since their populations are sensitive to the upwelling flux of infrared photons from the warm lower atmosphere.

(iii) LTE breaks down when the absorption of the strong solar radiation field is the prime process for the population of the excited levels. This occurs mainly for states which emit in the near-infrared part of the spectrum.

(iv) There are also situations where molecules are excited and de-excited by processes such as photochemical reactions, chemical recombination, electronic-vibrational and vibrational-vibrational energy transfers, dissociative recombination, and charged-particle collisions. Some examples of the non-LTE situations produced by these processes are given in Chapter 3. They are largely responsible for the atmospheric emissions which constitute the terrestrial dayglow, nightglow and auroral spectra, which historically have not been referred to as non-LTE emissions; however, since the populations of the excited states which produce these phenomena are far from a Boltzmann distribution at the local kinetic temperature they should come under this heading. Having said that, an extensive literature exists on airglow and aurorae and their detailed study is not an objective of the present book, where we are mainly concerned with non-LTE emissions which fall in the infrared part of the spectrum and those which require a detailed radiative transfer treatment.

1.5 The Importance of Non-LTE

The departure of the excited energy levels from a Boltzmann distribution has an important effect on the radiative cooling produced by the bands originating from them. The two principal examples in the atmosphere are the emission by the ν_2 bending mode of carbon dioxide near $15\,\mu\mathrm{m}$, which dominates all other sources of cooling in the upper mesosphere and lower thermosphere, and the non-LTE emission by nitric oxide at $5.3\,\mu\mathrm{m}$, which produces the main radiative cooling in the thermosphere. Both cooling rates would be much greater than they actually are if the upper levels of these bands were in LTE at the high kinetic temperatures found in the thermosphere.

Heating as well as cooling is affected by non-LTE. For example, the heating (i.e. the conversion of solar radiative energy into kinetic energy of the atmospheric molecules) by bands such as CO_2 at 4.3 and $2.7\,\mu\mathrm{m}$ is much smaller than if the upper levels of these transitions were completely thermalized locally. The net absorption in the solar pumped bands is much less for LTE. Also the re-emission may be fluorescent rather than resonant, which contributes to a smaller heating than expected in LTE. The heating rate can be further reduced by collisional transfers. In general, the collision processes that result in energy being transferred to the thermal reservoir are rather complex. For example, molecules such as O_2 and O_3 which absorb in the visible and UV, can transfer energy to $CO_2(v_3)$, with the $N_2(1)$ state acting as an intermediate agent, with subsequent re-emission to space.

Such cooling and heating processes are important for the determination of equilibrium temperature profiles in the atmosphere, and for creating the horizontal pressure gradients which drive atmospheric motions, so increasingly sophisticated parameterizations of radiative transfer, particularly in CO_2, are now a necessary part of most 2D and 3D dynamical models.

Since the advent of high resolution, high sensitivity instruments capable of being mounted on satellite platforms, remote sensing of the atmosphere has become an increasingly important tool in meteorology and other applied sciences. For example, limb sounding is a powerful technique used to derive the temperature structure and the minor constituent abundances of the middle and upper atmosphere. Non-LTE effects play an important role in the remote sensing of these upper regions, where the excited states which produce the infrared emission being measured depart from their Boltzmann populations. By limiting, or enhancing, the strength of the signal, non-LTE effects often impose the upper height limit at which remote sensing is possible. Well below that, non-LTE models are required to correctly retrieve the atmospheric temperature and species abundances. For example, the derivation of the kinetic temperature above $70\,\mathrm{km}$ from measurements of the limb radiance at $15\,\mu\mathrm{m}$, the most commonly-used technique, requires correction for non-LTE. The derivation of ozone abundances from measurements of the $9.6\,\mu\mathrm{m}$ limb radiance requires correction for non-LTE in O_3 and spectrally-overlapping species, especially

CO_2, above about 50–60 km. Similarly, deriving water vapour concentrations above around 60 km requires corrections for non-LTE in H_2O and possibly NO_2 as well. The derivation of CO from measurements of the limb radiances near 4.6 μm must incorporate non-LTE populations at altitudes above the stratopause, and the retrieval of NO from limb radiance measurements at 5.3 μm in the stratosphere and mesosphere requires a knowledge of the non-LTE populations of the NO(1) level in these regions and in the thermosphere. Similar considerations apply to the interpretation of spectral data on the atmospheres of planets other than the Earth, and of comets.

1.6   Some Historical Background

Having defined the subject and discussed its current importance, we now look briefly at its recent history.

Historically, the solution of non-LTE situations has been linked to the development of radiative transfer theory and to the description of the radiating properties of matter. The concept of non-LTE was introduced in the context of stellar atmospheres by Milne in 1930. After this first treatment, this subject has been extensively studied in astrophysics, mainly to interpret stellar spectra and to study line formation in hot stars (see 'References and Further Reading' section).

Fig. 1.7   Edward Arthur Milne, who first introduced the non-LTE concept.

In 1949, Spitzer first pointed out the possibility that at the low pressures found in the upper atmosphere, the radiative field could upset the state of LTE. No quantitative treatment was given, however. The first application of a non-LTE formulation in the terrestrial atmosphere was by Curtis and Goody in their study in 1956 of the CO_2 15 μm cooling rate in the mesosphere. They formulated the problem for the simplest case of a two-level transition, including only thermal collisions and radiative processes. At the same time, coinciding with the development of the first electronic computers, Curtis devised a linear parameterization of the radiative transfer equation to calculate the cooling rates of the CO_2 15 μm bands. Although this was devised for LTE conditions, it can also be applied to non-LTE situations by replacing the Planck function by the source function. This was the origin of the Curtis matrix method, as it is now called, which is extensively used in studies of atmospheric infrared bands under non-LTE and is described in detail in Sec. 5.7.

The basic principles of the solution of the system of coupled equations have remained essentially unchanged since this first formulation. Their application to atmospheric infrared bands has continued vigorously, facilitated by the availability of ever more powerful, although never completely adequate, computers. Most researchers have paid particular attention to the relative contributions of the many possible energy transfer paths, with a view to obtaining computational economy as well as insight into the molecular processes involved. These included Houghton, and Kuhn and London, both in 1969, and Dickinson, who established the importance of weak bands in the radiative cooling of the atmosphere and developed very comprehensive models for the CO_2 levels and bands in the Venusian and Martian atmospheres. Shved became another of the pioneers in this field in 1975, when he developed a general treatment for linear molecules in the upper atmospheres of planets and cool stars. He and his colleagues later applied the same treatment to all of the major infrared-active atmospheric molecules, making their group one of the reference centres in the non-LTE field.

Kumer and co-workers also made significant progress in the seventies with their studies of the CO_2 4.3 μm bands, the excitation mechanisms of these levels and the analysis of rocket measurements, including observations under auroral conditions. This effort continued in the eighties and nineties with at the Air Force Research Laboratory where these studies were advanced with further modelling and analysis of rocket and spacecraft measurements of CO_2 at 15, 4.3 and 2.7 μm.

One of the long-term, driving aims in non-LTE research has been to develop accurate and computationally efficient algorithms for the calculation of cooling rates (particularly for CO_2 at 15 μm) in global models. Many investigators have paid attention to this problem using different techniques and parameterizations (see Chapter 5 for more details) and currently achieve good parameterizations that, with the continued rapid growth of very fast computers, are constantly being updated.

One of the important advances in the late seventies/eighties was the extension of the theoretical treatment of non-LTE to deal with situations where both vibrational-translational (thermal) and vibrational-vibrational collisions were important. The latter were included to account for the vibrational relaxation or collisional reordering of states excited mostly under daytime conditions, either by solar radiation, photochemical processes or from collisions with excited molecules. Originally, only two extreme situations were considered: (i) when only thermal collisions and radiative losses were at work, e.g., the classical non-LTE case treated by Curtis and

Goody; or (ii) daytime conditions where direct solar excitation was the only source. The case of CO_2 at $15\,\mu m$, under daytime conditions, has an additional complexity: the absorption of solar radiation by the near-IR bands at 4.3 and $2.7\,\mu m$, and subsequent vibrational relaxation, has important effects on the populations of the v_2 levels. This problem was first addressed by Kumer (1977), who used an iterative scheme, and later on by López-Puertas *et al.* (1986) who solved the coupled system of radiative transfer and statistical equilibrium equations for the CO_2 energy levels and bands at 2.7, 4.3, and $15\,\mu m$. Subsequently, several other groups treated this problem. For example, Solomon *et al.* (1986) extended the formulation of non-LTE with the incorporation of vibrational-vibrational collisions, to study the O_3 $10\,\mu m$ emission measured by the LIMS satellite instrument.

Early in the nineties, a major advance was achieved, and reported by three groups of investigators using different sources of measurements. This was an appreciation of the very efficient excitation and de-excitation of the CO_2 bending mode which results from collisions with atomic oxygen. Although this effect had, in fact, been previously predicted by Crutzen in 1970, the discovery dramatically changed our view of the energy balance of the upper atmospheres of the Earth, Mars, and Venus.

Other recent non-LTE studies, reviewed in Chapter 8, have focussed on the effects of non-LTE populations on the radiance emitted from the atmospheric limb by excited molecules, and the use of measurements to understand the processes at work. Transmission, as opposed to emission, measurements, using solar occultation at high spectral resolution, have also proved very useful for understanding non-LTE processes. Both emission and transmission measurements have utilised our knowledge of non-LTE to retrieve atmospheric abundances. The next step is to implement non-LTE theory directly into retrieval schemes used with measurements taken by infrared instruments on board space platforms to retrieve accurate temperature and composition measurements of the mesosphere and lower thermosphere. Another direction of continued interest is the development of accurate cooling/heating algorithms to be implemented in global models of the upper atmospheres of the terrestrial and giant planets.

1.7 Non-LTE Models

One purpose of the present book is to describe how radiative transfer in the middle and upper atmosphere can be computed in spite of the complications introduced by the effects of non-LTE. As we have seen, LTE breaks down when collisions between molecules are infrequent enough that the radiative lifetimes of the important vibrational energy states of a molecule become comparable to the mean time between collisions. Then the populations of the states are no longer determined by simple Boltzmann statistics, but rather by a mixture of collisional interactions and the exchange of quanta. The former can involve both vibrational-translational (V–T) and vibrational-vibrational (V–V) exchanges, and both can involve a multiplicity of levels and of exchange partners. When all of these are considered it is no simple matter to perform reliable calculations of the middle atmosphere temperature structure, to incorporate the non-LTE energy exchange into dynamical models of the region, or to calculate populations of excited states in order to retrieve temperature or atmospheric constituents from measurements of the limb radiance.

The non-LTE model for CO_2 we describe in Chapter 6 is developed from that published by the present authors and our colleagues in 1986. It includes the vibrational-vibrational exchange of energy between different states of a single or different molecule(s) and treats radiative transfer using a modified version of the original Curtis matrix method, which allows the inclusion of the V–V collisional processes. The most important bands of CO_2 at 15, 4.3 and 2.7 μm, and the energy exchange between the v_2 (bending quanta) and the v_3 (asymmetric stretching quanta) reservoirs are included, thus allowing the daytime case to be treated. The output is not only the populations of the different states, including those of interest for temperature sounding of the upper atmosphere, but also the thermal cooling and heating rates needed to investigate the energetics of carbon dioxide. The model incorporates the CO 4.7 μm state, the H_2O 6.3 and 2.7 μm states, and the vibrational energy exchange between the H_2O levels with vibrationally-excited oxygen. The large concentration of $O_2(1)$ means it acts as an efficient intermediate agent in transferring energy between the H_2O, CO_2, CH_4 and CO states.

Results which are discussed in later chapters include populations for the CO(1) state, and the bending, symmetric- and asymmetric-stretch modes of H_2O. These results also include the populations of the first vibrational level of N_2 and O_2, which act as reservoirs for the CO_2 v_3 and H_2O v_2 levels, respectively. In addition to these molecules, the model is applied to other species emitting in the infrared, including (i) O_3, for which the nascent vibrational distribution in its formation reaction and subsequent collisional relaxation are key processes; (ii) CH_4 and N_2O, which are rather similar to H_2O in that collisions with $O_2(1)$ also play a major role and little is known about the non-LTE populations of their vibrational levels; (iii) NO_2, for which non-LTE processes are also poorly known and present some similarities with the ozone molecule; and (iv) NO, for which radiative transfer is not very important but which exhibits rotational non-LTE populations in the thermosphere, where its emission represents one of the major sinks of energy. Of the other nitrogen compounds important in stratospheric chemistry, nitric acid does not exhibit significant non-LTE effects, since its abundance is only significant up to the stratopause, where collisions are fast.

The main deficiencies in non-LTE models at the present time are: (i) their computational complexity, which limits the extent to which they can be included in retrieval codes and in 2D and 3D dynamical models; (ii) the limited amount of experimental data elucidating the processes in the atmosphere; (iii) uncertainties in the rates of the energy transfer in vibrational-translational and vibrational-vibrational collisional processes; and (iv) incomplete databases on spectroscopic parameters such as line strengths, especially for certain species such as O_3 and NO_2 in their vibrationally excited modes. All four areas are gradually being improved by faster computers, new satellite programmes, and studies in the laboratory. The most dramatic progress is likely to come, however, from simultaneous and co-located measurements of the non-LTE emissions by infrared instruments, on one hand, together with independent 'non-LTE free' measurements (e.g. using occultation or microwave techniques) of the kinetic temperature and species abundances under consideration on the other.

1.8   Experimental Studies of Non-LTE

The experimental determination of the various rate coefficients and other molecular parameters which are used in the formulation of non-LTE radiative transfer models is a complex topic which is largely beyond the scope of this book. We will use the best values currently available, and provide references to the original work for the reader who wishes to delve into how the values were actually obtained.

It should be noted, however, that it is quite difficult to simulate the non-LTE behaviour of the atmosphere in the laboratory, because of the difficulties involved in reproducing the long path lengths required at the appropriate temperatures and pressures, and in introducing the appropriate solar irradiation. In practice, therefore, the coefficients used in atmospheric models are, as often as not, extrapolated from conditions which are well outside the range of parameters within which we apply them. Furthermore, the extrapolation is usually based, at best, on theoretical assumptions which are often quite naive. This is the best we can do at the present time, and it needs to be constantly kept in mind that the possibility for error and uncertainty is considerable.

The experimental study of non-LTE in the real atmosphere began with infrared radiance measurements from rocket payloads, but nowadays it more often involves instruments on orbiting satellites. The data can be used to improve non-LTE models, by clarifying the processes involved and reducing the uncertainties described in the previous paragraph. Numerous examples of this will be found in the following chapters, especially 6 and 7. In Chapter 8, the emphasis is more on the situation where the non-LTE model is assumed to be reliable and is used to interpret measured radiances in terms of basic atmospheric properties such as temperature or composition. Clearly, there is interplay between the two, and situations will arise where it is difficult to say whether discrepancies between models and measurements are due to unusual (and potentially interesting) atmospheric conditions, or more prosaically to the limitations of the model and its input parameters.

Experiments are constantly being devised to obtain better molecular constants, in the laboratory and in the field, so that models, and their value as a tool for probing the large-scale behaviour of the atmosphere, are steadily improving.

1.9 Non-LTE in Planetary Atmospheres

Non-LTE situations are not limited to the Earth's atmosphere. The fact that the atmospheres of the other terrestrial planets, Mars and Venus, consist of nearly pure carbon dioxide makes the study of non-LTE processes in that molecule of crucial importance for the understanding of the thermal structure of their upper atmospheres. Also, the different physical and chemical conditions in these atmospheres with respect to the Earth offer scenarios where different combinations of processes may be at work, and comparison with the terrestrial case can provide deeper insights into the non-LTE problem. For example, the weak hot and isotopic bands of CO_2 are much more important for radiative cooling in the Venusian atmosphere than on the Earth. In the $15\,\mu$m bands, LTE applies up to much lower pressures (higher altitudes) than on the Earth. The larger concentration of CO_2 means that these bands are optically thicker at a given pressure level, less radiative energy is lost, and a slower thermal collisional rate is enough to keep their populations in LTE.

Differences in the compositions of the terrestrial planet atmospheres are important in other ways. The major atmospheric constituent in Earth's atmosphere, N_2, acts as an intermediary through its first vibrationally excited state to redistribute the vibrational energy of CO_2 v_3 levels excited by absorption of solar radiation. The second most abundant species, O_2, plays a similar role for distributing the vibrational energy of water vapour and methane. On Mars and Venus it is the CO_2 itself which, through much faster near-resonant vibrational-vibrational energy exchange, distributes the absorbed solar energy internally and makes possible efficient thermal cooling at $15\,\mu$m.

The atmospheres of the giant planets, and some of their satellites, especially Titan, contain radiatively active gases such as methane and other hydrocarbons. Terrestrial experience suggests that these gases should play an important role in the thermal balance of their upper atmospheres, but the application of non-LTE methods to these cases is in its infancy. Also, relatively few measurements are available for comparison to test the preliminary models which have been produced, or to derive accurate concentrations for the active gases. The situation is similar concerning the behaviour and concentrations of water vapour, carbon monoxide, methane and other gases in the very tenuous atmospheres which make up the comae of comets. Non-LTE in all these atmospheres is discussed in Chapter 10.

1.10 References and Further Reading

For more background to the basic properties of the atmosphere and atmospheric processes (dynamics, chemistry and radiation), see Houghton (1986), Brasseur and Solomon (1986), Andrews *et al.* (1987), Chamberlain and Hunten (1987), Rees (1989), Goody (1995), Salby (1996), and Andrews (2000). Wayne (1985) treats the chemistry of the planets and satellites in detail as well as that of the Earth. Current knowledge of the lower and middle atmosphere, its chemistry and the possible relevance to global change is described in Brasseur *et al.* (1999). Trends in CO_2 measurements from Mauna Loa can be found in http://cdiac.esd.ornl.gov/trends/co2/-nocm-ml.htm (Keeling and Whorf, 2000).

Atmospheric remote sounding is treated in Houghton *et al.* (1984), and more recently by Rodgers (2000). Hanel *et al.* (1992) cover the topic with emphasis on planetary atmospheres in the infrared, with a good basic treatment of radiative transfer. Chapter 4 in Brasseur and Solomon (1986) and Chapter 2 in Andrews *et al.* (1987) also give brief introductions to atmospheric radiation.

More comprehensive texts on atmospheric radiation include those by Kondratiev (1969), Coulson (1975), and Liou (1980). Goody and Yung (1987) is recommended as the most complete text on the subject, at least for the infrared part of the spectrum. Liou (1992) has a detailed treatment of radiation with emphasis on radiation and clouds.

The historical development of the study of non-LTE may be traced through a selection of seminal papers, including Milne (1930), Curtis and Goody (1956), Houghton (1969), Kuhn and London (1969), Williams (1971), Dickinson (1972, 1984), Kumer and James (1974), Shved (1975), López-Puertas *et al.* (1986a,b), and Wintersteiner *et al.* (1992). Sections on non-LTE may be found in the general atmospheric physics books by Goody and Yung (1989), Chapter 5 in Houghton (1986), Chapter 4 in Brasseur and Solomon (1986), Chapter 2 in Andrews *et al.* (1987), and Chapter 13 in Thorne (1988).

Most of the existing literature on non-LTE actually deals with stellar atmospheres, e.g., Chandrasekhar (1960), Feautrier (1964), Rybicki (1971), Athay (1972), Mihalas (1978), and Simonneau and Crivellari (1993). The molecules of interest and the physical conditions are, of course, quite different to those which concern the Earth and planetary atmospheric scientist, but interesting comparisons, especially with mathematical and computational techniques and approximations can be made; for overviews see e.g. Mihalas (1978) and Rybicki and Lightmann (1979).

Chapter 2

Molecular Spectra

2.1 Introduction

In this chapter we consider the energy levels of an individual molecule, and then how the populations of those levels are determined in a gas in LTE and in non-LTE situations, by interactions between colliding molecules and between molecules and photons.

As mentioned in Chapter 1, the statistics of energy level populations in LTE are the results of the translational energy of the gas reaching dynamic equilibrium with the internal vibrational and rotational energy states of the individual molecules, so that the mean population of each level is a function only of the kinetic temperature of the gas. In non-LTE situations, on the other hand, important processes include the exchange of vibrational quanta between molecules during collisions, chemical effects, and the absorption and emission of photons travelling to and from sources at temperatures different from the local kinetic temperature in the neighbourhood of the molecule. We shall look at each of these effects in the chapters which follow. First, we need to explain the energy levels of individual molecules, the way in which they interact with individual photons, and the transitions which result.

To do this we use models of the molecules of interest and calculate their energy states using basic quantum theory. In general, we want to use the simplest model which accounts, with sufficient accuracy, for the more readily observable features of the infrared spectrum of the molecule. We start, therefore, by calculating the energy levels of the simplest possible model of the simplest non-trivial molecule, the semi-classical 'rigid rotor' representation of a diatomic like CO, and then proceed gradually to more complex models and more complex molecules.

The atmosphere is composed mainly of diatomic molecules of nitrogen and oxygen. However, these are homonuclear molecules which lack a permanent dipole moment, and so do not interact with photons in the way asymmetric, and therefore polarized, molecules do. In fact, most of the interesting species for atmospheric radiation are triatomic molecules —the most important for infrared radiative transfer being CO_2, H_2O and O_3. After these three, the commonest gases which are important for radiative transfer in the atmosphere are the diatomics CO and NO, and the pentatomic CH_4.

2.2   Energy Levels in Diatomic Molecules

Molecules, if they contain two or more atoms, can store internal energy as rotations and vibrations about the common centre of gravity. As with the electronic states of an atom or molecule, molecular vibrational and rotational energy levels are quantized. The separation between vibrational levels is much less than for electronic states, while adjacent rotational levels are much closer than adjacent vibrational levels (Fig. 2.1). A molecule which is in an excited vibrational or rotational state can relax to a lower energy state by emission of a photon, and the latter is generally of such an energy that its wavelength is in the infrared region of the spectrum, and, conversely, an infrared photon of a wavelength which matches an allowed transition can be absorbed by a molecule by excitation of one or more vibrational or rotational states.

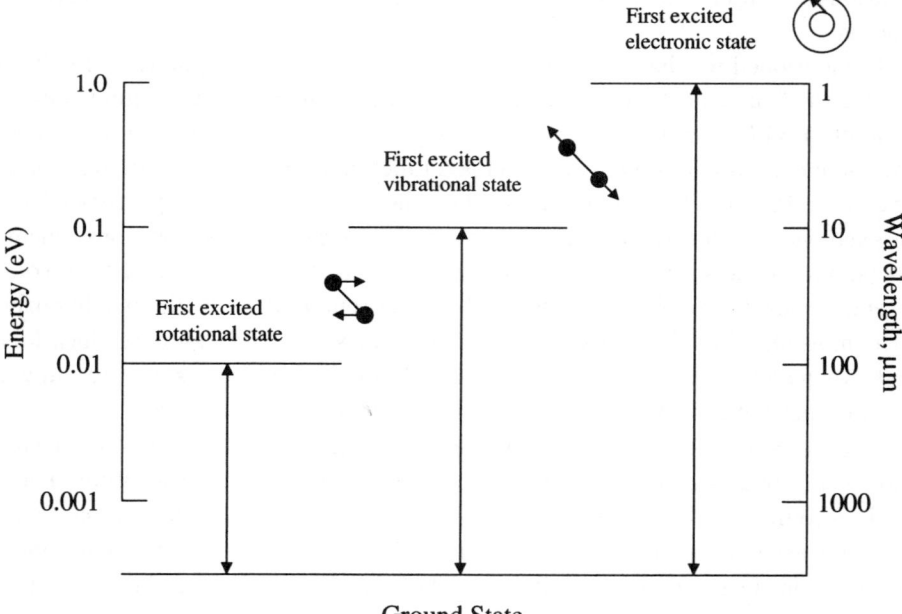

Fig. 2.1   Energy level diagram for a diatomic molecule, illustrating the three types of transition which can occur, and their relative magnitudes.

2.2.1   *The Born–Oppenheimer approximation*

The energy of the first excited state relative to the ground state for each of the three kinds of internal energy is typically an order of magnitude or more greater for pure vibrational than for pure rotational transitions (see Fig. 2.1). Electronic transitions are in turn, one or more orders of magnitude more energetic than vibrational transitions. In terms of wavelength, electronic transitions give rise to (or absorb) visible or UV photons, while vibrational and rotational transitions involve photons in the near to middle infrared, and far infrared to microwave parts of the electromagnetic spectrum respectively.

To first order, the different kinds of internal energy of a molecule can be treated separately, as if they do not interact with each other (the *Born–Oppenheimer* approximation). The energy of the first excited vibrational state, for example, will then be independent of the state of rotational excitation of the molecule, and so on. With the help of this simplification, we can examine the interactions between molecules and photons emitted from those levels using semi-classical models, starting with the simplest. We will return later to the question of what happens when the Born–Oppenheimer approximation breaks down, as it does most noticeably when large quantum numbers, i.e. highly excited states, are involved.

2.2.2 *Rotation of diatomic molecules*

Diatomic molecules rotate about their centre of mass. In the simplest model, the bond joining the two atoms is rigid. This is a reasonably good approximation, although for rapid rates of rotation the bond can stretch, changing the molecular moment of inertia and shifting the energy levels relative to those predicted by the rigid bond model. A refinement in which the bond is modelled as a linear (Hooke's law) spring, a better but still not exact approximation to reality, gives good results.

2.2.2.1 *The rigid rotator model*

In this model we assume that the molecule consists of two unequal masses m_1 and m_2, joined by a rigid bond of fixed length r (Fig. 2.2).

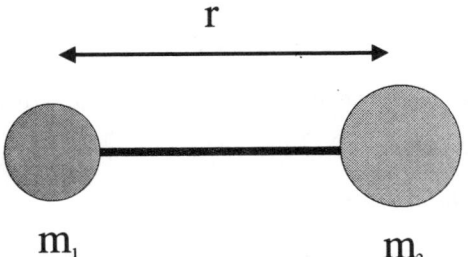

Fig. 2.2 The rigid rotator model of a diatomic molecule.

The system possesses energy $E = I\omega^2/2$ and angular momentum $I\omega$ due to the rotation of the molecule about its centre of mass, where the moment of inertia is $I = mr^2$, $m = m_1 m_2/(m_1 + m_2)$ is the *reduced mass*, and ω is the angular velocity of rotation. The time–independent Schrödinger equation for this system is

$$\frac{h^2}{8\pi^2}\left(\frac{1}{m_1}\nabla_1^2 + \frac{1}{m_2}\nabla_2^2\right)\Psi + E\,\Psi = 0$$

with solution

$$\Psi_{J,M} = \sqrt{\frac{(2J+1)(J-|M|)!}{4\pi(J+|M|)!}}\ e^{im\phi}\ P_J^{|M|}(\cos\theta), \qquad (2.1)$$

where J and M are integral quantum numbers representing the total angular momentum, and its component along the polar axis, respectively ($|M|\leq$ J). The terms $P_J^{|M|}(\cos\theta)$ are associated Legendre polynomials. The corresponding energies of the rotating molecule are

$$E_J = BJ(J+1) \qquad\qquad J = 0,1,2,3,\ \ldots \qquad (2.2)$$

where h is Planck's constant, $\hbar = h/(2\pi)$, and $B = \hbar^2/(2I)$ is called the *rotational constant*. Since the energy of the system is $E = I\omega^2/2$, it follows that

$$I\omega = \hbar\sqrt{J(J+1)}. \qquad (2.3)$$

$I\omega \approx \hbar J$ for large J, so the angular momentum of the system is approximately proportional to the value of J, while the separation of the energy levels increases linearly with increasing J (Fig. 2.3).

Only certain transitions between states are allowed. The selection rule for a transition between a state J'' and one of higher energy J' follows from the form of the wavefunctions given above. The interaction energy between two states is proportional to the relevant element of the electric dipole moment matrix, which is

$$\mu_{J'M'J''M''} = \int \Psi_{J'M'}^* \,\mu\, \Psi_{J''M''}\, d\varphi,$$

where φ is the solid angle and the integral is non-zero only if $\Delta J = J' - J'' = \pm 1$.

Thus, the energy involved in a transition from one level to another is

$$E_{J+1} - E_J = 2B(J+1), \qquad (2.4)$$

from which it follows that the lines in the spectrum are equally spaced by $2B$ wavenumbers.

2.2.2.2 *Non-rigid rotator model*

The deficiencies of the simple rigid rotator model are exposed if we construct a table of the values of the atomic separation r which is obtained by measuring the separation of several pairs of adjacent lines. The example in Table 2.1 is for the diatomic molecule hydrogen fluoride (HF). (Note the convention whereby in describing a transition between rotational states, J'' is used to denote the rotational quantum number of the lower, and J' of the upper state, respectively).

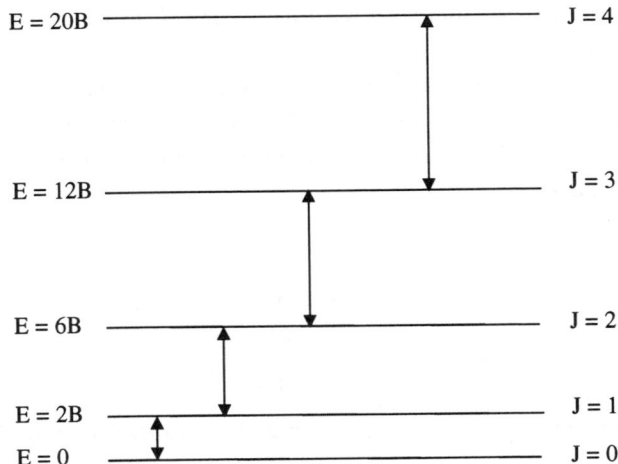

Fig. 2.3 Energy level diagram for a rigid rotator, showing allowed transitions between rotational levels.

Table 2.1 Atomic separation in HF.

Lines use $(J'' - J')$	Atomic separation* (Å)
0–1 and 1–2	0.929
1–2 and 2–3	0.931
2–3 and 3–4	0.932
...	...
...	...
10–11 and 11–12	0.969

*Determined from the separation of several different pairs of transitions.

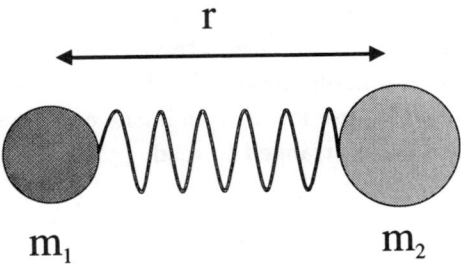

Fig. 2.4 The non-rigid rotator model of a diatomic molecule.

We see that the inferred value of r increases systematically with J. The classical interpretation of this behaviour is that the interatomic bond is not in fact rigid, but stretches in response to the centrifugal force due to the rotation of the molecule.

The stretching, of course, increases with increasing J, as the molecule rotates faster to store more energy.

In accordance with this discovery, we adopt a more sophisticated model in which the rigid bond is replaced by a linear spring of force constant k (Fig. 2.4).

Equating the restoring force due to bond stretching to the centrifugal force due to rotation we have

$$k(r - r_0) = mr\omega^2 \qquad (2.5)$$

from which we see that the length of the distorted bond is

$$r = \frac{k}{(k - m\omega^2)}\, r_0. \qquad (2.6)$$

The energy of the system is the sum of that due to rotation and that stored in the spring, i.e.

$$E = \frac{1}{2}I\omega^2 + \frac{1}{2}k(r - r_0)^2 = \frac{(I\omega)^2}{2I} + \frac{(I\omega)^4}{2kr^2I^2} \qquad (2.7)$$

but

$$\frac{1}{2I} = \frac{1}{2mr^2} \approx \frac{1}{2mr_0^2} - \frac{2(I\omega)^2}{2kr_0^2I^2}$$

hence, writing $I\omega = \hbar\sqrt{J(J+1)}$ as before,

$$E \approx \frac{\hbar^2 J(J+1)}{2mr_0^2} - \frac{\hbar^4 J^2(J+1)^2}{2km^2r_0^6}, \qquad (2.8)$$

collecting the constants together we get

$$E = BJ(J+1) - DJ^2(J+1)^2, \qquad (2.9)$$

where B is the same as before and we have defined the *centrifugal distortion constant* D. While B is typically of the order of $10\,\mathrm{cm}^{-1}$, D is much smaller at around $10^{-3}\,\mathrm{cm}^{-1}$. Hence the second term does not become comparable to the first until J is of order 100, and can often be neglected for transitions near the ground state. This explains why a reasonably good value for r can be obtained from the rigid rotator model, provided we use low-J lines in the experiment.

Equation (2.8) can be re-arranged to read

$$E = \frac{\hbar^2 J(J+1)}{2I_0}\left(1 - \frac{\omega_r^2}{\omega_v^2}\right), \qquad (2.10)$$

where $\omega_r = (\hbar/I)\sqrt{J(J+1)}$ is the frequency of rotation and $\omega_v = \sqrt{k/m}$ is the classical frequency of vibration of the model system, which is of course an harmonic oscillator. The ratio ω_r^2/ω_v^2 is typically about 10^{-2} or smaller. In this way, the vibrational frequency ω_v can be found from measurements of high-J line position in the purely rotational spectrum.

2.2.3 *Vibrations of diatomic molecules*

Since the interatomic bond in a diatomic molecule can stretch, we would expect that energy can also be stored as vibrations of the system. Clearly, we could compute very easily the energy stored in vibrations of the simple system shown in Fig. 2.4. However, although this model works quite well for a purely rotating and stretching molecule, we shall find below that it is not good enough to analyse the vibrations of the real molecule it represents, because the amplitude of the stretching is generally larger during vibrations and the non-linearities of the spring can no longer be neglected.

2.2.3.1 *Interaction energy*

Let us therefore consider the potential energy of the system as a function of the separation of the two atoms. The quantity shown in Fig. 2.5 is the *interaction energy* $V(r)$, defined as the potential energy (E_p) which the atoms possesses by virtue of being bound, i.e. the total E_p minus the sum of the E_p's which the two atoms would possess if separated to infinity.

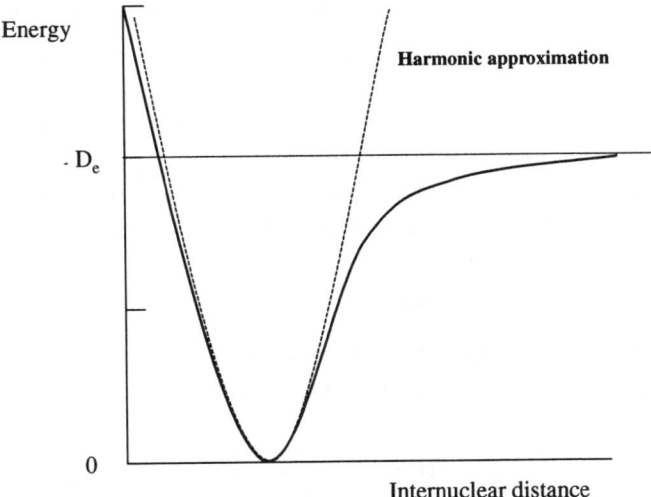

Fig. 2.5 The potential energy V of the HCl molecule as a function of internuclear distance r. The dashed curve shows the simple harmonic approximation to $V(r)$. After Herzberg.

The general appearance of the solid curve in Fig. 2.5, which of course is different in detail for different molecules (and for different electronic and vibrational levels of the same molecule) can be estimated from the following arguments:

(i) $V(r)$ must tend to infinity as r tends to zero, since the atoms cannot be moved infinitely close together. The smoothness of the curve as drawn recognises that the atoms are not rigid balls, but that we expect the resistance to bringing the two close together to increase rapidly for small r;

(ii) $V(r)$ must tend to a constant value of zero as r tends to infinity, by definition of $V(r)$;

(iii) $V(r)$ must have a minimum at some finite value of r, otherwise the molecule could not exist as a stable entity. It is convenient to redefine this minimum to be the zero of the energy scale; the value at $r = \infty$ then becomes the *dissociation energy*, D_e.

We need an expression for V in terms of r in order to calculate the energy of the vibrations. One way to do this is to model the actual interactions of the nuclei, and in this way to derive an expression for $V(r)$. Alternatively, it is possible to find an analytic function which approximately fits the shape of the molecular potential energy curve. One such is the *Morse* function

$$V(r) = D_e \left\{ 1 - \exp[-\beta\,(r - r_0)] \right\}^2 ,$$

where the constant D_e is the depth of the potential well (that is, the dissociation energy of the molecule) and the constant β is equal to $\sqrt{k/(2D_e)}$. When the Morse function is used, an explicit solution to the Schrödinger wave equation can be found in the form

$$E = \left(v + \frac{1}{2}\right)\nu_v - \left(v + \frac{1}{2}\right)^2 \nu_v \chi, \tag{2.11}$$

with selection rule $\Delta v = \pm 1, \pm 2$. The constant $\nu_v = \hbar\sqrt{k/m}$ is the *fundamental frequency* and $\chi = \nu_v/(4D_e)$ is known as the *anharmonicity constant*.

2.2.3.2   *The harmonic oscillator model*

A more physically insightful way to derive the same expression is to write the interaction energy in the form of a Maclaurin series, thus:

$$V(r - r_0) = V(0) + \left[\frac{\mathrm{d}V}{\mathrm{d}r}\right]_{r=r_0} (r - r_0) + \frac{1}{2}\left[\frac{\mathrm{d}^2 V}{\mathrm{d}r^2}\right]_{r=r_0} (r - r_0)^2 + \ldots \tag{2.12}$$

Now redefine the zero of potential energy to be that at the rest position $r = r_0$ where $\mathrm{d}V/\mathrm{d}r = 0$. Then $V(0)$ is also zero, and

$$V(r - r_0) = \frac{1}{2}\,k\,(r - r_0)^2 + \text{higher order terms.} \tag{2.13}$$

Retaining just the first non-zero term is the same as assuming that the potential energy curve can be approximated by a parabolic function (dashed curve in Fig. 2.5), i.e. that the spring is linear in restoring force as a function of displacement. This is therefore known as the *harmonic oscillator* approximation.

The solution of the wave equation for the harmonic oscillator gives us an expression for the quantized energy which is

$$E = \left(v + \frac{1}{2}\right)\hbar\sqrt{\frac{k}{m}} = \left(v + \frac{1}{2}\right)\nu_v. \tag{2.14}$$

Note that the system has ground state energy even when the vibrational quantum number v is zero. The selection rule is $\Delta v = \pm 1$, from which it follows that the energy of a transition is

$$E' - E'' = \left[\left(v' + \frac{1}{2}\right) - \left(v'' + \frac{1}{2}\right)\right]\nu_v = \Delta v\, \nu_v = \nu_v \tag{2.15}$$

and so all lines fall at a single frequency ν_v called the *fundamental frequency**. As already noted, vibrational quanta correspond in energy to photons with wavenumbers of the order of $1000\,\mathrm{cm}^{-1}$. We should therefore expect the spectrum to consist of a single strong line in the middle infrared.

In reality vibrational and rotational transitions can, and usually do, take place together and produce a *band* of lines. Any particular line in the band arises from the absorption of photons which have the right wavelength to correspond in energy to the sum or difference of a vibrational and a rotational energy level. Since the rotational energies are relatively small, the lines are grouped around the fundamental frequency, which is at the centre of the band and can be accurately determined from expressions such as (2.10) and the more refined versions to come below. Simultaneous vibrational and rotational transitions (*vibration-rotation bands*) will be dealt with in Sec. 2.4. For the moment, we will continue to treat vibrational transitions in isolation for simplicity.

For hydrogen chloride, HCl, the fundamental frequency is $2886\,\mathrm{cm}^{-1}$. It follows from Eq. (2.15) that the bond strength for this molecule corresponds to a force constant of $k = m\nu_v^2/\hbar^2$, or $480\,\mathrm{N\,m}^{-1}$. This is a very strong spring and it is an interesting exercise to consider what is the amplitude of the vibration for the first excited state ($v=1$). If we let the maximum value of $(r - r_0) = x$, then, in the harmonic approximation, we can equate the potential energy stored in the fully stretched spring to the kinetic energy at zero spring extension, thus:

$$\frac{1}{2}kx^2 = \frac{3}{2}\hbar\sqrt{\frac{k}{m}} \tag{2.16}$$

which can be solved for x given k and m. Alternatively, given that

$$\omega_r = \frac{\hbar\sqrt{J(J+1)}}{I} = \frac{\hbar\sqrt{2}}{mr^2} \quad \text{and} \quad \omega_v = \sqrt{\frac{k}{m}}$$

---

*This is also sometimes called 'harmonic frequency'.

for $J = 1$, then we can write

$$\frac{x}{r} \approx \sqrt{\frac{3}{\sqrt{2}} \frac{\omega_r}{\omega_v}}$$

which is of the order of 0.1 for a typical diatomic molecule.

2.2.3.3   *The anharmonic oscillator*

Laboratory measurements of spectra reveal that transitions forbidden according to the harmonic oscillator model, i.e. those corresponding to $\Delta v = \pm 2$, $\Delta v = \pm 3$ and so forth, do occur with significant probabilities. These arise as a result of non-linearities in the spring connecting the atoms and point to the need for a non-harmonic oscillator model.

To obtain an expression for the energy levels of the anharmonic oscillator we need a better expression for the interaction energy. At this point we obtain the benefit of having set up the expression for $V(r)$ as a series; we can simply keep additional terms in Eq. (2.13) until we have a satisfactory fit. Neglecting all but the first two non-zero terms gives

$$V(r - r_0) = \frac{1}{2}k(r - r_0)^2 + \frac{1}{6}k'(r - r_0)^3$$

whence

$$E = \left(v + \frac{1}{2}\right)\nu_v - \left(v + \frac{1}{2}\right)^2 \chi\,\nu_v \qquad (2.17)$$

with selection rule $\Delta v = \pm 1, \pm 2$. The fundamental frequency has the same value as before, i.e. $\nu_v = \hbar\sqrt{k/m}$ and the anharmonicity constant χ can be evaluated in terms of k', m and the fundamental constants. Terms after the first are generally written as a product so that the fundamental frequency can be factored out of the series, which makes it easier to see the relative size of successive terms. For instance, for HCl

$$E = \left(v + \frac{1}{2}\right)\nu_v\left[1 - 0.0179\left(v + \frac{1}{2}\right) + \ldots\right].$$

Thus we can see how, as for the case of rotation, the harmonic approximation can be useful for transitions which take place near the ground state, but becomes increasingly poor with increasing v. Note that the term added for anharmonicity is negative; this is as we would expect classically since the spring becomes proportionately weaker at higher amplitudes of stretch. Retaining progressively higher order terms admits the progressively weaker harmonic transitions $\Delta v = \pm 3, \pm 4$, etc. The existence of overtone bands like this, and the hot bands already described (see also Sec. 2.4.2), means that, unlike the case for rotational levels, transitions between any pair of vibrational levels are normally allowed.

2.2.4 Breakdown of the Born–Oppenheimer approximation

Vibration and rotation can no longer be independent in the anharmonically oscillating molecule. Classically, we picture the effect of increasing the rate of rotation as stretching the non-linear spring representing the inter-atomic bond to a different value of k, or the effect of non-harmonic vibration as altering the mean value of r and hence the rate of rotation. Thus, the energy of the vibrational levels depends on J as well as v, and vice versa. In a full quantum mechanical treatment, the end result is to introduce a term in the expression for the energy levels which depends on both quantum numbers in a simple linear manner. It is convenient to write this *interaction term* in the form $a_r(v + 1/2)J(J + 1)$, where a_r is called the *interaction constant*. Now we have

$$E_r = BJ(J + 1) - DJ^2(J + 1)^2 - a_r\left(v + \frac{1}{2}\right)J(J + 1) \qquad (2.18)$$

for the rotational levels. The interaction constants are generally quite small, less than $1\,\text{cm}^{-1}$, but cannot be neglected in precise work, especially if J or v is large. The negative sign before a_r in these expressions is to make the constant itself a positive number, showing explicitly that the effect of the interaction between vibration and rotation is to reduce the energy of the levels. Classically, this recognises that increasing v will increase the mean moment of inertia of the molecule and so reduce B.

We have already noted that vibrational and rotational transitions can, and do, occur together. Thus, the general expression for all of the energy levels in a diatomic molecule (in the electronic ground state) can be written

$$E_{vr} = \left(v + \frac{1}{2}\right)\nu_v - \left(v + \frac{1}{2}\right)^2\nu_v\chi + BJ(J + 1) - DJ^2(J + 1)^2 - a_r\left(v + \frac{1}{2}\right)J(J + 1).$$
$$(2.19)$$

Again using the example of HCl, values of the constants are $\nu_v = 2886\,\text{cm}^{-1}$, $\nu_v\chi = 52.02\,\text{cm}^{-1}$, $B = 10.59\,\text{cm}^{-1}$, $D = 0.004\,\text{cm}^{-1}$, and $a_r = 0.3019\,\text{cm}^{-1}$.

Additional, higher–order terms can of course be added to achieve greater precision as needed. However, their effect would be so small as to require extraordinary measurement techniques to observe them. Eq. (2.19) is the complete expression for the energy levels of a vibrating and rotating molecule that we set out to find.

2.3 Energy Levels in Polyatomic Molecules

2.3.1 General

The vibrations and rotations of polyatomic molecules can be extremely complicated, particularly for large molecules. We will consider some of the simpler cases, illustrated in each case by a commonly–encountered example. The basic approaches which we will use are extensions of those already applied to diatomic, and provide an introduction to the treatment of complex molecules which can be found in specialised textbooks on the subject.

2.3.2 *Rotation of polyatomic molecules*

The analysis of the rotation of polyatomic molecules of arbitrary complexity proceeds by choosing a coordinate system of three perpendicular axes about which the moment of inertia can be determined. In general, the three values, I_A, I_B, and I_C, may all be different, such molecules are known as asymmetric tops. One common example is the water vapour molecule, H_2O. Other important classes possess symmetry such that an axis can be chosen to make one or more moments of inertia zero, or to make one or more of the three moments equal. Molecules with only one non-zero moment of inertia obviously are linear; CO_2 is one such. Pyramid-shaped molecules like ammonia (NH_3) have two identical non-zero moments, i.e. $I_A = I_B \neq I_C$, and are known as symmetric tops. If all three moments are the same, we have a spherical top; we shall be considering methane (CH_4) as an example of this.

2.3.2.1 *Linear molecules – Carbon dioxide*

One of the most common triatomic, carbon dioxide, is symmetrical (Fig. 2.6) and therefore has no permanent dipole moment and no pure rotational spectrum. Nitrous oxide, on the other hand, is not symmetrical (Fig. 2.6) and so does have a rotational spectrum. In either case, the expression for the energy of rotation is exactly as for diatomic molecules, i.e. in the rigid rotor approximation

$$E_J = \frac{\hbar^2}{2I}J(J+1) = BJ(J+1). \qquad (2.20)$$

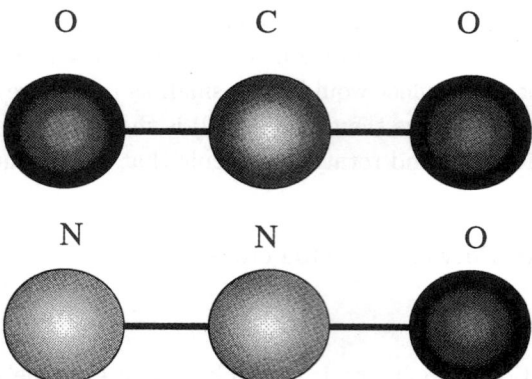

Fig. 2.6   Two linear triatomic molecules – carbon dioxide and nitrous oxide.

The moment of inertia $I = mr^2$, is now calculated with the reduced mass m

given by

$$m^{-1} = m_1^{-1} + m_2^{-1} + m_3^{-1}. \tag{2.21}$$

Obviously, I will in general be larger for triatomic than for diatomic molecules, so the rotational constant B tends to be smaller. For example, $B = 0.2\,\mathrm{cm}^{-1}$ for carbonyl sulphide (OCS), about one order of magnitude smaller than the corresponding value for CO.

2.3.2.2 *Symmetric top molecules – Ammonia*

An example of a *prolate* symmetric top is ammonia, NH_3. The molecule is shaped like a squat pyramid, with the N atom poised above the three H atoms, which form a triangle in a plane (Fig. 2.7). Co-ordinates can be chosen so that such a structure has two equal moments of inertia $I_B = I_C$, and a third moment I_A which is different from the other two. The component of angular momentum around each axis is $p_A = I_A \omega_A$, etc. and the total angular momentum is quantized so that

$$p_A^2 + p_B^2 + p_C^2 = J(J+1)\hbar^2 \tag{2.22}$$

and the energy is

$$E = \frac{p_A^2}{2I_A} + \frac{p_B^2}{2I_B} + \frac{p_C^2}{2I_C}. \tag{2.23}$$

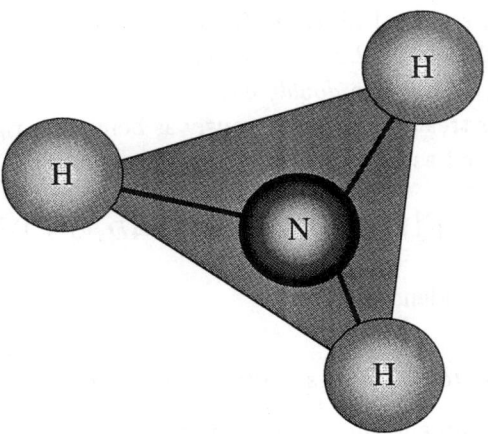

Fig. 2.7   The symmetric top molecule NH_3.

However, the angular momentum along the axis of symmetry is also quantized, with quantum number K:

$$p_A = K\hbar \qquad K = 0, \pm 1, \ldots, \pm J \tag{2.24}$$

and the result, (if the molecule is rigid) is

$$E_J = \frac{\hbar^2 J(J+1)}{2I_B} + K^2\hbar^2\left(\frac{1}{2I_A} - \frac{1}{2I_B}\right) \tag{2.25}$$

or

$$E_J = BJ(J+1) + (A-B)K^2 \tag{2.26}$$

with

$$B = \frac{h}{8\pi c I_B}, \quad A = \frac{h}{8\pi c I_A} \quad [\text{cm}^{-1}].$$

And selection rules (with the dipole moment along the figure axis):

$$\Delta J = 0, \pm 1; \Delta K = 0, K \neq 0$$
$$\Delta J = \pm 1; \Delta K = 0, K = 0.$$

Hence the lines are at wavenumbers given by

$$\nu = 2B(J+1) \quad [\text{cm}^{-1}] \tag{2.27}$$

i.e. they are *independent* of K. Classically, the reason for this is that rotations about the unique axis do not produce a change in angular momentum. Therefore, transitions between states of different K are not allowed. K can take the values $J, J-1, J-2, \ldots, 0, \ldots, 2-J, 1-J, -J$, but $E(K) = E(-K)$, therefore we should expect $J+1$ levels, all but one *doubly degenerate*.

If the molecule is treated more rigourously as being non-rigid, the corresponding formula can be derived as

$$\nu = 2B(J+1) - 2D_{JK}K^2(J+1) - 4D_J(J+1)^2 \quad [\text{cm}^{-1}] \tag{2.28}$$

revealing a weak dependence on K.

2.3.2.3 *Asymmetric top molecules – Water vapour and Ozone*

Asymmetric tops represent the completely general case of molecules with three unequal moments of inertia ($I_A \neq I_B \neq I_C$). The most important examples are water vapour (H_2O) and ozone (O_3). In this case J is the only good quantum number, but the lines are ($2J+1$)-fold degenerate, corresponding to the different values of K which all have the same energy.

The mathematical treatment of these molecules is extremely complicated and we shall not attempt it here: see advanced texts such as Herzberg.

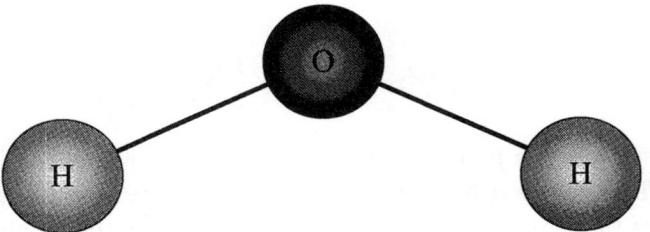

Fig. 2.8   Structure of the asymmetric top molecule H_2O.

2.3.2.4   *Spherical top molecules – Methane*

This apparently simple class of molecules has three equal moments ($I_A = I_B = I_C$); an example is methane, CH_4. There is no change with rotation in the dipole moment presented to an incoming photon, hence no rotational spectrum under normal conditions. (Under high–pressure conditions, such as are found in the atmospheres of the outer planets, sufficient collisional distortion of the molecule occurs to endow methane, and other molecules, with observable rotational spectra).

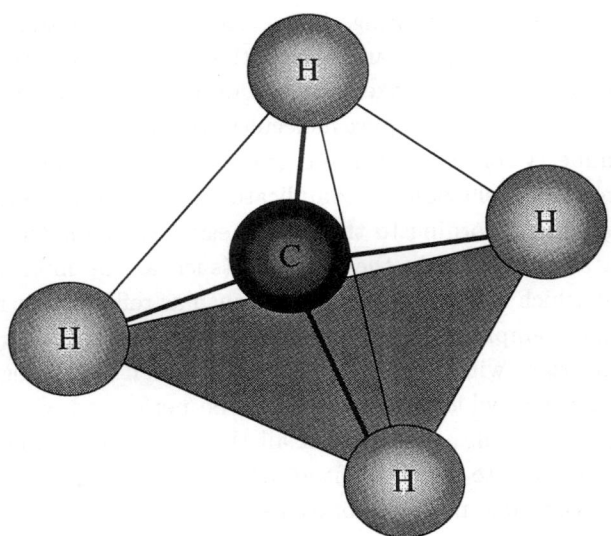

Fig. 2.9   Structure of the spherical top molecule CH_4.

2.3.3   *Vibration of polyatomic molecules*

Again we choose axes of symmetry which allow molecules to be separated into classes. Any mode of vibration can be treated as the sum of *normal vibrations*, such that

$$E_p = \frac{1}{2} \sum_i f_i Q_i^2; \qquad E_k = \frac{1}{2} \sum_i \left(\frac{\mathrm{d}Q_i}{\mathrm{d}t}\right)^2, \qquad (2.29)$$

where E_p and E_k are the potential and kinetic energies, and the Q_i's are *normal coordinates*. These can be defined, for molecules with more than 2 atoms, as linear combinations of $(3N-6)$ cartesian coordinates, since this is the number of internal degrees of freedom in a molecule made up of N atoms (of the $3N$ degrees of freedom possessed by the atoms individually, 3 are used to specify the position and 3 the orientation of the assembly in space, leaving $3N-6$ for internal modes).

The f_i's are force constants, the roots of the secular equation. The energy of the system is

$$E = \sum_i \left(v + \frac{1}{2}\right) h\nu_i, \qquad \text{where } \nu_i = \frac{\sqrt{f_i}}{2p_i}. \qquad (2.30)$$

The ν_i's are the $3N-6$ *normal vibrations*, and any vibration of the molecule may be considered as being made up of a sum of these. Even in a simple molecule, it can be very difficult to determine which modes are present, as an examination of a model of a CO_2 molecule executing all three of its fundamental modes simultaneously, for example, rapidly brings home. At least in this simple case the axes of symmetry are readily defined, and the motion can be analysed in terms of oscillations perpendicular to these axes and the amplitude of each mode determined. It is not easy to tell in more complex molecules which motions constitute the normal modes. Symmetry considerations and group theory assist in determining these. The vibration-rotation behaviour of complicated molecules is studied by assigning them to *point groups*, according to their symmetry. A *symmetry element* is some definable point, axis or plane in the molecule (such as the internuclear axis in a diatomic), about which a *symmetry operation*, such as reflection or rotation, can be defined. Thus, for example, an axis C_3 can be defined in NH_3, where C_p means the vibrations are invariant with respect to rotation through $360/p$ degrees. Ammonia also has three planes in which a reflection can be performed without altering the appearance of the molecule. The sum of all the symmetry elements a particular molecule has determines to which point group it belongs. All of the members of a point group have the same normal vibrations.

2.3.3.1   *Fundamental vibrations of CO₂*

Let us consider vibrations of this important linear triatomic molecule in more detail. There are three fundamental modes, as shown in Fig. 2.10, and so any vibration of the molecule can be considered to be a linear combination of these. Fundamental modes of vibration are designated ν_1 to ν_n where the ν-numbers are used consistently to designate a particular type of motion, for example, ν_1 is always a symmetric stretching mode. We also called the bands arising from these modes of vibration by ν, e.g., the ν_2 or the ν_3 band. Vibrational levels are represented by v, e.g., (v_1, v_2, v_3) for the level excited in the vibrational modes ν_1, ν_2, and ν_3 (see Appendix C for more details on this terminology).

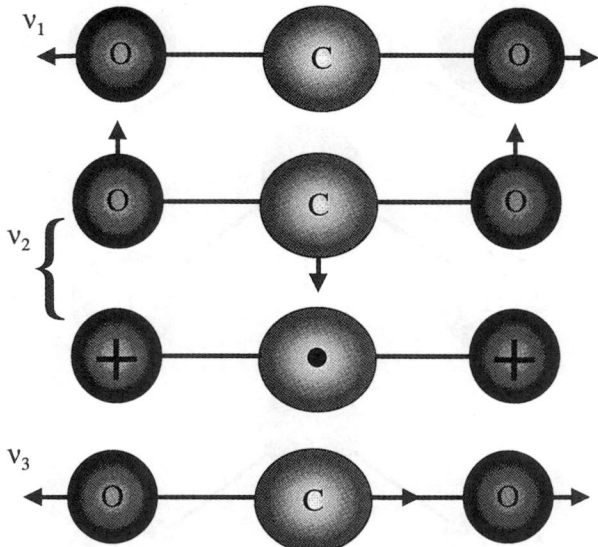

Fig. 2.10   Fundamental vibrations of the CO_2 molecule. The solid dot and the '+' denote approaching and receding movements perpendicular to the plane of the page, respectively.

In CO_2, the ν_1 vibration, like pure rotation, is inactive for absorption and emission of radiation as there is no change of dipole moment associated with the motion. For the other two fundamentals, the selection rules are $\Delta J = 0, \pm 1$ for the ν_2 band, and $\Delta J = \pm 1$ for the ν_3 band. The asymmetric mode, ν_3, is also called a *parallel* mode, because there is no change of dipole moment in a direction perpendicular to the internuclear axis. Purely vibrational transitions (i.e. without accompanying rotational transitions) are forbidden in this case. The bending mode ν_2 (a *perpendicular* band) has no such restriction, but does have the complexity that the mode is doubly degenerate (because the bending can take place in either of two perpendicular directions). The degeneracy is removed by the rotation of the molecule, giving rise to *l–type* doubling; each line in the spectrum consists of a closely–spaced pair or *doublet*.

2.3.3.2   *Water and ozone vibrational modes*

The three fundamental modes of vibration for nonlinear triatomic molecules like water and ozone are shown in Fig. 2.11. In this case, none of the modes is degenerate. The energy levels, in the harmonic approximation with no coupling between modes, are found as described above, with each atom executing simple harmonic motion. The more realistic anharmonic case is again a problem to be found in specialised

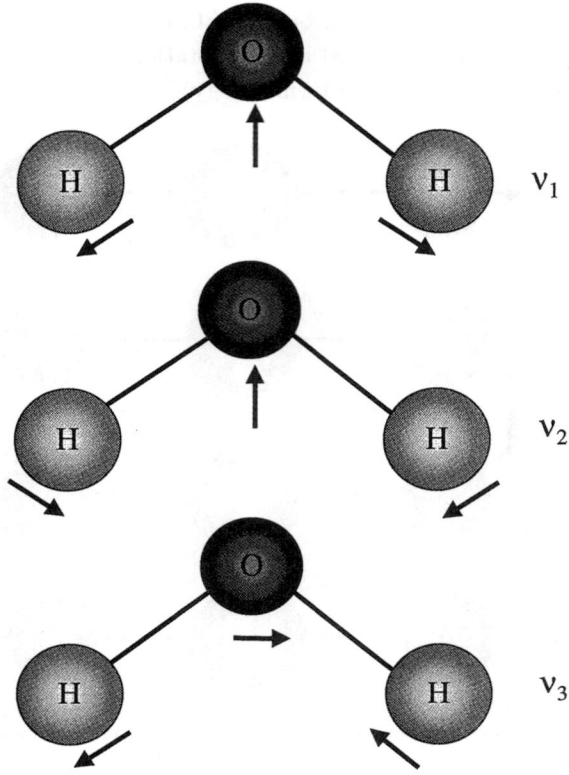

Fig. 2.11   Fundamental vibrations of water vapour.

textbooks.

2.4   Transitions and Spectral Bands

2.4.1   *The vibration-rotation band*

We have seen that the energy E of a photon, incident on a diatomic molecule, can be absorbed if E matches the separation between any of the levels defined by Eq. (2.9) when J'' and J' differ by ± 1. This selection rule for rotational transitions is observed rigourously by real molecules, unaffected by anharmonicity; for vibrational transitions, on the other hand, the selection rule $\Delta v = \pm 1$, which arises in the harmonic approximation, does not hold and any change in v is possible. As noted above, $v'' = 0$ to $v' = 1, 2, 3, \ldots$ are the most important transitions under LTE conditions, but only because the ground state alone contains a significant population at ordinary temperatures (see Sec. 2.5.1). The energy levels of the system for this example, and the allowed transitions, are shown in Fig. 2.12. The vertical lines at the bottom of the figure shows the relative position of the spectral lines on a scale with wavenumber increasing from left to right. In the harmonic oscillator model, the positions are given by

$$\nu_R = \nu_0 + 2B(J+1) \quad \text{for lines with} \quad \Delta J = +1 \quad \text{(the } R \text{ branch)} \quad (2.31)$$

$$\nu_P = \nu_0 - 2BJ \quad \text{for lines with} \quad \Delta J = -1 \quad \text{(the } P \text{ branch).} \quad (2.32)$$

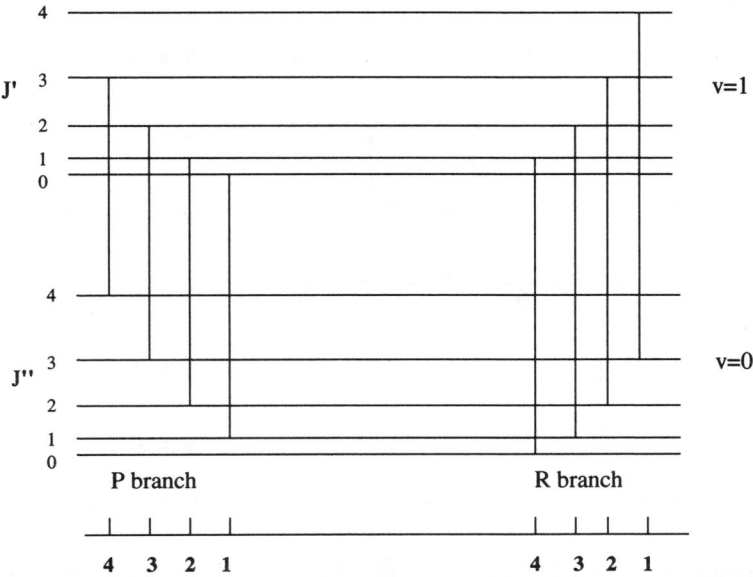

Fig. 2.12 Energy level diagram showing transitions for the 1–0 band of a typical diatomic molecule.

The names P and R branch are of historical origin, but still in common use. In polyatomic molecules, and some unusual diatomics of which nitric oxide, NO, is the commonest example, we can encounter Q branches for which $\Delta J = 0$, but in most diatomics such transitions are forbidden by the selection rules. Thus, the statement above that rotational and vibrational transitions *can* occur together is not quite right for most diatomics; they *must* occur together if a vibrational transition is involved at all.

The factor which determines whether a Q branch occurs (i.e. whether vibrational transitions can occur in isolation) is whether the component of the dipole moment along the axis of symmetry changes when the vibrational state changes. This explains why diatomics do not normally exhibit a Q branch. In the case of CO_2, the ν_2 band, arising from the bending mode, does have a Q branch, while the ν_3 band, arising from the asymmetric stretching mode along the main axis, does not.

2.4.2 *Hot bands*

The selection rule $\Delta v = \pm 1$, which controls the harmonic oscillator, means that the transitions shown in Fig. 2.13 are allowed. In the laboratory, it is found that although all do occur, they have very different strengths. The transition from $v = 0$ to $v = 1$ (or vice versa), sometimes written (0–1) or (1–0) and called the *fundamental*, is by far the strongest. The reason is simply that most of the molecules in a gas at standard temperature and pressure are in the vibrational ground state so, statistically, transitions which originate in this state are more likely to be observed.

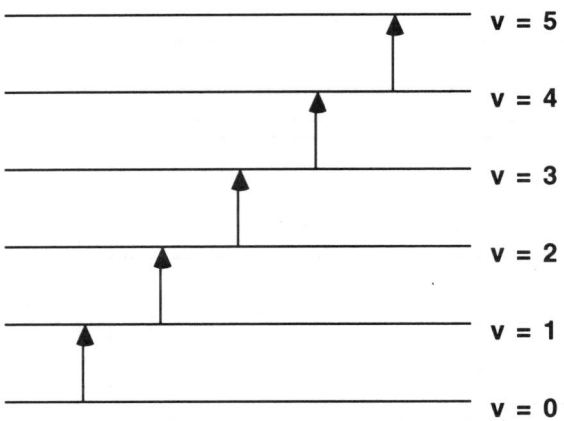

Fig. 2.13   Energy level diagram, showing allowed transitions between vibrational levels.

We shall see later (Sec. 2.5.1) that a gas like CO, in thermodynamic equilibrium at standard temperature and pressure (STP), has only about one in 10^5 of its molecules in the state $v = 1$, and even fewer in the more highly excited vibrational states. Therefore the initial conditions for (2–1), (3–2), etc., seldom exist. The situation can change at high temperatures or at low pressures in the atmosphere. Again, as we shall see, gases in equilibrium at higher temperatures have more molecules in excited vibrational states, because the increased kinetic energy of the heated gas is available, during collisions, to excite vibrations. Because the bands with lower level $v'' > 0$ occur profusely in hot gases (for example, in the atmosphere of the Sun), they are often known as *hot bands*. In gases at low pressures (and normal temperatures) the higher vibrational levels can become densely populated by non-LTE processes, as will be discussed further below, and the hot bands then have increased significance.

2.4.3   *Overtone bands*

Laboratory measurements of spectra reveal that transitions forbidden according to the harmonic oscillator model, i.e. those corresponding to $\Delta v = \pm 2, \pm 3$ and so forth, do occur with significant probabilities. They are called *overtone* bands. Like hot bands, they normally have considerably reduced strength relative to the fundamental. The fact that they occur at all is explained classically by recognising that the bond is not in fact a linear spring, but something more complex which does not vibrate with a single harmonic. This is equivalent to saying that the potential energy is a higher–order function of atomic separation than the parabolic one we have assumed. Table 2.2 summarises the terminology which is used to describe the various vibrational bands which are observed.

Table 2.2   Terminology for vibrational bands.

Upper state	Lower state	Δv	Name
$v' = 1$	$v'' = 0$	1	1–0 or fundamental
$v' = 2$	$v'' = 1$	1	2–1 or 1st hot band
$v' = 3$	$v'' = 2$	1	3–2 or 2nd hot band
			etc.
$v' = 2$	$v'' = 0$	2	2–0 or 1st overtone or 2nd harmonic
			etc.
$v' = 3$	$v'' = 0$	3	3–0 or 2nd overtone or 3rd harmonic
			etc.
$v' = 3$	$v'' = 1$	2	3–1 or 2nd harmonic of first hot band
			etc.

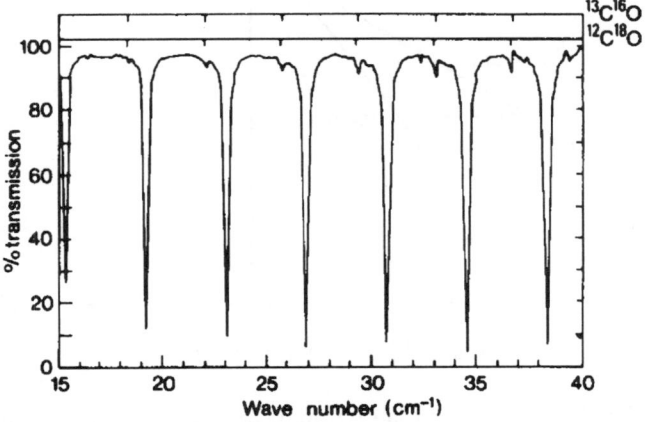

Fig. 2.14   CO spectrum.

2.4.4   *Isotope bands*

We have seen (c.f. Eq. 2.3) that during rotation of a molecule the quantity which is quantized is the angular momentum, $I\omega$. This implies that a heavier isotope will have a slower rate of rotation, and a smaller value for B, than a lighter one of the same species. The spectral lines will then have a slightly closer spacing, and will fall to the long–wavelength side of the lighter isotope. This effect can be seen in the CO spectrum in Fig. 2.14, where lines due to $C^{13}O^{16}$ and $C^{12}O^{18}$ appear (albeit blended together by the limited resolution of the spectrometer used for this measurement) in addition to the more common isotope $C^{12}O^{16}$.

Since we must have $I'\omega' = I\omega$ and $k'/m' = k/m$, the net effect is that vibrational frequencies are proportional to $1/\sqrt{m}$ and rotational to $1/m^2$.

The different apparent line strengths in the spectrum of Fig. 2.14 reflect the relative abundances of the isotopes in the sample, and not a real difference in the strength of the same line in different isotopes.

2.4.5   *Combination bands*

The carbon dioxide laser provides a useful illustration of the concept of combination bands. These involve changes in more than one vibrational mode as a result of the absorption or emission of a single photon.

The CO_2 laser is an extremely powerful device, generating 1 GW for 2 ns or several kW of continuous power. It relies for its operation on transitions between energy levels in the molecule corresponding to different types of vibrational mode, as shown in Fig. 2.15.

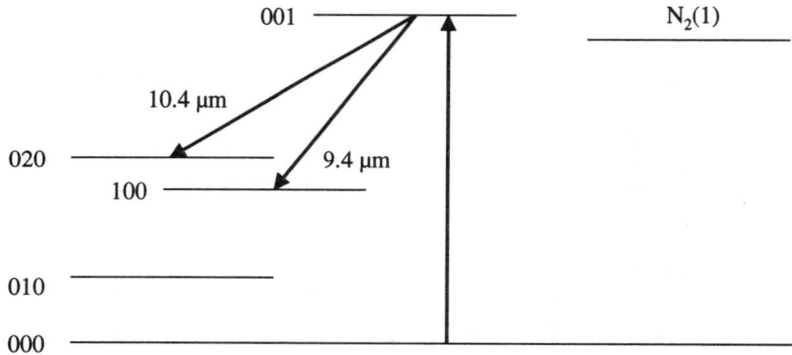

Fig. 2.15   The CO_2 laser bands.

The upper level of the system is one quantum of the asymmetric stretch mode ν_3. This is populated by an electric discharge through the gas. An admixture of N_2 is found to enhance the output of the laser, due to the fact that its fundamental vibration is in near resonance with ν_3 of CO_2 and cannot relax to the ground state by radiation except via CO_2. (This is an example of vibration-vibration energy transfer). In this way it is produced an inversion of the populations of the band, i.e., we have more molecules in the upper state of the transition $(0,0^0,1)$ than in the lower, $(1,0^0,0)$ or $(0,2^0)$ (see Fig. 2.15). Thus, this situation favours the de-excitation of $(0,0^0,1)$ by induced emission too, producing the amplification of the emission. That is why these bands are called laser bands. The laser lines themselves are at wavelengths of 9.4 and 10.4 μm. The most intense transitions occur for $J \approx 18$. These vibrational levels are sufficiently long–lived that they tend to relax to the ground state mainly as a result of collisions, in which their vibrational energy is converted into an enhancement of the kinetic energy of the colliding molecules, i.e. into heating of the gas. These laser bands occur naturally in the atmospheres of Mars and Venus (see Sec. 10.4.3.3).

2.5   Properties of Individual Vibration-Rotation Lines

Individual lines in a spectrum are not perfectly sharp, as the discussions of the previous sections assume. In fact, a molecule in a given state with nominal energy E can have any of a range of energies $E + \Delta E$ as a consequence of the uncertainty principle. Furthermore, not all transitions occur with the same probability. These concepts together imply that a measured spectral line has the attributes of strength (proportional to the probability that a photon will be absorbed or emitted, following a given transition path, per photon–molecule interaction) and width or shape (the distribution of transmitted or emitted radiance with frequency) which depend on the properties of the molecule concerned and its physical state (temperature, pressure and collision partners).

Spectral lines can, of course, be observed either in absorption or emission depending on whether the corresponding transition is upward or downward in energy. Lines are defined as being observed in emission or absorption depending on whether they appear bright or dark, respectively, against the background (or continuum, as it is usually known in spectroscopy). In absorption spectroscopy, the continuum is just the unaltered source radiance at wavelengths outside the spectral lines. In emission, it depends on the nature of the background. Of course, these considerations are relative, since in any collection of molecules absorption and emission will be going on together, and the appearance of the lines to the observer will depend on whether the background emits more or less strongly than the gas. If both are at the same temperature, the lines cannot be seen at all, although transitions are still taking place.

2.5.1   *Spectral line strength*

Consider a path of gas of length l and temperature T observed in absorption using a source of temperature T_s where $T_s \gg T$. This is the normal arrangement for absorption spectroscopy in the laboratory. Let the radiance entering the gas sample be $L_0(\nu)$ and that leaving be $L(\nu)$. Then the expression

$$L(\nu) = L_0(\nu) \exp[-k(\nu)m], \qquad (2.33)$$

where m is the absorber amount, $m = lp$, where l is the physical path length, and p is the pressure of the gas, assumed pure, defines the absorption coefficient $k(\nu)$, and

$$S = \int_0^\infty k(\nu)\, d\nu \qquad\qquad (2.34)$$

defines the strength of the line, S. If the integration takes place over all of the lines in a band, then the quantity is known as the *band strength* (see Sec. 3.6.3).

The main factor determining the strength of a spectral line or band is the population of the lower state, in other words the number of molecules available to make that particular transition. The relative population of two vibrational levels separated by E_v is determined by Boltzmann statistics, i.e. is proportional to $\exp[-E_v/(kT)]$. For HCl, with $E_1 - E_0 = 2143\,\mathrm{cm}^{-1}$, we find that

$$\frac{n(v=1)}{n(v=0)} \approx 3.4 \times 10^{-5} \quad \text{at} \quad T = 296\,\mathrm{K}.$$

The exponential dependence of population on the energy of a state, and the fact that most first vibrational states are several times greater in energy than kT, explains why 'hot bands' like 2–1 ($v' = 2 \rightarrow v'' = 1$) are so weak at room temperature. Nearly all of the molecules are in the ground state at any given time, and for most species only about one molecule in a million or so is available to contribute to a transition which starts out from the first excited state.

From similar considerations for rotation levels, the number in the level with quantum number J is

$$n(J) = \frac{n_a}{Q_r}(2J+1)\exp\left(-\frac{BJ(J+1)hc}{kT}\right), \qquad\qquad (2.35)$$

where n_a is the total number of molecules in the sample and Q_r is a normalization factor called the partition function (sometimes the *rotational state sum*). This is defined as

$$Q_r = \sum_J (2J+1)\exp\left(-\frac{BJ(J+1)hc}{kT}\right). \qquad\qquad (2.36)$$

It can be shown, by treating J as a continuous variable and integrating Eq. (2.36), that $Q_r \approx kT/(hcB)$, actually valid only for linear molecules. The term $(2J+1)$ is a statistical weight, arising from the $(2J+1)$-fold degeneracy of the energy levels, itself due to the quantization not only of total angular momentum J, but also of angular momentum M in the polar direction. There are $2J+1$ ways in which a total value of J can be oriented so as to give an integral value of M. These integral values are $J, J-1, \ldots, 0, \ldots, 1-J, -J$.

The expression (2.35) explains the characteristic shape of a P or R branch in a rotation–vibration band (Fig. 2.16). Line strength tends to decrease with increasing J due to the exponential term, higher energy levels always having smaller Boltzmann population factors. Set against this is the greater multiplicity of the higher J levels, tending to increase the number of molecules in the lower state, and hence the probability of a transition, by increasing the number of degenerate states. Thus, the intensities of lines at first increase and then decrease with increasing J (Fig. 2.16). By differentiating Eq. (2.35), we find that the peak, which depends only on T for a given species, occurs at

$$J = 0.589\sqrt{\frac{T}{B}} - \frac{1}{2}. \tag{2.37}$$

This expression has some practical applications. For example, in astronomy, where bands in the light from stars or planets maybe observed only at low spectral resolution, the temperature can be determined by finding the wavenumber at which the P or R branch has the greatest strength. Individual lines do not need to be resolved so long as the rotational constant B is known.

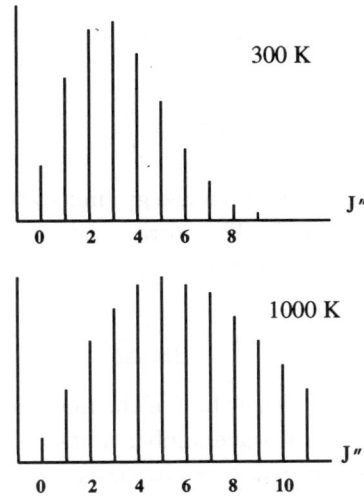

Fig. 2.16 Relative populations of rotational levels as a function of quantum number J for the HCl molecule at two temperatures: 300 K and 1000 K.

For theoretical work on atmospheric radiative transfer, such as that which concerns us in most of this book, it is necessary to know the absolute value of the spectral line strengths, and not just their relative strengths. In principle it is possible to calculate these from transition probabilities, but in practice they are usually obtained by measurement. The simplest method is to obtain a spectrum of the band of interest in a cell containing gas at known values of temperature, pressure and path length, and then to calculate the spectrum using the theoretical relative line intensities, all multiplied by the same unknown factor. The value of this factor (the band strength) which gives the best fit can then be determined. Note that the individual lines do not have to be resolved, and in fact an equally valid (although generally less accurate) alternative approach is to measure the area under the band envelope observed at low resolution, under conditions where the lines are unsaturated. This gives the band strength directly.

2.5.2   *Widths and shapes of spectral lines*

The fact that spectral lines are not infinitely narrow, but have a characteristic shape, means that photons having a small range of energies around the specific values calculated in the sections above are also valid, although less probable, partners in an energy exchange with a molecule. One way of looking at this is to picture the energy states themselves as blurred, rather than sharp, levels. This blurring arises from several different mechanisms, of which two are important in the atmosphere, but it is instructive to consider three.

2.5.2.1   *Natural broadening*

The natural width of a line arises from the uncertainty principle in the following form:

$$\Delta E \Delta t \sim \frac{h}{2\pi}, \tag{2.38}$$

where Δt is the transition time between the ground state and the state with energy E, and ΔE is the uncertainty in E. The corresponding uncertainty in the frequency of the photon required to excite the transition is

$$\alpha_N = \frac{1}{2\pi c \Delta t}. \tag{2.39}$$

The line width α_N is the half width at half of the maximum value of k and, as noted earlier, is commonly expressed in wavenumber units.

The lifetime Δt of a vibrational or rotational state is generally quite long compared to the mean time between collisions for a molecule under most atmospheric conditions. Furthermore, α_N is usually negligible in comparison to the broadening induced by the thermal agitation of the molecules. In practice, therefore, the natural width of a line cannot be observed except under special observing conditions (such as extremely low temperature and pressure) since it is swamped by accompanying Doppler broadening or pressure broadening.

2.5.2.2   *Doppler broadening*

In any path through a gas a range of velocities is observed, resulting in a range of Doppler shifts which broadens the lines. Molecular motions follow Maxwell–Boltzmann statistics, so that the number of molecules with a line-of-sight velocity v and mass m is given by

$$n(v) \, dv = N \sqrt{\frac{m}{2\pi kT}} \exp\left[-\frac{mv^2}{2kT}\right] dv. \tag{2.40}$$

Since the observed frequency of light emitted or absorbed by a molecule moving with velocity v is changed by a factor $(1 - v/c)$, we find that

$$k(\nu) = \frac{S}{\alpha_D \sqrt{\pi}} \exp\left[-\frac{(\nu - \nu_0)^2}{\alpha_D^2}\right], \qquad (2.41)$$

where the Doppler width α_D is given by

$$\alpha_D = \frac{\nu_0}{c} \sqrt{\frac{2RT}{M}}, \qquad (2.42)$$

R is the universal gas constant, M is the molecular weight, and T is temperature. The quantity $f(\nu) = k(\nu)/S$ is called the normalized line shape or just the line shape. Note that this depends on the line centre frequency ν_0; Doppler broadening is more important for shorter wavelengths. The symbol α_D is sometimes used for a slightly different quantity, the half width at half height (or half maximum) (HWHM), which is smaller by a factor of $\sqrt{\ln 2}$, $\alpha_{\text{HWHM}} = \sqrt{\ln 2}\, \alpha_D$. An example of a Doppler line profile is shown in Fig. 2.17.

2.5.2.3 *Pressure broadening*

Collisions in a gas shorten the lifetime of an excited state, and t_0, the mean time between collisions at STP, replaces the natural lifetime in the expression for ΔE, Eq. (2.38), giving

$$\alpha_L(\text{STP}) = \frac{1}{2\pi c t_0}. \qquad (2.43)$$

The profile of the line is the same as for natural broadening, since the two mechanisms have the same effect, being two different ways of limiting the lifetime of the excited state. The absorption coefficient k_ν arising as a result of the transition centred on frequency ν_0 has a frequency distribution (the line shape) with a characteristic width of α_L. The shape of the distribution can be found by obtaining the Fourier transform of a sine wave of finite duration t_0, and integrating the product of this (squared, to give intensity rather than amplitude) with the probability of a given value of t_0 (from Eq. 2.43). The result is the standard 'dispersion' shape

$$k(\nu) = \frac{S}{\pi} \frac{\alpha_L}{(\nu - \nu_0)^2 + \alpha_L^2}, \qquad (2.44)$$

where k is the absorption coefficient, S is the strength of the line and the normalization factor $(1/\pi)$ enters since the integral of k must equal S by definition.

According to simple kinetic theory, the rate at which collisions take place is proportional to the pressure and inversely proportional to the square root of the temperature. Thus, if the sample of gas is at other than STP, then

$$\alpha_L(p, T) = \alpha_L(\text{STP}) \frac{p}{p_0} \sqrt{\frac{T_0}{T}}. \qquad (2.45)$$

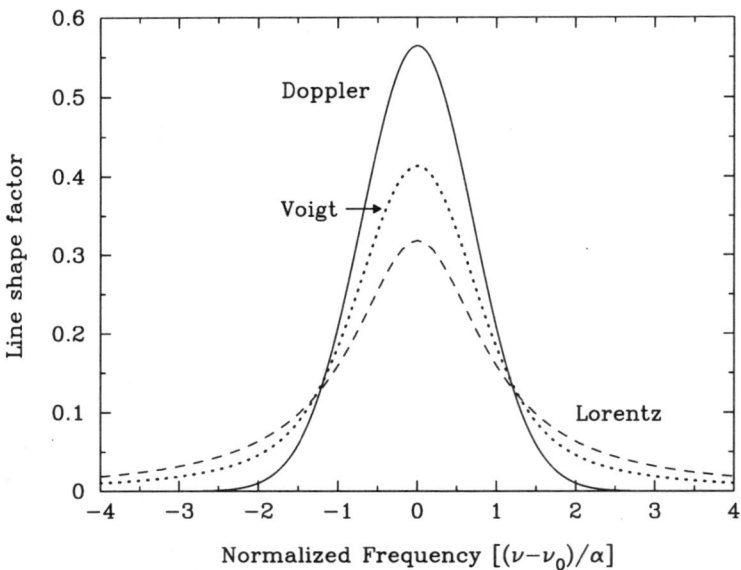

Fig. 2.17 Doppler and Lorentz line shapes for the same strength and line widths. The corresponding Voigt profile for the same strength and half line width is also shown.

In practice, the process of collision broadening is very complicated and the simple line shapes and temperature dependences described here are not followed by all molecules or lines in the real cases. In particular, the temperature dependence of line width seldom follows the square root law predicted in Eq. (2.45), and the spectral databases usually provide a different value for the exponent which has been obtained by fitting measurements or from complex theories of molecular potentials and their interactions during collisions. These coefficients often vary not only with molecule and band, but also with transition, having, for example, a systematic dependence on the value of the rotational quantum number J.

2.5.2.4 *Combining different broadening effects*

Except in gases where the pressure is extremely low, α_N can nearly always be neglected in comparison to α_L since natural lifetimes of states are long compared to typical times between collisions. α_N is also small compared to α_D, except for extremely low temperatures. These two facts taken together mean that natural broadening can be ignored in most practical situations.

The relative importance of Doppler broadening and pressure broadening depends on the mass of the molecule in question, its internal structure, the temperature, and the pressure; but as a rule of thumb Lorentzian line widths of common gases at STP are of the order of $0.1\,\mathrm{cm}^{-1}$ while the Doppler widths are between one and two orders of magnitude smaller. Then the latter may often be neglected, leaving just the relatively simple Lorentz line shape and width to be used in calculating the spectral line profile.

When great accuracy is required, or when α_L and α_D are more comparable (for example, for a gas at a low pressure, such as the atmosphere above an altitude of about 30 km) then the combined effects of Doppler and Lorentz broadening need to be considered. The result of convolving the two different line shapes leads to the so-called *Voigt* line profile, given by

$$k(\nu - \nu_0) = \frac{S}{\alpha_D \sqrt{\pi}} \frac{y}{\pi} \int_{-\infty}^{\infty} \frac{e^{-t^2} \, dt}{y^2 + (x - t)^2}, \qquad (2.46)$$

where $y = \alpha_L/\alpha_D$, and $x = (\nu - \nu_0)/\alpha_D$. This integral cannot be solved analytically, but computer subroutines which calculate it are widely available. Fig. 2.17 shows a comparison between lines of Doppler, Lorentz, and Voigt shapes. At most atmospheric pressures of interest for non-LTE studies, the Voigt profile has a Doppler core and Lorentzian wings.

2.6 Interactions between Energy Levels

Energy levels often relate to each other in ways other than by simple radiative transitions. For example, if two otherwise unconnected levels have the same energy, they tend to have the same populations as well because they can exchange quanta during collisions much more readily than well-separated levels. Levels couple in a variety of different ways, some of which will be important for the non-LTE models of the atmosphere developed in later chapters.

2.6.1 *Fermi resonance*

Two otherwise dissimilar vibrational modes which happen to have nearly the same value of energy may be in Fermi resonance. The effect of this is to raise the higher of the two levels in frequency, and depress the lower; there is also some sharing of line strengths, i.e. the weaker of the two transitions becomes stronger, at the expense of the other which becomes weaker.

An example is to be found in CO_2, where ν_1 is expected to fall at $1330 \, \text{cm}^{-1}$ and $2\nu_2$ at $1334 \, \text{cm}^{-1}$. In fact, they are observed at about 1285 and $1385 \, \text{cm}^{-1}$ respectively. Furthermore, they are observed (in the Raman spectrum, since ν_1 is inactive in absorption) to have about the same strength, whereas the fundamental would, without the effect of Fermi resonance, be expected to be much stronger than the overtone. This is an extreme case; normally, the amount of line strength 'borrowing' is much less, and sometimes zero, even for transitions which are very close in energy. Generally, this amount, and the wavenumber shifts, have to be determined empirically as the relevant theory is complex.

2.6.2   *Coriolis interaction*

Modes can be coupled, i.e. one mode can excite another, especially if they are close in energy. An example of this which is exhibited by CO_2 is the Coriolis interaction. An inspection of Fig. 2.18 shows how this operates.

The individual atoms in a linear triatomic molecule executing the ν_3 vibration will, if the molecule is also rotating, experience Coriolis forces in the directions shown by the ν_2 set of arrows. The net effect of these forces is to tend to induce the ν_2 vibration, although the transition is most unlikely to occur unless the quantum of energy is the same, or nearly the same, for each mode.

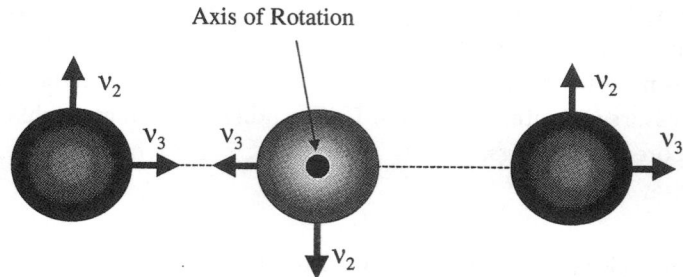

Fig. 2.18   The Coriolis interaction in a linear triatomic molecule. Imagine the molecule rotating in the plane of the page; the Coriolis forces associated with the motions corresponding to the ν_2 vibration tend to excite the ν_3 mode, and vice versa.

2.6.3   *Vibration-vibration transfer*

This process involves the exchange of vibrational energy quanta between colliding molecules, without going through kinetic energy as an intermediary. Such processes are important in the atmosphere, especially in non-LTE situations. For example, the deactivation of $CO_2(0,2,0)$ by $CO_2(0,0,0)$, in which both molecules end up in the $(0,1^1,0)$ state, can be shown to be responsible for 25% of the total relaxation at altitudes below 90 km. As another example, an effective excitation mechanism of the asymmetric (ν_3) mode of the less common CO_2 isotopes is V–V energy transfer from the main isotope, although the rate of these processes is not well known.

The near-resonant reaction between the CO_2 isotopes and the first vibrational level of N_2 plays a very important role in determining the populations of the CO_2 $(0,0^0,1)$ states. Collisions with N_2 are frequent, so the molecule, although itself not active in the infrared spectrum, acts effectively as a reservoir for v_3 quanta, which are swapped back and forth. The effectiveness of N_2 as a reservoir is enhanced since it cannot relax radiatively, but only by V–V exchange. There is also an important transfer of energy from vibrational and electronically excited molecules, radicals and atoms, such as $OH(v \leq 9)$ and $O(^1D)$ to N_2 quanta and subsequent transfer to CO_2 v_3 quanta. Amongst processes not involving CO_2 directly, one of the most important is the near resonant V–V energy exchange between $H_2O(0,1,0)$ and $O_2(1)$.

2.7 References and Further Reading

There are many books on basic molecular spectroscopy. A useful introduction at the simplest level is Banwell and McCash (1994), or the paper by Schor and Teixeira (1994), which covers diatomic spectra, while the quantum mechanical formulation is treated in most graduate level modern physics texts, such as Eisberg and Resnik (1985). For more sophisticated, but still basic, treatments of molecular spectroscopy, Penner (1959), Steinfeld (1985), Atkins and Friedman (1996), and Bunke and Jensen (1998) are recommended.

The tomes by Herzberg (1945, 1950) are old, but still the most complete reference books on the subject.

Chapter 3

Basic Atmospheric Radiative Transfer

3.1 Introduction

Following the philosophy expressed earlier, we will look at simplified cases which establish the basic principles and provide a feel for the quantities involved before we graduate to the full complexities of the non-LTE radiative transfer problem. To compute the transfer of energy as radiation within the atmosphere is not a simple task, even when LTE can be assumed. As in other branches of Physics, the complexity of the problem has given rise to the widespread use of *models*. These are simplifications, which represent interesting or important features of the matter under study with enough accuracy to be useful while neglecting unnecessary or unwanted detail. Obviously, the choice of model will depend on why one is making the calculation or study in the first place, and careful track has to be kept of the possible errors being introduced by the approximations made.

In the introductory chapter, we said that we will be dealing with non-LTE mostly in molecular vibration-rotation levels that emit and absorb electromagnetic radiation principally in the infrared. Thus we will concentrate on the absorption and emission terms of the radiative transfer equation, and not deal with scattering processes. In the presence of clouds, scattering could be important, specially at near-infrared wavelengths, but not usually when we also have non-LTE emissions, since clouds are restricted to the lower atmosphere where LTE prevails for essentially all transitions. However, radiation scattered by tropospheric clouds might, under certain conditions, reach and be partially absorbed in the mesosphere and lower thermosphere where non-LTE situations most frequently appear. This situation can be treated adequately in models by adding an additional flux at the lower boundary, which is much simpler than including the complexities of scattering in the computation.

3.2   Properties of Radiation

We turn now to the general situation where photons of all wavelengths are present, travelling in all directions, at a point P in space. We say that a radiation *field* is present at P. To describe this field consider an infinitesimal area $d\sigma$ at P which has its normal unit vector **n** oriented in an arbitrary direction (Fig. 3.1). Consider also an arbitrary direction along the unit vector **s** at point P that forms an angle θ with **n**. The *radiance* (sometimes called *specific intensity*) at point P in an arbitrary direction **s**, $L_\nu(P, \mathbf{s})$, is defined by

$$dE_\nu = L_\nu(P, \mathbf{s}) \cos(\theta)\ d\sigma\ d\omega\ d\nu\ dt, \tag{3.1}$$

where dE_ν is the energy transported by the radiative field across the element of area $d\sigma$ in all directions confined to an element of solid angle $d\omega$, during the time dt. Integrating $L_\nu(P, \mathbf{s})$ over the frequency interval $(\nu, \nu + d\nu)$ yields the spectrally-integrated radiance with dimensions of $\mathrm{W\,m^{-2}\,sr^{-1}}$, since the units of σ are $\mathrm{m^2}$ and those of frequency are Hz (see Appendix C). If we then integrate again over solid angle, we obtain the *flux*, which has units of $\mathrm{W\,m^{-2}}$. This quantity is sometimes referred to as *flux density*.

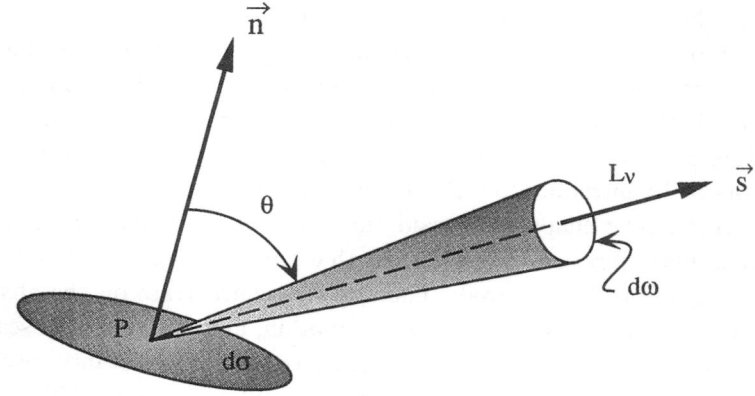

Fig. 3.1   The radiance, or specific intensity of radiation, at a point P.

From the definition of the radiance it follows that the energy transported by a radiation field can vary not only from point to point but also with direction at any point. Thus the radiance depends on the three cartesian coordinates defining the point P and the three direction cosines needed to describe the direction of **s**. Further parameters may be needed to characterise the state of polarization of the radiation field, but these are not generally important in the infrared and will be ignored here (see Chandrasekhar, 1960, for a discussion of polarization effects in radiative transfer). A radiation field that is independent of direction at a point is said to be *isotropic*. If it is the same at all points, then is said to be *homogeneous*.

Equation (3.1) gives the radiant energy flowing across $d\sigma$ confined to an element of solid angle $d\omega$ in the direction that forms an angle θ to its outward normal in the interval $\nu, \nu + d\nu$. The radiative *flux* along the $\mathbf{n}$ direction, $F_{\nu,n}(P)$ is obtained by integrating the travelling pencils of radiation contained in $d\omega$ over all $\mathbf{s}$ directions:

$$F_{\nu,n}(P) = \int_{\omega} L_{\nu}(P,\mathbf{s}) \cos\theta \, d\omega. \qquad (3.2)$$

The integration over ω can be divided into the half-spheres (2π steradians) above and below $d\sigma$, i.e., to obtain the intensities of radiation flowing outward and coming inward through $d\sigma$ (or propagating upwards and downwards, if $d\sigma$ is a horizontal surface in the atmosphere). For this reason $F_{\nu,n}(P)$ is sometimes called *net* flux instead of just flux. It is clear from the definition that the *flux* depends on the direction of the normal vector of the surface considered, $\mathbf{n}$, i.e., it is itself a vector quantity. Assuming a three-dimensional orthogonal system with (x, y, z) axes and respective unit vectors $\{\mathbf{i}, \mathbf{j}, \mathbf{k}\}$, then

$$\cos\theta = \mathbf{n} \cdot \mathbf{s} = (\mathbf{n} \cdot \mathbf{i})(\mathbf{s} \cdot \mathbf{i}) + (\mathbf{n} \cdot \mathbf{j})(\mathbf{s} \cdot \mathbf{j}) + (\mathbf{n} \cdot \mathbf{k})(\mathbf{s} \cdot \mathbf{k}).$$

Including this expression in Eq. (3.2) and taking into account the definition of the flux for each of the orthogonal directions, i.e.,

$$F_{\nu,x}(P) = \int_{\omega} L_{\nu}(P,\mathbf{s}) \, (\mathbf{s} \cdot \mathbf{i}) \, d\omega$$

we have

$$F_{\nu,n}(P) = F_{\nu,x}(P)(\mathbf{n} \cdot \mathbf{i}) + F_{\nu,y}(P)(\mathbf{n} \cdot \mathbf{j}) + F_{\nu,z}(P)(\mathbf{n} \cdot \mathbf{k}) \qquad (3.3)$$

which can be written in the vector form

$$\mathbf{F}_{\nu}(P) = F_{\nu,x}(P) \, \mathbf{i} + F_{\nu,y}(P) \, \mathbf{j} + F_{\nu,z}(P) \, \mathbf{k}. \qquad (3.4)$$

The divergence of the radiative flux gives us the net rate at which the energy per unit volume of the radiation field is increased, or, changing its sign, the net rate per unit volume of energy gained by matter interacting with the radiation field. The latter, a scalar quantity, h_{ν}, is defined by

$$h_{\nu} = -\nabla \cdot \mathbf{F}_{\nu}. \qquad (3.5)$$

The divergence of the radiative flux, $-h_{\nu}$, is usually referred to as the cooling rate, or heating rate (h_{ν}). We should emphasise, however, that they are not always identical concepts. The cooling or heating rate is defined as the rate of loss or gain of energy from the thermal reservoir, i.e. the net kinetic energy of the molecules. If, for example, V–V transitions are involved, then when flux is absorbed it may not all end up as heat; some may go into the excitation of a higher vibrational state. As another example, if flux is emitted when sources of chemical energy are present, then all of the energy may not have originated in the thermal reservoir. We will normally refer to $-\nabla \cdot \mathbf{F}_\nu$ as a heating rate regardless of this distinction, calling attention where necessary to the fact that it may be wholly or partially *potential*, rather than *actual*, heating.

Inserting in Eq. (3.5) the definition of the flux we have

$$
\begin{aligned}
h_\nu &= -\left(\frac{\partial F_{\nu,x}}{\partial x} + \frac{\partial F_{\nu,y}}{\partial y} + \frac{\partial F_{\nu,x}}{\partial z}\right) = -\int_\omega \left[\frac{\partial L_\nu}{\partial x}\mathbf{s}\cdot\mathbf{i} + \frac{\partial L_\nu}{\partial y}\mathbf{s}\cdot\mathbf{j} + \frac{\partial L_\nu}{\partial z}\mathbf{s}\cdot\mathbf{k}\right]\mathrm{d}\omega \\
&= -\int_\omega \frac{\mathrm{d}L_\nu}{\mathrm{d}s}\,\mathrm{d}\omega.
\end{aligned}
\tag{3.6}
$$

The mean radiance of the radiation field, $\bar{L}_\nu(P)$, is defined as the average of the radiance over all solid angles:

$$
\bar{L}_\nu(P) = \frac{1}{4\pi}\int_\omega L_\nu(P,\mathbf{s})\,\mathrm{d}\omega.
\tag{3.7}
$$

3.3 The Radiative Transfer Equation

With these definitions in place we go on to discuss in this section how the absorption coefficient and the source function can be defined, and introduced into the equation of radiative transfer, from a consideration of the processes of interaction of radiation with matter.

We usually distinguish two kinds of interactions between radiation and matter; those which decrease the energy in the radiation field, extinction (either scattering or absorption) processes, and those which give rise to an increase, emission processes. *Lambert's law* states that the rate of change (reduction) of the radiance experienced in passing through matter is proportional to the amount of radiation, L_ν, and to the amount of matter*. The change (decrease) of radiance when travelling a path $\mathrm{d}s$ across a medium is given by

$$
\mathrm{d}L_\nu = -e_\nu n_a L_\nu\,\mathrm{d}s,
\tag{3.8}
$$

where n_a is the number density of absorbing molecules (or atoms) and e_ν is the molecular extinction coefficient (including both absorption and scattering), a constant that characterises the radiation-matter interaction. Applying the same argument to emission, the increase in the radiance when crossing the path $\mathrm{d}s$ is given by

$$
\mathrm{d}L_\nu = j_\nu n_a\,\mathrm{d}s,
\tag{3.9}
$$

where the proportionality coefficient now is j_ν, the emission coefficient, which characterises the *emission* properties, including scattering, of the molecules in the path.

---

*In planetary atmospheres, any deviation from this fundamental law is normally completely negligible, except when induced emission takes place. In that case, we usually keep the same equation but add a negative term to the extinction coefficient, see Sec. 3.6.2. For a discussion on the applicability of Lambert's law see Rybicki and Lightmann (1979), p. 11.

As we will see in Sec. 3.5, under thermodynamic equilibrium Kirchhoff's law states that the emission and absorption coefficients, j_ν and e_ν, are related by a universal function depending solely on temperature (later demonstrated to be the Planck function),

$$\frac{j_\nu}{e_\nu} = f_\nu(T). \tag{3.10}$$

This relation is extended to situations other than thermodynamic equilibrium (or even local thermodynamic equilibrium as we see later) by defining a general source function as

$$J_\nu = \frac{j_\nu}{e_\nu},$$

whence Eq. (3.9) is expressed as

$$dL_\nu = e_\nu n_a J_\nu \, ds. \tag{3.11}$$

If there is no other kind of interaction, extinction and emission must combine to give a change in radiance governed by

$$\frac{dL_\nu(P, \mathbf{s})}{ds} = -e_\nu n_a \left[L_\nu(P, \mathbf{s}) - J_\nu(P, \mathbf{s})\right]. \tag{3.12}$$

This is the *radiative transfer equation*, originally derived in this form by Schwarzschild and Milne in the early part of the 20th century. The equation itself does not reveal much about the physics of the processes involved, since these are implicit in the extinction coefficient and the source function which account for all interactions between the radiation field and matter. For later use, we can express the heating rate in terms of these new quantities by integrating Eq. (3.12) over all solid angles and writing Eq. (3.6) as

$$h_\nu(P) = 4\pi e_\nu n_a \left[\bar{L}_\nu(P) - \bar{J}_\nu(P)\right], \tag{3.13}$$

where the mean emission, $\bar{J}_\nu(P)$, is defined analogously to the mean radiance in Eq. (3.7). Now the heating rate is given in terms of the difference between the mean radiance and the mean emission from any matter present. Monochromatic radiative equilibrium, defined as $h_\nu(P) = 0$, exists (as a concept) at frequency ν when the radiance averaged over all directions equals the mean emission. In the real world, we must consider broad spectral intervals, perhaps the whole spectrum from 0 to infinity.

Chapter 2 dealt with the interaction processes between radiation and molecules, the latter under the conditions found in planetary atmospheres where compounds have electronic, vibrational and rotational internal energy. The physical processes of interest for us are those which can alter these modes of internal energy, particularly the populations of vibrational and rotational energy levels.

In the absorption process, a photon causes the matter to be excited into one of its internal energy modes, while in the reverse process, emission, an excited state is deactivated by emitting a photon. The absorption process is often followed by the transfer of the energy of the excited internal energy mode to translational energy in a subsequent matter-matter interaction, but this is not inevitable. The energy of the photon may be emitted, or transferred to some other internal state of the same molecule or one of its collision partners. For example, under some circumstances the absorption and emission of photons by matter occur much faster than any other matter-matter interactions, so that very little energy of the absorbed photons is actually transferred into translational energy or to other internal energy states. Such cases are conceptually similar to the 'simple scattering' process, in which the photon changes direction without changing the internal energy state of the molecule, and are sometimes referred to as 'scattering' by analogy. The two types of interaction are different, however, in that 'real' scattering is characterised by a phase function which relates the directions of the absorbed and emitted photons, whereas the 'absorption followed by re-emission' process is normally isotropic. This process is normally called *resonant fluorescence*.

Another process of interest is that in which the photon is absorbed and an internal energy mode is excited, then one or more photons are emitted so that their total emitted energy is smaller than that of the photon absorbed. This process is called, in general, *fluorescence*. For example, CO_2 absorbs $2.7\,\mu m$ photons from the Sun, giving rise to the excitation of the (v_1+v_3) vibrational combination state. When the CO_2 subsequently emits only the ν_3 photon at $4.3\,\mu m$ (the rest of the energy may be relaxed in collisions) we call it the fluorescence emission at $4.3\,\mu m$. The alternative path in which a $2.7\,\mu m$ photon is absorbed and two ν_2 (with energy approximately equal to ν_1) and one ν_3 photons are emitted is also an example of florescence although it is also described in some texts as *incoherent scattering*.

If 'simple scattering' is ignored, then the extinction coefficient is the same as the absorption coefficient, i.e.

$$e_\nu = k_\nu \qquad\qquad (3.14)$$

and the emission term will be given by

$$e_\nu n_a J_\nu = k_\nu n_a J_\nu, \qquad\qquad (3.15)$$

where J_ν excludes the 'simple scattering' contribution. There is no dependence of k_ν and J_ν on the angular distribution of photons, i.e., we assume that both absorption and emission are isotropic. In most cases this is a good approach for planetary atmospheres in the infrared part of the spectrum.

3.4   The Formal Solution of the Radiative Transfer Equation

Now we introduce the integral form of the radiative transfer equation. Although it is a mere mathematical manipulation, this form is very useful when the source function is known. This is the case for many common problems, for example, the calculation of the radiance and the cooling rates under LTE conditions (for which the source function is given by the Planck function); and the calculation of the atmospheric emission measured by any instrument, particularly by those used in satellite remote sensing, both for LTE and for non-LTE (once the source function has been previously calculated).

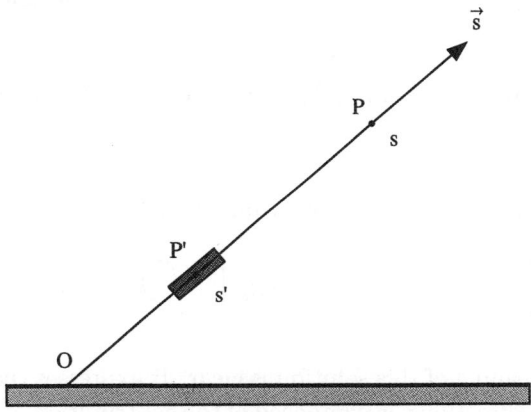

Fig. 3.2   Optical path.

Consider the optical path in Fig. 3.2. Introducing the optical thickness $\bar{\tau}_\nu$[†], between two points s' and s along the direction $\mathbf{s}$ by the definition

$$\bar{\tau}_\nu(s', s) = \int_{s'}^{s} k_\nu(s'') n_a(s'') \, \mathrm{d}s'',\tag{3.16}$$

the radiative transfer equation (3.12) at the point P' at distant s' can be written as

$$\frac{\mathrm{d}L_\nu(s', \mathbf{s})}{\mathrm{d}\bar{\tau}_\nu} = L_\nu(s', \mathbf{s}) - J_\nu(s', \mathbf{s}).\tag{3.17}$$

Note that $\bar{\tau}_\nu$ has been defined as a positive quantity, since $s > s'$. The change of sign from Eq. (3.12) to (3.17) in the right-hand term comes from the negative sign of $\mathrm{d}\bar{\tau}_\nu/\mathrm{d}s'$, which is a consequence of the definition of $\bar{\tau}_\nu$ as a positive quantity, and the arbitrary choice of the origin of $\bar{\tau}_\nu$ at point P near the top of the atmosphere[‡].

[†]We keep the symbol τ_ν for the optical depth, i.e., defined from s to ∞, $\tau_\nu(s, \infty)$ (see Eq. 4.11 in Sec. 4.2.1).

[‡]The actual choice of an upper limit becomes more meaningful when considering a specific planetary atmosphere, and finding where the atmospheric density, and hence the optical thickness, become vanishing small.

As the equation of transfer (Eq. 3.17) is a linear first-order differential equation with constant coefficients, it must admit an integrating factor which, as can be easily guessed, is $\exp(-\bar{\tau}_\nu)$. Multiplying Eq. (3.17) by this factor we have

$$\frac{d\left[L_\nu(s',\mathbf{s})\exp(-\bar{\tau}_\nu)\right]}{d\bar{\tau}_\nu} = -\exp(-\bar{\tau}_\nu)\,J_\nu(s',\mathbf{s}) \qquad (3.18)$$

and integrating from s, where $\bar{\tau}_\nu = 0$, to the origin s_0

$$L_\nu(s,\mathbf{s}) = L_\nu(s_0,\mathbf{s})\exp\left[-\bar{\tau}_\nu(s_0,s)\right] + \int_0^{\bar{\tau}_\nu(s_0,s)} J_\nu(s',\mathbf{s})\exp\left[-\bar{\tau}_\nu(s',s)\right]\,d\bar{\tau}_\nu. \quad (3.19)$$

If we substitute the definition of $\bar{\tau}_\nu$ (Eq. 3.16) into Eq. (3.19), we have the alternative formal solution

$$\begin{aligned} L_\nu(s,\mathbf{s}) &= L_\nu(s_0,\mathbf{s})\exp\left[-\int_{s_0}^s k_\nu(s')n_a(s')\,ds'\right] + \\ &\quad \int_{s_0}^s k_\nu(s')n_a(s')J_\nu(s',\mathbf{s})\exp\left[-\int_{s'}^s k_\nu(s'')n_a(s'')\,ds''\right]\,ds'. \quad (3.20) \end{aligned}$$

The physical meaning of this solution is clear. It expresses that the radiance at a given point $P(s)$ in direction $\mathbf{s}$ is composed of the contribution at the boundary radiance $L_\nu(s_0,\mathbf{s})$ attenuated by the absorbing material between the boundary and the point being considered, plus the emission from all the volume elements at positions s' along the path, $k_\nu(s')n_a(s')J_\nu(s',\mathbf{s})$, also attenuated by the absorbing material between the position of the emitting elements s' and the point under consideration at s.

Equation (3.19) (or 3.20) solves the problem of radiative transfer if the source function at a point P' and direction $\mathbf{s}$, $J_\nu(P',\mathbf{s})$, is known, for example under LTE conditions when, as we show in the next section, the source function is given by the Planck function. Non-LTE is more complicated because the source function depends, in general, on the radiance at other points (P''), directions $(\mathbf{s}')$, and frequencies ν', $L_\nu(P'',\mathbf{s}',\nu')$, and it is then necessary to solve the radiative transfer equation together with the equation that gives us the source function, i.e., the statistical equilibrium equation. In the next section we explain the situation of local thermodynamic equilibrium (LTE), a very important case where the solution of the radiative transfer equation is greatly simplified. Afterwards, in Sec. 3.6, we derive the general expression for the source function, required for the solution of the radiative transfer equation under non-LTE conditions. We devote the next two chapters entirely to practical solutions of the radiative transfer equation in the atmosphere, starting with the simpler case of LTE (Chapter 4) and then following with the more difficult non-LTE case (Chapter 5).

3.5 Thermodynamic Equilibrium and Local Thermodynamic Equilibrium

The radiative properties of a gas parcel in thermodynamic equilibrium, inside an isothermal enclosure at temperature T, were first described by Kirchhoff at the end of the 19th century, as follows:

(1) The radiation inside the enclosure is homogeneous, unpolarized, and isotropic;
(2) The source function is equal to the radiance; and
(3) The radiance is a universal function of the temperature within the enclosure.

A search for the analytical function which could explain experimental measurements of the radiance in the enclosure as function of temperature and frequency motivated Planck to postulate the quantification of radiant energy, and then to develop an expression for what we now call Planck's function:

$$B_\nu(T) = \frac{2h\nu^3}{c^2} \frac{1}{[\exp(h\nu/kT) - 1]}, \tag{3.21}$$

where h is Planck's constant, and k Boltzmann's constant. This fundamental expression describes not only the radiance in the enclosure under thermodynamic equilibrium, but also the radiance which would be emitted from a small hole in the enclosure, assumed to be so small as to have a negligible effect on the radiative balance inside. This is the radiance corresponding to a temperature T and an emissivity of unity. A small hole in a large cavity with highly-absorbing interior walls is the nearest thing in the real world to the concept of a perfect *blackbody*, one which absorbs any photon which falls on it. The fact that the hole must be a perfect emitter can perhaps be better appreciated by considering its properties as an absorber. Any photon falling on the hole from outside must be absorbed by the internal walls, either on the first encounter or after multiple internal reflections. Since the hole occupies a negligibly small fraction of the surface area of the enclosure, the photon cannot escape. A perfect absorber must also be a perfect emitter, according to Kirchhoff's law. Flat or grooved surfaces painted black are black bodies but not perfect blackbodies, i.e. they are approximations to this ideal condition. For reasons which will now be obvious, $B_\nu(T)$ is also known as the blackbody function, and the radiance in the cavity and leaving the hole is called blackbody radiation. The importance of Eq. (3.21) for our present purposes is that it gives us the expression we seek for the function $J_\nu(T)$ for a system in thermodynamic equilibrium. It is worth emphasising that, at a given frequency, the source function depends only on the temperature T if the gas is in thermodynamic equilibrium. The properties of different gases, and of solid surfaces, emitting and absorbing radiation at frequency ν are contained in their absorption coefficient k_ν.

In reality, no system is completely closed and so true thermodynamic equilibrium never applies. However, the simplicity of the expressions for the absorption coefficient and, particularly, the source function for the case of thermodynamic equilibrium lead us to look for realistic situations where, although not strictly true, the ideal case of thermodynamic equilibrium could be applied without a significant loss of accuracy. In the astrophysical community, between 1900 when Planck developed his blackbody function and Milne's analysis of non-LTE in 1930, there was a general belief that the Planck function could be considered to be the source function for matter under any conditions. It was thought that it was an intrinsic property of matter, without realising that both collisional and absorption/emission processes also play a role in maintaining the conditions under which Planck's function applies.

Planetary atmospheres are not in thermodynamic equilibrium —for example, there are temperature gradients, while in thermodynamic equilibrium temperature should be the same everywhere— but the redistribution of the kinetic energy of molecules occurs very quickly, much more quickly than the redistribution with other energy forms such as radiation or internal energy. Exchanges of translational energy to and from other energy forms are not sufficient to force a given atmospheric parcel away from translational equilibrium. That is, the kinetic temperature may change when these transfers occur (as actually happens since the kinetic temperature varies from one region to another), but translational equilibrium, i.e., the establishment of a local kinetic temperature, is reached in practically the whole atmosphere. In fact, this can be applied up to the exosphere of the planetary atmospheres and up to the mid part of the cometary coma.

Thus, a Maxwellian distribution of molecular velocities for a local kinetic temperature $T(z)$ at a height z can be assumed with a high degree of accuracy at any atmospheric height from the surface up to the exosphere. This is defined as $T = 2E_k/(3k)$, where k is the Boltzmann constant and E_k the mean kinetic energy of molecules at height z. Having defined a local kinetic temperature $T(z)$, we say that a given state (electronic, vibrational or rotational) is in local thermodynamic equilibrium (LTE) when its population is given by Boltzmann's law at this local kinetic temperature. As we will show later in Secs. 3.6.3 and 3.6.4, this situation will occur if thermal collisions between the atmospheric molecules are faster than any other collisional or radiative process. For a given excited state in LTE, it then follows that the radiating properties of transitions (the source function and the absorption coefficient) associated with this state depend only on the kinetic temperature. In particular, the source function is described by the Planck function at the local kinetic temperature, $T(z)$.

The origin of the term 'local' thermodynamic equilibrium should now be clear; in the atmosphere, neighbouring parcels of gas are described as if they were in thermodynamic equilibrium, each at whatever kinetic temperature prevails locally. Thus:

(a) In LTE we first assume that a kinetic temperature can be defined and then that the excited internal states, which interact with and emit radiation, are coupled with it by collisions.

(b) LTE can apply to individual forms of internal energy, and not necessary to all internal modes at the same time. Thus, LTE may prevail for rotational but not for vibrational levels, or for one vibrational mode but not another, to give two common examples.

(c) In true thermodynamic equilibrium the radiative field is blackbody radiation ($L_\nu = B_\nu$) and the source function is given by the Planck function ($J_\nu = B_\nu$). In LTE the source function is still given by Planck's function, $J_\nu = B_\nu(T)$, but the radiance L_ν can differ from B_ν.

(d) A direct consequence of point (c) and Eq. (3.13) is that LTE, unlike true thermodynamic equilibrium, is compatible with a net gain or loss of radiative energy by the gas. In other words, it is compatible with a non-zero heating rate. The only requirement is that collisions should be fast enough to transfer the net absorbed or emitted radiative energy into kinetic energy.

The terms 'thermodynamic equilibrium' or 'local thermodynamic equilibrium' should not be confused with 'radiative equilibrium' (see Sec. 4.2.4). Thermodynamic equilibrium implies radiative equilibrium (net flux is zero) but the opposite is not generally true. If we have radiative equilibrium for the spectral range of a vibrational band, either at each individual frequency (monochromatic equilibrium, $h_\nu = 0$) or over the band interval $h_{\Delta\nu} = 0$, then, when all other non-thermal processes are negligible, LTE applies. On the other hand, as we have seen above, we can have LTE without radiative equilibrium.

Since the forms of the absorption coefficient and particularly the source function are so simple for LTE, compared to fully non-equilibrium situations, it is very useful to know when LTE is expected to occur. To see, in a general way, under which conditions the population of an excited state, or the emission of a band arising from this state, is that of LTE, we need to know the level population or the source function of the band. This is obtained from the equation of statistical equilibrium which takes into account all the microscopic processes that affect the population, and is derived and described in detail in the next section. Then, later in Sec. 3.7, we mention some general ideas about when and where we expect LTE to occur and the most common non-LTE cases that we can find in the atmosphere.

3.6 The Source Function in Non-LTE

We have written the radiative transfer equation in a form that involves macroscopic quantities, and seen that, for LTE conditions, the quantities that express the radiation-matter interactions and, in particular, the emission term (the source function), depend on temperature only. We look now for the form of the source function for the general situation, which is usually expressed as a function of the number densities of the populations of the upper and lower energy levels of the transition involved. We should note however that this may not give much information for solving a particular problem, since we are merely replacing one unknown quantity, the source function, by another unknown quantity, the upper level number density. For a solution to the problem we further need to state the equation that governs the populations of the levels involved in each transition of interest. This is given by the balance between all microscopic processes that affect these populations, expressed in statistical equilibrium equations, and illustrated in Fig. 3.3. Also, it is useful to relate the absorption coefficient, which we introduced 'ad hoc' into the radiative transfer equation, to the parameters introduced at microscopic levels in Einstein's formulation, as we show in Sec. 3.6.2.

It is fundamental to realise that, in non-LTE, the upper state population of a transition itself depends on the radiative field in which the molecule resides. The reason for this is that the molecule will be excited by the absorption of photons, and this affects its population. Of course, this also occurs in LTE, but then collisions occur sufficiently frequently to cause the distribution of populations to adjust to the Boltzmann distribution corresponding to the kinetic temperature of the gas. In non-LTE, thermal collisions are less important and may be entirely negligible, so that excited states relax by some other process (or combination of processes, perhaps including some contribution from collisions) altogether. Then, the populations of the energy levels have little or no relation to the kinetic temperature of the gas, and we have to introduce the concept of vibrational (T_v) or rotational (T_r) temperature to describe them. These are the temperatures which when introduced in the Boltzmann exponential factor give the observed populations, e.g.,

$$\frac{n_{v,r}}{n_0} = \frac{g_{v,r}}{g_0} \exp\left(-\frac{E_{v,r}}{kT_{v,r}}\right), \qquad (3.22)$$

where $n_{v,r}$ is the number density of the upper (vibrational or rotational) state, $E_{v,r}$ is its energy, n_0 is the number density of the lower state, $g_{v,r}$ and g_0 are their respective degeneracies, and k is the Boltzmann constant. It follows that if the vibrational or rotational temperature of a level differs from the local kinetic temperature, the level is in vibrational or rotational non-LTE.

Non-LTE situations also exist for the electronic transitions of atoms. The latter are more important in astrophysical situations than atmospheric, and indeed are not usually referred to as non-LTE emissions in the atmospheric literature at all. The derivation of the source function for atomic transitions has been extensively treated in the astrophysical literature (see, e.g., Thomas, 1965; Ivanov, 1973).

3.6.1 *The two-level approach*

Let us consider first an atmospheric molecule with just two vibrational levels, that with the larger excitation energy being referred to as the 'upper', and the other as the 'lower' state. The latter would most commonly be the ground state, since most molecules at atmospheric temperatures are normally in the ground state, but we will consider the general case where it is also an excited state. Each vibra-

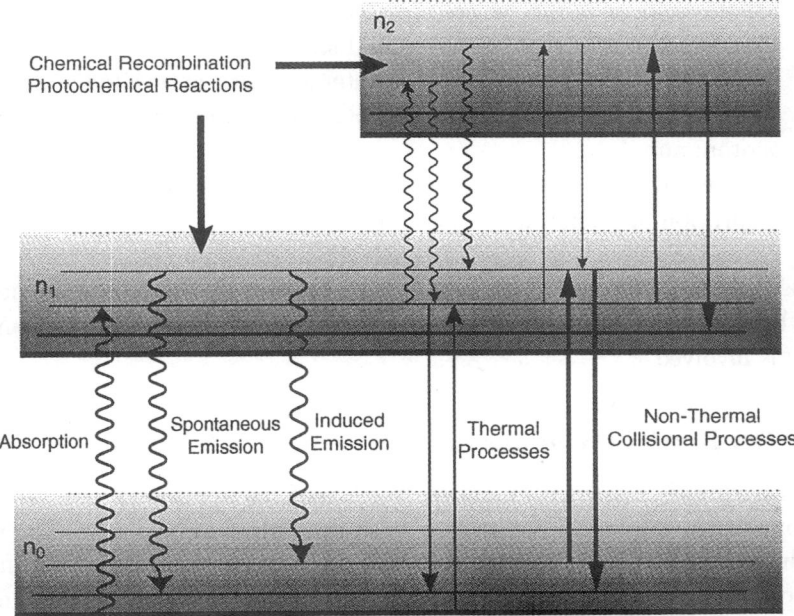

Fig. 3.3   Processes affecting the populations of vibrational levels.

tional state has a rotational fine structure that gives rise to the typical shape of a vibration-rotation band (Chapter 2). These two vibrational levels are connected by radiative and collisional processes. We distinguish explicitly between two kinds of collisional processes. Thermal (or vibrational-translational) processes are those where the vibrational energy of the upper state, relative to that of the lower state, is converted to or taken from the translational energy of a collision partner. If the energy involved in a collisional interaction is transferred to (or taken from) a non-translational source of energy we say that the molecule has suffered a non-thermal collision. Thermal processes will lead the population states to LTE, while non-thermal, in general, will drive them out of LTE. Examples of non-thermal processes are vibrational-vibrational (V–V) energy transfer processes, and electronic-vibrational energy transfer. As we will see in Chapters 6 and 7, in most situations these processes produce non-LTE emissions. Under some less common circumstances, however, they can also cause some vibrational levels to have LTE populations. This happens, for example, to the CO_2 (0,2,0) and (0,3,0) levels in the Martian mesosphere (Chapter 10). We have also considered non-LTE processes in which the upper level is not excited from a molecule in a lower-energy state but the molecule is directly formed in the excited state. Examples of this kind are the chemical recombinations and photochemical reactions.

The interaction between radiation and matter for the vibration-rotation transitions between the two levels of this system can be described in terms of three fundamental processes:

(i) *Spontaneous emission*, where a molecule is de-excited producing a photon of energy equal to that of the excited state;

(ii) *Induced emission*, where a photon interacts with an excited state inducing another photon identical in phase, polarization and direction to the incident photon; and

(iii) *Absorption*, in which a ground-state molecule becomes excited after the absorption of a photon.

Thus, the molecules are de-excited by spontaneous and induced emissions, and excited by absorption of photons. In each case, the appearance or disappearance of photons is involved.

3.6.2   The Einstein relations

Consider a sample of gas, with number densities of the upper and lower vibrational state populations n_2 and n_1 interacting with the radiation field $L_\nu{}^*$. The rate per unit volume and frequency interval at which the molecules are de-excited from the upper to the lower level by spontaneous emission is proportional to the number of molecules excited in the upper level. Considering all rotational transitions between the J' rotational states in the upper vibrational level and the J'' rotational states in the lower, the rate of spontaneous emission is given by

$$\text{rate of spontaneous emission} \quad = \quad \sum_{J'} n_{2,J'} A_{J'},$$

where the sum extends over all rotational levels J' in the upper state 2, $A_{J'}$ includes the spontaneous emission rate for all allowed transitions from level J' to any level J'', and $n_{2,J'}$ is the number of molecules excited in the J' rotational level of the upper vibrational state.

If we now introduce the Einstein coefficient for the vibration-rotation band (2–1), A_{21}, by

$$A_{21} = \sum_{J'} n_{2,J'} A_{J'} / n_2, \tag{3.23}$$

and the normalized factor for the rotational states distribution

$$q_{r,J'} = n_{2,J'} / n_2, \tag{3.24}$$

with $n_2 = \sum_{J'} n_{2,J'}$, the rate per unit volume and frequency interval at which the molecules are de-excited from level 2 to 1 by spontaneous emission is given by

$$\text{rate of spontaneous emission} = n_2 A_{21} q_{r,s}(\nu - \nu_0). \tag{3.25}$$

---

*We avoid the use of subscript 0 for the lower level, since it is commonly used for the ground vibrational state, which does not have necessarily to be the lower state.

In the normalized factor $q_{r,s}$ we have replaced index J' by $(\nu - \nu_0)$, where ν is the frequency of the absorbed or emitted photons and ν_0 is the frequency of the vibrational transition or band centre. As shown in Chapter 2, each vibration-rotation line has in turn a frequency dependency around its line centre, the line shape. This is not considered here in $q_{r,s}(\nu - \nu_0)$ but can be included later, when required, for example, by including the full frequency dependence of the absorption coefficient (Eq. 3.34). The rotational levels are normally considered to be in LTE among themselves at a given temperature T, hence $q_{r,s}(\nu - \nu_0)$ has the form of a normalized Boltzmann exponential at T. This is a good assumption in all but the most rarified atmospheric regions because far fewer collisions are required than for vibrational levels. This rotational non-LTE is discussed in detail in Sec. 8.9.

Following a similar procedure, and assuming also that the number of molecules de-excited by induced emission is proportional to n_2 and to the mean radiance $\bar{L}_\nu$ incident from all directions, and that the rate at which they are excited by absorption is proportional to the number of molecules in the lower state n_1 and again to the mean radiance $\bar{L}_\nu$, we have

$$\text{rate of induced emission} \ = \ n_2 B_{21} \bar{L}_\nu q_{r,i}(\nu - \nu_0), \tag{3.26}$$

$$\text{and rate of absorption} \ = \ n_1 B_{12} \bar{L}_\nu q_{r,a}(\nu - \nu_0). \tag{3.27}$$

Note that we have assumed different rotational state distributions for the upper and lower vibrational states.

The Einstein coefficients depend only on the quantum-mechanical properties of molecules and transitions; they are independent of the radiative field and of the thermodynamic state of the molecular gas, but are related to each other. Because of this, we can derive the relationship between them by making any assumption about the thermodynamic state, the simplest being thermodynamic equilibrium, and the result will apply for any state, including non-LTE.

In thermodynamic equilibrium, Boltzmann's law states that the ratio of the populations of two levels is determined by the temperature T, according to

$$\frac{\bar{n}_2}{\bar{n}_1} = \frac{g_2}{g_1} \exp\left(-\frac{h\nu_0}{kT}\right), \tag{3.28}$$

where the overbars indicate the equilibrium values. Another consequence of equilibrium conditions is that the rate of photons absorbed per volume in the frequency interval ν to $\nu + d\nu$ equals that of emitted photons, so that

$$\bar{n}_2 A_{21} q_{r,s}(\nu - \nu_0) + \bar{n}_2 B_{21} \bar{L}_\nu q_{r,i}(\nu - \nu_0) = \bar{n}_1 B_{12} \bar{L}_\nu q_{r,a}(\nu - \nu_0). \tag{3.29}$$

Finally, thermodynamic equilibrium means that the mean radiance $\bar{L}_\nu$ is determined by the Planck function B_ν. Assuming that Eq. (3.29) should be fulfilled for any frequency and temperature (the so-called principle of detailed balance), it follows from Eqs. (3.21), (3.28) and (3.29) that

$$\frac{A_{21}q_{r,s}}{B_{21}q_{r,i}} = \frac{2h\nu^3}{c^2} \tag{3.30}$$

and

$$\frac{B_{12}q_{r,a}}{B_{21}q_{r,i}} = \frac{g_2}{g_1}\exp\left(\frac{h(\nu - \nu_0)}{kT}\right). \tag{3.31}$$

Note that if we assume the photons of the different rotational transitions to have the same energy $\nu = \nu_0$ and that the rotational states distributions in the upper and lower vibrational levels are the same, ($q_{r,s} = q_{r,a}$ and $q_{r,i} = q_{r,s}$), we then have the usual relationships for the Einstein coefficients[†]

$$\frac{A_{21}}{B_{21}} = \frac{2h\nu_0^3}{c^2} \quad \text{and} \quad \frac{B_{12}}{B_{21}} = \frac{g_2}{g_1}. \tag{3.32}$$

This is the 'two-level' approximation, in which the rotational structure superposed on the vibrational levels is ignored. In this limit, the Einstein coefficients have a constant relation to each other, independent of the physical state of the gas, as originally anticipated. For some bands which do not extend over a wide spectral interval ($\sim 100\,\mathrm{cm}^{-1}$), that approximation is plausible. For the H_2O bands, which extend over several hundreds of cm^{-1}, it is less appropriate. In the more exact case, where those terms are not neglected, what is happening is that absorbed photons, exciting rotational levels in the upper vibrational state, can be equilibrated by collisions before relaxing back to the ground vibrational state, and they can be relaxed by emission of photons at a different frequency. Thus, although still a useful approximation, 'detailed balance' does not in fact apply rigourously to vibration-rotation transitions.

3.6.3   *Radiative processes*

Let us now calculate the balance of absorbed and emitted photons for a general case when radiation interacts with matter. The net (emission minus absorption) rate of emitted photons is proportional to the change in the radiance of the radiative field. The change of the radiance in a distance $\mathrm{d}s$, within the solid angle $\mathrm{d}\omega$, $[\mathrm{d}L_\nu/\mathrm{d}s]\mathrm{d}\omega$, is given by the net emission of photons within the volume of a cylinder of length $\mathrm{d}s$ and cross section $\mathrm{d}\sigma$ in the solid angle $\mathrm{d}\omega$.

[†]Some authors give different expressions for the relationships between the spontaneous emission coefficient and the induced and absorption coefficients, e.g., $A_{21}/B_{21} = 8\pi\nu^2/c^3$ (Goody and Yung, 1989) or $A_{21}/B_{21} = 8\pi h\nu^3/c^3$ (Lenoble, 1993). The reason is the different definitions of the induced and absorption coefficients, B_{21} and B_{12}, used by different authors. For example, Goody and Yung defined them in terms of the energy density, and the radiative energy is considered in units of $h\nu$. We use here the mean radiance $\bar{L}_\nu$. The ratio between the energy density u_ν and $\bar{L}_\nu$ is $u_\nu/\bar{L}_\nu = 4\pi/c$; and considering the $h\nu$ units of u_ν we get the $4\pi/(hc\nu)$ ratio between the A_{21}/B_{21} relationship given by Goody and Yung and Eq. (3.32) above.

Thus,

$$\frac{\mathrm{d}L_\nu}{\mathrm{d}s}\frac{\mathrm{d}\omega}{h\nu} = n_2 A_{21} q_{r,s}\frac{\mathrm{d}\omega}{4\pi} + n_2 B_{21} q_{r,i} L_\nu \frac{\mathrm{d}\omega}{4\pi} - n_1 B_{12} q_{r,a} L_\nu \frac{\mathrm{d}\omega}{4\pi}, \qquad (3.33)$$

where the factor $1/(h\nu)$ in the first term has been introduced to convert the energy units of L_ν into number of photons, and the factor $\mathrm{d}\omega/(4\pi)$ in the right hand terms has been included to express the number of photons produced in $\mathrm{d}\omega$. Using the relations between the Einstein coefficients (Eqs. 3.30 and 3.31) and Boltzmann's equation (3.28), rearranging the terms in Eq. (3.33) and comparing with the radiative transfer equation for absorption processes (Eqs. 3.12, 3.14, and 3.15) we find, for the absorption coefficient k_ν and the source function J_ν:

$$k_\nu = \frac{h\nu}{4\pi}\frac{n_1}{n_a}B_{12}q_{r,a}\left[1 - \frac{n_2}{n_1}\frac{\bar{n}_1}{\bar{n}_2}\exp\left(-\frac{h\nu}{kT}\right)\right] \qquad (3.34)$$

$$J_\nu = \frac{2h\nu^3}{c^2}\left[\frac{n_1}{n_2}\frac{g_2}{g_1}\exp\left(\frac{h(\nu - \nu_0)}{kT}\right) - 1\right]^{-1}. \qquad (3.35)$$

Using the expression for k_ν (Eq. 3.34) and the definition of B_ν (Eq. 3.21), J_ν can also be written as

$$J_\nu = B_\nu \frac{n_2}{\bar{n}_2}\frac{\bar{k}_\nu}{k_\nu}. \qquad (3.36)$$

This expression for the source function reduces to Planck's function for conditions of thermodynamic equilibrium (or local thermodynamic equilibrium), as required, and the expression for the absorption coefficient is then

$$\bar{k}_\nu = \frac{h\nu}{4\pi}\frac{\bar{n}_1}{n_a}B_{12}q_{r,a}\left[1 - \exp\left(-\frac{h\nu}{kT}\right)\right]. \qquad (3.37)$$

We have so far obtained the absorption coefficient at a particular frequency ν of a rotational line. To obtain the absorption coefficient of a complete vibration-rotation band, the so-called *band strength*[‡], S, we integrate Eq. (3.34) over the frequency interval $\Delta\nu$ over which the band extends:

$$S = \int_{\Delta\nu} k_\nu \,\mathrm{d}\nu = \frac{B_{12}}{4\pi}\frac{n_1}{n_a}\int_{\Delta\nu} h\nu q_{r,a}\left[1 - \frac{n_2}{n_1}\frac{g_1}{g_2}\exp\left(-\frac{h(\nu - \nu_0)}{kT}\right)\right]\,\mathrm{d}\nu. \qquad (3.38)$$

This expression can be simplified for most atmospheric molecular bands. One assumption often made is that since the spectral interval over which the rotational lines in a vibration-rotation band extend is fairly narrow (usually a few tens of wavenumbers, or a few percent of the frequency ν_0 at which the band centre lies, see Chapter 2), $h\nu$ and $\exp(-h\nu/kT)$ change slowly across $\Delta\nu$ and can be taken

---

[‡]The term 'band strength' is commonly used in optics and radiative transfer textbooks to mean its value in LTE conditions only, $\bar{S}$. We use here the term for any conditions, independently if n_2 and n_1 are in LTE or not. S is also sometimes called *band intensity*.

outside the integral with ν_0 replacing ν. The expressions for the source function and for the band strength are then

$$J_\nu = \frac{2h\nu_0^3}{c^2} \left[\frac{n_1}{n_2} \frac{g_2}{g_1} - 1 \right]^{-1} \tag{3.39}$$

and

$$S = \frac{h\nu_0}{4\pi} \frac{n_1}{n_a} B_{12} \left[1 - \frac{n_2}{n_1} \frac{g_1}{g_2} \right], \tag{3.40}$$

where we have made use of the normalization property of $q_{r,a}$. This approximation is not accurate for every band, for example the extensive vibrational bands of water vapour near $6.3\,\mu$m. The expression of the strength of a band in local thermodynamic equilibrium with the same assumptions is given by:

$$\bar{S} = \frac{h\nu_0}{4\pi} \frac{\bar{n}_1}{n_a} B_{12} \left[1 - \exp\left(-\frac{h\nu_0}{kT} \right) \right]. \tag{3.41}$$

Another common approximation is to drop the exponential term $\exp(-h\nu_0/kT)$ in these expressions to get

$$\bar{S} = \frac{h\nu_0}{4\pi} \frac{\bar{n}_1}{n_a} B_{12}. \tag{3.42}$$

This exponential is much smaller than unity for most atmospheric molecular bands, although it becomes significant for longer wavelengths and higher temperatures. For example, in the CO_2 bands around $15\,\mu$m, that factor is only 0.0083 for a mean atmospheric temperature of $220\,$K, but it reaches a value of 0.041 at $300\,$K, so if vibrational temperatures or cooling rates in the thermosphere are needed with an accuracy better than $\sim 4\%$, it should not be neglected. Also it should be included for the study of the atomic oxygen $63\,\mu$m emission in the thermosphere where, for a temperature of $500\,$K, it has a value of 0.63. This exponential factor also determines the ratio of the number densities in the upper and lower states of the transition in thermodynamic equilibrium (Eq. 3.28). So this discussion, as we said in Sec. 2.5.1, is simply reinforcing the fact that, under LTE conditions, the fraction of atmospheric molecules which are excited in high vibrational states is usually negligible compared to those in the ground state.

It is worth noting that a similar approximation for the band strength can be used for some non-LTE situations when the $n_2 g_1/n_1 g_2$ ratio is negligible compared to unity, either because the system is not very far from local thermodynamic equilibrium or, although far from equilibrium, because the upper level is under-, rather than over-, populated with respect to its LTE value. Then, the band strength is given by

$$S = \frac{h\nu_0}{4\pi} \frac{n_1}{n_a} B_{12}. \tag{3.43}$$

In fact, this expression is applicable to most of the vibrational bands in the atmosphere, since it is unusual for the non-LTE populations of excited states to be very much greater than their LTE equivalents. At high altitudes, in non-LTE conditions, collisions are not fast enough to compensate for the spontaneous emission, and absorption is also very weak, so the usual effect is a general depletion of the vibrational states with respect to LTE. There are however situations when that assumption should not be made. A good example is the CO_2 $10\,\mu$m bands in the upper atmosphere of the terrestrial planets (particularly Venus and Mars) during daytime conditions, when n_2 is very similar to n_1 or even becomes greater, giving rise to an inversion of populations so that the induced emission term becomes very important. (These are the bands which are active in the powerful carbon dioxide gas laser, and are often called the 'laser' bands, even in atmospheric applications).

If we are considering a *fundamental* transition, i.e., that whose lower level is the ground vibrational state, and assume that most molecules are in the ground state, then $n_1 = n_0 \simeq n_a$ and

$$S = \frac{h\nu_0}{4\pi} B_{12}. \qquad (3.44)$$

An important conclusion for these bands under these conditions is that the band strength has the same expression for LTE and for non-LTE.

For many purposes, it is useful to have the relationship between the Einstein spontaneous emission coefficient, A_{21}, and the absorption coefficient k_ν or the band strength, S. Eliminating B_{12} in Eq. (3.34) by using Eqs. (3.30) and (3.31)

$$A_{21}q_{r,s} = \frac{8\pi\nu^2}{c^2} \frac{n_a}{n_1} \frac{g_1}{g_2} \exp\left[-\frac{h(\nu - \nu_0)}{kT}\right] \left[1 - \frac{n_2}{n_1}\frac{g_1}{g_2}\exp\left(-\frac{h(\nu - \nu_0)}{kT}\right)\right]^{-1} k_\nu. \qquad (3.45)$$

Integrating over the molecular spectral band $\Delta\nu$, using the normalization property of $q_{r,s}(\nu - \nu_0)$, and replacing ν by ν_0 we have

$$A_{21} = \frac{8\pi\nu_0^2}{c^2} \frac{n_a}{n_1} \frac{g_1}{g_2} \left[1 - \frac{n_2}{n_1}\frac{g_1}{g_2}\right]^{-1} S. \qquad (3.46)$$

At this point it is useful to introduce the vibrational partition function, Q_{vib}, defined by

$$n_0 = \frac{n_a f_{\text{iso}}}{Q_{\text{vib}}}, \qquad (3.47)$$

where n_0 is the number density of molecules in the *ground* vibrational level, $v = 0$, and f_{iso} is the isotopic ratio of the isotope under consideration.[§] Introducing

---

[§]In the definition of Q_{vib}, f_{iso} is normally taken as unity. We are assuming here that n_a is the *total* number density of the molecular species under consideration, including its isotopic fractions, as in the HITRAN compilation. We also recall that Q_{vib} is different for each isotope.

Eq. (3.47) into (3.46) we have

$$A_{21} = \frac{8\pi\nu_0^2}{c^2} \frac{Q_{\text{vib}}}{f_{\text{iso}}} \frac{n_0}{n_1} \frac{g_1}{g_2} \left[1 - \frac{n_2}{n_1} \frac{g_1}{g_2}\right]^{-1} S. \qquad (3.48)$$

It is also useful[¶] to express this relation by replacing the populations of the upper and lower levels, n_2 and n_1 by their respective vibrational temperatures defined by Eq. (3.22), obtaining

$$A_{21} = \frac{8\pi\nu_0^2}{c^2} \frac{Q_{\text{vib}}}{f_{\text{iso}}} \frac{g_0}{g_2} \frac{\left[1 - \exp\left(-\left(\dfrac{E_2}{kT_{v,2}} - \dfrac{E_1}{kT_{v,1}}\right)\right)\right]^{-1}}{\exp\left(-\dfrac{E_1}{kT_{v,1}}\right)} S, \qquad (3.49)$$

where E_2 and E_1 are the vibrational energy of the upper and lower levels referred to the ground vibrational level, and $T_{v,2}$ and $T_{v,1}$ their respective vibrational temperatures (equal to the kinetic temperature if in LTE).

The source function can be obtained using the same approximations given above for the band strength. In particular, if: (i) ν can be replaced by ν_0; (ii) the exponential term $\exp(-h\nu_0/kT)$ is much smaller than unity; and (iii) the system is not very far from equilibrium (or in nonequilibrium if the upper level is under-populated with respect to LTE), then the source function reduces to

$$J_\nu = B_\nu \frac{n_2}{\bar{n}_2} \frac{\bar{n}_1}{n_1}. \qquad (3.50)$$

The simplest non-LTE case is when, in addition to these approximations, we are considering a fundamental transition. For this, the source function is simply expressed by

$$J_\nu = B_\nu \frac{n_2}{\bar{n}_2}. \qquad (3.51)$$

If ν can be replaced without loss of accuracy by ν_0 in the expression for the band strength, we can also replace J_ν and B_ν by J_{ν_0} and B_{ν_0}.

We have now obtained general (e.g. valid for LTE as well as non-LTE) expressions for the absorption coefficient k_ν and the source function J_ν. However, they are still dependent on unknown quantities, in particular the number densities of the populations of the upper and lower states of the transition, n_2 and n_1. The other quantities in the expressions for k_ν and J_ν, e.g., $\bar{n}_2$, $\bar{n}_1$ and B_ν can be obtained from the kinetic temperature which we assume to be known. The number densities of the states are governed by the statistical equilibrium equations, which are developed below.

¶This will become clear when describing the actual calculation of the non-LTE populations of levels in Chapters 6 and 7.

3.6.4 *Thermal collisional processes: the statistical equilibrium equation*

To obtain the population of n_2 and n_1, or the ratio between them, we consider now the statistical equilibrium equation, i.e., the principle of detailed balance as applied to collisions. The populations of the upper and lower levels are governed primarily by radiative and by collisional processes. We have already discussed the radiative processes and will now consider the collisional processes, which are of two kinds, thermal and non-thermal. In *thermal* collisions, the upper state of the transition is excited (or de-excited) with the energy taken from (or going to) the kinetic energy of the colliding molecule. In *non-thermal* collisions a significant part, possibly all, of the vibrational energy is taken from (or goes to) the internal energy of the colliding partner, e.g., the vibrational-vibrational and electronic-vibrational collisions, and chemical recombination.

Consider the vibrational-translational (V–T) process:

$$k_t : \quad n(2) + M \rightleftharpoons n(1) + M + \Delta E, \tag{3.52}$$

where M is any air molecule, and $\Delta E = E_2 - E_1 = h\nu_0$ is the energy difference of the upper and lower levels. As for radiative processes, we assume that the rate of loss of molecules in the upper state is proportional to the number density n_2 and to the number density of the collision partners, [M]. Introducing the proportionality coefficient k_t, called the *rate coefficient* of process (3.52) in the forward direction (in units of $cm^3 \, mol^{-1} \, s^{-1}$), the rate of loss of molecules per volume is given by $k_t[M]n_2$, or, $l_t n_2$, where $l_t = k_t[M]$ (in units of s^{-1}) is the *specific loss* of n_2. Analogously, the rate of production of n_2 is given by $P_t = p_t n_1$, where $p_t = k'_t[M]$ and k'_t is the rate coefficient of process (3.52) in the reverse direction. Under steady state conditions, the net production of molecules excited in the upper state is given by

$$P_{\text{net},2} = k'_t[M] \, n_1 - k_t[M] \, n_2. \tag{3.53}$$

As for radiative processes, k_t and k'_t are not independent. In the particular case of thermodynamic equilibrium, the effects of transitions in both directions cancel out, so that they do not provide a net excitation of the upper state. It then follows from Eq. (3.53) that

$$\frac{k'_t}{k_t} = \frac{\bar{n}_2}{\bar{n}_1} = \frac{g_2}{g_1} \exp\left(-\frac{h\nu_0}{kT}\right). \tag{3.54}$$

Analogously to the Einstein relationships, we obtain a relationship between k_t and k'_t which solely depends on temperature and the energy difference of the states, independently of the type of collision or the molecules involved. So it is a universal expression, valid for both LTE and non-LTE, whose only requirement is that a temperature T can be defined.

Consider now a case out of thermodynamic equilibrium where excitation and de-excitation by radiative (or collisional) processes alone do not balance. Assuming steady state conditions for the populations n_2 and n_1, the net production rate of

n_2 by radiative processes must be cancelled by the net removal rate by collisional processes. Hence, from Eqs. (3.25–3.27), integrating over the spectral interval of the transition, we have the statistical equilibrium equation (SEE)

$$n_1 B_{12} \bar{L}_{\Delta\nu} - n_2 A_{21} - n_2 B_{21} \bar{L}_{\Delta\nu} + k_t'[\mathrm{M}] n_1 - k_t[\mathrm{M}] n_2 = 0, \qquad (3.55)$$

where we have assumed that the induced and absorption line shapes are the same, and have introduced the mean radiance averaged over the spectral band, $\bar{L}_{\Delta\nu}$, defined by

$$\bar{L}_{\Delta\nu} = \int_{\Delta\nu} \bar{L}_\nu \, q_{r,a}(\nu) \, d\nu = \frac{1}{4\pi} \int_\omega \int_{\Delta\nu} L_\nu \, q_{r,a}(\nu) \, d\nu \, d\omega,$$

which is equivalent to the more common definition of

$$\bar{L}_{\Delta\nu} = \frac{1}{4\pi S} \int_\omega \int_{\Delta\nu} L_\nu \, k_\nu \, d\nu \, d\omega. \qquad (3.56)$$

Equation (3.55) can be expressed as

$$\frac{n_2}{n_1} = \frac{B_{12} \bar{L}_{\Delta\nu} + k_t'[\mathrm{M}]}{A_{21} + B_{21} \bar{L}_{\Delta\nu} + k_t[\mathrm{M}]}. \qquad (3.57)$$

We then have the desired expression for n_2/n_1 which includes a dependence on the radiation field. Inserting this expression into the equation for the source function (3.36), making use of the relations between the Einstein coefficients (3.30 and 3.31), using Eqs. (3.21) and (3.54), and defining ϵ as

$$\epsilon = \frac{l_t}{A_{21}} \left[1 - \exp(-h\nu_0/kT) \right] \qquad (3.58)$$

we obtain

$$J_{\nu_0} = \frac{\bar{L}_{\Delta\nu} + \epsilon B_{\nu_0}}{1 + \epsilon}. \qquad (3.59)$$

We now have the source function in a form suitable to be included in the radiative transfer equation prior to integration, since it is expressed as function of the unknown $\bar{L}_{\Delta\nu}$ and known parameters. This equation is often called the *integral equation* for the source function, particularly in the Astrophysical literature, since $\bar{L}_{\Delta\nu}$ depends on J (Eq. 3.20). From the solution of both equations, (3.59 and 3.20), we can derive the two unknowns, J and $\bar{L}_{\Delta\nu}$.

For some formulations of radiative transfer, e.g., the Curtis matrix method, it is useful to express the source function, J, in terms of the heating rate, instead of the mean integrated radiance, $\bar{L}_{\Delta\nu}$. Integrating Eq. (3.13) over the spectral band, and making use of Eqs. (3.14) and (3.56) we get for the heating rate, h_{12}:

$$h_{12} = 4\pi S n_a [\bar{L}_{\Delta\nu} - J_{\nu_0}], \qquad (3.60)$$

and eliminating $\bar{L}_{\Delta\nu}$ from (3.59) and (3.60),

$$J_{\nu_0} = B_{\nu_0} + D\, h_{12}, \tag{3.61}$$

where

$$D = \frac{1}{4\pi S n_a \epsilon}. \tag{3.62}$$

This equation is more conveniently written for computational purposes, by replacing the definition of ϵ and using the relation between A_{21} and S (Eq. 3.46), as

$$D = \frac{2\nu_0^2}{c^2} \frac{g_1}{g_2} \frac{1}{l_t n_1} \left[1 - \frac{n_2 g_1}{n_1 g_2} \right]^{-1} \left[1 - \exp\left(-\frac{h\nu_0}{kT} \right) \right]^{-1}. \tag{3.63}$$

The equation of transfer, in the form of (3.12) or (3.13), is then formally solved, and the source function is determined by the statistical equilibrium equation in the form of (3.59) or (3.61).

Strictly speaking, we have not completely replaced the population of the upper level by the source function, since the factor $[1 - (n_2 g_1)/(n_1 g_2)]$ in the expression of D still depends on n_2. Recall, however, that the term n_2/n_1 is much smaller than unity for most infrared bands in planetary atmospheres when the upper state is not far from LTE or when its population is smaller than that corresponding to LTE. Also, for these bands, the term $\exp(-h\nu_0/kT)$ is much smaller than unity. Under these circumstances, ϵ and D can be approximated by the simple expressions

$$\epsilon \simeq \frac{l_t}{A_{21}} \quad \text{and} \quad D \simeq \frac{2\nu_0^2}{c^2} \frac{g_1}{g_2} \frac{1}{l_t n_1}. \tag{3.64}$$

When these approximations are not accurate enough, then an initial estimate of n_2/n_1 can be obtained from these approximate equations and iterated using the accurate versions.

3.6.5 *Non-thermal processes*

We have seen in the previous section the most common case of non-LTE, sometimes called the classical case, in which, because of a paucity of thermal collisions, the radiative processes lead to a non-Boltzmann distribution of populations, i.e., a distribution not determined by the local kinetic temperature, in which the upper state is usually depleted compared to LTE. In practice, this occurs mainly under night-time conditions. Particularly during the daytime, several non-thermal processes also occur in the upper atmosphere, and these contribute to non-LTE. In such cases, the excited states often have populations larger than those corresponding to LTE. The following is a summary of the most important.

(1) Vibrational-vibrational (V–V) energy transfer, in which the vibrational excitation energy of a level is exchanged during a collision with one or more

different vibrational states. These processes play a fundamental role in the redistribution of the energy absorbed from solar radiation or from other non-LTE sources. Typical examples involve the exchange of v_2 and v_3 quanta between CO_2 molecules in the terrestrial planet atmospheres, and of v_3 quanta between $CO_2(v_1, v_2, v_3)$ and $N_2(1)$, and of v_2 quanta between atmospheric water vapour and $O_2(1)$, in the Earth's atmosphere.

(2) Electronic to vibrational energy transfer, as illustrated by the activation of the nitrogen molecule in its first vibrational mode by atomic oxygen electronically excited in the $O(^1D)$ state.

(3) Chemical recombination or chemiluminescence, in which the molecule is formed in excited vibrational states. The non-LTE excitation comes from the chemical energy of the molecules. Typical examples are the excitation of the O_3 infrared bands near $10\,\mu m$, $O+O_2+M \rightarrow O_3{}^*(v_1, v_2, v_3)+M$, and those of NO_2 near $6.2\,\mu m$, $O_3+NO \rightarrow NO_2{}^*(v_1, v_2, v_3)+O_2$.

(4) Photochemical reactions, such as the daytime O_2 infrared systems at $1.27\,\mu m$ and $1.58\,\mu m$, which are principally produced by the photolysis of ozone in the Hartley band.

(5) Dissociative recombination, of which $O_2{}^+ + e^- \rightarrow O^* + O$ is the most familiar example in the upper atmosphere which produces the red oxygen multiplet airglow, and,

(6) The collisions of atmospheric atoms and molecules with fast charged particles, for example during auroras and proton events. These give rise to a number of excited states that emit at rates far from the Planck function at the local temperature.

We describe in the next two sections the non-LTE source function for two important cases: non-thermal collisional processes, and chemical recombination (sometimes called chemiluminescence processes).

3.6.5.1 *Collisional processes*

In addition to thermal processes like those in (3.52), the upper state n(2) (or n_2) can also be excited by non-thermal collisional processes of the types (1) and (2) mentioned above:

$$k_{vv}: \quad M(v) + n(1) \rightleftharpoons n(2) + M(v') + \Delta E_v \qquad (3.65)$$

$$k_{ev}: \quad N^* + n(1) \rightarrow n(2) + N + \Delta E_{ev} \qquad (3.66)$$

at rates k_{vv} and k_{ev}, respectively. Note that process (3.66) has not been considered in the backwards direction because it is negligible, hence $l_{ev} = 0$. Here M(v) is an atmospheric molecule excited in vibrational level v before, and in v' (with $E_{v'} < E_v$) after, the collision; and N^* is either an excited atom or a molecule electronically or vibrationally excited. In general, M(v) and N^* have non-LTE populations. $\Delta E_v = E_v - E_{v'} - h\nu_0$ and $\Delta E_{ev} = E^* - h\nu_0$ are the energy exchanges, where E^* is the energy of N^* and $h\nu_0 = E_2 - E_1$. Following a similar procedure to the thermal (V–T) case, we have

$$\frac{n_2}{n_1} = \frac{B_{12}\bar{L}_{\Delta\nu} + p_t + p_{nt}}{A_{21} + B_{21}\bar{L}_{\Delta\nu} + l_t + l_{nt}}, \qquad (3.67)$$

where $p_{nt} = k_{vv}[M(v)] + k_{ev}[N^*]$ is the specific production rate (i.e., per molecule in the lower state) of n_2 molecules due to the non-thermal processes (3.65) and (3.66), and $l_{nt} = k'_{vv}[M(v')]$ is the specific loss of n_2 due to the same processes. The concentrations $[M(v)]$, $[M(v')]$, and $[N^*]$ are assumed to be known in general. The case where $[M(v)]$, $[M(v')]$ are unknown is treated in a more general way in the next section for the multilevel case. Also, as discussed previously for process (3.52), k_{vv} and k'_{vv} are related by

$$k'_{vv} = k_{vv}\frac{g_1}{g_2}\frac{g_v}{g_{v'}}\exp\left(-\frac{\Delta E_v}{kT}\right). \qquad (3.68)$$

Following similar steps to the case of thermal collisions we find,

$$J_{\nu_0} = \frac{\bar{L}_{\Delta\nu} + \epsilon_1 B_{\nu_0}}{1 + \epsilon_2}, \qquad (3.69)$$

where

$$\epsilon_1 = \frac{p_T}{A_{21}}\frac{g_1}{g_2}\left[\exp\left(\frac{h\nu_0}{kT}\right) - 1\right], \qquad \epsilon_2 = \frac{l_T}{A_{21}}\left(1 - \frac{p_T}{l_T}\frac{g_1}{g_2}\right), \qquad (3.70)$$

and $p_T = p_t + p_{nt}$ and $l_T = l_t + l_{nt}$ are the total excitation and de-excitation rates of n_2 for all thermal and non-thermal processes.

Analogously, the relationship between the source function and the heating rate is also modified, having the expression

$$J_{\nu_0} = B'_{\nu_0} + D_{nt}h_{12}, \qquad (3.71)$$

where

$$B'_{\nu_0} = \epsilon_r B_{\nu_0}, \qquad (3.72)$$

$$\epsilon_r = \frac{\epsilon_1}{\epsilon_2} = \left[\exp\left(\frac{h\nu_0}{kT}\right) - 1\right]\left[\frac{l_T g_2}{p_T g_1} - 1\right] \qquad (3.73)$$

and

$$D_{nt} = \frac{1}{4\pi S n_a \epsilon_2} = \frac{2\nu_0^2}{c^2}\frac{g_1}{g_2}\frac{1}{l_T n_1}\left[1 - \frac{n_2 g_1}{n_1 g_2}\right]^{-1}\left[1 - \frac{p_T g_1}{l_T g_2}\right]^{-1}. \qquad (3.74)$$

Note that when the contribution of non-thermal processes is not significant, $p_T \simeq l_T(g_2/g_1)\exp(-h\nu_0/kT)$, which leads to $\epsilon_1 = \epsilon_2 = \epsilon$, $\epsilon_r = 1$, and $D_{nt} = D$. Under these situations, the general expressions for the source function (Eqs. 3.69 and 3.71) reduce to those when only thermal collisional processes are present (Eqs. 3.59

and 3.61). Similarly, when $M(v)$ has an LTE population and the contribution of process k_{ev} is negligible, invoking Eq. (3.68), the expressions for J also reduce to those for thermal processes, as expected.

Neglecting the induced emission term by assuming $\exp(-h\nu_0/kT) \ll 1$, and assuming that the non-thermal collisional excitation rate does not exceed by much that corresponding to thermal excitations, so that $p_T/l_T \ll 1$, the expressions for the source function have the forms given above but with simplified coefficients, i.e.,

$$\epsilon_1 \simeq \frac{p_T}{A_{21}} \frac{g_1}{g_2} \exp\left(\frac{h\nu_0}{kT}\right), \quad \epsilon_2 \simeq \frac{l_T}{A_{21}}, \quad \epsilon_r \simeq \frac{p_T}{l_T} \frac{g_1}{g_2} \exp\left(\frac{h\nu_0}{kT}\right), \quad (3.75)$$

and

$$D_{nt} = \frac{2\nu_0^2}{c^2} \frac{g_1}{g_2} \frac{1}{l_T n_1}. \quad (3.76)$$

We should note that when non-thermal processes are present the divergence of the radiative flux, h, no longer has the meaning of 'heating' in the sense that the radiative energy is converted into (or taken from) the kinetic energy of the atmospheric molecules. Now, part of the energy (nearly all in some cases) can be taken from (or, less frequently, converted to) the internal energy of the colliding molecule, and therefore a balance in the radiative energy does not necessarily coincide with the net thermal balance of the atmospheric parcel under consideration. This is not relevant when computing the populations of the excited energy levels but should be taken into account when computing the heating or cooling that a given transition can produce in the atmosphere.

3.6.5.2   *Chemical recombination*

In the previous section we have considered the formulation for non-thermal collisional processes in general. There is however a kind of non-thermal process, the chemical recombination or chemiluminescence, which has the peculiarity that the rate of production of n_2 is *not* directly proportional to the number of molecules in the lower state of the transition n_1. This kind of processes can be represented by

$$k_c: \quad \text{A} + \text{B} + \text{M} \rightarrow \text{n}(2) + \text{M}, \quad (3.77)$$

where A and B are atmospheric atoms or molecules which, after recombination, give rise to the excited molecule of density n(2). M is a third-body constituent involved in some reactions. The rate of production of n_2 by this chemical process is given by $P_c = k_c \varphi [\text{A}][\text{B}]$ or expressed as a specific production rate, $p_c = P_c/n_1 = k_c \varphi [\text{A}][\text{B}]/n_1$, where we have included the efficiency factor φ to account for the yield of reaction (3.77) in producing n(2).

Following a similar procedure as for the other non-thermal processes, we obtain the statistical equilibrium equation as

$$\frac{n_2}{n_1} = \frac{B_{12}\bar{L}_{\Delta\nu} + p_t + p_{nt} + P_c/n_1}{A_{21} + B_{21}\bar{L}_{\Delta\nu} + l_t + l_{nt}}, \tag{3.78}$$

where p_{nt} does not include excitation by chemical recombination. The loss of n_2 by the reverse of reaction (3.77) is not generally important and has been neglected. Following the procedure in the previous section we obtain the same expressions for the source function and for the radiative transfer equation but with p_T replaced by $p_T + P_c/n_1$ in the formulae for ϵ_1, ϵ_2, ϵ_r, and D_{nt}.

These equations, however, do not fulfil our objective of having expressions with coefficients independent of the populations of the levels n_2 and n_1, since we still have the term P_c/n_1. A general solution is to set up the system of equations for statistical equilibrium for all the excited states, and of radiative transfer for all the important radiative transitions, and solve it for all of these at once. This is formulated in the next section for the multilevel case. Some assumptions, however, can be made which greatly simplify the solution of the problem. First, for most of the cases found in the atmosphere, radiative excitation can be neglected for most of the levels except for the very few low-energy ones, that is, the upper levels of fundamental transitions. Secondly, the recombination energy is usually deposited in the higher energy levels which then relax in cascade by collisions or spontaneous emissions to the lower levels, resulting in many highly excited molecular levels. Assuming that chemical recombination is the only non-thermal process under consideration, the population for the highly energetic levels, n_u, can then be approximated by

$$n_u \simeq \frac{P_t + P_c + \sum_{j>u} \alpha_j P_{c,j}}{A + l_t}, \tag{3.79}$$

where $\sum_{j>u} \alpha_j P_{c,j}$ accounts for the production of n_u from the collisional and radiative relaxations of the upper states. α_j are then assumed as known functions of the kinetic relaxation rates and the Einstein coefficients for radiative cascading to lower levels.

The procedure for getting the populations of all of the excited levels is to start from the uppermost one and to solve for each of the less energetic levels down to the first vibrationally excited level. Note that the inclusion of radiative excitation of the lower lying levels through fundamental transitions is not difficult, since for fundamental transitions n_l coincides very closely with the number of molecules in the ground state, which is usually known beforehand from the consistent (i.e., when all productions and losses of species A, B, C, and N are considered) output of chemical-dynamical models or measurements. So, for the fundamental transitions, Eq. (3.78) can be used directly.

Although much less common, we can also find cases where the levels are not excited from higher energy states cascading to lower levels, but in the reverse direction. For these situations the presence of the term P_c/n_1 in the coefficients ϵ_1, ϵ_2, ϵ_r, and D_{nt} of Eqs. (3.60) and (3.71) is not a difficulty since we can start by obtaining the population of the lowest level, knowing the population of the ground state, and then proceed to higher levels one step at a time.

3.6.6   *The multilevel case*

We extend the discussion now to the case where the n_2 and n_1 levels are affected by radiative and collisional processes involving other energy levels. Let us assume the general case of levels j with energy lower than E_1 $(E_j < E_1)$ and levels k with energy higher than E_2 $(E_k > E_2)$ (Fig. 3.4).

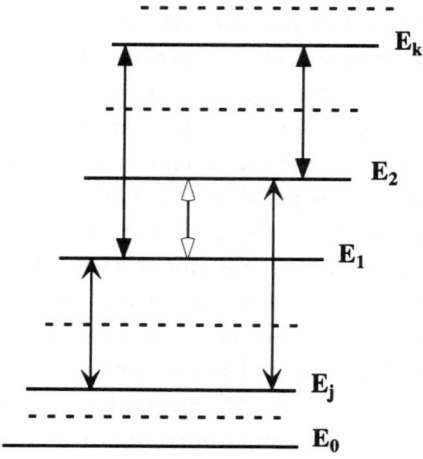

Fig. 3.4   Energy levels and transitions for the multilevel case.

The production rate of level 2 due to radiative processes is given by

$$B_{12}\bar{L}_{12}\, n_1 + \sum_{j<1} B_{j2}\bar{L}_{j2}\, n_j + \sum_{k>2}(A_{k2} + B_{k2}\bar{L}_{k2})\, n_k$$

and the radiative losses are

$$\left(A_{21} + B_{21}\bar{L}_{21} + \sum_{j<1}(A_{2j} + B_{2j}\bar{L}_{2j}) + \sum_{k>2} B_{2k}\bar{L}_{2k}\right) n_2,$$

where the sums over j and k extend over the levels with energy lower than level 1 and higher than level 2 (see Fig. 3.4), and the sub-indexes in $\bar{L}_{ij}$ can be exchanged with no effect. The productions and losses due to collisional processes are given by

$$P_2 = p_{21}n_1 + \sum_{m\neq 1,2} p_{2m}n_m = \sum_{m\neq 2} p_{2m}n_m, \quad \text{and}$$

$$\mathcal{L}_2 = \left(l_{21} + \sum_{m\neq 1,2} l_{2m}\right)n_2 = \sum_{m\neq 2} l_{2m}n_2,$$

where levels j and k have been merged together under index m.

Considering the continuity equation for level n_2, e.g., equating its radiative plus collisional productions to all its losses, and using the expression for the heating rates of the j–2 and 2–k transitions in the form

$$\frac{h_{j2}}{h\nu_{j2}} = B_{j2}\bar{L}_{j2}n_j - (A_{2j} + B_{2j}\bar{L}_{2j})n_2 \tag{3.80}$$

and

$$\frac{h_{2k}}{h\nu_{2k}} = B_{2k}\bar{L}_{2k}n_2 - (A_{k2} + B_{k2}\bar{L}_{k2})n_k \tag{3.81}$$

we may obtain, after some manipulation, the statistical equilibrium equation for level 2,

$$B_{12}\bar{L}_{12}n_1 + \sum_{j<1}\frac{h_{j2}}{h\nu_{j2}} - \sum_{k>2}\frac{h_{2k}}{h\nu_{2k}} + p_{21}n_1 + \sum_{m\neq1,2}p_{2m}n_m =$$
$$\left(A_{21} + B_{21}\bar{L}_{21} + l_{21} + \sum_{m\neq1,2}l_{2m}\right)n_2, \tag{3.82}$$

where we have taken into account that, according to its definition, the heating rates of transitions between level 2 and lower levels, j–2, produce a net excitation of level 2, while those taking place to higher levels, 2–k, contribute a net de-excitation. ν_{ij} defines the energy difference between levels j and i and, as for $\bar{L}_{ij}$, the sub-indexes can be exchanged with no effect. This type of equation can also be derived from (3.39) and (3.60), using the Einstein equations and the relation of the spontaneous emission coefficient, A_{21}, with the band strength, S, Eq. (3.46).

If we include the heating rate for the 2–1 transition and group all the collisional productions and losses, we have the simplified equation

$$\sum_{j\leq1}\frac{h_{j2}}{h\nu_{j2}} - \sum_{k>2}\frac{h_{2k}}{h\nu_{2k}} + \sum_{m\neq2}p_{2m}n_m = \sum_{m\neq2}l_{2m}n_2. \tag{3.83}$$

This expression relates the population of level 2 to those of the other levels m which significantly interact with it, and to the heating rate of all transitions in which level 2 is involved both as a lower or as an upper state. The system is completely represented if we write down similar continuity equations for the other excited levels, and the corresponding radiative transfer equations for each of the transitions involved. Since the statistical equilibrium equation relates the concentrations of at least two levels, that is, for a two-level system we have one equation but two unknowns, the previous system requires an additional equation to close it. This is obtained by requiring the conservation of the total number of absorbing molecules n_a (or $n_a f_{\mathrm{iso}}$), independent of the level in which they are excited, which is normally introduced by using the vibrational partition function Q_{vib} (see Eq. 3.47). In practice, this equation might not be needed since most molecules are in the ground state and its population can be taken as known, e.g., equal to the total number of absorbing molecules. The radiative heating is usually computed from Eq. (3.60) instead of Eq. (3.80), where the source function J can be replaced by its expression as a function of the ratio of the upper and lower levels (Eq. 3.35) or its simplified subsequent equations. The rest of the parameters are known with the exception of the induced emission term in the expression for S (Eq. 3.40), which is usually neglected. Thus, the problem is formally solved for the multilevel case. The different ways in which the equations are re-arranged to find a solution give rise to the different methods used in non-LTE models and calculations. These are treated in detail in Chapter 5.

Some methods use the source function and heating rates of the bands involved as unknowns, as in Eq. (3.71) but generalized for the multilevel case. Thus, from Eq. (3.82) we have

$$\frac{n_2}{n_1} = \frac{\left(\dfrac{h_{12}}{h\nu_0} + \sum_{j<1}\dfrac{h_{j2}}{h\nu_{j2}} - \sum_{k>2}\dfrac{h_{2k}}{h\nu_{2k}}\right)\dfrac{1}{n_1} + p_{21} + \sum_{m\neq1,2} p_{2m}\dfrac{n_m}{n_1}}{l_{21} + \sum_{m\neq1,2} l_{2m}}. \tag{3.84}$$

A given m level may be the upper or lower state of several transitions, therefore when replacing the n_m unknown populations in the latter equation by the unknown source functions we have to choose one among all the possible transitions. Usually n_m is replaced by the source function of the strongest transition originating from the m-th level. Assuming that approach and neglecting the induced emission in the expression of the source function, Eq. (3.84) can be re-written as

$$J_{21} = B'_{21} + \sum_{m\neq2} \mathcal{A}_{2m} J_m + D\left(h_{12} + \sum_{j<1}\frac{\nu_{12}}{\nu_{j2}}h_{2j} - \sum_{k>2}\frac{\nu_{12}}{\nu_{2k}}h_{k2}\right), \tag{3.85}$$

where

$$J_m = \frac{2h\nu_{lm}^3}{c^2}\frac{n_m}{n_{l,m}}\frac{g_{l,m}}{g_m}, \tag{3.86}$$

with $n_{l,m}$ being the lower level of the stronger transition arising from the m-th level, $g_{l,m}$ is its statistical weighting factor, and the $\mathcal{A}_{2m}$ coefficients are given by

$$\mathcal{A}_{2m} = \sum_{m\neq2}\frac{\nu_0^3}{\nu_{lm}^3}\frac{n_{l,m}}{n_1}\frac{g_1}{g_2}\frac{g_m}{g_{l,m}}\frac{p_{2m}}{l_T}. \tag{3.87}$$

B'_{21} and D are given by their respective expressions for the two-level approach, Eqs. (3.72–3.74), which, under the assumption of neglecting the induced emission considered in the three equations above (Eqs. 3.85–3.87), simplify to Eqs. (3.75) and (3.76). We should note that p_T in those expressions represents the collisional productions of n_2 *only* from n_1 and not from any other level. The other collisional terms are included in the $\mathcal{A}_{2m}$ coefficients. However, l_T in those equations and in Eq. (3.87) includes the collisional losses from *all* levels considered, $l_T = l_{21} + \sum_{m\neq1,2} l_{2m}$.

Equation (3.85) relates the source function for the transition 2–1 with those of the energy levels connected with it and with the heating rates of all transitions in which level 2 is introduced as an upper state. If we express this equation for all excited levels, and the radiative transfer equation for each of the transitions, in a form that relates the heating rate to the source function (see the discussion of the Curtis matrix method in Sec. 5.7), we have a system of equations to solve in which the source function and heating rates are the unknowns.

3.7 Non-LTE Situations

Now that we have derived the mathematical expressions for the source function of an excited level we are in a better position to consider in some detail when and where we might expect non-LTE situations to occur. We use here the expressions we have obtained in the sections above for the simpler case of the two-level model.

First, it is worthwhile to show that the expressions we have derived for non-LTE situations also incorporate the LTE case. Where only thermal collisional processes are important, i.e. where collisions are very fast, then from Eqs. (3.58), (3.59) and $\epsilon \gg 1$ it follows that $J_{\nu_0} = B_{\nu_0}$. Using Eqs. (3.35) and (3.21) it is seen that the ratio of the populations of levels 2 and 1 is given by the Boltzmann relation (Eq. 3.28). The same conclusion can be reached from Eqs. (3.61) and (3.62), which also tell us that LTE conditions are compatible with a net heating or cooling rate.

3.7.1 *The classical case of non-LTE*

This is the situation where only thermal collisions and radiative processes are assumed to be important. We consider separately the cases where weak and strong radiation fields are present.

3.7.1.1 *Weak radiative field*

Consider first the situation where the radiation field is so weak that the molecules are not significantly excited by absorption of radiation (losses by induced emission are also expected to be negligible). Neglecting the absorption and induced radiation terms in Eqs. (3.57) and (3.59) and using (3.54), the population of the upper state and the source function then simplify to

$$\frac{n_2}{n_1} = \frac{\bar{n}_2}{\bar{n}_1} \left(\frac{1}{1 + A_{21}/l_t} \right) \quad \text{and} \quad J_{\nu_0} = \frac{B_{\nu_0}}{1 + A_{21}/l_t}, \tag{3.88}$$

where we have made $\epsilon = l_t/A_{21}$ in the expression for the source function. Under these conditions we expect non-LTE to set in at that altitude where the collisional de-excitation rate, l_t, is significant or smaller than the spontaneous emission rate, A_{21}. The density of n_2 will be then smaller than that corresponding to LTE.

Another parameter to be considered in a crude estimation of the height at which non-LTE occurs is the energy of the transition. For those with small energy jumps the average number of collisions ($k_t[M]$) required to keep the levels in equilibrium is also smaller, so they can be in LTE up to higher altitudes in the atmosphere. As an example we can cite the CO_2 4.3 μm band, which starts to break down from LTE at $\sim$40 km; while the O(^3P) states, which emit at 63 μm, and the rotational levels of most atmospheric molecules are in LTE up to high altitudes ($\sim$200 km) in the thermosphere. Rotational non-LTE is discussed in more detail in Sec. 8.9.

Permitted electronic transitions have radiative lifetimes which are much shorter than those of vibrational transitions, and also the energy gaps between states are much larger. These transitions are therefore in non-LTE over most of the atmosphere. On the other hand, the magnetic dipole (or fine structure) transitions of atomic oxygen have a much longer radiative lifetime, and the energy difference between them is much smaller than for the vibrational transitions. They are expected to be in LTE up to high altitudes in the thermosphere.

The method discussed above for predicting the height of LTE breakdown is not accurate when the transition under consideration is optically thick around that altitude. In those cases, LTE usually prevails up to higher altitudes because photons cannot escape to space, the effective radiative losses are very small, and fewer collisions are needed to provide the level with an LTE population. A much better approximation of the LTE departure altitude is given then by the level at which the collisional thermal losses are equal to the spontaneous emission rate, multiplied by the probability of escape of the photons to space (see Sec. 5.2.1.2). Typical examples of these situations are the CO_2 15 μm emission in the upper atmosphere of Mars and Venus. In these atmospheres this band departs from LTE at levels where the pressure is at least an order of magnitude lower than those where the collisional and radiative deactivations are equal. These situations are unique in the sense that we still have an LTE situation even when the radiative processes are much faster than the collisional ones.

3.7.1.2 *Strong radiative field*

In situations where excitation by radiative processes dominates over the thermal collisional excitation, it follows from Eqs. (3.57) and (3.59), and the Einstein relationships (Eq. 3.32) that the population of the upper state and the source function are approximately given by

$$\frac{n_2}{n_1} = \frac{g_2}{g_1} \frac{c^2}{2h\nu_0^3} \frac{\bar{L}_{\Delta\nu}}{(1 + l_t/A_{21})} \quad \text{and} \quad J_{\nu_0} = \frac{\bar{L}_{\Delta\nu}}{1 + l_t/A_{21}}, \tag{3.89}$$

where we have neglected the induced emission term in the left equation (the term $c^2\bar{L}_{\Delta\nu}/(2h\nu_0^3)$ in the denominator) and have approximated ϵ by l_t/A_{21} in the source function. If, in addition, radiative relaxation is faster than collisional de-excitation, the source function is reduced to $\bar{L}_{\Delta\nu}$.

When we have such a strong radiative field, whether from the surface and lower atmosphere below (earthshine) or external (i.e. solar radiation), the strong absorption generally causes n_2 to be larger than its LTE value in some atmospheric regions. When the external radiation source is the Sun, the source function can be simplified to $J_{\nu_0} = \bar{L}_{\Delta\nu,\odot}$. For transitions at wavelengths shorter than $\sim 5\,\mu$m, this is usually much larger than in LTE in the middle atmosphere, and is the reason why non-LTE situations are more common in a daytime atmosphere than at night. Also, solar radiation dissociates some species, thus providing the reactants for the chemiluminescence and recombination processes, which are another significant source of non-LTE emissions.

The absorption of radiation from a distant source can be of importance even for night-time conditions, however. For example, optically thin transitions at mesospheric levels, where the local temperature is low and hence the thermal collision excitations are infrequent, can absorb significant amounts of emission from the warmer stratopause region or the surface, leading again to non-LTE populations larger than those corresponding to LTE. Most of the atmospheric infrared bands in the night-time upper mesosphere show this condition.

3.7.2 *Non-classical non-LTE situations*

When excitation by non-thermal processes (non-thermal collisions, or chemical or photochemical reactions) dominate over that from thermal collisions and radiative processes, a non-LTE population of the upper level usually results. For these situations, the population of the upper state and the source function are given, according to Eqs. (3.78) and (3.69), by

$$\frac{n_2}{n_1} \simeq \frac{p_t + p_{nt} + P_c/n_1}{A_{21} + l_t} \quad \text{and} \quad J_{\nu_0} \simeq \frac{p_t + p_{nt} + P_c/n_1}{p_t(1 + A_{21}/l_t)} B_{\nu_0}, \qquad (3.90)$$

where we have assumed $\exp(-h\nu_0/kT) \ll 1$, $p_T/l_T \ll 1$, and that losses by nonthermal collisional processes are negligible. We have also kept the productions by thermal processes. These might be negligible at high altitudes but it is safe to keep them since they will dominate at lower altitudes. Hence, the vibrational level or the transition originating from it will be in non-LTE if the production term of the non-thermal process, p_{nt}, or the corresponding term P_c/n_1 if we are considering chemical recombination, is of the same order or larger than the production by thermal collisions. These processes again usually lead the excited state to have populations much larger than those corresponding to LTE. The height at which this occurs is that at which the thermal excitation rate is of a similar magnitude to the non-LTE excitation source. For several bands, non-LTE is important at low atmospheric heights, i.e., NO at $5.3\,\mu$m can be in non-LTE as low as the stratosphere, and O_3 and NO_2 are in non-LTE in the low mesosphere (see Chapter 7).

3.8   References and Further Reading

Basic treatments of radiation in the atmosphere are presented by Liou (1980) and
Coulson (1975).   Chapter 4 in Brasseur and Solomon (1986) and Chapter 2 in
Andrews *et al.* (1987) also give brief but comprehensive introductions.   Goody and
Yung (1989) is possibly the most complete and comprehensive book on atmospheric
radiation, in particular for the infrared part of the spectrum.   Kondratiev (1969)
is an older, but also very detailed, text, while Liou (1992) and Lenoble (1993)
offer particular emphasis on modern treatments of radiative transfer in clouds. The
classic work on scattering processes is Chandrasekhar (1960), which can also be
consulted for a discussion of polarization effects in radiative transfer.   Another
classic monograph on scattering processes is Sobolev (1975).

    Good descriptions of LTE and non-LTE can be found in Milne (1930), Curtis and
Goody (1956), Chapter 13 in Thorne (1988), and Goody and Yung (1989). A brief
history of the derivation of the Planck function can be found in a modern book about
infrared remote sensing by Hanel *et al.* (1992). More detailed derivations, including
the Rayleigh-Jeans and Wien approximations, can be found in the published lectures
of Planck (1913), or in quantum mechanics textbooks such as Eisberg and Resnik
(1985). The derivations of the source function and absorption coefficients in non-
LTE are treated in Milne (1930), Kuhn and London (1969), Chapter 2 in Andrews
et al. (1987), and Goody and Yung (1989). A recent work on the general behaviour
of the source function for a vibration-rotation band in planetary atmospheres is
given by Shved and Semenov (2001).   The derivation of the source function for
atomic transitions includes some interesting mathematical treatments and has been
extensively treated in the astrophysical literature, for example, by Thomas (1965)
and Ivanov (1973).

Chapter 4

Solutions to the Radiative Transfer Equation in LTE

4.1 Introduction

Chapter 3 covered the basic formulation of radiative transfer both for LTE and non-LTE conditions. We now proceed to the actual integration of the radiative transfer equation (RTE), including simplified as well as more general cases, for LTE (in this chapter) and non-LTE conditions (in the next chapter). It is convenient to use the simpler LTE case to introduce useful ways of using the radiative transfer equation, including methods for integrating over the spectral range of a molecular band and over all directions and solid angles, as required for many applications, such as calculating atmospheric heating rates.

4.2 Integration of the Radiative Transfer Equation over Height

This section introduces some useful approximations for the solution of the RTE which are valid for a number of practical cases. The first neglects the curvature of the Earth so that the atmosphere can be treated as a stack of layers with negligible horizontal variations, separated by parallel, horizontal surfaces. Next we assume that the layers are thin enough that the properties of the atmosphere do not vary significantly within the layers (see Fig. 4.1). Both assumptions are widely used in atmospheric studies to simplify the RTE, by reducing the derivatives over the optical thickness in any of the three spatial directions to that over the geometric altitude above the surface of the planet, and to make the derivatives easy to calculate, by summing over layers. When these two conditions are applied we say that are using the plane-parallel or stratified approach.

4.2.1 The RTE in the plane-parallel approach

Because of the symmetry of a plane-parallel model atmosphere it is convenient to use a coordinate system with the z-axis in the upward vertical direction, perpendicular to the atmospheric layers, and to express the angular dependence of the radiation

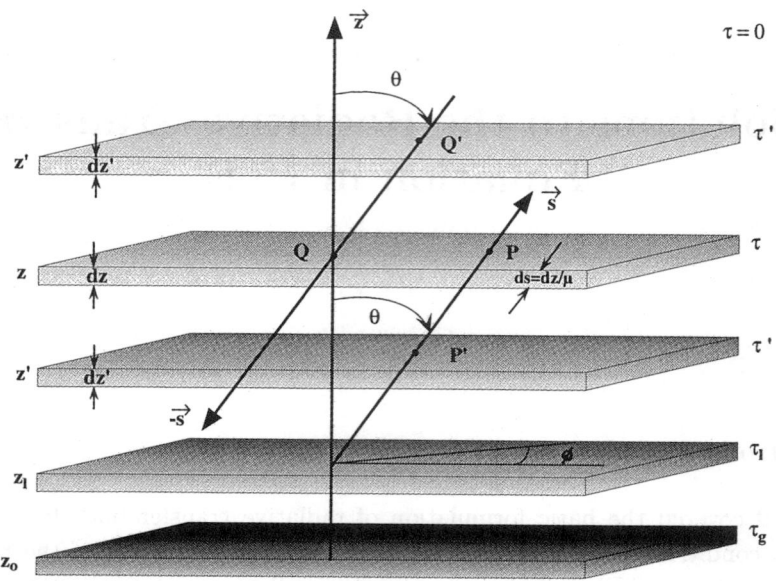

Fig. 4.1   An illustration of the plane-parallel approximation, in which the atmosphere is divided up into a number of homogeneous layers for the purpose of radiative transfer calculations. The vertical coordinates are height above the surface, z, and optical depth (measured from space), τ.

field in terms of the polar coordinates (θ, ϕ) referenced to the z-axis, with θ being measured from the vertical. In this system the element of solid angle is given by $d\omega = \sin\theta \, d\theta \, d\phi$. Under these conditions the radiance does not depend on the azimuth angle ϕ and the vector flux consists only of the z-component given by

$$F_\nu(z) = 2\pi \int_0^\pi L_\nu(z, \theta) \, \sin\theta \, \cos\theta \, d\theta = 2\pi \int_{-1}^1 L_\nu(z, \mu) \, \mu \, d\mu, \qquad (4.1)$$

where $\mu = \cos\theta$. We may define the upward flux, $F_\nu^\uparrow$, as the radiance flowing across a surface in the x-y-plane into the half–space above (2π steradian), and the downward flux, $F_\nu^\downarrow$, as the radiance flowing downwards across the surface into the half-space below by

$$F_\nu^\uparrow(z) = 2\pi \int_0^{\pi/2} L_\nu^\uparrow(z, \theta) \, \sin\theta \, \cos\theta \, d\theta = 2\pi \int_0^1 L_\nu^\uparrow(z, \mu) \, \mu \, d\mu \quad \text{and} \quad (4.2)$$

$$F_\nu^\downarrow(z) = 2\pi \int_\pi^{\pi/2} L_\nu^\downarrow(z, \theta) \, \sin\theta \, \cos\theta \, d\theta = 2\pi \int_0^{-1} L_\nu^\downarrow(z, \mu) \, \mu \, d\mu, \qquad (4.3)$$

where μ takes positive values in $F_\nu^\uparrow(z)$ and negative in $F_\nu^\downarrow$. The total or *net* flux F_ν is then given by

$$F_\nu(z) = F_\nu^\uparrow(z) - F_\nu^\downarrow(z). \qquad (4.4)$$

The heating rate is simply:

$$h_\nu = -\frac{\mathrm{d}F_\nu(z)}{\mathrm{d}z}. \tag{4.5}$$

Considering the direction **s** along the vertical coordinate in the atmosphere, z, we have $\mathrm{d}s \equiv \mathrm{d}z/\mu$ (Fig. 4.1). Then, from Eqs. (3.12) and (3.14) the expression for the RTE for a plane-parallel atmosphere when considering absorption processes only (i.e., excluding scattering) is given by

$$\mu \frac{\mathrm{d}L_\nu(z,\mu)}{\mathrm{d}z} = -k_\nu n_a \left[L_\nu(z,\mu) - J_\nu(z) \right], \tag{4.6}$$

where k_ν is the absorption coefficient, n_a is the number density of absorbing molecules, and we have also assumed that the source function is isotropic and hence independent of θ.

We now look for the formal solution of the RTE for a plane parallel atmosphere, including the appropriate boundary conditions at the top and bottom of the atmosphere. Strictly speaking, in the solution for the coordinate z, we should have a boundary condition for each value of μ. In practice each boundary condition can be simplified to two values, one for the upward and another for the downward radiance. For the radiation travelling upwards, $L_\nu^\uparrow(0 < \mu \leq 1)$, we assume a blackbody radiance at a given temperature T_g at the lower boundary z_0 (usually the surface or cloud top temperature),

$$L_\nu^\uparrow(z_0, \theta) = B_\nu(T_g) \qquad \text{for} \qquad 0 < \mu \leq 1. \tag{4.7}$$

Note that since blackbody emission is isotropic, we are assuming the same value for the upward radiance in all directions. For the radiation travelling downwards, $L_\nu^\downarrow(-1 \leq \mu < 0)$, the boundary condition is simply provided by the incident radiation from outer space. For night-time conditions, this is negligible in all downward directions, i.e.

$$L_\nu^\downarrow(\infty, \theta) = 0 \qquad \text{for} \qquad -1 \leq \mu < 0. \tag{4.8}$$

The daytime case with solar radiation present is considered in the next section. It can be added directly as an additional term to the downward radiance (Eq. 4.10 below), assuming that the incident radiance at the top of the atmosphere is normally negligible in all directions except within the small solid angle subtended by the Sun.

Taking into account these boundary conditions, (3.20) has the form

$$L_\nu^\uparrow(z,\mu) = B_\nu(T_g) \exp\left[-\int_{z_0}^{z} k_\nu(z')\, n_a(z')\, \frac{\mathrm{d}z'}{\mu} \right] +$$

$$\int_{z_0}^{z} k_\nu(z')\, n_a(z')\, J_\nu(z') \exp\left[-\int_{z'}^{z} k_\nu(z'')\, n_a(z'')\, \frac{\mathrm{d}z''}{\mu} \right] \frac{\mathrm{d}z'}{\mu} \tag{4.9}$$

for $1 \geq \mu > 0$. Analogously, for downward radiance, integrating now from infinity down to point z

$$L_\nu^\downarrow(z,\mu) = \int_\infty^z k_\nu(z')\, n_a(z')\, J_\nu(z') \exp\left[-\int_{z'}^z k_\nu(z'')\, n_a(z'')\, \frac{\mathrm{d}z''}{\mu}\right] \frac{\mathrm{d}z'}{\mu}, \qquad (4.10)$$

for $-1 \leq \mu < 0$. Since μ only takes negative values in Eq. (4.10), the negative sign cancels out and the downward radiance is positive. Also note that $z' > z$ which makes the integral in the exponent negative, thus giving the appropriate absorbing factor.

These expressions are often written in a more concise form using the concept of *optical depth*. This simplifies some problems, like the calculation of the temperature profile of an atmosphere in radiative equilibrium (Sec. 4.2.4). From the definition of optical thickness (Eq. 3.16), considering the case of a plane-parallel atmosphere with the direction **s** along the vertical coordinate, and integrating along the vertical trajectory from a given z to infinity (i.e. to the top of the atmosphere) we have the expression for the optical depth

$$\tau_\nu \equiv \tau_\nu(z,\infty) = \int_z^\infty k_\nu(z')\, n_a(z')\, \mathrm{d}z'. \qquad (4.11)$$

With this definition, the radiative transfer equation for a plane-parallel atmosphere is simply

$$\mu \frac{\mathrm{d}L_\nu(\tau_\nu,\mu)}{\mathrm{d}\tau_\nu} = L_\nu(\tau_\nu,\mu) - J_\nu(\tau_\nu), \qquad (4.12)$$

where the change of sign relative to Eq. (4.6) comes from setting the zero of τ_ν at the top of the atmosphere (Fig. 4.1).

The upward and downward radiances can then be expressed in terms of the optical depth by using

$$\tau_{\nu,g} \equiv \tau_\nu(z_0,\infty) = \int_{z_0}^\infty k_\nu(z')\, n_a(z')\, \mathrm{d}z', \qquad \text{and}$$

$$\tau_\nu' \equiv \tau_\nu(z',\infty) = \int_{z'}^\infty k_\nu(z'')\, n_a(z'')\, \mathrm{d}z''$$

to give

$$L_\nu^\uparrow(\tau,\mu) = B_\nu(T_g) \exp\left[-\frac{\tau_{\nu,g} - \tau_\nu}{\mu}\right] + \int_\tau^{\tau_g} J_\nu(\tau') \exp\left[-\frac{\tau_\nu' - \tau_\nu}{\mu}\right] \frac{\mathrm{d}\tau_\nu'}{\mu} \qquad (4.13)$$

for $1 \geq \mu > 0$, and

$$L_\nu^\downarrow(\tau,\mu) = -\int_0^\tau J_\nu(\tau') \exp\left[-\frac{\tau_\nu' - \tau_\nu}{\mu}\right] \frac{\mathrm{d}\tau_\nu'}{\mu}, \qquad (4.14)$$

for $-1 \leq \mu < 0$.

Introducing these equations into those for the upward and downward fluxes, (4.2) and (4.3), we obtain

$$F_\nu^\uparrow(\tau) = 2\pi B_\nu(T_g) \int_0^1 \exp\left[-\frac{\tau_{\nu,g} - \tau_\nu}{\mu}\right] \mu \, d\mu \; +$$

$$2\pi \int_\tau^{\tau_g} J_\nu(\tau') \int_0^1 \exp\left[-\frac{\tau_\nu' - \tau_\nu}{\mu}\right] d\mu \, d\tau_\nu' \quad \text{and} \quad (4.15)$$

$$F_\nu^\downarrow(\tau) = 2\pi \int_0^\tau J_\nu(\tau') \int_0^1 \exp\left[-\frac{\tau_\nu - \tau_\nu'}{\mu}\right] d\mu \, d\tau_\nu'. \quad (4.16)$$

These expressions can also be written in a more concise form by using the exponential integral of order n,

$$E_n(x) = \int_1^\infty \frac{e^{-xt}}{t^n} dt$$

to obtain

$$F_\nu^\uparrow(\tau) = 2\pi B_\nu(T_g) \, E_3(\tau_{\nu,g} - \tau_\nu) + 2\pi \int_\tau^{\tau_g} J_\nu(\tau') \, E_2(\tau_\nu' - \tau_\nu) \, d\tau_\nu', \quad \text{and} \quad (4.17)$$

$$F_\nu^\downarrow(\tau) = 2\pi \int_0^\tau J_\nu(\tau') \, E_2(\tau_\nu - \tau_\nu') \, d\tau_\nu'. \quad (4.18)$$

The total flux obtained from these expressions is given by

$$F_\nu(\tau) = 2\pi B_\nu(T_g) \, E_3(\tau_{\nu,g} - \tau_\nu) + 2\pi \int_\tau^{\tau_g} J_\nu(\tau') \, E_2(\tau_\nu' - \tau_\nu) \, d\tau_\nu' \; -$$

$$2\pi \int_0^\tau J_\nu(\tau') \, E_2(\tau_\nu - \tau_\nu') \, d\tau_\nu', \quad (4.19)$$

and the heating rate can then be obtained directly from Eqs. (4.5) and (4.19).

4.2.2   *Solar radiation*

The radiation from the Sun can be separated into two components: the direct beam, and diffuse solar radiation which has been scattered by the atmosphere. As already noted, true scattering is usually negligible in the absence of cloud or aerosol particles, and our concern is with the process where photons from the Sun are absorbed and then immediately re-emitted at the same wavelength by the atmospheric molecules. This differs from aerosol scattering in that the re-emission occurs isotropically, so some photons are scattered (re-emitted) backwards towards the direction of the Sun. We will consider this component later when calculating the source function of the emitting levels and the flux divergence produced by an atmospheric transition.

From the formal solution for the downward radiance, assuming no source within the atmosphere, and with the solar radiance at the top of the atmosphere as the boundary condition, we have

$$L_\nu(z, \mu_\odot) = L_{\nu,\odot}(\infty) \exp\left[-\frac{\tau_\nu(z, \infty)}{\mu_\odot}\right], \qquad (4.20)$$

where $\mu_\odot = \cos\chi$, being χ the solar zenith angle, and $L_{\nu,\odot}(\infty)$ is the monochromatic radiance of the Sun at frequency ν at the top of the atmosphere.

From the relationship between radiance and radiative flux (Eq. 3.2), integrating over the solid angle subtended by the Sun, we obtain the monochromatic flux of direct solar radiation along the vertical coordinate given by

$$F_{\odot,\nu}(z) = -\mu_\odot F_{\odot,\nu}(\infty) \exp\left[-\frac{\tau_\nu(z, \infty)}{\mu_\odot}\right], \qquad (4.21)$$

where the solar flux outside the atmosphere on a surface normal to the solar radiation beams is $F_{\odot,\nu}(\infty) = (\pi R_\odot^2/d^2)L_{\nu,\odot}(\infty)$, with $R_\odot$ being the solar radius, and d the mean Earth–Sun distance. The rate at which energy is gained by the atmosphere from the absorption of solar radiation, is obtained by calculating the divergence of the solar flux:

$$h_{\odot,\nu} = \mu_\odot F_{\odot,\nu}(\infty) \frac{d\left[\exp[-\tau_\nu(z, \infty)/\mu_\odot]\right]}{dz}. \qquad (4.22)$$

This is generally referred to as the solar heating rate at frequency ν, but of course since it is a monochromatic quantity it has to be integrated over the solar spectrum to get the true heating rate.

4.2.3 *Atmospheric sphericity*

The plane-parallel atmosphere approximation is generally valid for most non-LTE studies. Horizontal inhomogeneities are greatest in the troposphere, while the middle and upper atmosphere where non-LTE situations are generally encountered are more stratified and homogeneous. An exception is found under auroral conditions, where the proper solution requires an account of the spatial variability of the auroral energy deposition. The presence of noctilucent clouds might be another exception. There are two other common cases where the atmospheric layers can be considered to be homogeneous, but the concept of plane parallel geometry is not applicable. These are:

(a) When the direct solar radiation is incident at zenith angles larger than ∼80° (note that this can be larger than 90°), and

(b) When the radiation is emerging from the atmosphere at directions which are nearly tangential to the surface (Fig. 4.2, case b), as is frequently the case in satellite remote sensing, where limb measurements are used to obtain good vertical resolution.

In the former case, the curvature of the atmosphere means that the solar zenith angle χ is not the same at all altitudes (see Figs. 4.2 and 4.3). Then the optical depth cannot be obtained using the $1/\cos\chi$ factor, but instead has to be computed using the complete expression

$$\tau_\nu(z, \infty) = \int_z^\infty \frac{k_\nu(z')\, n_a(z')}{\mu(z')}\, \mathrm{d}z'. \tag{4.23}$$

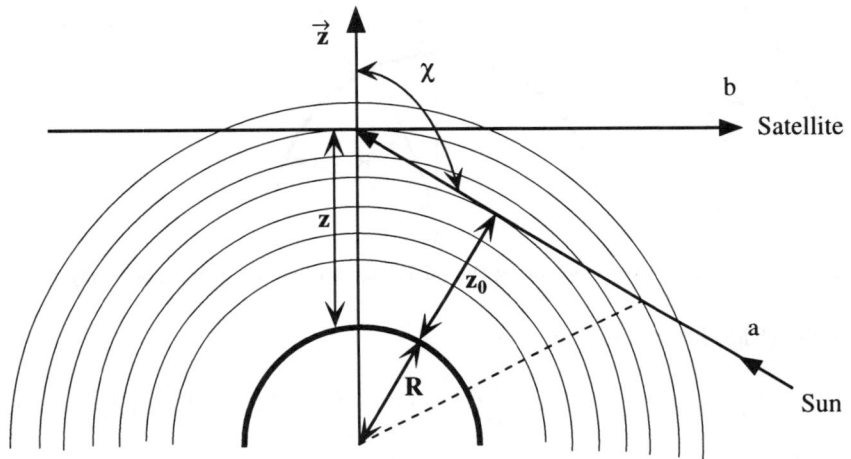

Fig. 4.2 a) Geometry for calculating the absorption of solar radiation at solar zenith angles greater than 90°; and b) the atmospheric limb radiance, i.e. the emission from the atmosphere in directions parallel to the tangent of the Earth's surface. Not to scale.

There are integrated forms for Eq. (4.23) in which the $\mu(z')$ function is taken out of the integral and replaced by the *Chapman function* $\mathrm{Ch}(x, \chi)$, which has the form

$$\mathrm{Ch}(x, \chi) = x \sin\chi \int_0^\chi \csc^2 t \, \exp(x - x \sin\chi \, \csc t)\, \mathrm{d}t, \tag{4.24}$$

where $x = R_\oplus + z$, $R_\oplus$ is the radius of the Earth, and t is a dummy variable of integration.

This equation, which is easy to compute, has the advantage of treating solar zenith angles larger than $\sim 80°$ in the same way as the smaller ones, just by replacing $1/\mu$ by $\mathrm{Ch}(x, \chi)$. The approximation is only valid, however, if the concentration of the molecular species $n(z')$ varies exponentially with altitude, which requires that the constituent has a constant volume mixing ratio. Since this condition is not satisfied by all molecules of interest, the Chapman function is not always useful and the optical path of the solar rays often has to be calculated numerically. For χ greater than $\sim 80°$ but smaller than 90° (Fig. 4.3) we use Eq. (4.23) with

$$\frac{1}{\mu(z')} = \frac{R + z'}{\sqrt{(R + z')^2 - (R + z)^2 \sin^2\chi}}. \tag{4.25}$$

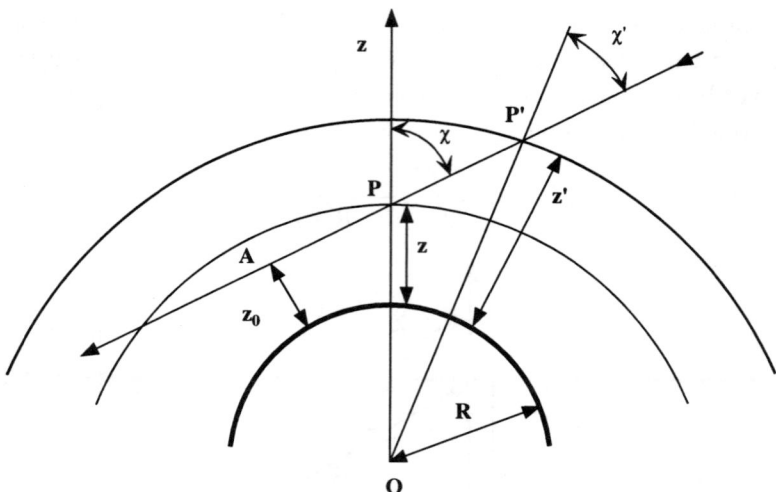

Fig. 4.3   Geometry for calculating the path length in curved atmospheric layers.

For $\chi > 90°$ (Fig. 4.2, case a) we use

$$\tau_\nu(z, \infty) = 2 \int_{z_0}^{z} \frac{k_\nu(z') \, n_a(z')}{\mu(z')} \, dz' + \int_{z}^{\infty} \frac{k_\nu(z') \, n_a(z')}{\mu(z')} \, dz', \qquad (4.26)$$

where $z_0 = R(\sin\chi - 1) + z\sin\chi$, and $\mu(z')$ is given by Eq. (4.25).

The function $1/\mu(z')$ (Eq. 4.25) has a discontinuity for $\chi = 90°$ and values of z' close to z. For these cases we use the integrated form, which gives the geometric distance between two adjacent atmospheric layers at z' and $z' + \Delta z'$, $\Delta s = \Delta z'/\mu$, by

$$\Delta s = \sqrt{(R + z' + \Delta z')^2 - (R + z)^2 \sin^2 \chi} - \sqrt{(R + z')^2 - (R + z)^2 \sin^2 \chi}. \quad (4.27)$$

The limb sounding (Fig. 4.2, case b) is equivalent to solar absorption for a zenith angle of 90° so the integration along the line of sight is also given by Eq. (4.27).

4.2.4   *The radiative equilibrium temperature profile*

An important application of the radiative transfer equation is the calculation of radiative equilibrium temperature profiles. The principle is to calculate the energy gained and lost by each layer by radiative exchange between it and all the other layers, and with the surface, the Sun, and cold space, then to obtain the temperature for which all of these positive and negative contributions balance. The result should, if all of the spectral bands have been included and the calculations performed accurately, resemble the real atmosphere in the stratosphere and above, where radiative balance is the main factor determining the temperature. In the troposphere, the radiative equilibrium profile is unstable and vertical convection of heat is the dominant effect.

Taking the radiative transfer equation for a stratified atmosphere in the form of Eq. (4.12), integrating over all directions, using the definition of the mean radiance (Eq. 3.7) and of the flux for a stratified atmosphere (Eq. 4.1), we have

$$\frac{\mathrm{d}F_\nu(\tau_\nu)}{\mathrm{d}\tau_\nu} = 4\pi \left[\bar{L}_\nu(\tau_\nu) - J_\nu(\tau_\nu) \right], \tag{4.28}$$

where we have assumed that the source function is isotropic. This equation relates $F_\nu(\tau_\nu)$ to $\bar{L}_\nu(\tau_\nu)$, assuming for the moment that $J_\nu(\tau_\nu)$ is known. To close the system we need another equation. This is obtained by multiplying both sides of the RTE (Eq. 4.12) by the direction cosine $\mu = \cos\theta$, and integrating again over all directions, to calculate the second moment of the RTE. Such an integration, if carried out for any general angular distribution of the radiation field, does not give a useful analytic equation. It becomes useful to invoke an approximation to the angular distribution (the full treatment will be given in Sec. 4.4), such as the two-stream approach. This assumes that the radiance travelling upwards and downwards is approximately isotropic, independently in both hemispheres, i.e.,

$$L_\nu(\mu, \tau_\nu) = L_\nu^\uparrow(\tau_\nu) \quad \text{for} \quad 0 < \mu < 1, \quad \text{and} \tag{4.29}$$
$$L_\nu(\mu, \tau_\nu) = L_\nu^\downarrow(\tau_\nu) \quad \text{for} \quad -1 < \mu < 0. \tag{4.30}$$

Note that we are using the optical depth τ_ν, measured from space in the downward direction, as the vertical coordinate. Then, from this equation and the definitions in (Eq. 3.7) and (Eq. 4.1), the mean radiance and the radiative flux at τ_ν are found to be

$$\bar{L}_\nu(\tau_\nu) = \frac{L_\nu^\uparrow(\tau_\nu) + L_\nu^\downarrow(\tau_\nu)}{2} \tag{4.31}$$

and

$$F_\nu(\tau_\nu) = \pi[L_\nu^\uparrow(\tau_\nu) - L_\nu^\downarrow(\tau_\nu)], \tag{4.32}$$

and the relationship between them is

$$\bar{L}_\nu(\tau_\nu) = L_\nu^\uparrow(\tau_\nu) - F_\nu(\tau_\nu)/2\pi = L_\nu^\downarrow(\tau_\nu) + F_\nu(\tau_\nu)/2\pi. \tag{4.33}$$

Multiplying the RTE (4.12) by μ, integrating over all directions and using the approximations given by Eqs. (4.29) and (4.30), the left-hand side of Eq. (4.12) can be written as

$$\int_{4\pi} \mu^2 \frac{\mathrm{d}L_\nu(\tau_\nu, \mu)}{\mathrm{d}\tau_\nu} \, \mathrm{d}\omega = 2\pi \frac{\mathrm{d}}{\mathrm{d}\tau_\nu} \left[\int_{-1}^{+1} \mu^2 L_\nu(\tau_\nu, \mu) \, \mathrm{d}\mu \right] = \frac{4\pi}{3} \frac{\mathrm{d}\bar{L}_\nu(\tau_\nu)}{\mathrm{d}\tau_\nu}, \tag{4.34}$$

where we have used Eq. (4.31) in the last equality. After these manipulations, the right-hand side of Eq. (4.12) is equal to $F_\nu(\tau_\nu)$, where we have used the definition of the flux (Eq. 4.1) and again assumed the source function to be isotropic. So we get

$$\frac{\mathrm{d}\bar{L}_\nu(\tau_\nu)}{\mathrm{d}\tau_\nu} = \frac{3}{4\pi}F_\nu(\tau_\nu) \tag{4.35}$$

or, introducing Eq. (4.28) in the latter,

$$\frac{\mathrm{d}^2 F_\nu(\tau_\nu)}{\mathrm{d}\tau_\nu^2} = 3F_\nu(\tau_\nu) - 4\pi\frac{\mathrm{d}J_\nu(\tau_\nu)}{\mathrm{d}\tau_\nu}. \tag{4.36}$$

Combining Eq. (4.28) with (4.33) gives

$$\frac{\mathrm{d}F_\nu(\tau_\nu)}{\mathrm{d}\tau_\nu} = 4\pi\left[L_\nu^\uparrow(\tau_\nu) - J_\nu(\tau_\nu)\right] - 2F_\nu(\tau_\nu) = 4\pi\left[L_\nu^\downarrow(\tau_\nu) - J_\nu(\tau_\nu)\right] + 2F_\nu(\tau_\nu). \tag{4.37}$$

To find the radiance and the radiative flux with this approach we need to assume boundary conditions. First we look at the solution where the ground is a blackbody at temperature T_g and there is no incident radiation at the top, as at night. These conditions fit the two-stream approach, since the blackbody at the ground is isotropic by definition, and the radiance at the top is also isotropic since it is zero for all downward directions.

With these boundary conditions, we have for the upward radiance at the ground

$$L_\nu^\uparrow(\tau_{\nu,g}) = F_\nu(\tau_{\nu,g})/2\pi + J_\nu(\tau_{\nu,g}) + \frac{1}{4\pi}\left[\frac{\mathrm{d}F_\nu(\tau_\nu)}{\mathrm{d}\tau_\nu}\right]_{\tau=\tau_g} = B_\nu(T_g) \tag{4.38}$$

and for the downward radiance at the top of the atmosphere ($\tau_\nu = 0$)

$$L_\nu^\downarrow(0) = -F_\nu(0)/2\pi + J_\nu(0) + \frac{1}{4\pi}\left[\frac{\mathrm{d}F_\nu(\tau_\nu)}{\mathrm{d}\tau_\nu}\right]_{\tau=0} = 0. \tag{4.39}$$

It is useful to illustrate the solution of this equation for conditions of *radiative equilibrium*. The concept means that the divergence of the net flux is zero, i.e., the net flux is constant. The heating rate (Eq. 3.5) is zero under these conditions since there is no *net* exchange of energy between the radiation field and the molecules. This concept usually applies to the whole spectrum, and does not mean that the divergence of the net flux is necessarily zero at a given frequency or in a limited spectral interval, since in principle fluxes in, say, the infrared and the ultraviolet, could cancel out. However, the idealised case of *monochromatic radiative equilibrium* illustrates some simple solutions and will help with the more realistic case of net radiative equilibrium which covers the whole spectrum. So, making $\mathrm{d}F_\nu(\tau_\nu)/\mathrm{d}\tau_\nu = 0$ in Eq. (4.28) and assuming LTE, $\bar{L}_\nu(\tau_\nu) = J_\nu(\tau_\nu) = B_\nu(\tau_\nu)$ and introducing this result into Eq. (4.35), we have

$$\frac{\mathrm{d}B_\nu(\tau_\nu)}{\mathrm{d}\tau_\nu} = \frac{3}{4\pi}F_\nu(\tau_\nu). \tag{4.40}$$

The upward radiance at the ground for monochromatic radiative equilibrium is, from Eq. (4.38),

$$L_\nu^\uparrow(\tau_{\nu,g}) = B_\nu(T_1) + F_\nu/2\pi = B_\nu(T_g) \tag{4.41}$$

and the downward radiance at the top of the atmosphere from Eq. (4.39) is

$$L_\nu^\downarrow(\tau_\nu = 0) = B_\nu(T_0) - F_\nu/2\pi = 0, \tag{4.42}$$

where T_1 and T_0 are the air temperatures at $\tau_{\nu,g}$ and $\tau_\nu = 0$, respectively, and F_ν is constant with τ_ν.

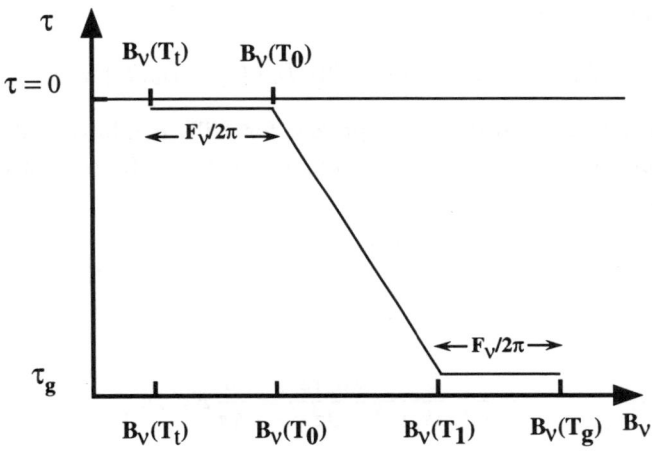

Fig. 4.4   The blackbody function at atmospheric temperature against the optical depth for the atmosphere in monochromatic radiative equilibrium. In the absence of an incident downward flux at the top of the atmosphere $B_\nu(T_t) = 0$, and $B_\nu(T_0) = F_\nu/2\pi$.

The solution of Eq. (4.41) requires a discontinuity in the Planck function, and hence in temperature, at the ground. Since the flux is positive (Eq. 4.42), the temperature of the air is smaller than that of the surface, $T_1 < T_g$ (see Fig. 4.4). Also, the temperature T_0 at the top of the atmosphere, which is given by

$$B_\nu(T_0) = F_\nu/2\pi = B_\nu(T_g) - B_\nu(T_1) \tag{4.43}$$

is not zero. The net flux of radiative energy leaving the atmosphere at the top is given by

$$F_\nu^\uparrow(0) = F_\nu = 2\pi B_\nu(T_0), \tag{4.44}$$

which is twice that emitted by a blackbody at temperature T_0.

The solution for $B_\nu(\tau_\nu)$, from Eq. (4.40) and the boundary conditions above, is

$$B_\nu(\tau_\nu) = B_\nu(T_0)(1 + 3\tau_\nu/2). \tag{4.45}$$

From this equation we can see that the total radiative flux leaving the atmosphere is characteristic of the thermal radiation at $\tau_\nu = 2/3$.

Now let us include a representation of the mean solar flux, by assuming that the atmosphere is irradiated by a blackbody at temperature T_t, i.e. $L_\nu^\downarrow(\tau_\nu = 0) = B_\nu(T_t)$. Then we have

$$B_\nu(T_g) + B_\nu(T_t) = B_\nu(T_1) + B_\nu(T_0) \tag{4.46}$$

and

$$F_\nu/2\pi = B_\nu(T_g) - B_\nu(T_1) = B_\nu(T_0) - B_\nu(T_t) \tag{4.47}$$

and the solution at a given τ (see Fig. 4.4) is

$$B_\nu(\tau_\nu) = B_\nu(T_0) + (3/2)\big[B_\nu(T_0) - B_\nu(T_t)\big]\tau_\nu. \tag{4.48}$$

Next, consider the two-stream approach for LTE conditions and with radiative equilibrium over the whole spectrum. Using the definition of optical depth, Eqs. (4.28) and (4.35) can be written as

$$-\frac{1}{n_a(z)k_\nu(z)}\frac{dF_\nu(z)}{dz} = 4\pi\big[\bar{L}_\nu(z) - B_\nu(z)\big] \tag{4.49}$$

and

$$-\frac{1}{n_a(z)k_\nu(z)}\frac{d\bar{L}_\nu(z)}{dz} = \frac{3}{4\pi}F_\nu(z). \tag{4.50}$$

We want to use the monochromatic radiative equilibrium solution to look for an equation similar to (4.40) for quantities integrated over ν. This is not possible in general, but can be obtained using several approximations. Consider the case of a *grey* atmosphere, defined as one where the absorption coefficient is constant over the whole spectrum, i.e. it does not depend on ν although it can vary with height. Integrating Eq. (4.49) over frequency, and assuming radiative equilibrium, so that $F(z) = \int_0^\infty F_\nu(z)\,d\nu$ is equal to a constant independent of z, we have $\bar{L}(z) = B(z)$, where

$$\bar{L}(z) = \int_0^\infty \bar{L}_\nu(z)\,d\nu, \quad \text{and} \quad B(z) = \int_0^\infty B_\nu(z)\,d\nu.$$

Integrating Eq. (4.50) over frequency and including $\bar{L}(z) = B(z)$, we have

$$-\frac{1}{n_a(z)k(z)}\frac{dB(z)}{dz} = \frac{3}{4\pi}F \quad \text{or} \quad \frac{dB(z)}{d\tau} = \frac{3}{4\pi}F, \tag{4.51}$$

where $d\tau = -n_a(z)\,k(z)\,dz$. Thus, a grey atmosphere in radiative equilibrium has the same solution as that for monochromatic radiative equilibrium described above.

We can now obtain the temperature profile for a grey atmosphere in LTE and in radiative equilibrium. Equation (4.51) has a solution of the form of (4.45) for the integrated blackbody radiance, which, according to the Stefan–Boltzmann law, gives

$$T^4(\tau) = T_0^4(1 + 3\tau/2). \tag{4.52}$$

From this equation the air temperature at the ground, $T_1 = T(\tau_g)$ is given by

$$T_1^4 = T_0^4(1 + 3\tau_g/2),$$

and from Eq. (4.43), the temperature of the ground itself is given by

$$T_g^4 = T_1^4 + T_0^4 = T_0^4(2 + 3\tau_g/2), \tag{4.53}$$

where T_0 is usually called the *skin temperature*. This can be related to the *planetary effective temperature*, T_e, a useful quantity which is defined as the temperature of the equivalent blackbody which emits the same total radiative energy to space as the planet. The effective temperature is obtained assuming that the stratosphere is radiated from below by absorption of the atmospheric radiation at the effective temperature, and cooled by emission in all directions (both hemispheres). Since the ratio of the solid angles is $1/2$, then according to the Stefan–Boltzmann law, $(T_0/T_e)^4 = 1/2$, and $T_g^4 = T_e^4(1 + 3\tau_g/4)$. This illustrates how large temperatures can be reached at the ground when the optical depth is large, a well-known phenomenon commonly called the *greenhouse effect*.

The grey atmosphere assumption is often used in the astrophysical literature because stellar spectra show a strong absorption continuum over a large spectral range, making the approximation more valid than in the planetary case, where most of the opacity is contained in discrete absorption bands. The grey solution, however, can be useful to describe atmospheric radiation under certain conditions (including, in particular, the opaque and transparent limits) by defining an appropriate mean or effective absorption coefficient for the whole, or a large part, of the spectrum. For example, integrating $\mathrm{d}F_\nu(z)/\mathrm{d}z$ over frequency in Eq. (4.49) and assuming radiative equilibrium we obtain

$$\int_0^\infty \bar{L}_\nu(z)\, k_\nu(z)\, \mathrm{d}\nu = \int_0^\infty B_\nu(z)\, k_\nu(z)\, \mathrm{d}\nu.$$

Under some circumstances, including the limits mentioned above, it can be assumed that $\bar{L}_\nu(z)$ and $B_\nu(z)$ have the same frequency dependence at any altitude, e.g., $\bar{L}_\nu(z) = B_\nu(z)$. Then, from Eq. (4.50), we have a solution similar to the grey case:

$$-\frac{1}{n_a(z)\bar{k}(z)} \frac{\mathrm{d}B(z)}{\mathrm{d}z} = \frac{3}{4\pi} F, \tag{4.54}$$

where

$$\frac{1}{\bar{k}(z)} = \frac{\int_0^\infty \frac{1}{k_\nu(z)} \frac{dB_\nu(z)}{dz} \, d\nu}{\int_0^\infty \frac{dB_\nu(z)}{dz} \, d\nu}. \tag{4.55}$$

If we additionally assume that $k_\nu(z)$ does not depend on z, then it can be introduced into the left-hand side of Eq. (4.50) to give a solution similar to Eq. (4.54) but with $\bar{k}$ given by

$$\frac{1}{\bar{k}} = \frac{1}{B(z)} \int_0^\infty \frac{B_\nu(z)}{k_\nu(z)} \, d\nu. \tag{4.56}$$

The mean absorption coefficients of Eqs. (4.55) and (4.56) are called the *Rosseland mean* absorption coefficients.

We could use alternative means to reach a solution similar to (4.54). With the assumption that $\bar{L}_\nu(z) = B_\nu(z)$, moving $k_\nu(z)$ to the right-hand side in (4.28) before integrating over ν, we have a similar solution to (4.54) where now $\bar{k}(z)$ is given by

$$\bar{k}(z) = \frac{1}{F} \int_0^\infty F_\nu(z) \, k_\nu(z) \, d\nu, \tag{4.57}$$

which is known as the *Chandrasekhar mean* absorption coefficient.

4.2.5 *Heating and cooling rates*

We describe in this section the *exchange integral* solution for the heating rate in a stratified atmosphere, and then in the next two sections the heating rate for two specific, extreme LTE cases.

The heating rate of a vibrational band for a stratified atmosphere can be obtained by integrating the monochromatic heating rate (Eq. 4.5) over the spectral interval occupied by the band $\Delta\nu$,

$$h = -\frac{dF}{dz} = -\frac{d(F^\uparrow - F^\downarrow)}{dz}, \tag{4.58}$$

where we have used the definition of the net flux (Eq. 4.4).

The band-integrated upward and downward fluxes, $F^\uparrow$ and $F^\downarrow$, can be obtained by integrating the expressions (4.15) and (4.16) over frequency. This is what numerical line-by-line integration schemes usually do. However, it is instructive, and can be useful as we will see below, to proceed by separating the integration over frequency from that over height. We describe this alternative method for obtaining $F^\uparrow$ and $F^\downarrow$ below.

We have seen in Sec. 4.2.1 that the optical thickness for a stratified atmosphere is given by

$$\bar{\tau}_\nu(z',z,\mu) = \int_{z'}^{z} \frac{k_\nu(z'')\,n_a(z'')}{\mu}\,\mathrm{d}z''$$

and, according to its definition (Eq. 3.16), we have for the upward radiance ($0<\mu \leq 1$, or $z > z'$, see Fig. 4.1)

$$\frac{\mathrm{d}\bar{\tau}_\nu(z',z,\mu)}{\mathrm{d}z'} = -\frac{k_\nu(z')\,n_a(z')}{\mu}.$$

Introducing the definition of the *monochromatic transmission*, $\mathcal{T}_\nu(z',z,\mu)$, by

$$\mathcal{T}_\nu(z',z,\mu) = \exp\left[-\bar{\tau}_\nu(z',z,\mu)\right] \tag{4.59}$$

we get

$$\frac{\mathrm{d}\mathcal{T}_\nu(z',z,\mu)}{\mathrm{d}z'} = \frac{k_\nu(z')\,n_a(z')}{\mu}\exp\left[-\bar{\tau}_\nu(z',z,\mu)\right],$$

so the formal solution for the upward radiance (Eq. 4.9) can be re-written as

$$L_\nu^\uparrow(z,\mu) = B_\nu(T_g)\,\mathcal{T}_\nu(z_0,z,\mu) + \int_{z_0}^{z} J_\nu(z')\,\frac{\mathrm{d}\mathcal{T}_\nu(z',z,\mu)}{\mathrm{d}z'}\,\mathrm{d}z'. \tag{4.60}$$

We introduce now the *flux transmission* as the average of transmission weighted by $\mu = \cos\theta$ (z-component of the flux) over solid angle in the upper half-hemisphere (the definition is equally valid for the lower hemisphere), i.e.,

$$\mathcal{T}_{F,\nu}(z',z) = \frac{\int_0^{2\pi} d\phi \int_0^{\pi/2} \sin\theta\,\cos\theta\,\mathcal{T}_\nu(z',z,\theta)\,\mathrm{d}\theta}{\int_0^{2\pi} d\phi \int_0^{\pi/2} \sin\theta\,\cos\theta\,\mathrm{d}\theta} = 2\int_0^1 \mathcal{T}_\nu(z',z,\mu)\,\mu\,\mathrm{d}\mu =$$

$$2\int_1^\infty \frac{\exp\left[-\tau_\nu(z',z,\eta)\right]}{\eta^3}\,\mathrm{d}\eta = 2\,\mathrm{E}_3\left[\tau_\nu(z',z)\right], \tag{4.61}$$

where E_3 is the third exponential integral, and $\eta = \sec\theta$. We also define the *average or mean transmission* over the spectral interval of the band of interest, $\Delta\nu$, by*

$$\mathcal{T}(z',z) = \frac{1}{\Delta\nu}\int_{\Delta\nu} \mathcal{T}_\nu(z',z)\,\mathrm{d}\nu, \tag{4.62}$$

or

$$\mathcal{T}_F(z',z) = \frac{1}{\Delta\nu}\int_{\Delta\nu} \mathcal{T}_{F,\nu}(z',z)\,\mathrm{d}\nu. \tag{4.63}$$

---

*We use $\mathcal{T}_\nu$, with subscript ν, for monochromatic transmission and $\mathcal{T}$, without the subscript ν, for the average transmission over $\Delta\nu$.

Then, introducing the upward radiance (Eq. 4.60) into the expression for the upward flux (Eq. 4.2), integrating over the spectral interval $\Delta\nu$, and making use of the definitions above, we find that

$$F^\uparrow(z) = \pi\Delta\nu \left[B(T_g, \nu_0)\, \mathcal{T}_F(z_0, z) + \int_{z_0}^z J(z', \nu_0)\, \frac{\mathrm{d}\mathcal{T}_F(z', z)}{\mathrm{d}z'}\, \mathrm{d}z' \right], \qquad (4.64)$$

where we have assumed that the radiance at the lower boundary (the Planck function in this case) and the source function vary with frequency much more slowly than the absorption coefficient, so $B_\nu(T_g)$ and J_ν can be taken out of the integral over ν, and replaced by their average values over the band interval $\Delta\nu$ $[B(T_g, \nu_0)$ and $J(z, \nu_0)$ respectively] without significant loss of accuracy. Note that the angular and frequency integrations are now confined to the quantities in Eqs. (4.61) and (4.63), respectively, and we have an equation for the radiative flux (4.64) that depends only on height. Methods for carrying out the integrations over frequency and solid angle are described in detail in Secs. 4.3 and 4.4.

Now we look for a similar expression for the downward flux. Exchanging the limits of the integral over z', and changing μ from a negative-definite to a positive-definite quantity in Eq. (4.10), since these two changes of sign cancel out, the equation can be re-written as

$$L_\nu^\downarrow(z, \mu) = L_\nu^\downarrow(\infty, \mu) \exp\left[-\int_z^\infty \frac{k_\nu(z')\, n_a(z')}{\mu}\, \mathrm{d}z' \right] +$$
$$\int_z^\infty \frac{k_\nu(z')\, n_a(z')}{\mu}\, J_\nu(z') \exp\left[-\int_z^{z'} \frac{k_\nu(z'')\, n_a(z'')}{\mu}\, \mathrm{d}z'' \right] \mathrm{d}z',$$

where we have included the solar radiation incident at the top of the atmosphere.

The optical thickness for downward fluxes is, again,

$$\bar{\tau}_\nu(z, z', \mu) = \int_z^{z'} \frac{k_\nu(z'')\, n_a(z'')}{\mu}\, \mathrm{d}z'',$$

where now $z' > z$ $(-1 < \mu < 0)$ (see Fig. 4.1), and then, from Eq. (4.59),

$$\frac{\mathrm{d}\mathcal{T}_\nu(z, z', \mu)}{\mathrm{d}z'} = -\frac{k_\nu(z')\, n_a(z')}{\mu} \exp\left[-\bar{\tau}_\nu(z, z', \mu) \right].$$

Introducing the latter equation for the downward radiance into the equation for the downward flux (Eq. 4.3), and using the expressions above for $\bar{\tau}_\nu(z, z', \mu)$ and $\mathrm{d}\mathcal{T}_\nu(z, z', \mu)/\mathrm{d}z'$, we have

$$F^\downarrow(z) = \pi\Delta\nu \left[\bar{L}_{\Delta\nu}^\downarrow(\infty)\, \mathcal{T}_F(z, \infty) - \int_z^\infty J(z')\, \frac{\mathrm{d}\mathcal{T}_F(z, z')}{\mathrm{d}z'}\, \mathrm{d}z' \right]. \qquad (4.65)$$

Note that the second term on the right-hand side gives a positive contribution since $\mathrm{d}\mathcal{T}_F(z, z')/\mathrm{d}z'$ is negative.

Introducing these expressions (Eqs. 4.64 and 4.65) for the upward and downward fluxes in the equation for the heating rate (4.58) we obtain

$$
\frac{h}{\pi \Delta \nu} = -\frac{d}{dz} \left[B(T_g)\, \mathcal{T}_F(z_0, z) + \int_{z_0}^{z} J(z') \frac{d\mathcal{T}_F(z', z)}{dz'}\, dz' - \bar{L}^{\downarrow}_{\Delta \nu}(\infty)\, \mathcal{T}_F(z, \infty) + \int_{z}^{\infty} J(z') \frac{d\mathcal{T}_F(z, z')}{dz'}\, dz' \right].
\tag{4.66}
$$

Carrying out the derivative over z we find

$$
\frac{h}{\pi \Delta \nu} = -B(T_g) \frac{d\mathcal{T}_F(z_0, z)}{dz} - \int_{z_0}^{z} J(z') \frac{\partial^2 \mathcal{T}_F(z', z)}{\partial z \partial z'}\, dz' - \int_{z}^{\infty} J(z') \frac{\partial^2 \mathcal{T}_F(z, z')}{\partial z \partial z'}\, dz' + \bar{L}^{\downarrow}_{\Delta \nu}(\infty) \frac{d\mathcal{T}_F(z, \infty)}{dz} - J(z) \left[\frac{d\mathcal{T}_F(z', z)}{dz'} - \frac{d\mathcal{T}_F(z, z')}{dz'} \right]_{z'=z}.
\tag{4.67}
$$

If we consider that

$$
\int_{z_0}^{z} J(z) \frac{\partial^2 \mathcal{T}_F(z', z)}{\partial z \partial z'}\, dz' = J(z) \left[\left(\frac{d\mathcal{T}_F(z', z)}{dz} \right)_{z'=z} - \frac{d\mathcal{T}_F(z_0, z)}{dz} \right],
$$

$$
\int_{z}^{\infty} J(z) \frac{\partial^2 \mathcal{T}_F(z, z')}{\partial z \partial z'}\, dz' = J(z) \left[\frac{d\mathcal{T}_F(z, \infty)}{dz} - \left(\frac{d\mathcal{T}_F(z, z')}{dz} \right)_{z'=z} \right]
$$

and that $d\mathcal{T}_F(z', z)/dz' = -d\mathcal{T}_F(z, z')/dz'$, then, by subtracting the left-hand sides and adding the right-hand sides of those equations to the right-hand side of (4.67), this can be re-written as

$$
\frac{h}{\pi \Delta \nu} = -[B(T_g) - J(z)] \frac{d\mathcal{T}_F(z_0, z)}{dz} - \int_{z_0}^{z} [J(z') - J(z)] \frac{\partial^2 \mathcal{T}_F(z', z)}{\partial z \partial z'}\, dz' - \int_{z}^{\infty} [J(z') - J(z)] \frac{\partial^2 \mathcal{T}_F(z, z')}{\partial z \partial z'}\, dz' - [J(z) - \bar{L}^{\downarrow}_{\Delta \nu}(\infty)] \frac{d\mathcal{T}_F(z, \infty)}{dz}.
\tag{4.68}
$$

This solution for the heating (or cooling) rate is called the *exchange integral* formulation and is equally valid for LTE and non-LTE conditions, although there is a difference in how it is applied. In obtaining the equation for the general non-LTE case it was assumed that the integration over frequency is carried out over a single vibrational band, because different bands usually have different source functions. Under LTE conditions, all bands have B_ν as their source function and the frequency integration in $\mathcal{T}_F(z, z')$ applies for any spectral interval which is sufficiently narrow so that $B_\nu(T)$ can be assumed constant over the interval.

One of the main advantages of this formulation is that, like the formal solution for the radiance, the physical meaning of each term is clear. To illustrate this, let us consider LTE conditions, i.e., $J_\nu(z) = B_\nu(z)$. The first term in the right-hand side of (4.68) represents the energy exchange of the atmospheric layer at z with the lower boundary. When the temperature at the lower boundary is larger than $T(z)$, since $d\mathcal{T}_F(z_0, z)/dz < 0$, the lower boundary contributes some heating to the z-layer. The second and third terms give the net exchange of radiation with all layers below and above, respectively. Since $\partial^2 T_F(z, z')/\partial z \partial z'$ and $\partial^2 T_F(z', z)/\partial z \partial z'$ are always negative (or zero), they contribute to heating the z-layer whenever $T(z') > T(z)$. The last term on the right-hand side represents the exchange with the upper boundary. Since $\mathrm{d}T_F(z, \infty)/\mathrm{d}z \geq 0$, if we assume that the input radiance is zero at the top of the atmosphere (night-time conditions) then this term represents cooling to space.

The Sun is approximately a blackbody at a temperature close to 6000 K. For most atmospheric layers the solar input during the day, $\bar{L}_{\nu,\odot}(\infty)$, is usually larger than $B(z)$ or $J(z)$ and a net heating occurs. The rate of energy absorbed from the solar beam, given by $\bar{L}_{\Delta\nu}^{\downarrow}(\infty)[\mathrm{d}T_F(z, \infty)/\mathrm{d}z]$, is not completely thermalized in the middle and upper atmosphere, so a calculation of the flux divergence does not give us a reliable value for the heating of the atmosphere. In fact, in the mesosphere and lower thermosphere the source function $J(z)$ is far from LTE and usually much larger than $B(z)$, approaching $\bar{L}_{\Delta\nu}^{\downarrow}(\infty)$. Hence, the heating rate induced is not nearly as large as the LTE value $\bar{L}_{\Delta\nu}^{\downarrow}(\infty)[\mathrm{d}T_F(z, \infty)/\mathrm{d}z]$. We will return to this point in Sec. 5.2 and in Chapter 7.

4.2.6 *The 'cooling-to-space' approximation*

In optically thin regions sufficiently far from the lower boundary, where $T_F(z_0, z)$ changes very slowly and $T_F(z, \infty)$ changes rather rapidly, the exchange of photons with the adjacent layers is small, and the incident radiation from space is negligible (i.e. during the night), then Eq. (4.68) reduces to

$$h \simeq -\pi \Delta\nu \, \frac{\mathrm{d}T_F(z, \infty)}{\mathrm{d}z} \, B(z). \tag{4.69}$$

Since there is now only one important term, this expression for the heating rate is called the *'cooling-to-space'* approximation. It can be quite accurate for the major bands, except obviously in the troposphere, and where there is a strong temperature gradient. It is most accurate in the middle and upper atmosphere where the transmission to space approaches unity, but the assumption that $J(z) = B(z)$ has not yet broken down. From the definition of $T_F(z, \infty)$ (Eq. 4.63) and assuming optically thin conditions (very small optical depth), the change with altitude of $T_F(z, \infty)$ can be approximated by $2Sn_a/\Delta\nu$, where S is the band strength, so

$$h \simeq -2\pi \, S \, n_a(z) \, B(z). \tag{4.70}$$

Under some conditions where this approximation is useful, the number of photons emitted downwards can be much larger than those absorbed from the underlying layers. Then it may be better to assume that the photons emitted at z are not only lost into the hemisphere above but also into the hemisphere below. The resulting cooling rate would then be twice that of the 'cooling-to-space' approximation, i.e., $h \simeq -4\pi \, S \, n_a(z) \, B(z)$. This approximation is sometimes called the *transparent* or *'total-escape'* approximation.

Along with the 'cooling-to-space' approximation, the concept of *Newtonian cooling* is often introduced. If an atmospheric parcel at z with an equilibrium temperature T experiences a dynamical or other disturbance perturbing the temperature to T', the change in the heating rate is given by

$$\Delta h \simeq -\pi \Delta \nu \, \frac{\mathrm{d}\mathcal{T}_F(z, \infty)}{\mathrm{d}z} \, \Delta B(z)$$

and, assuming the change in temperature $\Delta T = T' - T$ is small, then non-linear effects can be neglected and

$$\Delta h \simeq -\pi \Delta \nu \, \frac{\mathrm{d}\mathcal{T}_F(z, \infty)}{\mathrm{d}z} \left[\frac{\mathrm{d}B}{\mathrm{d}T} \right] \Delta T. \tag{4.71}$$

The rate of temperature change that this variation in the heating rate would produce is given by

$$\left[\frac{\partial T}{\partial t} \right] \simeq -\frac{\pi \Delta \nu}{c_p \rho} \frac{\mathrm{d}\mathcal{T}_F(z, \infty)}{\mathrm{d}z} \left[\frac{\mathrm{d}B}{\mathrm{d}T} \right] \Delta T,$$

where ρ is the air density and c_p is the specific heat of air at constant pressure. This linear approximation allows the heating rate for different temperature profiles to be calculated rapidly and efficiently once it has been calculated for a standard profile. This is why the Newtonian cooling approach is widely used for computing the radiative contribution to the energy balance in atmospheric general circulation models.

4.2.7 *The opaque approximation*

Now consider the other limit, an optically thick atmosphere in which the mean free path of photons is much smaller than the typical scale of other atmospheric processes. Assuming that $B(z)$ is a continuous function of z, a Taylor series expansion of the Planck function for a height z' close to z gives

$$B(z') = B(z) + (z' - z)\frac{\mathrm{d}B}{\mathrm{d}z} + \frac{(z' - z)^2}{2!}\frac{\mathrm{d}^2 B}{\mathrm{d}z^2} + \frac{(z' - z)^3}{3!}\frac{\mathrm{d}^3 B}{\mathrm{d}z^3} + \dots .$$

We assume that we are dealing here only with regions which are unaffected by the boundaries, where $B(z)$ can have discontinuities (Sec. 4.2.4); under the optically thick assumption this requires only a small separation.

With this expansion for $B(z)$, Eq. (4.68) can be written for LTE conditions as

$$\frac{h}{\pi \Delta \nu} = -\left[B(T_g) - B(z)\right]\frac{\mathrm{d}\mathcal{T}_F(z_0, z)}{\mathrm{d}z} - \left[B(z) - \bar{L}^\downarrow_{\Delta \nu}(\infty)\right]\frac{\mathrm{d}\mathcal{T}_F(z, \infty)}{\mathrm{d}z} +$$
$$\sum_{n=1}^{\infty} \left[(-1)^n \mathcal{L}^\downarrow_n + \mathcal{L}^\uparrow_n\right]\frac{1}{n!}\frac{\mathrm{d}^n B}{\mathrm{d}z^n}, \tag{4.72}$$

where the functions $\mathcal{L}_n^{\downarrow}$ and $\mathcal{L}_n^{\uparrow}$, called the downward and upward radiation lengths, are given by

$$\mathcal{L}_n^-(z) = -\int_{z_0}^z |z' - z|^n \, \frac{\partial^2 \mathcal{T}_F(z', z)}{\partial z \partial z'} \, \mathrm{d}z',$$

and

$$\mathcal{L}_n^+(z) = -\int_z^\infty (z' - z)^n \, \frac{\partial^2 \mathcal{T}_F(z, z')}{\partial z \partial z'} \, \mathrm{d}z'.$$

Under optically thick conditions, not only the contribution from the surface but also those from any layer not very close to z is negligible, hence the first moments $\mathcal{L}_1^{\downarrow}$ and $\mathcal{L}_1^{\uparrow}$ are very similar, and the term in $\mathrm{d}B/\mathrm{d}z$ vanishes. Then the heating rate becomes

$$\frac{h}{\pi \Delta \nu} \simeq \left(\frac{\mathcal{L}_2^{\downarrow} + \mathcal{L}_2^{\uparrow}}{2} \right) \frac{\mathrm{d}^2 B}{\mathrm{d}z^2} = \left(\frac{\mathcal{L}_2^{\downarrow} + \mathcal{L}_2^{\uparrow}}{2} \right) \frac{\mathrm{d}B}{\mathrm{d}T} \frac{\mathrm{d}^2 T}{\mathrm{d}z^2}, \tag{4.73}$$

where a term in $(\mathrm{d}^2 B/\mathrm{d}T^2)(\mathrm{d}T/\mathrm{d}z)^2$ is usually neglected. This is the *opaque approximation*. It is sometimes also called the *diffusion or diffusivity approximation*, but we will reserve the latter name for the approach by which the integral over μ in the flux transmittance (Eq. 4.61) is replaced by a single multiplicative 'diffusivity factor' (see Sec. 4.4).

Equation (4.72) can be used for other regimes. For example, for most atmospheric bands a height can be defined at which the atmosphere is optically thick downwards but optically thin upwards. This falls in the upper mesosphere for the CO_2 15 μm fundamental band and in the lower mesosphere for the hot and isotopic bands. In this situation, the term in the first derivative cannot be neglected since the downward radiation length is much larger than the upward one. The reverse situation also occurs, where the atmosphere is more optically thick upwards than downwards and the higher-level contribution dominates, e.g. for the O_3 9.6 μm band in the lower stratosphere and for CO 4.7 μm in the upper stratosphere.

4.3 Integration of the Radiative Transfer Equation over Frequency

Section 4.2.1 discussed the formal solution of the radiative transfer equation for a stratified atmosphere at a single frequency or wavelength. For LTE, it is relatively simple since the source function is equal to the Planck function, and the absorption coefficient k_ν has a simple form. The problem is more difficult for non-LTE situations, but is still quite straightforward if the source function and absorption coefficient are known or can be calculated. The major problem remaining to be tackled in either case is how to carry out the integration over the spectrum properly. This is essential for nearly all practical problems, for example, when we want to calculate the radiance emitted to space in the bandpass measured by an instrument on a satellite (see Chapter 8), the heating or cooling rate, or the populations of a level in non-LTE.

Having obtained an expression for the monochromatic radiance L_ν, in general we need to form the integral $\int_{\Delta\nu} L_\nu \, d\nu$ to calculate the mean radiance over the spectral interval $\Delta\nu$. For the heating rate, we have already shown in Sec. 4.2.5 that if B_ν and L_ν do not change significantly over $\Delta\nu$, the integration over frequency is limited to the computation of the mean transmittance (Eq. 4.63) for all optical paths (z, z') required. We now deal with the more general case.

The frequency integration is simple in principle, since all that is required is to evaluate L_ν at an appropriate number of points, N, spaced $\delta\nu$ apart, where $\delta\nu$ is small enough that the sum $\delta\nu \sum_N L_\nu$ is a good approximation to the integral $\int L_\nu \, d\nu$. However, the variation of absorption coefficient with frequency is complex for atmospheric molecular bands (see Chapter 2), and so this type of numerical step-by-step integration (usually called 'line-by-line' integration) can consume inordinate amounts of time even on modern computers. The problem is exacerbated by the need to sum the contributions of all of the spectral lines, with evaluation of the line shape function for each line, quite complex in itself. Where atmospheric models are concerned, we are dealing with quantities which vary in all three spatial dimensions and in time, as well as in wavelength, and it is usually necessary to integrate over the angular distribution as well. While modern machines may be up to the task in most cases, applications where speed is more important than accuracy, or where the error is limited more by input parameters such as rate coefficients than by the radiative transfer calculation, various approximations are still widely used. These include special algorithms to perform the integration over ν efficiently, and of *band models*, which avoid the step-by-step integration altogether.

4.3.1 *Special line-by-line integration techniques*

Since the width of a typical pressure-broadened line is around $0.07\,p\,\mathrm{cm}^{-1}$, where p is pressure in bars, the increment $\delta\nu$ in a direct numerical integration of the RTE has to be less than this; in practice, $\delta\nu \sim 10^{-3}\,\mathrm{cm}^{-1}$ or smaller. However, because k_ν varies rapidly only near the line centre, and very slowly over other parts of the spectral range, large savings are possible by using an 'adaptive' integration scheme. Many of these exist; we shall describe here only the general principles and refer the reader to specialised publications for detailed descriptions, methods and codes (see the 'References and Further Reading' section).

Consider, for example, the problem of finding the total radiance $L_{\Delta\nu}^{\uparrow}(\infty, \mu)$ leaving the top of an atmosphere in the frequency interval $\Delta\nu$ at a zenith angle θ, with $\mu = \cos\theta$. From Eq. (4.60), integrating over ν and assuming LTE conditions, it follows that this involves evaluating the expression

$$L_{\Delta\nu}^{\uparrow}(\infty,\mu) = \int\limits_{0}^{\Delta\nu} \int\limits_{z_0}^{\infty} B(z',\nu) \frac{\mathrm{d}\mathcal{T}_{\nu}(z',z,\mu)}{\mathrm{d}z'} \, \mathrm{d}z' \, \mathrm{d}\nu,$$

where we have dropped the contribution of the lower boundary for simplicity. The discretization of this equation in the altitude z and frequency ν leads to

$$L_{\Delta\nu}^{\uparrow}(\infty,\mu) = \sum_{j} \sum_{i} B_{i,j} \left(\mathcal{T}_{i+1,j} - \mathcal{T}_{i,j} \right) \delta\nu_{j}, \qquad (4.74)$$

where the sum over j replaces the integral over frequency, and that over i the integral over height. $\mathcal{T}_{i,j}$ is the transmission from the i-th layer to space at frequency j, given by

$$\mathcal{T}_{i,j} = \exp\left(-\sum_{i'=i}^{\infty} \sum_{m} k_{i',j,m} \, n_{a,i',m} \, \Delta z_{i'}/\mu \right), \qquad (4.75)$$

where the sum over the m different atmospheric species is included, and the sum over i' from layer i to the top of the atmosphere converts layer transmissions into that for the whole column. The line shape of the absorption coefficient k is given by the Voigt formula.

Not all of the spectral lines need to be included: if these have been read in from a database in the usual way, then some of the weaker lines may have a negligible effect on the spectrum and can be left out. Most adaptive schemes include a simple test for this, for example by setting a lower limit on the product of line strength multiplied by absorber amount. They should also include a test for the range of influence of each line; strong or broad lines influence the spectrum at a much greater distance from their centre than do weak or narrow lines. Similarly, it is not efficient to make every layer in the atmospheric model the same physical thickness; the higher layers may contain very little absorber mass relative to the lower ones, and a judicious choice can reduce the total number of layers in the summation.

Most of the saving in an adaptive scheme comes from the frequency integration, however. An effective scheme is one in which the frequency interval (the sum over j in Eq. 4.74) is set equal to some fraction of the half width of the narrowest line. As the altitude increases the line width decreases and the number of frequency steps increases. This avoids the use of too fine a frequency grid in the lower atmosphere, where the lines are broad. However, it does not allow for the fact that the strongest lines may be saturated, so that the transmission is zero there regardless of the precise value of k_{ν}. The most efficient schemes actually examine the difference between adjacent frequency points and use a finer mesh where the gradient is steepest. Across regions where the transmission is zero, or unity, a very coarse frequency mesh will obviously suffice.

In practice, the procedure is to evaluate the transmission on a coarse mesh first, and then to compare adjacent values. If they are the same within a specified limit, say 0.001, then the frequency resolution is adequate. If the difference is larger, a new point is taken midway between the existing ones and the cycle repeated. In this way, the program automatically fits a fine mesh over the sharp features in the spectrum, and a coarse one over smooth regions, and the user can trade off accuracy against computation time to suit the problem being addressed. Various refinements to this basic approach have been developed and programs to carry out the integration are available from the literature, or in some cases commercially (see 'References and Further Reading').

In order to obtain the radiative flux F, we require in addition the integration of the RTE over solid angle. We then have an equation similar to (4.74) but with an additional sum over μ or, alternatively, $\mathcal{T}_{i,j}$ can be replaced in that equation by $2E_3(\mathcal{T}_{i,j})$, where E_3 is the third exponential integral (see Eq. 4.61). The computation of the heating rates, as discussed in Sec. 4.2.5, requires an additional difference in height, i.e. Eq. (4.74) is used but with a double difference over index i of $\mathcal{T}_{i,j}$. This approach is also required when computing the non-LTE populations, as we will see in the next chapter.

4.3.2   *Spectral band models*

Band models represent the complex distribution of lines in a spectral interval by some simplified concept, for which the averaged transmission over the interval can be calculated from a simple formula instead of the tedious integrations described in the previous section. The accuracy achieved obviously will depend on whether the model chosen is a good representation of reality, but it is generally found that measured and calculated transmissions disagree by something of the order of 10% when using band models, whereas a good line-by-line integration scheme can achieve a performance of the order of 1%. Band models are most useful for problems where very high accuracy is not required, or where the computation is so daunting that lower accuracy has to be accepted, for example in calculating heating rates in general circulation models.

4.3.3   *Independent-line and single-line models*

The calculation of transmission is greatly simplified if we can assume that the spectral lines involved are essentially independent of each other, i.e. do not overlap. The simplest assumption of all is that in which a band of lines is represented by a single line of the same effective strength as the total band. The *equivalent width* W of a line (that is, the width of a square-sided line that has the same integrated absorption as the line in question) is defined as:

$$W = \int_{-\infty}^{\infty} (1 - \mathcal{T}_\nu) \, d\nu = \int_{-\infty}^{\infty} \left[1 - \exp(-S \, f(\nu) \, m)\right] d\nu, \qquad (4.76)$$

where $m = \int_z^{z'} n_a(z'') \, dz''$ is the absorber amount; S the band strength, and $f(\nu)$ the normalized line shape.

In the limits of strong and weak absorption, most line shapes have simple expressions for the equivalent width. These limits are determined by the asymptotic behaviour as the optical depth at the line centre, $Sf(\nu_0)m$, tends to zero or to infinity. The weak limit is independent of the line shape because as $Sf(\nu_0)m \to 0$, $W \to Sm$.

If the origin of the frequency scale is taken for convenience to be at the line centre, the equivalent width of a Lorentz line is given by

$$W_L = \int_{-\infty}^{\infty} \left\{1 - \exp\left[-\frac{Sm\alpha_L}{\pi(\alpha_L^2 + \nu^2)}\right]\right\} d\nu = 2\pi \, \alpha_L \, \mathcal{L}\left(\frac{Sm}{2\pi\alpha_L}\right), \qquad (4.77)$$

where α_L is the Lorentz width (in the same units as ν!). $\mathcal{L}$ is known as the *Ladenberg–Reiche* function and can be expressed in terms of modified Bessel functions:

$$\mathcal{L}(y) = y \exp(-y) \left[I_0(y) + I_1(y)\right].$$

The strong limit is determined by $Sm/(2\pi\alpha) \to \infty$, and so the equivalent width in this case is determined by the behaviour of the Bessel functions as $y \to \infty$. However, it is easier to see what is happening physically by studying Eq. (4.77). The term α_L^2 can be neglected far from the line centre where $\nu^2 \gg \alpha_L^2$, and also close to the line centre where the numerator tends to ∞, i.e. the line is blacked out. Thus, dropping the term altogether, we have

$$W_L = \int_{-\infty}^{\infty} \left[1 - \exp\left(-\frac{Sm\alpha_L}{\pi\nu^2}\right)\right] d\nu = \sqrt{Sm\alpha_L/\pi} \int_{-\infty}^{\infty} \left[1 - \exp\left(-x^{-2}\right)\right] dx$$

$$= 2\sqrt{Sm\alpha_L}. \qquad (4.78)$$

A useful simple approximation to the equivalent width of a Lorentz line is

$$W_L = S \, m \left[1 + Sm/(4\alpha_L)\right]^{-1/2}. \qquad (4.79)$$

This has the same strong and weak limits, and is found to deviate by less than 8% for all values of the parameters. Other approximations of varying accuracy and complexity can be devised; some are discussed in the 'References and Further Reading' section.

The equivalent width of a Doppler line is given by

$$W_D = \int_{-\infty}^{\infty} \left\{1 - \exp\left[-\frac{Sm}{\alpha_D\sqrt{\pi}} \exp\left(-\frac{\nu^2}{\alpha_D^2}\right)\right]\right\} d\nu. \qquad (4.80)$$

This expression cannot be integrated using standard functions, but it can be written in terms of a single variable

$$W_D = \alpha_D F_D \left(\frac{Sm}{\alpha_D \sqrt{\pi}} \right),$$

(4.81)

where

$$F_D(y) = \int_{-\infty}^{\infty} \left\{ 1 - \exp\left[-y \exp(-x^2) \right] \right\} \, \mathrm{d}x.$$

(4.82)

A numerical approximation to the function $F_D(y)$ is given by

$$W_D = 2\,S\,m \left[\frac{1 + \ln(1 + Sm/\alpha_D \sqrt{\pi})}{4 + \pi (Sm/\alpha_D \sqrt{\pi})^2} \right]^{1/2},$$

(4.83)

and the strong limit of the equivalent width of a Doppler line is of the form

$$W_D = 2\,\alpha_D \sqrt{\ln(Sm/\alpha_D \sqrt{\pi})}.$$

(4.84)

In most real problems, the line shape of interest is not the pure Lorentz or Doppler case, but the composite of the two represented by the Voigt function. The equivalent width of this cannot be represented by simple functions, and generally it is necessary to resort to a subroutine which calculates the function explicitly. It can, however, be approximated to within a maximum error of 8% by

$$W_V = \sqrt{W_L^2 + W_D^2 - \left(\frac{W_L W_D}{Sm} \right)^2}.$$

(4.85)

A single-line band model may be used to approximate transmission in bands where the spectral lines do not overlap appreciably, i.e. when they are well-spaced and/or the pressure is low. The equation for the transmission uses the approximate expressions for the equivalent width derived above, or more precise values computed numerically, and is simply:

$$\mathcal{T} = 1 - \frac{\sum_i W_i}{\Delta \nu},$$

(4.86)

where the sum is over all of the lines in the interval of interest, $\Delta \nu$. Despite its attractive simplicity, the single-line model is mainly of academic interest since the need to make vast savings in computing time (or to estimate transmissions without the aid of a computer at all) is no longer as relevant as it was in the days when the theory was first worked out, and the accuracy which can be achieved with such a simple representation is very limited in most cases.

4.3.4   *Band models with overlapping lines*

This is the most general case, and a large variety of models has been developed, each with its advantages and disadvantages, for particular cases. Often the hardest part of using a band model for a particular application is deciding which of the various candidates to adopt. Sometimes the model will have a physical basis, for example the assumption of regularly-spaced (or alternatively randomly-spaced) lines, which has a recognisable correlation with the band under study. More often, however, tests have to be carried out in which sample cases are compared with more exact line-by-line techniques in order to assess the error being introduced by the approximations inherent in the model. We will now look in more detail at some of the more popular band models in use.

4.3.5   *The regular or Elsasser model*

This was one of the first models developed, proposed by Elsasser in 1938. It assumes that the lines are all identical, and equally spaced (but overlapped) by a distance δ (Fig. 4.5).

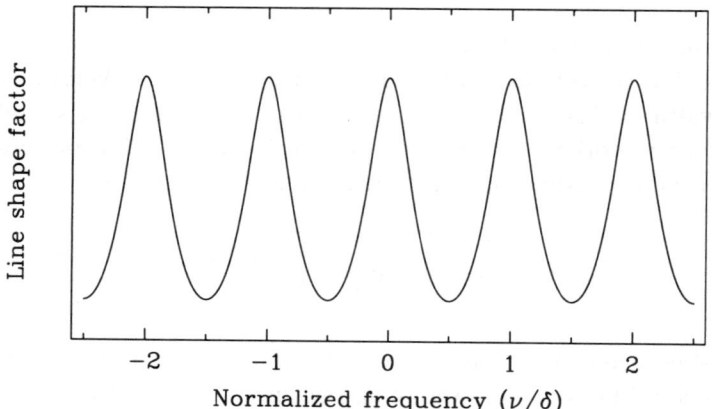

Fig. 4.5   An idealised spectral band, as represented by the Elsasser model.

Elsasser showed that the line contour is given by

$$f_E(\nu) = \frac{1}{\delta} \frac{\sinh(2\pi\alpha/\delta)}{\cosh(2\pi\alpha/\delta) - \cos(2\pi\nu/\delta)},$$

and the transmission is

$$\mathcal{T}_E = \frac{1}{\delta} \int_\delta \exp\left[-S\,f_E(\nu)\,m\right] \mathrm{d}\nu. \tag{4.87}$$

This expression cannot be integrated in terms of elementary functions, but it can be computed or held in a look-up table. It also reduces to simple expressions in the strong and weak limits. For strong lines $(Sm \gg 2\pi\alpha)$ we have

$$\frac{W_E}{\delta} = \mathrm{erf}\left(\frac{\sqrt{Sm\pi\alpha}}{\delta}\right),\tag{4.88}$$

where W_E is the equivalent width in an interval of width δ. In the weak line limit, $Sm \ll 2\pi\alpha$, and $W_E = Sm$, as usual.

4.3.6 *Random band models*

Several different models exist which are based on the assumption that the positions and the strengths of lines are randomly distributed. While this is not strictly true, bands of many important species such as water vapour and carbon dioxide do have a quasi-random appearance (see, for example, Fig. 4.6) and the model has been shown to work well. It also follows that lines calculated for different species occupying the same spectral interval —a common situation— can be multiplied together to obtain the total transmission in that interval. This 'multiplication property' is not valid in general for transmissions averaged over finite spectral intervals; it is strictly valid only for monochromatic transmissions, but again has been shown to work well for random bands, in the laboratory as well as in simulations.

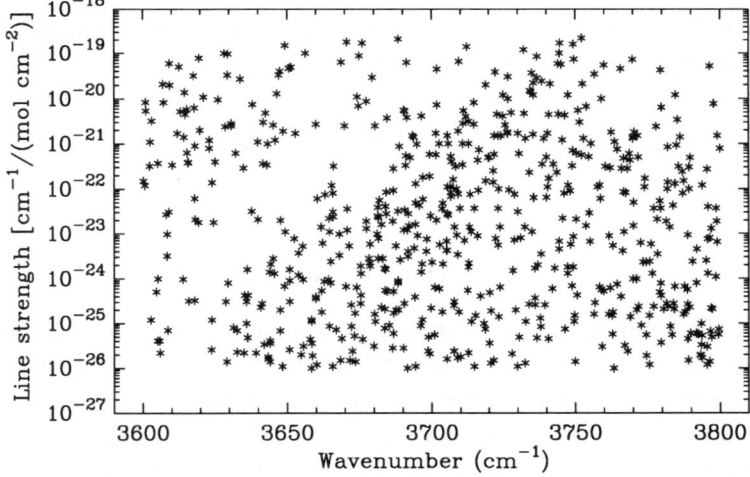

Fig. 4.6   Line strength distribution of water vapour in the 2.7 μm bands, illustrating the quasi-random characteristics. The line strengths are from HITRAN 1992.

The proof of the multiplication property is as follows. Consider a spectral interval $\Delta\nu$ having a monochromatic transmission $\mathcal{T}_{0,\nu}$, and therefore an average transmission given by

$$\mathcal{T}_0 = \frac{1}{\Delta\nu}\int_{\Delta\nu}\mathcal{T}_0(\nu)\,\mathrm{d}\nu.\tag{4.89}$$

If a line is introduced at a position ν' having a monochromatic transmission $\mathcal{T}_1(\nu - \nu')$ and an equivalent width W_1, much smaller than $\Delta\nu$, $W_1 \ll \Delta\nu$, then the total frequency averaged transmission will become

$$\mathcal{T} = \frac{1}{\Delta\nu} \int_{\Delta\nu} \mathcal{T}_0(\nu)\, \mathcal{T}_1(\nu - \nu')\, d\nu.$$

If this is averaged over all possible positions of the new line, it becomes

$$\mathcal{T} = \frac{1}{\Delta\nu} \int_{\Delta\nu} \left[\frac{1}{\Delta\nu} \int_{\Delta\nu} \mathcal{T}_0(\nu)\, \mathcal{T}_1(\nu - \nu')\, d\nu \right] d\nu'$$

which, on integration, gives simply

$$\mathcal{T} = \mathcal{T}_0\, \mathcal{T}_1,$$

where $\mathcal{T}_1$ is the frequency-averaged transmission of the new line. In general, if the spectral lines are spaced at least approximately randomly in frequency, it is possible to multiply frequency-averaged transmissions without serious error.

The expression for the general random band model follows from a consideration of a single line whose equivalent width W_i gives a transmission in the spectral interval $\Delta\nu$ of

$$\mathcal{T}_i = 1 - W_i/\Delta\nu \approx \exp(-W_i/\Delta\nu),$$

provided that $W_i \ll \Delta\nu$. Using the multiplication property, the transmission of N lines in the interval $\Delta\nu$, averaged over all positions, is

$$\mathcal{T} = \prod_{i=1}^{N} \exp\left(-W_i/\Delta\nu\right) = \exp\left(-\sum_{1}^{N} W_i/\Delta\nu\right),$$

thus

$$\mathcal{T} = \exp(-W/\Delta\nu), \tag{4.90}$$

where W is the total equivalent width of all lines in the interval assumed to be independent. This derivation has assumed that $W_i \ll \Delta\nu$, but in fact it is possible to derive the same expression for the general case provided that it can be assumed that the interval $\Delta\nu$ is embedded in a wider array of random lines with the same statistical properties. Then, any absorption due to lines inside $\Delta\nu$ which falls outside the interval is offset by absorption inside $\Delta\nu$ due to lines whose centres fall outside.

The general random band model, in spite of its simple formulation, is not particularly easy to use because the equivalent widths of all of the lines have to be evaluated. If we assume that the distribution of line strengths is given by some known function $N(S)$, i.e. that the number of lines in the interval with strengths between S and $S+dS$ is $N(S)\, dS$, then a simpler expression for the transmission can be found. Three distribution functions have been found which are often used. These are named after their originators, Goody, Godson and Malkmus, and illustrated in Fig. 4.7.

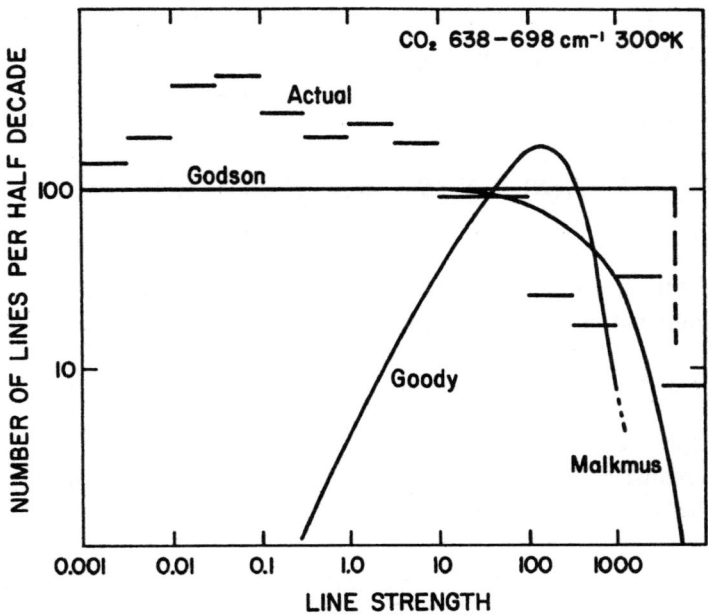

Fig. 4.7   Comparison of an actual distribution of line strengths with the three model distributions discussed in the text.

The Goody model assumes that

$$N(S) = \frac{N_0}{k} \exp(-S/k),$$

where the distribution has a total of N_0 lines of mean strength k. The total equivalent width is given by

$$W = N_0 \overline{W} = \int_0^\infty N(S)\, W(S)\, \mathrm{d}S =$$
$$\int_0^\infty \frac{N_0}{k} \exp(-S/k) \left[\int_{-\infty}^\infty \left\{ 1 - \exp\left[- S\, f(\nu)\, m \right] \right\} \mathrm{d}\nu \right] \mathrm{d}S,$$

where $f(\nu)$ is the line shape function. Integrating over S, we obtain

$$W = N_0 \int_{-\infty}^\infty \frac{k\, m\, f(\nu)}{1 + k\, m\, f(\nu)}\, \mathrm{d}\nu = N_0 \int_{-\infty}^\infty \frac{\bar\tau(\nu)}{1 + \bar\tau(\nu)}\, \mathrm{d}\nu, \qquad (4.91)$$

where $\bar\tau(\nu)$ is the optical thickness for a line of strength k. For the Lorentz shape, in particular,

$$W = \frac{N_0\, k\, m}{\sqrt{1 + k\, m/(\pi \alpha_L)}}.$$

The transmission for the Goody random model with Lorentz lines is usually written

$$T = \exp\left(-\frac{k\,m}{\delta\sqrt{1 + k\,m/\pi\alpha_L}}\right), \tag{4.92}$$

where $\delta = \Delta\nu/N_0$ is the mean line spacing.

The random model due to Godson has N_0 lines for each factor of e in strength, an infinite number altogether. Then

$$N(S) = N_0/S \qquad \text{for} \quad S < k, \quad \text{and}$$
$$N(S) = 0 \qquad \text{for} \quad S > k. \tag{4.93}$$

For a general line shape there is no simple expression for W, but for the Lorentz shape it can be written in terms of modified Bessel functions:

$$W = 2\pi\,\alpha_L\,N_0\left[e^{-y}I_0(y) + 2y\,e^{-y}\left[I_0(y) + I_1(y)\right] - 1\right], \tag{4.94}$$

where $y = km/(2\pi\alpha_L)$.

The most realistic random band model is that by Malkmus, who combined the Goody and Godson approaches to get

$$N(S) = \frac{N_0}{S}\exp(-S/k). \tag{4.95}$$

The total equivalent width for a general line shape may be written as

$$W = \int_0^\infty \frac{N_0}{S}\exp(-S/k)\left[\int_{-\infty}^{+\infty}\left\{1 - \exp\left[-S\,f(\nu)\,m\right]\right\}d\nu\right]dS =$$
$$N_0\int_{-\infty}^{+\infty}\ln\left[1 + \bar\tau(\nu)\right]d\nu = -N_0\int_{-\infty}^{+\infty}\frac{\nu}{1 + \bar\tau(\nu)}\frac{d\bar\tau(\nu)}{d\nu}\,d\nu \tag{4.96}$$

which, for a Lorentz line can be integrated to become

$$W = \frac{\pi\alpha_L N_0}{2}\left[\sqrt{1 + 4km/(\pi\alpha_L)} - 1\right]. \tag{4.97}$$

In order to use these models we have to introduce values for the various parameters. This can be done by considering again the values to which W reduces in the weak and strong limit, and fitting the known quantities (the individual line strengths and widths) to those. We find the limits by putting $km/\pi\alpha_L \to 0$ or ∞ in the expressions for W. The results are as listed in Table 4.1. For the Malkmus model, for example, we would set

$$N_0\,k = \sum_i S_i, \qquad \text{and} \qquad N_0\sqrt{k\pi\alpha_L} = \sum_i \sqrt{S_i\alpha_{L,i}}.$$

The expressions for transmission are

$$T = \exp\left[-(w^{-2} + s^{-2})^{-1/2}\right] \tag{4.98}$$

Table 4.1   Equivalent widths of band models.

Model	Weak limit	Strong limit
General	$\sum_i S_i m$	$2 \sum_i \sqrt{S_i \alpha_{L,i} m}$
Goody	$N_0 k m$	$N_0 \sqrt{k \pi \alpha_L m}$
Godson	$N_0 k m$	$4 N_0 \sqrt{k \alpha_L m}$
Malkmus	$N_0 k m$	$2 N_0 \sqrt{k \pi \alpha_L m}$

for the Goody model, and

$$\mathcal{T} = \exp \left[-\frac{1}{(2w)^{-1} + \left[(2w)^{-2} + s^{-2} \right]^{1/2}} \right] \qquad (4.99)$$

for Malkmus, where

$$w = \frac{1}{\Delta \nu} \sum_i S_i m \quad \text{and} \quad s = \frac{2}{\Delta \nu} \sum_i \sqrt{S_i \alpha_{L,i} m}.$$

Strictly these are valid only for Lorentz lines, but reasonable results are obtained under most conditions for Voigt lines. A decision as to what is acceptable accuracy has to be made by the researcher by running tests against the full line-by-line solution, and comparing the difference to his or her requirements.

4.3.7   *Empirical models*

Another way to use band models is to fit them to measured spectra. In that way, the constants in the expression for transmission can be determined without the need to know any of the individual line positions, strengths or widths. In particular, the laboratory data used in this approach can be of quite low resolution, and so relatively easy to acquire, especially at low temperatures and pressures characteristic of the Earth's atmosphere (and at the high temperatures and pressures found, for instance, on Venus).

As an example of how models can be employed in this way consider the Goody model, Eq. (4.98), in the form

$$(-\ln \mathcal{T})^{-2} = (\ln \mathcal{T}_W)^{-2} + (\ln \mathcal{T}_S)^{-2}, \qquad (4.100)$$

where $\mathcal{T}_W$ is the transmission in the weak limit, and $\mathcal{T}_S$ is the strong limit, both of which can be determined from data. The fit in the regions of intermediate strength can be arbitrarily improved by adding extra terms and multipliers, but it has to be kept in mind that the model is only reliable for the range of temperatures, pressures, and absorber amounts over which it was derived. In other words, empirical models can be used for interpolation, but not reliably for extrapolation, within this range of parameters. A very direct approach due to Smith generalizes the simple empirical model

$$T = \exp(-k\, m^a\, p^b\, T^c),$$

to the form

$$\ln(-\ln T) = \ln k + a \ln m + b \ln p + c \ln T + \dots, \qquad (4.101)$$

where T is temperature. The accuracy of this model can be improved indefinitely by adding more terms.

4.3.8   The 'sum of exponentials' method

This approach is particularly valuable as a way of representing broad-band transmissions in a scattering medium, for example when analysing remote sensing measurements of a cloudy atmosphere, because the transmission function is represented as a series of pseudo-monochromatic terms, and the multiplication property holds for each term. They can then be used in one of the many multiple-scattering algorithms which are available for computing radiative transfer in clouds. A variation of this method, known as *correlated-k*, is also in common usage.

The transmission as a function of absorption coefficient k and absorber amount m is represented by

$$T(m) = \sum_i a_i \exp(-k_i m), \qquad (4.102)$$

and the coefficients a_i and k_i are chosen by a numerical fitting procedure so that a good fit is obtained to T with the smallest number of terms.

A similar method involves defining an 'inverse transmission function' $f(k)$ such that its Laplace transform is the transmission:

$$T(m) = \int_0^\infty f(k) \exp(-k\, m)\, \mathrm{d}k. \qquad (4.103)$$

It can be seen that $f(k)$ corresponds to the a_i coefficients in the discrete version, and can be obtained from tables of inverse Laplace transforms.

4.3.9   Inhomogeneous optical paths

For most atmospheric problems, pressure and temperature will vary along the paths for which we want to calculate the transmission. Sometimes, if the spectroscopically-active gas is not uniformly mixed, the absorber amount will also vary, in a way which is not simply proportional to the pressure.

The use of band models in these common situations, where the line shape varies along the path, requires further approximations. The most common of these is the

Curtis-Godson approximation, which defines a homogeneous path with the same equivalent widths in the weak and strong limits as the real path.

Consider first the situation in which the temperature is constant, so that the equivalent homogeneous path is defined by two parameters, an effective pressure $\overline{p}$, and an absorber amount $\overline{m}$. The weak limit correspondence requires $\int S \, dm = S \overline{m}$, and the strong Lorentz limit $\int S \, \alpha \, dm = S \overline{\alpha} \, \overline{m}$, where $\overline{\alpha} = \alpha_0 \overline{p}$. We can eliminate S and α_0 to give the simplest form of the Curtis-Godson approximation,

$$\overline{m} = \int dm, \quad \text{and} \quad \overline{m} \, \overline{p} = \int p \, dm. \tag{4.104}$$

It is tempting also to define by analogy, an effective mass weighted temperature $\overline{T}$ as

$$\overline{m} \, \overline{T} = \int T \, dm, \tag{4.105}$$

the justification being mainly intuitive, although it is often found that this simple approach gives quite good results in many cases. As with band models themselves, it is necessary to carry out tests in which approximations are compared to more exact numerical integrations in order to estimate what accuracy is being achieved in any particular case. The literature is full of specific cases, including derivations of more sophisticated approaches to the definition of equivalent paths which have been found to be useful for a variety of realistic problems. In practice the Curtis-Godson approximation is used in two rather distinct cases. One, within monochromatic models, where it is used to find equivalent conditions for each path segment, spanning a small range of p and T conditions. The complete path is then represented by the multiplied transmittances of each of these segments. Another case in within band models, where the approximation is used to find equivalent conditions for the complete path to some point in the atmosphere, spanning a much larger range of p and T conditions.

4.4 Integration of the Radiative Transfer Equation over Solid Angle

We have seen that obtaining the source function under non-LTE conditions, and the atmospheric heating rate, requires angular integration of the radiance. There are several situations where the last integral in Eq. (4.61) can be performed explicitly, for example in both the weak and the strong limits of a Lorentz line. However, it cannot be done in the general case of the transmission over a finite spectral interval. Like the integration over frequency, this can be done by direct numerical integration, but is so costly in computer time that even using band models it is very desirable in most cases to approximate the angular integration as well. The solution is the use of quadrature schemes, which require the evaluation of the transmittance for only a few angles (the quadrature points). A weighted mean is then formed using a set of quadrature weights. The choice of scheme depends on the situation being modelled and the accuracy required; in the radiative transfer literature Gaussian quadrature is the most common. Tables of angles and weights for any number of points are available (see 'References and Further Reading'). Four points are usually sufficient for most atmospheric infrared emission calculations, but if higher accuracy is needed, the number of points in the quadrature can be increased indefinitely.

Situations which require large numbers of points are usually those which include particulate scattering, i.e. where the atmosphere is cloudy or hazy. For typical non-LTE situations where this does not apply, further simplifications may be appropriate. A simple and very common approach, sometimes called the *diffusivity approximation*, is essentially a one-point quadrature in which the integration over μ in the flux transmittance (Eq. 4.61) is replaced by the transmittance evaluated at a certain angle θ_d, such that, $\beta = 1/\cos\theta_d$, where β is called the diffusivity factor, i.e.,

$$\mathcal{T}_{F,\nu}(z', z) = 2 \int_0^1 \mathcal{T}_\nu(z', z, \mu)\, \mu \, d\mu \simeq \mathcal{T}_\nu(z', z, \mu = 1/\beta) \qquad (4.106)$$

or, equivalently,

$$E_3\big[\tau_\nu(z', z)\big] \simeq \frac{1}{2}\exp\big[-\beta\,\tau_\nu(z', z)\big], \qquad (4.107)$$

where E_3 is the exponential integral. In this approach we are actually replacing the diffuse radiation (i.e., in all directions) by a beam of radiation at angle $\theta_d = \arccos(1/\beta)$. In practice we do not wish to use different β values for each frequency-grid point, but a single value for the frequency interval in use. Then Eq. (4.106) can be used with the mean, instead of the monochromatic, transmission.

In principle, the parameter β can be adjusted to give more accurate results for any particular case. Let us consider the extreme cases of the weak and strong transmission limits, to gain an idea of its range of possible values. As shown in Sec. 4.3.3, the equivalent width of a line (or of a spectral region) in the weak limit is given by $W = Sm$. It is easy to show that in this case $\beta = 2$, and we would expect this value to give accurate cooling rates in the thin upper atmosphere. For the strong limit of a Lorentz line, the equivalent width is proportional to $\sqrt{Sm}$ and, including this result in the definition of β (Eq. 4.106), it can be shown that $\beta = 16/9 = 1.778$. The same result is obtained for the strong limit of the Goody, Godson and Malkmus random models with Lorentz lines. For the strong limit of the Doppler line, taking into account its equivalent width (Eq. 4.84), we get a value of $\beta = \sqrt{e}$, close to 1.649.

The diffusivity approximation for the angular integration is often used in the calculation of heating rates. In Eq. (4.66), the heating rate is proportional to the derivative of the flux transmittance $d\mathcal{T}_F(z', z)/dz$, so it is tempting to obtain the β value from this instead of the transmittance. From the definitions of flux transmittance, optical thickness, and the exponential integral,

$$2 \int_{\Delta\nu} k_\nu(z) \, E_2(\tau_\nu) \, d\nu = \beta \int_{\Delta\nu} k_\nu(z) \, \exp(-\beta\,\tau_\nu) \, d\nu \qquad (4.108)$$

for a spectral range $\Delta\nu$, or

$$2 \, E_2(\tau_\nu) = \beta \, \exp(-\beta\,\tau_\nu) \qquad (4.109)$$

for the monochromatic case. The values of β derived from these equations for the extreme cases of the weak and strong limits of both Lorentz and Doppler lines coincide with those obtained from Eq. (4.106). β values that give exact results for Lorentz and Doppler line shapes for intermediate optical depth can be found in the literature (see 'References and Further Reading').

Elsasser found empirically that the band model named after him gave good results in the calculation of cooling rates with the approximation $\beta = 5/3 = 1.667$. This factor has been generally adopted for the computation of cooling rates in the troposphere and stratosphere where it has been found to give accuracies of 1–2% in most situations. However, Apruzese found that a diffusivity factor of 1.81 gave better results for the cooling rate of an isothermal atmosphere.

4.5 References and Further Reading

The original derivation of the Chapman function was presented by Chapman (1931). A tabulation of this function was given later by Wilkes (1954), and an useful approximation by Titheridge (1988). A general formulation of optical paths for large solar zenith angles in the Earth's curved atmosphere is given by Wang *et al.* (1981).

Several general purposes radiance/transmittance line-by-line models have been developed. Some of the most common are: FASCODE (Clough *et al.*, 1989; Chetwynd, 1994); GENLN2 (Edwards, 1992; Edwards *et al.*, 1993; 1998); ARC (Wintersteiner *et al.*, 1992); LINEPAK (Gordley *et al.*, 1994); RFM (Dudhia, 2000; http://www.atm.ox.ac.uk/RFM/sum.html), and KOPRA (Stiller *et al.*, 1998; Höpfner *et al.*, 1998, http://www-imk.fzk.de:8080/imk2/ame/publications/kopra_docu).

A discussion of the line-by-line integration technique may be found in Mitzel and Firsov (1995). The best review of spectral band models is that of Goody and Yung (1989), but see also Liou (1992). A very good summary of approximate methods for calculating the transmission by bands of spectral lines, including inhomogeneous optical paths, can be found in Rodgers (1976). For integrations using different quadratures see Chandrasekhar (1960) and Abramowitz and Stegun (1970).

The paper by Edwards and Francis (2000) provides a good overview of the current status of fast integrations methods including the correlated-k method.

Of the various compilations of sources of spectral data in the infrared part of the spectrum, the most commonly used is HITRAN, which covers the region 0–23000 cm^{-1} and contains some data for rotational lines. The HITRAN database

(1986) edition is described by Rothman *et al.* (1987) but there have been corrections and updates in 1997, 1998, and the current edition HITRAN 2000 (see http://www.hitran.com and links within). The GEISA database covers the 0–23000 cm^{-1} region and contains a mixture of HITRAN and other data of use in the treatment of planetary atmospheres other than Earth. The latest version is from 1997 (Jacquinet-Husson *et al.*, 1999; see http://www.ara.polytechnique.fr/-alexei_index.html). The ATMOS database, compiled for the analysis of that instrument, covers the region from 0–10000 cm^{-1}. It includes an updated version of HITRAN 1992, some supplementary line data and some cross-section data (Brown *et al.*, 1995). The EPA database covers the infrared region only. It includes measured and calculated (based on HITRAN) spectra and provides the necessary information to calculate cross-sections, with emphasis on organic molecules (see http://earth1.epa.gov/ttnemc01/ftir/welcome.html).

A useful survey of the diffusivity factor, β, is given by Armstrong (1968). For applications of this for the computation of cooling rates in the troposphere and stratosphere, including justification for claims of an accuracy of 1 to 2%, see Elsasser (1938), Hitschfeld and Houghton (1961), Rodgers and Walshaw (1966), Ellingson and Gille (1978) and Apruzese (1980). Values for β that give exact results for Lorentz and Doppler line shapes for intermediate optical depth can be found in Shved and Bezrukova (1976).

Chapter 5

Solutions to the Radiative Transfer Equation in Non-LTE

5.1 Introduction

Chapter 3 described the basic formulation of radiative transfer both for LTE and non-LTE conditions and Chapter 4 the integration of the radiative transfer equations when LTE applies. In Chapter 4 we also illustrated how the integrations over altitude, frequency and solid angle can be carried out, much of which is equally applicable to non-LTE. In this chapter we describe some simplified solutions of the non-LTE radiative transfer equations which are applicable under certain conditions (i.e. some atmospheric regions and particular molecular bands), leading up to the full solution using the Curtis matrix method. Since the integrations over frequency and solid angle are similar to those for LTE, we focus mostly on the integration over height in this chapter.

5.2 Simple Solutions for Radiative Transfer under Non-LTE

In Sec. 3.7 we discussed when and where the more common non-LTE situations are expected to occur in the atmosphere. Now we consider some non-LTE cases in which the source function, the population of the states, and the heating/cooling rates can be computed with useful accuracy using simple expressions, without the need to solve the complete radiative transfer equation. We begin with several versions of the so-called *classical* non-LTE situation, which have in common the fact that exchange of photons between layers can be neglected, and only thermal collisional and local radiative processes are important.

5.2.1 *Weak radiative field*

In this case there is no pumping of the emitting level by the absorption of solar, terrestrial or atmospheric radiation, and therefore radiative processes are important only as a loss mechanism. This applies for most near-infrared bands in the upper atmosphere at night, and for moderate and strong mid-infrared bands, like the CO_2

$15\,\mu$m fundamental band, in the lower thermosphere, by both night and day. At $15\,\mu$m, solar radiation is weaker than terrestrial radiation and, in the thermosphere, the excitation of CO_2 vibrational states by collisions with atomic oxygen is more effective than absorption of stratospheric or tropospheric radiation. We consider two ways to treat the effective radiative losses: the 'escape-to-space' (or its variant known as 'total-escape') and the 'cooling-to-space' approaches.

5.2.1.1 *Escape to space*

This simply neglects excitation by absorption of radiation and the losses due to induced emissions. We already found in Sec. 3.7 that, neglecting the terms containing $\bar{L}_{\Delta\nu} = 0$ in Eqs. (3.57) and (3.59), the population of the excited state n_2 and the source function are given by

$$\frac{n_2}{n_1} = \frac{\bar{n}_2}{\bar{n}_1}\left(\frac{\epsilon}{1+\epsilon}\right),\tag{5.1}$$

and

$$J_{\nu_0} = \frac{\epsilon}{1+\epsilon}\,B_{\nu_0},\tag{5.2}$$

respectively, with $\epsilon = l_t/A_{21}$. This expression for the source function applies for LTE as well as the night-time non-LTE cases of two levels connected by an optically thin band, or at very low pressures where $A_{21} \gg l_t$ and $J \ll B$.

The cooling rate can be obtained from Eq. (3.60) by making $\bar{L}_{\Delta\nu} = 0$ and using Eq. (5.2) for the source function. Thus we obtain

$$h = -\frac{4\pi S n_a \epsilon}{1+\epsilon}\,B_{\nu_0} = -\frac{4\pi S n_a}{1 + A_{21}/l_t}\,B_{\nu_0}.\tag{5.3}$$

This assumes 'total-escape', where radiation emitted in the downward and upward directions is lost with no compensating absorption. For lower altitude non-LTE regions it is sometimes more realistic to assume that the radiative losses are uncompensated in the upward, but zero in the downward, hemisphere. Then the net value is $A_{21}/2$ instead of A_{21} and we have the 'escape-to-space' approach. Note that we do not require that the photons 'recovered' are strictly from the layers below, just that half of the photons emitted in all directions are recovered by absorption from all directions. It can be shown that under these circumstances the population and source function are given by Eq. (5.1) and (5.2) with $\epsilon = 2l_t/A_{21}$. For the cooling rate, however, one must be cautious since we can no longer assume $\bar{L}_{\Delta\nu} = 0$ in Eq. (3.60). We assume in this approach that the absorption compensates for half of the spontaneous losses. With this, it can be shown that the heating rate is

$$h = -\frac{2\pi S n_a}{1 + A_{21}/2l_t}\,B_{\nu_0}.\tag{5.4}$$

The heating rate can easily be expressed in terms of the population of the upper state of the transition (instead of the source function) when there is no significant radiative excitation. From Eqs. (3.60) and (3.35), making use of the relation between the Einstein coefficient A_{21} and the band strength S (Eq. 3.46), we have

$$h_{12} = -n_2 A_{21} h\nu_0 \left[1 - \frac{n_2 g_1}{n_1 g_2}\right]^{-2} \simeq -n_2 A_{21} h\nu_0. \qquad (5.5)$$

A further approximation can be assumed when the thermal collisional losses are much smaller than the radiative emissions, i.e., $A_{21} \gg l_t$. From Eqs. (5.1) and (5.5), remembering that $l_t = k_t [\text{M}]$, we have

$$h_{12} \simeq -n_1 k_t [\text{M}] \frac{g_2}{g_1} \exp\left(-\frac{h\nu_0}{kT}\right) h\nu_0, \qquad (5.6)$$

where the heating rate (which is actually cooling) is solely controlled by the rate of thermal collisions, k_t, independent of the value of A_{21}, and therefore independent of whether the radiative losses are A_{21} or $A_{21}/2^*$. This situation also occurs when the band is not optically thin, i.e. $\mathcal{T}^*$ is not very close to unity (see next section), provided that $A_{21}\mathcal{T}^*/2 \gg l_t$.

5.2.1.2 *Cooling to space*

The previous approach does not give good results for the layers which fall at heights intermediate between those in LTE and those in the total 'escape-to-space' situation, where $A_{21} \sim l_t$, particularly when the band is optically thick. For these conditions, a better approach is available, still without the need to solve the full radiative transfer problem. This is the 'cooling-to-space' approximation, which assumes that the net radiative losses are due only to those photons emitted in the upward hemisphere that are able to escape directly to space. These are smaller than the net losses in either the 'escape-to-space' or 'total-escape' approximations, since they include absorption from the layers below equal to all the photons emitted downwards, plus an additional absorption from the layers above equal to those photons emitted upwards and absorbed in the hemisphere above.

In this approach, the cooling rate is given by Eq. (4.69) but with the Planck function replaced by the source function J to accommodate non-LTE, i.e.

$$h \simeq -\pi\Delta\nu \frac{d\mathcal{T}_F(z,\infty)}{dz} J(z).$$

Introducing this expression into the equation that relates J and h, Eq. (3.61), we

---

*Remember that n_1 is the density of the lower state of the transition. If the band considered is a fundamental, n_1 would be very close to the number density of the absorbing molecules, $n_1 = n_0 \simeq n_a$.

find

$$J_{\nu_0} = \frac{B_{\nu_0}}{1 + \dfrac{\Delta\nu}{4Sn_a\epsilon}\dfrac{\mathrm{d}T_F(z,\infty)}{\mathrm{d}z}}.$$

From the definition of $T_F(z,\infty)$ (Eqs. 4.61 and 4.63) and defining the probability of a photon emitted at any frequency in the band escaping to space in any direction in the upper hemisphere, T^*, by

$$T^*(z) = \frac{1}{S}\int_{\Delta\nu}\int_0^1 k_\nu(z)\exp\left[-\tau_\nu(z,\infty,\mu)\right]\mathrm{d}\mu\,\mathrm{d}\nu, \tag{5.7}$$

we have

$$\frac{\mathrm{d}T_F(z,\infty)}{\mathrm{d}z} = \frac{2Sn_a\,T^*(z)}{\Delta\nu},$$

and hence

$$J_{\nu_0} = \frac{B_{\nu_0}}{1 + (A_{21}/2l_t)\,T^*}, \tag{5.8}$$

where we have also assumed that $\exp(-h\nu_0/kT)$ is much smaller than unity.

Inserting this equation into the previous expression for the heating rate, this becomes

$$h = -\frac{2\pi Sn_a\,T^*}{1 + (A_{21}/2l_t)\,T^*}B_{\nu_0}. \tag{5.9}$$

The treatment of radiative transfer in this approach is then reduced to the calculation of the derivative of the transmission to space at each layer or, equivalently, the 'escape-to-space' function.

Note that Eq. (5.8) tends to the expression for J under the 'escape-to-space' approach (Eq. 5.2 with $\epsilon = 2l_t/A_{21}$) for the upper atmosphere where the bands are optically thin and T^* tends to unity. For optically thick conditions, $T^* \to 0$, and hence $J \to B$, and the band remains in LTE even though the radiative relaxation time may be significantly smaller than the time between thermal collisions.

We can write alternative expressions for the source function and heating rate if we assume the diffusivity approximation (see Sec. 4.4). For these conditions,

$$\frac{\mathrm{d}T_F(z,\infty)}{\mathrm{d}z} \simeq \frac{\beta\,S\,n_a\,\overline{T}^*(z,\beta)}{\Delta\nu},$$

where $\overline{T}^*$ is now defined as

$$\overline{T}^*(z,\beta) = \frac{1}{S}\int_{\Delta\nu} k_\nu(z)\exp\left[-\tau_\nu(z,\infty,\mu = 1/\beta)\right]\mathrm{d}\nu,$$

and hence,

$$J_{\nu_0} = \frac{B_{\nu_0}}{1 + (\beta/4)\,(A_{21}/l_t)\,\overline{T}^*}, \tag{5.10}$$

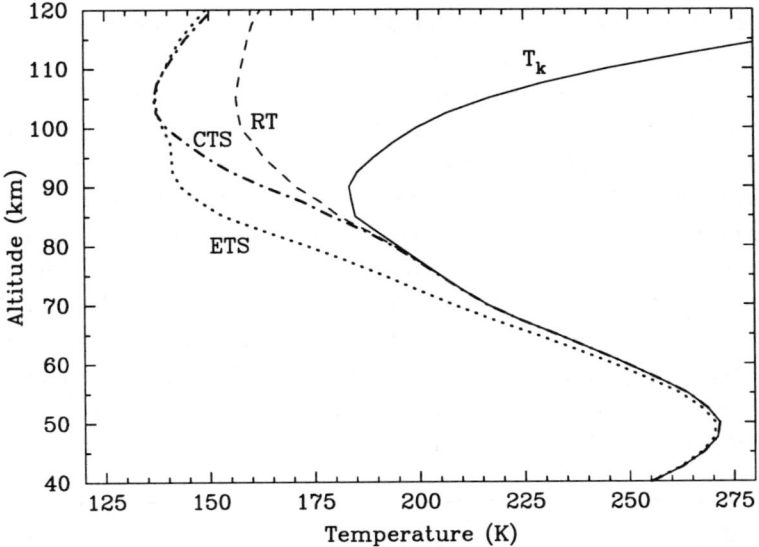

Fig. 5.1 The night-time vibrational temperature (see its definition in Eq. 3.22) of $CO_2(0,1^1,0)$ in the Earth's atmosphere calculated for the 'escape-to-space' (ETS) and 'cooling-to-space' (CTS) approximations, with an exact solution (RT) for comparison. All calculations use a slow rate coefficient for the thermal collisions of $CO_2(0,1^1,0)$ with atomic oxygen, $k_{CO_2\text{-}O}$. The kinetic temperature profile T_k is also shown.

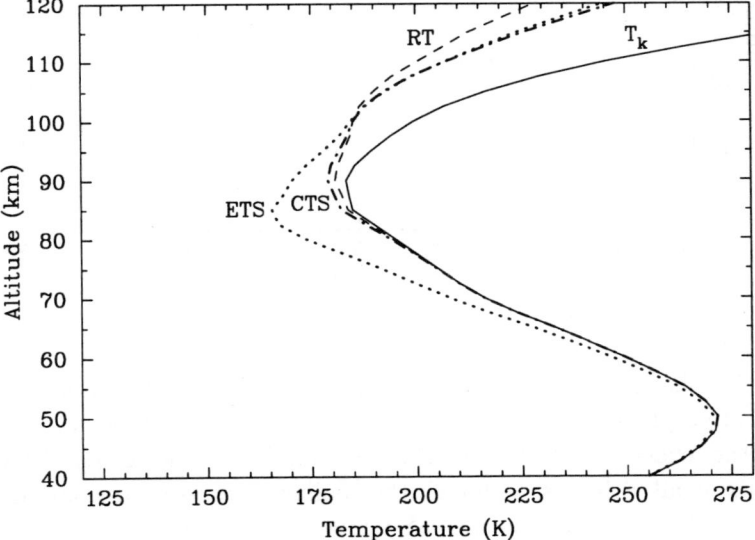

Fig. 5.2 As Fig. 5.1 but for a larger, more realistic value of the rate coefficient, $k_{CO_2\text{-}O}$. Note the smaller differences among the vibrational temperatures in comparison with previous figure.

and

$$ h = -\frac{\beta \pi S n_a \overline{T}^\star}{1 + (\beta/4)(A_{21}/l_t)\overline{T}^\star} B_{\nu_0}. \qquad (5.11) $$

An illustration of these two approaches, together with the 'exact' radiative transfer solution (see Sec. 5.7) for the source function, is shown in Figs. 5.1 and 5.2 for
the CO_2 $15\,\mu m$ fundamental band in the terrestrial atmosphere. Fig. 5.3 shows
the same comparison for Mars. Fig. 5.1 shows how the 'escape-to-space' approach,
with net radiative losses of $A_{21}/2$, gives a reasonable idea of the height at which the
band departs from LTE, but is not very accurate above that altitude. The 'cooling-
to-space' solution gives a better estimate of the LTE breakdown height, but still
deviates from the more accurate calculation in the region above. As expected, both
approaches predict much better solutions when the thermal collisional processes
are faster (Fig. 5.2), since radiative excitation is relatively less important. In the
case of the pure CO_2 atmosphere on Mars (Fig. 5.3) the 'escape-to-space' solution
does not even give a good estimate of the height of departure from LTE, while the
'cooling-to-space' approach does much better in this regard.

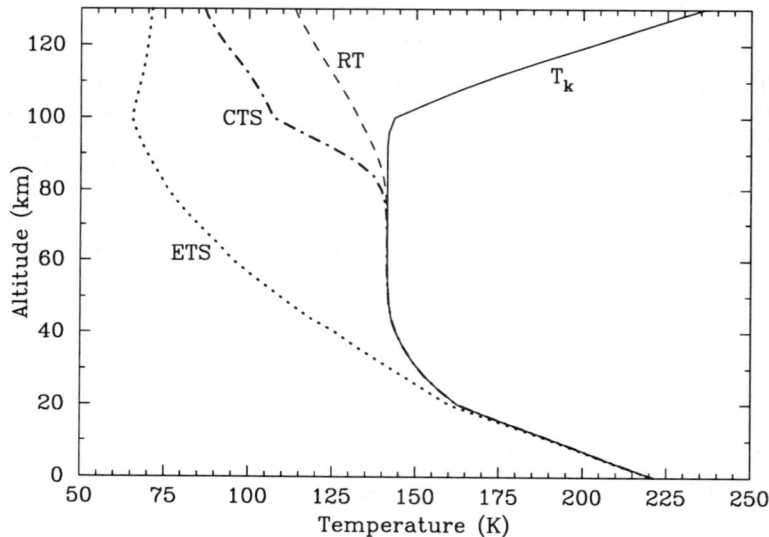

Fig. 5.3   As Figure 5.1, but for the Martian atmosphere.

Figure 5.4 shows the cooling rate of the CO_2 $15\,\mu m$ fundamental band in the
terrestrial atmosphere, using the 'total-escape' and 'cooling-to-space' approximations and the 'exact' non-LTE and LTE solutions. 'Cooling-to-space' gives fairly
good results at all altitudes, although the cooling rates are too small around the
stratopause. This is due to the warmer temperatures there, which mean that the
layers emit more radiation than they absorb, while the approximation assumes a
null net exchange. In the mesosphere it is the other way around, since absorption of
radiation from the warmer layers below is significant, thus offsetting the cooling to
space so that the approximation overestimates the cooling rate. As expected, the
'total-escape' approach, which ignores the large opacity of the lower atmosphere,
predicts an unrealistic large cooling in the stratosphere and mesosphere.

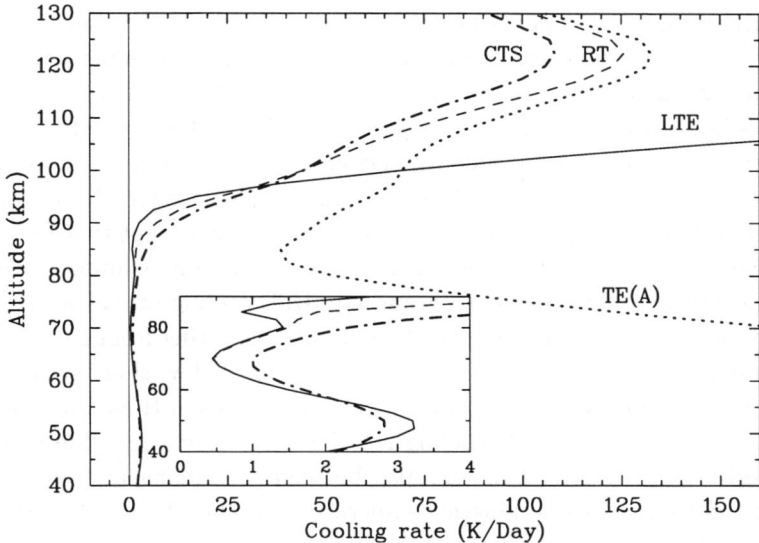

Fig. 5.4   Cooling rates of the CO_2 $15\,\mu m$ fundamental band in the terrestrial atmosphere, using the 'total-escape' (A_{21}) (TE) and 'cooling-to-space' (CTS) approximations and the 'exact' non-LTE solution for LTE and non-LTE (RT). A fast $k_{CO_2\text{-}O}$ coefficient (3×10^{-12} $cm^3 s^{-1}$) has been used.

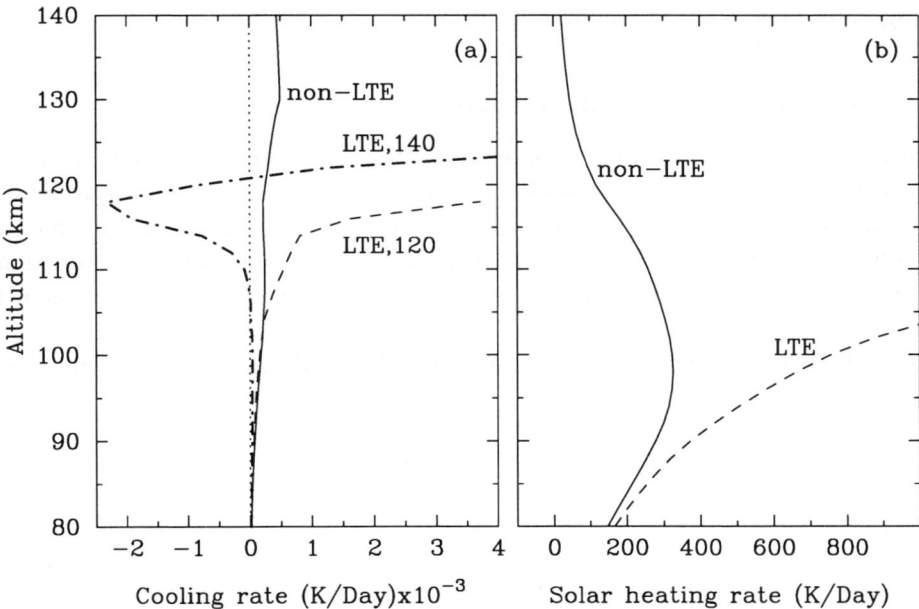

Fig. 5.5   Non-LTE effects on the cooling and heating rates in the Martian atmosphere. (a) The thermal cooling rate of the CO_2 $15\,\mu m$ fundamental band, for models assuming LTE with the upper boundary at 120 and 140 km, and for the more realistic non-LTE case. (b) The solar heating rate by CO_2 near-IR bands for LTE and non-LTE conditions.

In the thermosphere, 'cooling-to-space' underestimates the cooling because too few photons travelling upwards from below are absorbed to compensate for those emitted downwards at a rate of $A_{21}/2$. The 'exact' (RT) solution is intermediate between 'cooling-to-space' and 'total-escape' (TE), the difference between the two being, in fact, an estimate of the amount of energy absorbed from the lower regions. At higher altitudes this is smaller and the RT and TE solutions tend to converge. For bands other than CO_2 15 μm, similar results are expected, except that for optically thinner bands 'total-escape' will approach the exact solution at lower altitudes.

It is also worth noting that the LTE and non-LTE cooling rates are the same up to a few km below the height where the band starts departing from LTE ($\sim$95 km). The LTE rates are slightly smaller in a narrow region below that, because of the absorption of emission from the warmer regions above when these are assumed to be in LTE. Further up, the LTE cooling rate becomes very large, due to the high thermospheric temperature. This can also be seen by looking at Fig. 5.2, which shows that the thermospheric kinetic temperature is much larger than the actual vibrational excitation temperature.

A similar situation for the LTE/non-LTE cooling rate difference occurs in the Martian atmosphere, particularly when the upper boundary is set up at 140 km (Fig. 5.5a), where the downward contribution from the warm thermosphere is unrealistically large if LTE is assumed to apply at all levels. The effects on the Martian heating rates (Fig. 5.5b) of ignoring non-LTE is again huge.

5.2.2   *Strong external radiative source*

There are some situations where a source supplying radiation from outside the non-LTE region being considered (usually the Sun, but sometimes the surface, lower atmosphere, or a cloud top) is so strong that radiative exchange between layers is negligible by comparison. The total mean radiance, $\bar{L}_{\Delta\nu}$, and the source function and heating rates, can then be calculated easily and accurately from simplified expressions.

From Sec. 3.7, the population of the excited level and the source function are then

$$\frac{n_2}{n_1} = \frac{g_2}{g_1} \frac{c^2}{2h\nu_0^3} \frac{\bar{L}_{\Delta\nu}}{(1 + l_t/A_{21})},$$

and

$$J_{\nu_0} = \frac{\bar{L}_{\Delta\nu}}{1 + l_t/A_{21}}. \tag{5.12}$$

Note that we can no longer replace the net radiative losses by substituting A_{21} with $A_{21}/2$ or $A_{21}T^\star/2$ in these equations, since absorption and induced emission are already considered in full with the inclusion of the mean radiance, $\bar{L}_{\Delta\nu}$.

In general, a given energy state will be affected by more than one transition. It is quite common for a state to be excited by the absorption of solar radiation in a single band but de-excited in several bands. Many high-energy combination levels relax radiatively after solar pumping more quickly to intermediate vibrational states than to the ground level. For example, the CO_2 levels excited near $2.7\,\mu m$ have a higher probability of emitting a $4.3\,\mu m$ quanta than of re-emitting at $2.7\,\mu m$. These situations can be studied without considering the multilevel case explicitly, if we replace the radiative losses of the absorbing band, A_{21}, with a new term A_T which includes spontaneous emission of all possible transitions to lower levels. Under these conditions, Eqs. (3.60) and (5.12) give the heating rate as

$$h = 4\pi S n_a \, \frac{\bar{L}_{\Delta\nu}}{1 + A_T/l_t}. \tag{5.13}$$

A further particular case is when thermal losses are much smaller than the radiative relaxation, as is common in the upper atmosphere. Then, the source function is simply

$$J_{\nu_0} \simeq \bar{L}_{\Delta\nu}, \tag{5.14}$$

and the heating rate, given by

$$h = 4\pi S n_a \, \frac{l_t}{A_T} \, \bar{L}_{\Delta\nu}, \tag{5.15}$$

or by

$$h = \frac{c^2}{2\nu_0^2} \frac{g_2}{g_1} \, n_1 \, l_t \, \bar{L}_{\Delta\nu}. \tag{5.16}$$

To calculate the source function or the heating rate from these expressions, the mean radiance, $\bar{L}_{\Delta\nu}$, is required. We derive below expressions for this for the two usual sources of strong external fields, which are the Sun and the thermal emission from the lower boundary.

5.2.2.1 *Solar radiation*

Inserting solar radiation into the definition of $\bar{L}_{\Delta\nu}$, Eq. (3.56), by taking $L_\nu = L_{\nu_0,\odot}$, which is nearly independent of frequency over most spectral bands in the infrared, integrating over the solid angle subtended by the Sun ($\pi R_\odot/d$) being $R_\odot$ the solar radius and d the mean Earth-Sun distance, and including the atmospheric absorption (Eq. 4.20) we have

$$\bar{L}_{\Delta\nu}(z) = \frac{R_\odot^2}{4Sd^2} L_{\nu_0,\odot}(\infty) \int_{\Delta\nu} \exp\left[-\frac{\tau_\nu(z,\infty)}{\mu_\odot}\right] k_\nu \, d\nu \tag{5.17}$$

which, using the definition

$$T(z, \infty, \mu_\odot) = \frac{1}{\Delta\nu} \int_{\Delta\nu} \exp\left[-\frac{\tau_\nu(z, \infty)}{\mu_\odot}\right] d\nu,$$

can also be written as

$$\bar{L}_{\Delta\nu}(z) = \mu_\odot \frac{R_\odot^2}{d^2} \frac{L_{\nu_0,\odot}(\infty)\Delta\nu}{4Sn_a} \frac{d\left[T(z, \infty, \mu_\odot)\right]}{dz}. \qquad (5.18)$$

If excitation (but not losses) by thermal collisions is negligible, substituting the expression for $\bar{L}_{\Delta\nu}(z)$ above into Eqs. (5.12) and (5.13) gives for the source function and solar heating rates,

$$J_{\nu_0}(z) = \mu_\odot \frac{R_\odot^2}{d^2} \frac{L_{\nu_0,\odot}(\infty)\Delta\nu}{4Sn_a\,(1 + l_t/A_{21})} \frac{d\left[T(z, \infty, \mu_\odot)\right]}{dz}, \qquad (5.19)$$

and

$$h(z) = \mu_\odot \frac{F_\odot(\infty)}{(1 + A_{21}/l_t)} \frac{d\left[T(z, \infty, \mu_\odot)\right]}{dz}, \qquad (5.20)$$

where we have made use of $F_\odot(\infty) = (\pi R_\odot^2/d^2)\, L_{\nu_0,\odot}(\infty)\,\Delta\nu$.

In the lower atmosphere, the high pressures cause thermal collisional losses to be dominant. Then, from Eq. (5.20) we obtain

$$h(z) = \mu_\odot\, F_\odot(\infty) \frac{d\left[T(z, \infty, \mu_\odot)\right]}{dz},$$

i.e. all the radiation absorbed from the Sun (see also Eq. 4.22) is converted into kinetic energy. Frequently, the levels absorbing solar radiation are combination levels with high energies, absorbing in the near-IR where the solar radiation is strong. For them, collisions might be fast even in the upper atmosphere, especially V–V collisions which transfer the absorbed energy to other vibrational levels which then emit (e.g., M(v) in process 3.65 in Sec. 3.6.5.1). The 'heating rate', as formally defined, may then be given quite accurately by Eq. (5.20), but relatively little of it is thermalized. As noted above (Sec. 3.2), h is still measured in K/day or similar units, even when it is a potential rather than an actual heating rate.

In the highest atmospheric layers, where thermal losses are negligible compared to radiation and most of the bands are optically thin, $J_{\nu_0} \simeq \bar{L}_{\Delta\nu}$, and the source function, from (5.17), is

$$J_{\nu_0}(z) \simeq \frac{R_\odot^2}{4d^2} L_{\nu_0,\odot}(\infty). \qquad (5.21)$$

This represents resonant fluorescence, in which the radiation absorbed from the Sun within the solid angle $\pi R_\odot^2/d^2$ is redistributed isotropically in all directions (4π steradians). In this case, $l_t \to 0$, and from Eq. (5.20) it is found that the heating rate also tends to be zero. In other words, all of the radiation absorbed from the

Sun (Eq. 4.22) is re-emitted (but not necessary at the wavelength of absorption) and no net internal or kinetic energy is transferred to the atmospheric molecules.

5.2.2.2 *Terrestrial radiation*

The surface or lower boundary can be treated as a blackbody emitting at an effective temperature T_e located at height $z_b{}^\dagger$. From the definition of $\bar{L}_{\Delta\nu}$ in Eq. (3.56), with $L_\nu = B_\nu(T_e)$, again assumed to be nearly independent of frequency within any single spectral band, $B_\nu(T_e) \simeq B_{\nu_0}(T_e)$. Including the atmospheric absorption from the lower boundary z_b up to altitude z, (e.g., the first term in the right hand side of Eq. 4.9) we have

$$\bar{L}_{\Delta\nu}(z) = \frac{B_{\nu_0}(T_e)}{2S} \int_0^1 \int_{\Delta\nu} \exp\left[-\frac{\tau_\nu(z_b, z)}{\mu}\right] k_\nu \, d\nu \, d\mu, \qquad (5.22)$$

or, using the definition of the integrated flux transmittance (Eqs. 4.61 and 4.63),

$$\bar{L}_{\Delta\nu}(z) = \frac{B_{\nu_0}(T_e)\, \Delta\nu}{4Sn_a} \frac{d\big[\mathcal{T}_F(z_b, z)\big]}{dz}. \qquad (5.23)$$

As for the solar radiation, we can introduce this expression for $\bar{L}_{\Delta\nu}(z)$, into the equations for the source function and heating rates when only thermal collisions are important as losses, Eqs. (5.12) and (5.13), to obtain

$$J_{\nu_0}(z) = \frac{B_{\nu_0}(T_e)\, \Delta\nu}{4Sn_a(1 + l_t/A_{21})} \frac{d\big[\mathcal{T}_F(z_b, z)\big]}{dz}, \qquad (5.24)$$

and

$$h(z) = \frac{\pi B_{\nu_0}(T_e)\, \Delta\nu}{(1 + A_{21}/l_t)} \frac{d\big[\mathcal{T}_F(z_b, z)\big]}{dz}, \qquad (5.25)$$

where A_{21} can be replaced by A_T, if necessary.

When thermal losses are negligible and the blackbody radiation from the lower boundary is not significantly absorbed up to the layer in question, z, we have $J_{\nu_0}(z) = B_{\nu_0}(T_e)/2$. This again tells us that radiation emitted at z_b into the upward hemisphere and absorbed at height z, is re-emitted in all directions.

5.2.3 *Non-thermal collisional and chemical processes*

When collisions with molecules having non-LTE populations, chemical recombination or photochemical reactions dominate the excitation sources of the level in question (Sec. 3.7), the population of the excited state and the source function are

---

†The radiation from the surface and lower atmospheric layers travelling upwards in the atmosphere is often called *earthshine*.

given by (Eq. 3.90)

$$\frac{n_2}{n_1} = \frac{p_t + p_{nt} + P_c/n_1}{A_{21} + l_t},$$

and

$$J_{\nu_0} = \frac{\epsilon_1 B_{\nu_0}}{1 + \epsilon_2} \simeq \frac{p_t + p_{nt} + P_c/n_1}{p_t(1 + A_{21}/l_t)} B_{\nu_0},$$

where we have used the relations $\epsilon_1 \simeq (p_T/p_t)(l_t/A_{21})$ and $\epsilon_2 \simeq l_t/A_{21}$, and have assumed that non-thermal losses are negligible. Introducing the expression for the source function into that for the heating rate (Eq. 3.60), we have

$$h = -4\pi S n_a \frac{\epsilon_1}{1 + \epsilon_2} B_{\nu_0} \simeq -4\pi S n_a \frac{p_t + p_{nt} + P_c/n_1}{p_t(1 + A_{21}/l_t)} B_{\nu_0}. \qquad (5.26)$$

An alternative expression to compute the heating rate for situations where the excitation by radiation is negligible and the bands are optically very thin, was obtained previously in Sec. 5.2.1.1, Eq. (5.5). That approach is particularly useful when non-thermal populations are produced by chemical recombination. In these cases, Eq. (5.5), plus the expression for the population of the upper state (Eq. 3.78), neglecting the mean radiance, lead to a simple equation for the heating rate:

$$h = -\frac{P_t + P_{nt} + P_c}{1 + l_t/A_{21}} h\nu_0,$$

where we recall that $P_t = p_t n_1$ and $P_{nt} = p_{nt} n_1$. Again, when non-thermal processes are present, the divergence of the radiative flux (the formal 'heating rate') no longer corresponds to the usual meaning of 'heating' since a large fraction of the energy transferred is internal energy.

We can combine situations involving non-thermal processes with those where the radiative field is weak (Sec. 5.2.1), and consider various cases with the 'escape-to-space' or 'cooling-to-space' approaches to obtain expressions for the source function and heating rate directly. We can also study what happens when an important non-thermal process (collisional, chemical or photochemical) is involved and the absorption of radiation from an external source is much larger than the absorption of radiation from the atmospheric layers. The expressions for the source function and heating rates are obtained by substituting $\bar{L}_{\Delta\nu}(z)$ given in Eqs. (5.18) and (5.23) for the solar and terrestrial radiation, respectively, into Eqs. (3.69) and (3.60). For solar radiation and non-thermal processes we have:

$$J_{\nu_0}(z) = \frac{\mu_\odot F_\odot(\infty)}{4\pi S n_a(1 + \epsilon_2)} \frac{d[\mathcal{T}(z, \infty, \mu_\odot)]}{dz} + \frac{\epsilon_1}{1 + \epsilon_2} B_{\nu_0}, \qquad (5.27)$$

and

$$h(z) = \mu_\odot \frac{F_\odot(\infty)}{(1 + 1/\epsilon_2)} \frac{d[\mathcal{T}(z, \infty, \mu_\odot)]}{dz} - \frac{4\pi S n_a \epsilon_1}{1 + \epsilon_2} B_{\nu_0}. \qquad (5.28)$$

Analogously, for terrestrial radiation and non-thermal processes we have:

$$J_{\nu_0}(z) = \frac{B_{\nu_0}(T_e)\,\Delta\nu}{4Sn_a(1+\epsilon_2)} \frac{\mathrm{d}\big[T_F(z_b,z)\big]}{\mathrm{d}z} + \frac{\epsilon_1 B_{\nu_0}}{1+\epsilon_2}, \tag{5.29}$$

and

$$h(z) = \frac{\pi B_{\nu_0}(T_e)\Delta\nu}{(1+1/\epsilon_2)} \frac{\mathrm{d}\big[T_F(z_b,z)\big]}{\mathrm{d}z} - \frac{4\pi Sn_a\epsilon_1}{1+\epsilon_2} B_{\nu_0}, \tag{5.30}$$

where ϵ_1 and ϵ_2 are given approximately by $\epsilon_1 \simeq (p_T/p_t)(l_t/A_{21})$ and $\epsilon_2 \simeq l_t/A_{21}$.

5.3 The Full Solution of the Radiative Transfer Equation in Non-LTE

Finally, we come to the solution of the radiative transfer equation in the general case, when no major approximations are sufficiently valid to simplify the problem. We must then take account of (i) the radiative transfer of photons, between each layer and every other layer; (ii) the frequency variation of the atmospheric absorption coefficient, which includes a large number of vibration-rotation lines with shapes which vary with temperature and pressure; (iii) the different local thermal and non-thermal collisional processes which affect the populations of the internal energy levels of the molecules; and (iv) the coupling between some of the vibrational levels.

The magnitude of the problem is illustrated by the fact that the populations of the excited states, n_2, depends on the radiation field through the mean radiance, $\bar{L}$, which in turn depends on the number density of the excited level, and on the radiation at other atmospheric layers and, in some cases, on external radiation fields. Thus, the solution for n_2, or for the radiative heating or cooling produced by the band, requires first the integration of the radiative transfer equation over frequency and over solid angle, ν and μ, i.e. the calculations of the flux transmittances; and then the solution over altitude of the coupled system of radiative transfer and statistical equilibrium equations. When the V–V energy exchange between vibrational levels with non-LTE populations is very fast, the coupled equations of statistical equilibrium for all the affected levels and of radiative transfer for all bands originating from those levels have to be solved simultaneously in order to obtain accurate populations. The degree of coupling between levels, and how many of them are coupled, determine the numerical method to be used to solve the system: a matrix inversion or an iterative procedure, for example. Later, in Sec. 5.4.2, we discuss some ideas about how to make the choice.

The transmittance calculation problem is essentially the same as for the LTE case but with one major additional complexity: overlapping between lines of different bands cannot be easily taken into account since they have, in principle, different source functions. Less important complexities include the fact that, for hot bands, the transmission calculations require some previous knowledge of the populations of their lower-level states, because they might be in non-LTE. Also, when the induced emission and vibrational partition function factors are significant, the population being sought enters into the absorption term, leading to a non-linear system of equations. Fortunately, these effects are often found to be small, at least when the non-LTE deviations are not very large. In the calculation of quantities where differences of transmission are involved, e.g. in the cooling rate or in the limb radiance emerging from the atmosphere, the population of these levels can be treated as being in LTE for most practical cases. In some special cases, however, such as the CO_2 laser bands, the transmittance calculation of hot bands with lower levels in non-LTE can represent an additional difficulty.

5.4   Integration of the RTE in Non-LTE

This section surveys methods for calculating non-LTE radiative transfer in the atmosphere, starting with integration over frequency and solid angle and then describing the general approach to the integration over height. In Sec. 5.4.3 we briefly describe some formulations of non-LTE that have been developed for tackling specific problems.

5.4.1   *Integrations over frequency and solid angle*

The system of statistical equilibrium equation (SEE) (in the form of 3.55 or, equivalently, as in 3.61 or 3.71) and the radiative transfer equation (RTE) (e.g., as in 3.17 or, equivalently, 3.60) include integrations over solid angle and frequency, and over height. Most of the methods for their solution, with the exception of some of the line-by-line ones discussed below, assume these integrations can be separated. One of the most common approximations to allow for such separation is that the populations that enter into the transmission functions or radiances are known beforehand. Examples of these are the populations of the lower levels of the transitions for the hot bands, and those of the upper levels entering into the induced emission factors and into the vibrational partition functions. They are usually considered to be in LTE, or else their previous non-LTE values are used and a few iterations performed to re-calculate the transmission functions.

When separated from the integration over altitude, integration over frequency and solid angle can be carried out exactly as for LTE. The calculation of those integrals is equivalent to the computation of the radiative transfer terms, i.e., the exchange between layers and the cooling to space (see Sec. 4.2.5).

The various methods employed are usually given different names by different authors, but they have many similarities. Three methods may be distinguished: (i) the *flux formulations*; (ii) the *radiance formulations*; and (iii) those which formulate the problem in terms of the *escape probability functions*. In the first of these, the net exchange of radiation between atmospheric layers, which enters into the calculation of the heating rates and the excitation of the upper state by radiative exchange, is calculated from a double difference in the flux transmission times the difference in the source functions of the two layers. In the second, it is calculated directly from the average radiance, which involves the source function of only the source layer. In the third method, it is computed from a single difference in the escape probability function (Eq. 5.7) so that instead of carrying out the double difference in the flux transmittance numerically, it forms the first derivative analytically and then performs the second numerically. The flux formulations use the heating rate, or net absorption rate, as the fundamental quantity in their calculations, while the radiance methods calculate absolute absorption rates. The escape methods simply calculate the probability of photons not being absorbed within certain ranges.

In principle, calculating the 'exchange of radiation' term with the escape probability or the radiance in the *radiance* formulations should be more accurate, particularly when dealing with optically thin conditions, or very long and inhomogeneous optical paths. The double difference in the flux transmittances in the *flux* formulation can introduce significant round-off errors not once but twice, even in double precision arithmetic. On the other hand, calculations of the radiance or the escape probability functions intrinsically take more computer time than flux transmittances. In the end, comparisons of heating rates calculated by the radiance and flux formulations have shown very small differences, but the flux methods can make use of procedures developed for integrating over frequency and solid angle under LTE conditions.

The Curtis matrix method is usually seen as a typical example of a *flux* method; it is so described in detail in Sec. 5.7. However, its formulation is also compatible with the escape probability method, if it is used to calculate the Curtis matrix from a single difference in the escape probability functions (see Sec. 5.7), instead of the usual double difference in the flux transmittances. The Curtis matrix itself can be computed using band models, or using some other approximation for the integration over frequency, or it can be carried out using an accurate line-by-line frequency-integration code. It is thus a very flexible method.

5.4.2 *Integration over altitude*

The integration over altitude is significantly different in non-LTE calculations, compared to LTE. There are two ways to do it: by matrix inversion, or by iteration. Either can be combined with the *flux, radiance* or *escape probability function* methods. In general, matrix inversion is faster and more accurate for the most important atmospheric infrared bands. Dickinson found an iterative solution to be unstable when applied to some CO_2 bands in the Venusian atmosphere, and when he solved the problem with a different approach he needed hundreds of iterations for the strong bands to converge.

These procedures can be applied to a two-level system (only one statistical equilibrium equation and one radiative transfer equation) or to a multilevel system of M levels and N bands, with M statistical equilibrium equations and N radiative transfer equations. When many vibrational levels are collisionally coupled, the system includes a large number of equations which make it difficult to solve by matrix inversion. In such cases a mixture of both methods is generally used. Dickinson used matrix inversion for radiative transfer within each band, but iterated to couple the populations of the collisionally-linked levels. For CO_2 in the terrestrial, Martian and Venusian atmospheres, the system of statistical and radiative transfer equations has been solved using matrix inversion for groups of a few vibrational levels which are strongly coupled by V–V, switching to iteration for the exchange of quanta between the groups which are weakly coupled. If the strongly coupled levels are treated iteratively, the convergence is much slower and not always conclusive.

Induced emission factors, the vibrational partition function, and certain V–V collisional processes introduce non-linear terms into the statistical equilibrium equations. Those terms are easily included in the iterative method, and when matrix inversion is used, at least two iterations are required as well for an accurate solution. The chief advantage of the iterative method is, however, that the large number of equations no longer has to be solved simultaneously. This allows easier inclusion of the vibrational coupling between the different states.

5.4.3 *Specific non-LTE formulations*

Shved and his co-workers devised a method for calculating the populations of the vibration-rotation bands of linear molecules and applied it to CO_2 and CO in the atmospheres of the terrestrial planets, which was later extended for terrestrial O_3, and H_2O bands. They derived expressions for the source function of the vibration-rotation bands making use of astrophysical methods and obtaining equations similar to the integral equation described in Sec. 3.6.4 (Eq. 3.59). This formulation incorporated any collisional process (including non-thermal), and made some common assumptions including: (i) non-overlapping lines; (ii) rotational LTE; and (iii) negligible induced emission. The integral equation for the source function and the radiative transfer equation for $\bar{L}_{\Delta\nu}$ were solved by iteration.

The CO_2 $4.3\,\mu$m emissions in the Earth's atmosphere were studied by Kumer and his co-workers using a formulation based on the 'band transport' functions. They calculated the vibrational temperature of the $CO_2(0,0^0,1)$ state coupled with $N_2(1)$, formulating radiative transfer in the $CO_2(0,0^0,1-0,0,0)$ band by tabulating the 'band transport' functions, that is, (i) the flux equivalent width; (ii) the derivative of the equivalent width over the vertical coordinate (i.e., the band transmission function); and (iii) the solid-angle-integrated second order derivative of the equivalent width over the vertical coordinate (actually over two different altitudes in the atmosphere z and z'). The last is called Green's function, **H**, which gives the probability that a photon emitted at z is absorbed at z'. The similarity with the Curtis matrix method is clear, and in fact the elements of the **H** matrix are identical to the Curtis matrix elements. The difference is that those elements are computed from the expression resulting after obtaining the second order derivative of the flux transmittance analytically, instead of from the double difference of the flux transmittances.

Kumer also studied the more complicated problem of the joint solution of the CO_2 4.3 and 15 μm vibrational levels in an atmosphere illuminated by the Sun, using the same method for radiative transfer but, since the number of coupled equations is larger in this case, with an iterative procedure for solving the population equations for the CO_2 $(0,0^0,1)$ and $(0,v_2,0)$ states. He also solved the *time-dependent* statistical equation for a number of CO_2 levels, including collisional and radiative transfer processes. The intention was to analyse the CO_2 4.3 μm emissions under auroral conditions. In this problem, the continuity equations for a dozen states have to be solved, including not just radiative transfer but also time-evolution, because of the important temporal evolution of the auroral processes. However, it was necessary to incorporate some approximations either by treating radiative transfer with the 'cooling-to-space' approach (c.f. Sec. 5.2.1.2), or by an approximate treatment of the V–V collisional processes between the $CO_2(v_3)$ and $N_2(1)$ levels.

In 1991, Kutepov and his co-workers adapted the accelerated lambda iteration (ALI) technique, an iterative technique commonly used in solving multilevel non-LTE line formation problems in stellar atmospheres, to the non-LTE problem of molecular infrared bands in planetary atmospheres. Because convergence can be very slow, particularly for optically thick conditions, some strategies were devised to accelerate it by avoiding the full calculation of radiative transfer in each iteration. The lambda operator, Λ, is defined so that when applied to the source function, J_ν, we obtain the radiance, L_ν, given by the formal solution of the radiative transfer equation, e.g. Eq. (4.9) or (4.10),

$$L_\nu(z,\mu) = \Lambda_{\nu,\mu}[J_\nu], \qquad (5.31)$$

where the brackets mean 'applied to'. The ALI methods are based on splitting the lambda operator, i.e.

$$\Lambda_{\nu,\mu} = \Lambda^\star_{\nu,\mu} + (\Lambda_{\nu,\mu} - \Lambda^\star_{\nu,\mu}),$$

where the approximate lambda operator $\Lambda^\star_{\nu,\mu}$ is chosen so it can be quickly calculated while retaining most of the radiative exchange. With this approach, the iterative scheme is given by

$$L_\nu^i(z,\mu) = \Lambda^\star_{\nu,\mu}[J_\nu^i] + (\Lambda_{\nu,\mu} - \Lambda^\star_{\nu,\mu})[J_\nu^{i-1}], \qquad (5.32)$$

where J_ν^{i-1} is the source function from the previous iteration.

The various forms of the ALI procedure differ, essentially, by: (i) the choice of the approximate lambda operator $\Lambda^\star_{\nu,\mu}$ (which defines the radiative transfer quantities to be calculated at each step); and (ii) the way in which the level populations are calculated from the system of equations, e.g., the way by which the equations are linearized in terms of the populations using those from previous iterations. The approximate lambda operator used by Kutepov is just the diagonal elements of the full operator. That is, in each step only the *local* radiative transfer is calculated, and the exchange of radiation between different layers is excluded from the iteration process, except for the first step. We should note the similarity of this method to the iteration technique of Wintersteiner and his co-workers described below, in which the radiative transfer terms are calculated in the first step and then re-calculated only in a few later iterations. This version of ALI keeps the linearity of the statistical equilibrium and radiative transfer equations with respect to the population number densities in each step by introducing the densities from the previous calculation. Thus it exploits the advantages of an iterative method, as discussed above.

Another radiative transfer method commonly used in astrophysical problems is the 'diffusion approximation'. This has been applied to the CO_2 infrared emissions from the Venusian and Martian atmospheres, but is valid only for conditions close to LTE. It is useful, for example, for calculating the altitude where departure from LTE starts being significant. Another astrophysical formulation, the Feautrier method, was applied by Rodrigo and colleagues for calculating the upward and downward intensities of CO_2 at 15 and $4.3\,\mu m$ in the upper atmospheres of Mars and Venus for day and night-time conditions.

Kutepov and colleagues applied the Rybicki formulation, another well-known astrophysical technique[‡], to the study of the rotational non-LTE population of the $CO_2(0,0^0,1)$ vibrational level in an isothermal atmosphere of pure CO_2, resembling Mars and Venus. They later applied the method to study the non-LTE in the rotational and vibrational levels of $CO(1)$ and in the vibrational levels of CO_2. Their approach is particularly useful when studying the rotational populations in situations where the radiative transfer between layers is important. It consists essentially of the discretization of the equation of transfer, after including the statistical equation, with respect to three variables, frequency, solid angle and a vertical coordinate. There is a trade-off between the number of points in solid angle, in frequency, and in altitude, which are set up depending on the problem to be solved. For example, a finer mesh in solid angle is usually taken, at the expense of the frequency and altitude points, for problems involving scattering of solar radiation, while the opposite may apply when calculating spectra, i.e. a coarser mesh for solid angles and a finer grid for frequency.

Mlynczak and Drayson studied the infrared emissions from O_3, tackling the difficulty that the rate of excitation of the upper states is not proportional to the number density of O_3 in the ground vibrational state (see Sec. 3.6.5.2). They found that including only the absorption of an upwelling radiative flux (as described in Sec. 5.2.2.2), is sufficiently accurate for this case. Thus the equations for the non-

---

[‡]The Rybicki formulation is a particular version of the more general Feautrier method. See, e.g., Mihalas (1978).

LTE populations of the vibrational states included local processes only, making the solution much simpler.

A good representative example of the *radiance* formulations is the non-LTE radiative transfer algorithm developed at the Air Force Research Laboratory by Wintersteiner and colleagues. This model uses a line-by-line calculation of the radiative absorption rate which allows inclusion of the full altitude dependence of the absorption and emission line shapes, and hence calculates very accurately the exchange of photons among layers. The coupled integral equation for the source function (Eq. 3.59) and the expression for the mean integrated radiance (Eq. 3.56) are solved iteratively, starting with estimated populations for the radiating states, which give the emission rates everywhere in the atmosphere. It then calculates the absorption in each band at discrete altitudes from these emissions and, using this, recalculates the populations from the statistical equilibrium equation. Using a rapid nested iteration scheme, this continues until convergence criteria are satisfied. It should be noted that although the method is formulated for full iteration (like the lambda iteration method), in practice, the radiative transfer terms are not calculated in each iteration, but only in the first step and possibly one or two times more. In its practical application this method is very similar to a Curtis matrix method which solves the equations by iteration, instead of the usual inversion, and recalculates the Curtis matrix a few times at a later stage.

The fundamental quantity calculated in this method is the radiance, from which the populations of vibrational levels and the cooling rates of the bands that originate from them are ultimately calculated. As a result, the heating rate is a derived quantity. However, it is not obtained by introducing the calculated radiance in the expressions for the net, upward, and downward fluxes (Eqs. 4.4, 4.64, and 4.65, respectively), and then calculating its derivative over z. Instead, it is computed from the frequency-integrated radiative flux (Eq. 3.60). As discussed above, the calculation of the heating rate in the former way might introduce round-off errors twice: first in the difference between the upward and downward fluxes, and then in the derivative of the net flux. The calculation of the heating rate from Eq. (3.60) with $\bar{L}_{\Delta\nu}$ obtained from Eq. (3.56) only involves these round-off errors once; as in the escape probability function formulation.

This review of the most common methods helps us to realise that the iteration and inversion techniques have features in common, and that in practice, there are no 'pure' iteration or inversion methods but only mixtures of both, some putting more emphasis on iteration and others on inversion.

5.5   Intercomparison of Non-LTE Codes

Compared to radiative transfer codes for LTE, relatively few comparisons have been carried out between the different formulations of the non-LTE problem. A recent study for CO_2 in the terrestrial atmosphere found good agreement for vibrational temperatures and cooling rates under a range of conditions, with differences that are generally smaller than $2\,\mathrm{K}$ and $0.5\,\mathrm{K/day}$ respectively. These deviations are small, given the large differences between the methods: a line-by-line approach versus a code that calculates transmissions from analytical expressions for the equivalent width of histogrammed lines of the bands; an iterative algorithm versus matrix inversion; and over an extreme range of conditions of optical thickness and collisional parameters. In particular, the differences are much smaller than those introduced by the current uncertainties in the parameters used in the non-LTE models, such as the collisional rate constants or the different schemes for collisional coupling between the vibrational levels. Some of the differences found could be caused by model parameters such as the collisional rate constants and line spectral data, and not necessarily by the radiative formulations themselves. These comparisons tend to confirm that the use of approximate methods gives useful accuracy in many cases for which the use of sophisticated methods, in particular line-by-line integration, may not be justified, especially if there are significant uncertainties in the rate coefficients. We have, however, some clear exceptions as the CO(1) level discussed in Sec. 7.2, whose population is strongly dominated by radiative processes in the mesosphere; and the CO_2 populations in the atmospheres of Mars and Venus.

5.6   Parameterizations of the Non-LTE Cooling Rate

Models of the dynamics of the upper atmosphere need to include some treatment of radiative heating and cooling. This requirement has driven the development of fast and accurate non-LTE radiative transfer algorithms which do not place too heavy an additional computing requirement on codes which are generally already massive. The earlier parameterizations were extensions of those developed to study the effects of non-LTE on level populations and vibrational temperatures. Dickinson suggested in 1973 a parameterization based on an exact non-LTE calculation for a reference temperature profile, $h(T_0)$ plus an additional term accounting for the deviations from that profile given by the Newtonian cooling approach, e.g.,

$$h(z) = h(T_0) + \frac{h(T_0 + \delta T) - h(T_0)}{\delta T}\,(T - T_0), \qquad (5.33)$$

where $h(z)$ is the cooling rate at altitude z with temperature $T(z)$, and δT a small ($\sim 1\,\mathrm{K}$) temperature perturbation. This approach did not give accurate results. First, Newtonian cooling is expected to give a reasonable estimate of the 'cooling to space' term (see Sec. 4.2.5), since it depends only on the temperature of the layer considered, but not of the 'exchange between layers' term, which depends on the temperature of other layers. Also, if the contribution of hot bands is significant, as it is for CO_2 in the atmospheres of the terrestrial planets, their strong temperature dependence is not well represented by the linear Newtonian approach. Even when the starting heating rate is obtained from a non-LTE calculation, further errors are introduced because the cooling is not directly proportional to kinetic temperature, but to the vibrational excitation temperature of the emitting levels.

The 'cooling-to-space' approximation described in Sec. 5.2.1.2 has been widely used in many problems of radiative cooling in the atmospheres of the terrestrial planets. It has the advantage of being simple and fast to compute, including only a local term. It does not however give accurate results for very optically thick bands, and at places where vibrational-vibrational processes are involved, e.g., as in the case of the Martian and Venusian atmospheres.

Most of the parameterizations that have been developed use the Curtis matrix formulation, applied to a two-level transition where the Curtis matrix elements or, more precisely, the transmissions that go into the matrix elements, are parameterized. These can be obtained by interpolation from precomputed tabulated data, or calculated using suitable approximations. Generally, not all of the other layers contribute significantly to the heating rate of a given layer, so some parameterizations try to optimize the number of layers (or Curtis matrix elements) taken into account at different altitudes. An inverse Curtis matrix method has been also used in which the matrix elements are calculated from correlated-k coefficients derived from random band models.

The parameterization of Apruzese and his co-workers is quite different from the matrix methods. It uses the two-level formalism for the non-LTE source function and the two-stream solution for the radiative transfer equation, solved by an iterative procedure with an optimized but fixed absorption coefficient.

Kutepov developed a parameterization very similar to the second-order escape probability approximation, which essentially replaces the integral equation for the source function (Eq. 3.59), by an approximate first order linear differential equation relating the first derivative of the mean radiance, $d\bar{L}_{\Delta\nu}/d\tau$, to those for the source function and the 'escape-to-space' function (see Eq. 5.7), $dJ/d\tau$ and $dT^{\star}/d\tau$.

In the search for accurate and efficient schemes, practical questions of how to parameterize the heating rates have to be considered. For example, the altitude range and the altitude grid spacing, plus the number of parameters to be included at each level, can be varied. Within the latter, pressure and temperature are obviously essential, along with the species abundances (if the atmosphere is not completely mixed) and, for non-LTE, the collisional rate coefficients. Additional issues which have to be considered include how to introduce 'switching' from the LTE to the non-LTE regimes, where different approaches are normally used. In some cases, with more than one band contributing to the cooling, this involves accounting for the different source functions of the various bands. The 'References and Further Reading' section includes various publications which detail how different schemes have dealt with these and other practical issues.

5.7 The Curtis Matrix Method

We have already expressed the source function or, equivalently, the population of an upper state, in terms of the Planck function, the heating rate, and various coefficients which are generally found in radiative transfer problems. The radiative transfer equation completes the system. Now we derive one of the most common practical methods, the Curtis matrix, that solves the radiative transfer problem by expressing the heating rate at a given altitude in terms of the source function for other atmospheric layers and the radiance at the boundaries of the atmospheric region under consideration. We begin with the method for the two-level system, and extend it to the multilevel case later.

We first derive the expression for the net upward flux from the difference between the upward and downward fluxes for a stratified atmosphere, starting with those at level z derived from the formal solution in Sec. 4.2.5, (Eqs. 4.64 and 4.65). In those equations the upward and downward fluxes are expressed in terms of the source function, the flux transmissions, and the boundaries conditions, but the radiance no longer appears. The next step is to calculate the heating rate from these fluxes. For convenience, the transmission derivatives are first converted into derivatives of the source function by integrating by parts. From Eqs. (4.64) and (4.65), we have

$$\frac{F^{\uparrow}(z)}{\pi \Delta \nu} = J(z) + \left[\bar{L}^{\uparrow}_{\Delta \nu}(z_0) - J(z_0)\right]\mathcal{T}_F(z_0, z) - \int_{z_0}^{z} \mathcal{T}_F(z', z)\,\frac{dJ(z')}{dz'}\,dz', \quad (5.34)$$

and

$$\frac{F^{\downarrow}(z)}{\pi \Delta \nu} = J(z) + \left[\bar{L}^{\downarrow}_{\Delta \nu}(\infty) - J(\infty)\right]\mathcal{T}_F(z, \infty) + \int_{z}^{\infty} \mathcal{T}_F(z, z')\,\frac{dJ(z')}{dz'}\,dz'. \quad (5.35)$$

Next, some practical considerations about the conditions at the boundaries. The lower boundary is taken at the highest level below which the atmosphere is effectively opaque, and assumed to be black, so the radiance at the bottom of the atmosphere is the Planck function at the temperature T_g of that surface, $B(T_g)$. Since most atmospheric infrared bands are in LTE in the troposphere, their source functions at the lower boundary $J(z_0)$ can be replaced by $B[T(z_0)]$, where $T(z_0)$ is the air temperature at the lower boundary and may differ somewhat from T_g, especially if T_g is actually the temperature of the ground. Then Eq. (5.34) can be re-written as

$$\frac{F^{\uparrow}(z)}{\pi \Delta \nu} = J(z) + \left[B(T_g) - B[T(z_0)]\right]\mathcal{T}_F(z_0, z) - \int_{z_0}^{z} \mathcal{T}_F(z', z)\,\frac{dJ(z')}{dz'}\,dz'. \quad (5.36)$$

If T_g and $T(z_0)$ are equal, the second term on the right hand side vanishes.

The reason for sometimes taking the lower boundary to be some level above the surface becomes clear when we consider that the solution of the equation of transfer requires us to invert matrices of the order of $(N \times N)$ where N is the number of atmospheric layers. Raising the lower boundary reduces N and involves little loss of accuracy if the lower atmosphere is not participating in the solution at levels of importance for non-LTE because it is opaque.

We might wish to reduce the number of atmospheric layers even for bands which do not completely fulfil the previous assumption. If we know the gradient of the source function below z_0, we can use the previous solution and add a correction term of the form

$$\left[\frac{\mathrm{d}J(z')}{\mathrm{d}z'}\right]_{z'<z_0} \int_0^{z_0} \mathcal{T}_F(z',z)\,\mathrm{d}z'.$$

Note that Eq. (5.36) is also valid if the source function below z_0 does not change with altitude.

Considering now the upper boundary, we normally do not know the source function at the top of the atmosphere $J(\infty)$ beforehand, and it is often not practical to extend the integration up to very high altitudes where it vanishes. However, using the term

$$C_\downarrow = J(z_t)\mathcal{T}_F(z,\infty) - J(\infty)\mathcal{T}_F(z,\infty) + \int_{z_t}^\infty \mathcal{T}_F(z,z')\frac{\mathrm{d}J(z')}{\mathrm{d}z'}\,\mathrm{d}z' =$$

$$\int_{z_t}^\infty \left[\mathcal{T}_F(z,z') - \mathcal{T}_F(z,\infty)\right]\frac{\mathrm{d}J(z')}{\mathrm{d}z'}\,\mathrm{d}z', \tag{5.37}$$

we can rearrange equation (5.35) to obtain

$$\frac{F^\downarrow(z)}{\pi\Delta\nu} = J(z) + \left[L_\odot(\infty) - J(z_t)\right]\mathcal{T}_F(z,\infty) + \int_z^{z_t} \mathcal{T}_F(z,z')\frac{\mathrm{d}J(z')}{\mathrm{d}z'}\,\mathrm{d}z' + C_\downarrow, \tag{5.38}$$

where we have substituted the unknown $J(\infty)$ by $J(z_t)$ plus the derivative of $J(z)$ above z_t, and have introduced the correction term $C_\downarrow$. This is useful because we usually have a better knowledge about the derivative of $J(z)$ than about $J(\infty)$ itself. Actually, the source function usually departs strongly from the Planck function in the thermosphere, and for most bands is nearly constant or with only a small gradient above about 120 or 140 km. Thus, for many transitions the derivative $\mathrm{d}J(z')/\mathrm{d}z'$ is very small and the correction term $C_\downarrow$ can be neglected. Another good approach may be to evaluate the term $C_\downarrow$ by assuming $\mathrm{d}J(z')/\mathrm{d}z'$ is constant above z_t with the value at z_t.

The solar radiation term is included as a general case, but of course it vanishes at night, and for those atmospheric transitions that lie at wavelengths larger than about $5\,\mu$m.

Now, a numerical integration is performed which, as originally suggested by Curtis, allows us to treat the transmissions and source functions separately. The method is based on the replacement of the integration over height by a summation over the atmospheric layers, assumed equally spaced. Dividing the atmosphere from z_1 to z_t into N-1 layers equally spaced in geometric height by Δz, (Fig. 5.6), any height z is given by $z = z_1 + (i-1)\Delta z$ and $z' = z_1 + (j-1+\alpha)\Delta z$, with $0 \le \alpha \le 1$. Note that the j-th level in Fig. 5.6 represents the nearest level below

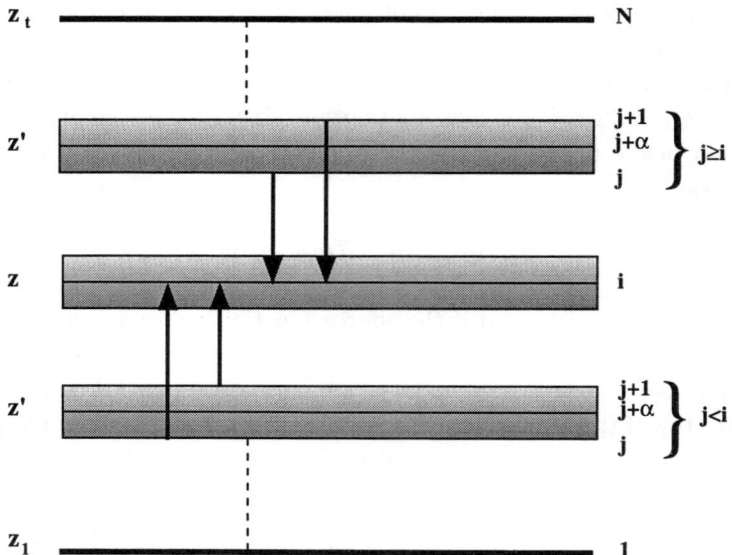

Fig. 5.6   Separation of the atmosphere into equally-spaced layers for the Curtis matrix method.

the variable height of integration z', for both the upward flux equation $(z' < z)$ and for the downward flux equation $(z' > z)$. Replacing the integral over altitude by summations over all levels, the upward flux at the i-th level, i.e., at altitude z_i, can be written as

$$\frac{F_i^\uparrow}{\pi \Delta \nu} = J(z_i) - \sum_{j=1}^{i-1} \int_0^1 \mathcal{T}_F(z_j + \alpha \Delta z, z_i) \frac{\mathrm{d}J(z_j + \alpha \Delta z)}{\mathrm{d}\alpha} \, \mathrm{d}\alpha. \qquad (5.39)$$

If the spacing is sufficiently narrow, we can assume that the source function varies linearly between two adjacent layers, so $\mathrm{d}J/\mathrm{d}\alpha = J(z_{j+1}) - J(z_j)$ (represented in short by $J_{j+1} - J_j$) can be assumed as constant within the limit of the integral over α. Then, by defining

$$\Gamma_{i,j} = \int_0^1 \mathcal{T}_F(z_j + \alpha \Delta z, z_i) \, \mathrm{d}\alpha \qquad \text{for} \qquad i > j, \qquad (5.40)$$

the equation for the upward flux becomes

$$\frac{F_i^\uparrow}{\pi \Delta \nu} = J_i - \sum_{j=1}^{i-1} \Gamma_{i,j} \, (J_{j+1} - J_j),$$

or, by re-arranging the summation, by

$$\frac{F_i^\uparrow}{\pi \Delta \nu} = \Gamma_{i,1} J_1 - \sum_{j=2}^{i-1} (\Gamma_{i,j-1} - \Gamma_{i,j}) J_j + (1 - \Gamma_{i,i-1}) J_i. \qquad (5.41)$$

With respect to the downward flux, neglecting the correction term $C^{\downarrow}$, we have

$$\frac{F_i^{\downarrow}}{\pi\Delta\nu} = (1 - \Gamma_{i,i})J_i + \sum_{j=i+1}^{N-1} \left[\Gamma_{i,j-1} - \Gamma_{i,j}\right]J_j + \left[\Gamma_{i,N-1} - \mathcal{T}_{F,\infty,i}\right]J_N + L_{\odot}(\infty)\mathcal{T}_{F,\infty,i},$$

$$(5.42)$$

where $\mathcal{T}_{F,\infty,i}$ is the flux transmission from layer i to the top of the atmosphere, and $\Gamma_{i,j}$ is now defined as

$$\Gamma_{i,j} = \int_0^1 \mathcal{T}_F(z_i, z_j + \alpha\Delta z)\,\mathrm{d}\alpha \qquad \text{for} \qquad j \geq i. \qquad (5.43)$$

The heating rate at layer i expressed in discrete form becomes, from (Eq. 4.58),

$$h_i = -\frac{(F_{i+1} - F_{i-1})}{2\Delta z} = -\frac{(F_{i+1}^{\uparrow} - F_{i+1}^{\downarrow} - F_{i-1}^{\uparrow} + F_{i-1}^{\downarrow})}{2\Delta z}, \qquad (5.44)$$

where i ranges from 2 to N$-$1.

Including Eqs. (5.41) and (5.42) into (5.44), the heating rate can be written as

$$h_i = \sum_j \mathcal{C}_{i,j}J_j + h_{l,i} + h_{u,i} \qquad \text{for} \quad 2 \leq i \leq N - 1, \qquad (5.45)$$

where the elements $\mathcal{C}_{i,j}$ forming the Curtis matrix depend only on the transmission properties of the atmosphere and are given by

$$\mathcal{C}_{i,j} = \frac{\pi\Delta\nu}{2\Delta z}\left[\Gamma_{i+1,j-1} - \Gamma_{i+1,j} - \Gamma_{i-1,j-1} + \Gamma_{i-1,j}\right] \text{ for } 2 \leq i,j \leq N-1. \quad (5.46)$$

From this equation the meaning of the Curtis matrix elements can be seen. They give directly the heating rate at a layer z per unit of source function change at layer z'. This method is conceptually similar to the exchange integral formulation described in Sec. 4.2.5. The diagonal elements of the Curtis matrix give us the 'cooling-to-space' term and the elements on the right- and left-hand sides of the diagonal represent the net exchange with the overlying and underlying layers, respectively.

The $h_{l,i}$ and $h_{u,i}$ terms are the contributions to the heating rate at layer i of the intensities incident at the lower and upper boundaries of the region, respectively, and are given, for $(2 \leq i \leq N - 1)$, by

$$h_{l,i} = \frac{\pi\Delta\nu}{2\Delta z}\left[\Gamma_{i-1,1} - \Gamma_{i+1,1}\right]J_1, \qquad (5.47)$$

and

$$h_{u,i} = \frac{\pi\Delta\nu}{2\Delta z}\left[\left[\Gamma_{i+1,N-1} - \Gamma_{i-1,N-1}\right]J_N + \left[\mathcal{T}_F(z_{i+1},\infty) - \mathcal{T}_F(z_{i-1},\infty)\right]\left[L_{\odot}(\infty) - J_N\right]\right]. \qquad (5.48)$$

With this method, the heating rate, h, and the source function, J, are calculated for heights ranging from 2 to N$-$1. Note that J_1 appears in the expression for h_l and J_N in that for h_u. J_1 is usually replaced by the Planck function at the lower boundary, and J_N can be replaced by J_{N-1} if we assume that the gradient of the source function at the top is negligible, or by $2J_{N-1} - J_{N-2}$ if the gradient of J is considered to have a non-zero but constant value. The heating rates at the bottom (z_1) and top (z_t) are usually calculated as extrapolations from their values in the adjacent layers once the system has been solved.

Williams (1971) assumed that the heating rates at the boundaries are linear extrapolations from their respective adjacent layers, e.g. $h_1 = \frac{1}{2}F_3 - 2F_2 + \frac{3}{2}F_1$ and $h_N = -\frac{3}{2}F_N + 2F_{N-1} - \frac{1}{2}F_{N-2}$, but before solving the system and without considering the boundary contributions for h_l and h_u (except the solar radiation). Hence, he considered a Curtis matrix of dimension N×N instead of (N−2)×(N−2). This generally works well when calculating heating rates but, because the Curtis matrix is then linearly dependent (the first and last rows are combinations of the adjacent rows), it does not have a unique inverse. Since this method calculates the source function by inverting the Curtis matrix, the solution will be unstable and generally gives a rapid-oscillating value for the source function (Murphy, 1985). The procedure described here avoids this problem by including the conditions at the lower and upper limits by the relations given for h_l and h_u, and using a Curtis matrix with dimensions of (N−2)×(N−2).

The major dependence of the heating rate on temperature comes principally from the source function since the transmissions involved in the Curtis matrix are much less dependent on temperature. This allows heating rates for different temperature profiles to be calculated quickly once a Curtis matrix has been calculated for a standard profile. If more accuracy is needed, a set of Curtis matrices can be pre-computed. Overall, this method provides a flexible procedure to compute heating rates efficiently and accurately.

The most important atmospheric heating and cooling rates are produced by the bands of CO_2, O_3 and H_2O which are not enormously variable in the middle and upper atmosphere. Some other minor species, e.g. NO, have seasonal and latitudinal variations of orders of magnitude but the infrared bands are so optically thin that the computation of their cooling rates do not require a full radiative transfer calculation and they can be accurately obtained by using the 'cooling-to-space' or total-escape approaches (see Secs. 5.2.1.1 and 5.2.1.2). Of course, the Curtis matrix method can be used to calculate heating rates under LTE conditions as well, and in fact it was originally derived for that purpose.

In matrix form we can write the expression (5.45) for the heating rate as

$$\mathbf{h} = \mathcal{C}\,\mathbf{J} + \mathbf{h_b}, \qquad (5.49)$$

where $\mathcal{C}$ is the Curtis matrix, a square matrix of order (N−2)×(N−2), and $\mathbf{h}$, $\mathbf{J}$, and $\mathbf{h_b} = \mathbf{h_l} + \mathbf{h_u}$ are vectors with components ranging from 2 to N−1. This expression relates the heating rates and the source functions in the atmospheric layers considered.

Expressing the equation for the source function (Eq. 3.61 or 3.71) in matrix form we have

$$\mathbf{J} = \mathbf{B}' + \mathcal{D}\,\mathbf{h}, \tag{5.50}$$

where $\mathcal{D}$ is a diagonal matrix, also of order $(N-2) \times (N-2)$, with the values of D at each altitude (see Eq. 3.63 or 3.74), and $\mathbf{B}'$, is a vector with components ranging from 2 to $N-1$. Thus we have a closed system of linear equations (5.50 and 5.49) for the unknowns $\mathbf{h}$ and $\mathbf{J}$. From the solution of these equations we have

$$\mathbf{h} = (\mathcal{I} - \mathcal{C}\mathcal{D})^{-1} [\mathcal{C}\,\mathbf{B}' + \mathbf{h_b}], \tag{5.51}$$

or

$$\mathbf{J} = (\mathcal{I} - \mathcal{D}\mathcal{C})^{-1} [\mathbf{B}' + \mathcal{D}\,\mathbf{h_b}], \tag{5.52}$$

where $\mathcal{I}$ is the unit matrix.

Finally, we consider the solution for the multilevel case where we have M vibrational levels among which N vibrational transitions take place. Let us consider first, for an easier description, the simplified case where only one radiative transition takes place from each of the M vibrational levels. In that case the number of transitions N is equal to number of levels M. Then we have one radiative transfer equation for each of the vibrational bands which, for the n-th band, following Eq. (5.49), has the form

$$\mathbf{h_n} = \mathcal{C}_n\,\mathbf{J_n} + \mathbf{h_{b,n}}. \tag{5.53}$$

Similarly, we have a statistical equilibrium equation for each of the source functions which has the n-th vibrational level as its upper state. Using Eq. (3.85), they can be expressed by

$$\mathbf{J_n} = \mathbf{B}'_n + \mathcal{D}_n\,\mathbf{h_n} + \sum_{k=1(\neq n)}^{N} \mathcal{A}_{n,k}\,\mathbf{J_k}. \tag{5.54}$$

In this case we have a closed system with N radiative transfer equations and N statistical equilibrium equations from which the N heating rates and N source functions can be obtained.

We consider now the case when more than one radiative transition can arise from a given energy level, so the number of heating rates and source functions (N) is larger than the number of vibrational levels (M). Eq. (5.53) for the heating rate is not affected, except that we now have more equations of this type, N ($>$M). Let us consider for each of the M vibrational levels a source function, $\mathbf{J_m}$. Since we can have more that one transition arising from a given vibrational level, we have several source functions to identify it. Usually that of the stronger transition originating from the considered level is taken. The statistical equilibrium equations (5.54) for the M vibrational levels have now the form

$$\mathbf{J_m} = \mathbf{B'_m} + \mathcal{D}_m\,\mathbf{h_m} + \sum_{k=1(\neq m)}^{N} \mathcal{A}_{m,k}\,\mathbf{J_k} + \sum_{l=1(\neq m)}^{N} \mathcal{D}_{m,l}\,\mathbf{h_l}, \qquad (5.55)$$

where the last term accounts for the additional radiative transitions, and we have replaced index n by m to denote that the number of these equations is M (<N).

In this situation, the number of source functions ($\mathbf{J_n}$) in Eq. (5.53), N, is larger than the M equations (5.54) we have for the source functions, $\mathbf{J_m}$. To close the system we need N−M equations. These are given by the relationships between the source functions which have a common upper level. That is, assuming that transition m takes place between levels k–j, transition n between k–i, and transition p between j–i (see Fig. 5.7), then, J_n, J_m, and J_p are related by

$$\frac{J_n}{J_m} = \frac{\nu_{0,n}^3}{\nu_{0,m}^3}\frac{c^2}{2h\nu_{0,p}^3}\frac{\gamma_m\gamma_p}{\gamma_n} J_p,$$

with the γ (induced emission) factors defined by

$$\gamma_p \equiv \gamma_{ji} = 1 - \frac{n_j g_i}{n_i g_j}.$$

The ratio of the γ-factors can be made equal to unity with no significant loss of accuracy.

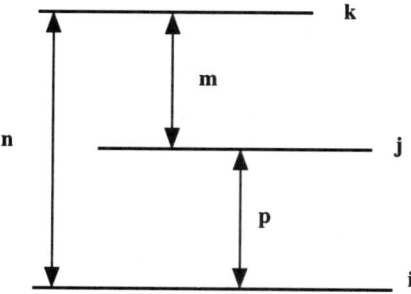

Fig. 5.7   Transitions arising from vibrational levels.

Although the solution of this system of equations seems very difficult, in practice it often can be greatly simplified, since not all energy levels involved are strongly coupled. Usually they are coupled on a one-to-one or at most a one-to-two basis. Sometimes it is possible to isolate subsets of energy levels which are strongly coupled among themselves and only weakly coupled to others. An example of such a subset would be the transitions which occur between the different energy levels within a particular vibrational mode. Then, the system for those can be solved with a few iterations and it is easier to solve the overall system because the number of equations is reduced.

When some levels are very close in energy, we can assume that they are in LTE among themselves, and the system of radiative and statistical equations reduced to just one of each. The savings can be applied, if desired, to incorporate a larger number of equations, for example those including higher energy levels. We shall see in Chapter 7 that in the case of O_3, for example, excited levels with vibrational quantum numbers up to 10 or higher play a role in the atmosphere. In other cases, such as CO_2 or H_2O, it is seldom necessary to include levels with v larger than 3 or 4.

5.8 References and Further Reading

The *flux* formulations have been used by Curtis (1956), Kuhn and London (1969), Williams (1971), Shved (1975), Houghton (1986), Kutepov and Shved (1978), Shved *et al.* (1978), Wehrbein and Leovy (1982), Murphy (1985), Solomon *et al.* (1986), López-Puertas *et al.* (1986a,b), and Andrews *et al.* (1987). A good description of the Curtis matrix method is given by Williams (1971). The *radiance* formulation was used by Kumer and James (1974) and Kumer (1977a,b); and is clearly described by Wintersteiner *et al.* (1992). A detailed description of the *escape function* method can be found in Dickinson (1972). This was used later by Dickinson (1976, 1984). A comparison of the cooling rates computed with a Curtis matrix *flux* method (López-Puertas *et al.*, 1986a,b; 1998a) and a *radiance* method (Wintersteiner *et al.*, 1992) have shown that the differences between both methods are very small (López-Puertas *et al.*, 1994).

For a detailed description of the resonant scattering theory for UV radiation see, e.g. Holstein (1947), Thomas (1963), Strickland and Donahue (1970), Finn (1971), and Strickland and Anderson (1977).

Radiative transfer was formulated in terms of the 'band transport' functions by Kumer and James (1974) for calculating the vibrational temperature of the $CO_2(0,0^0,1)$ state coupled with $N_2(1)$. Kumer (1977a) studied the joint solution of the CO_2 4.3 and $15\,\mu m$ vibrational levels in the daytime atmosphere for the first time. Kumer (1975, 1977b), in the first and so far the only reported study, solved the *time-dependent* statistical equation for a number of CO_2 levels, including collisional and radiative transfer processes. Kumer's method for studying the CO_2 infrared emissions in auroral conditions has been used by Winick *et al.* (1988).

For descriptions of the application of the accelerated lambda iteration (ALI) technique to solving multilevel non-LTE line formation problems in stellar atmospheres see Kudritzki and Hummer (1990) and Rybicki and Hummer (1991). The application of the ALI to solving multilevel vibration-rotation non-LTE problems for planetary atmospheres can be found in Kutepov *et al.* (1998). They also compared this technique with the matrix method.

The application of the Rybicki formulation to study the rotational non-LTE population of $CO_2(0,0^0,1)$ in an isothermal pure CO_2 atmosphere was carried out by Kutepov *et al.* (1985). More recently, the Rybicki formulation together with the ALI technique was applied by Kutepov *et al.* (1997) to the vibration-rotation non-LTE problem of CO(1) and by Shved *et al.* (1998) and Ogivalov *et al.* (1998) to the non-LTE vibrational populations of CO_2.

A description of the 'diffusion approximation' can be found in Mihalas (1978), and its application to the CO_2 infrared emissions in the Venusian, and Martian atmospheres in Battaner *et al.* (1982) and Rodrigo *et al.* (1982).

Among the parameterizations of the non-LTE cooling rates are those reported by Wehrbein and Leovy (1982), Fels and Schwarzkopf (1981) and Haus (1986), which use the Curtis matrix formulation with parameterizations of the Curtis matrix elements. Apruzese *et al.* (1984) used the two-level formalism for the non-LTE source function and the two-stream solution for the radiative transfer equation. Zhu (1990) and Zhu *et al.* (1992) proposed a parameterization of the CO_2 15 μm cooling based on the Curtis matrix interpolation, where the Curtis matrix elements were calculated from correlated-k coefficients for random band models.

Kutepov (1978) and Kutepov and Fomichev (1993) have developed a parameterization scheme based on a method very similar to the second-order escape probability approximation described in e.g., Frisch and Frisch (1975). In this method, the heating rate is calculated using a recurrence formula, which was applied later by Fomichev *et al.* (1993, 1998) for the computing the cooling by the CO_2 15 μm bands. The most recent work by Fomichev *et al.* (1998) presents a comprehensive review of the different schemes developed for LTE and non-LTE in the atmosphere.

Parameterizations of the CO_2 15 μm cooling rates based on the 'cooling-to-space' approach have been used by Gordiets *et al.* (1982) in the Earth's upper atmosphere and by Bougher *et al.* (1986) in the upper atmospheres of Mars and Venus.

See also the references on non-LTE radiative transfer, including the Curtis matrix method, in the 'References and Further Reading' sections of Chapters 6 and 7.

Chapter 6

Non-LTE Modelling of the Earth's Atmosphere I: CO_2

6.1 Introduction

We described in Chapters 3 and 5 the formulation of radiative transfer theory for non-LTE situations, and outlined the physics which is involved in the most important non-LTE processes. In this chapter we progress to the problem of incorporating all of this theory into a practical numerical model of the atmosphere which can then be used to calculate the populations of various important molecular energy levels and infrared absorption and emission rates of bands originating between them. We recap first the most important difficulties and provide some general guidelines for the development of non-LTE population models. We then proceed in this and the following chapter with the description of models for the most important atmospheric molecules. Of those, we pay most attention to carbon dioxide, CO_2, since this is the most important species for radiative transfer in the atmosphere of the Earth and the other terrestrial planets, Mars and Venus. The non-LTE model for CO_2 in the Earth's atmosphere is described in this chapter and for the atmospheres of Mars and Venus in Chapter 10. Non-LTE models for the population of the energy levels of other minor atmospheric constituents are covered in Chapter 7, and the experimental evidence for non-LTE in various atmospheric molecules is described in Chapter 8.

The major difficulties in formulating the non-LTE problem were described in Sec. 5.3. Usually, the population of the excited states depends on the radiation field which depends, in turn, on the number density of the excited states, on the radiative emission from the other atmospheric layers and, during the day, on the solar illumination conditions. The solution for the population of the excited states, or for the radiative heating or cooling produced by the bands which originate from them, requires a completely self-consistent simultaneous solution of a non-linear system of both the radiative transfer and statistical equilibrium equations.

Finding the solution is particularly complicated if:

(i) Radiative transfer between layers within the atmosphere makes an important contribution;

(ii) The absorption coefficient of the atmospheric bands has a very rapid frequency variation, and includes many lines of variable width and strength;

(iii) Several different non-thermal processes affect the population of the molecular vibrational levels, and have to be considered simultaneously; or

(iv) The strong collisional coupling between vibrational levels requires us to solve the problem for a relatively large number of energy levels all at once.

In the general situation, all of the above apply.

We are usually most interested in the non-LTE populations of those states which either give rise to an emission band which is measured by sensors to gain atmospheric information, or those which produce important contributions to the heating or cooling of the atmosphere. In either case we need information about the strengths, widths and ground-state energies of the spectral lines corresponding to the transitions involved.

There are several spectroscopic databases covering the infrared part of the atmospheric spectrum, of which the best known are HITRAN, GEISA, and ATMOS. These contain information about most of the important atmospheric bands. GEISA was compiled in France specifically for planetary atmospheres, and includes some bands of species like ammonia which are only important in the atmospheres of the giant planets. None of these compilations cover all of the hot bands originating from very highly energetic vibrational levels of some species, for example the ν_1, ν_2, ν_3, and $\nu_1+\nu_3$ hot bands of O_3, and the ν_3 hot bands of NO_2. For cases like these, the data required have to be specifically measured in suitably-equipped spectroscopy labs, or estimated theoretically, perhaps by analogy with other hot bands for which data do exist. Very accurate line data is not always essential for the calculation of non-LTE populations, unless the bands are very strong. In particular, the hot bands of O_3 and NO_2 mentioned above are not known with high accuracy, but this introduces only a small uncertainty to the non-LTE populations. High accuracy is needed, however, to compute the spectral radiances they emit, for comparison with measurements, especially if the latter are at high resolution.

Most calculations also require the number density of the lower state as an input. For transitions from the ground state, this is simply equivalent to knowing the atmospheric composition. For some molecules, e.g. O_3 and NO_2, which are vibrationally excited when formed in chemical reactions, the initial concentrations have to be calculated. For example, O_3 is vibrationally excited in its chemical formation in reaction $O_2+O+M \rightarrow O_3(v)+M$. Because the O_3 and O concentrations are not independent, since they are in photochemical equilibrium in the mesosphere, care must be taken when using O and O_3 abundances from different models or data sets which might not be fully consistent with each other. The same applies to the concentrations of NO, O_3, and NO_2 when modelling $NO_2(v_3)$ and $NO(1)$ populations (see Chapter 7).

The kinetic temperature is also assumed to be known in most cases, particularly when we are interested in determining non-LTE populations. This generally comes from measurements made at the same time as those of the non-LTE emissions under study or, in the case of theoretical studies, from models. For the Martian and Venusian atmospheres, on the other hand, non-LTE models have been developed and coupled with the energy budget equation to give the radiative equilibrium temperature profile as an output (see Chapter 10).

The next step is to consider which radiative and/or collisional processes are forcing the atmospheric region of interest out of LTE. Considering radiative processes first, spontaneous emission has, of course, to be considered, and the Einstein coefficients can be derived from spectroscopic data (see Eq. 3.46). Induced emission can be neglected for most cases as will be discussed below.

Next, we need to estimate if exchange of radiation between the atmospheric layers is important. We have already seen in the previous chapter that, if such exchange has to be incorporated, it substantially increases the complexity of the model. It can be neglected under certain circumstances, for example for weak bands where the source is solar pumping. In other cases, for example where energy from the warm troposphere is absorbed by weak bands in the mesosphere, it can be very important.

The definition of the lower boundary is important, for the same reason. It must be located low enough so that the assumption of LTE, as required by the Curtis matrix method, is fulfilled. Non-LTE models are normally developed for several bands of a given molecule, including weak and isotopic transitions. These are, in general, optically thin at the lower boundary, and they normally determine the level at which the lower boundary should be located. The region around the tropopause is in general a low enough boundary.

The upwelling flux at the lower boundary of the model (the tropospheric flux) in the spectral region of the band(s) under consideration is also required. This is particularly important for the optically thinner weak and isotopic bands, whose upper levels populations in the mesosphere are strongly affected by the tropospheric flux. For these, it should be accurately computed by using by a line-by-line radiance code and including all absorbers in the troposphere. The tropospheric flux depends on many parameters, such as the concentration of H_2O, the cloud coverage, and temperature. Measurements of surface temperatures, sky fractional coverage by clouds, and cloud top temperatures, by instruments on board satellites provides data from which the tropospheric flux can be derived. This flux is normally represented by an effective temperature (see Sec. 4.2.4) such that a blackbody emitting at this temperature would produce the same upward flux, at the frequency of interest. An example for the CO(1–0) $4.7\,\mu m$ band is given in Sec. 7.2.1.

Next we look for those collisional processes likely to affect the populations of the levels under consideration, distinguishing between thermal (i.e. vibrational-translational, or V–T) collisions, in which most of the vibrational internal energy is converted to or taken from the translational energy of the colliding molecules, and vibrational-vibrational (V–V) in which an important fraction of the internal energy is exchanged between the colliding molecules. More rarely, there can also be other collisional processes in which the internal energy is transferred in electronic form, or released when the molecule is formed in a chemical reaction. The rate coefficients and in some cases the efficiencies for many collisional processes are not well known, and this is often the main source of uncertainty in any calculated non-LTE populations.

Finally, we must consider the coupling of each relevant molecular state with another, of the same or a different molecule. This is particularly important when considering polyatomic molecules, e.g. CO_2, O_3. Only rarely is a useful model developed for just one state individually, and then it will usually belong to a diatomic, for example NO(1).

6.2   Useful Approximations

It will already be clear that non-LTE radiative transfer models represent a formidable computational task, and approximations need to be made whenever possible. Those that are often assumed include neglecting induced emission, assuming rotational LTE, grouping levels with very similar energies into resonant sets which can be treated as a single level, and neglecting overlapping between spectral lines of different bands. These are discussed further below.

6.2.1   *Induced emission*

We will not normally need to deal with wavelengths much longer than that of the longest CO_2 band, or about 15 μm, at which wavelength, and for typical atmospheric temperatures, induced emission is responsible for at most a few percent of the source function. Neglecting the effect simplifies the relationship between the population ratio of the upper and lower states of the transition and the corresponding source function (see Sec. 3.6.3 and ff.).

6.2.2   *Rotational LTE*

The rotational levels associated with a given vibrational level are generally assumed to be in LTE with respect to each other, because their separation in energy is so small that very few collisions are required to maintain a Boltzmann distribution among them. However, some exceptions, generally at very high altitudes, have been noted. Kutepov *et al.* (1991) have shown that the population of the $CO_2(0,0^0,1)$ vibrational level near the top of the Venusian atmosphere may be overestimated by 20% when rotational LTE is assumed. Also, Kutepov *et al.* (1997) have found a significant effect of rotational non-LTE on the population of the CO(1) level above about 110 and 140 km for night and daytime conditions, respectively, of the

Earth. The NO(1) level shows non-LTE rotational behaviour above about 120 km in the daytime terrestrial thermosphere, as do most of the OH vibrational levels. With these exceptions, it is generally acceptable cautiously to assume rotational LTE under all conditions in atmospheric non-LTE models. Rotational non-LTE is discussed in more detail in Sec. 8.9.

6.2.3 *Resonant levels*

Some states are very close in energy, and the coupling between them is very fast. For example, the Fermi resonant $(0,2^0,0)$ and $(1,0^0,0)$ states of CO_2 and the intermediate $(0,2^2,0)$ level in the region near $1335\,\mathrm{cm}^{-1}$ are about $100\,\mathrm{cm}^{-1}$ apart. The coupling between them is through V–T collisions of CO_2 with the most abundant constituents (N_2 and O_2 in the Earth and CO_2 itself in Mars and Venus). In such cases, we can assume that their relative populations are in LTE among themselves, and treat them as an equivalent single level with the appropriate statistical degeneracy; in the case of $(0,2,0)$ with a degeneracy of 4. Again, care is required, since this assumption might not be completely valid at the highest altitudes, particularly during daytime when $(0,2^0,1)$ and $(1,0^0,1)$, but not $(0,2^2,1)$, are solar pumped.

Other levels for which this approach can be assumed include the CO_2 $(0,3^1,0)$, $(1,1^1,0)$ (Fermi resonant) and $(0,3^3,0)$ levels situated near $2000\,\mathrm{cm}^{-1}$. These can be treated as one equivalent level, $(0,3,0)$, with a statistical degeneracy of 6. Again, this assumption does not introduce significant errors (at least below the thermosphere), since the number of collisions even at 120 km is still large enough to maintain LTE, given their small energy separation (about $40\,\mathrm{cm}^{-1}$) and the fast energy transfer rates in collisions with N_2 and O_2. Again, the six $(0,4^0,0)$, $(0,4^2,0)$, $(1,2^0,0)$, $(0,4^4,0)$, $(1,2^2,0)$, and $(2,0^0,0)$ states, which give rise to the weak 15 μm third hot bands around $2650\,\mathrm{cm}^{-1}$ (levels 9–14 in Table 6.1), can be assumed in LTE among themselves and treated as an equivalent level $(0,4,0)$ with a degeneracy of 9.

6.2.4 *Line overlapping*

In non-LTE models we normally evaluate the radiative transfer contribution to the population of an upper state by integrating over the band, setting up an independent calculation for each transition or band because they have different non-LTE source functions. This involves performing the frequency integration over the band in order to calculate the transmittances between the atmospheric layers (i.e. the Curtis matrix elements, see Sec. 6.3.3.1), with either a band model or with a line-by-line code. In the description of the CO_2 model, Sec. 6.3.3.1, we describe a histogramming quasi-line-by-line method, which is efficient, fast, and accurate for the CO_2 bands. This assumes that the individual spectral lines do not overlap each other, normally a good approach since, in the low-pressure region where non-LTE is usually important, the line broadening is much smaller than the line spacing. However, for some transitions, like CO(1–0), a line-by-line code which does allow for line overlap is found to be necessary to compute accurate non-LTE populations (see Sec. 7.2). While this involves more computing, it is not particularly problematical when dealing with the overlapping of lines within the same band.

What is much more difficult is the situation where there is overlapping between the lines of different bands. Because of the different source functions in each band, it then becomes necessary to evaluate the radiative transfer equation for each fine-mesh frequency point. So far, no non-LTE model has been developed which tackles this formidable computing task.

6.3   Carbon Dioxide, CO_2

CO_2 is the most important molecule in the atmosphere where radiative transfer is concerned. It is also a good example of one having many vibrational levels whose populations are interdependent over extended atmospheric regions, requiring a complex and interactive model. The one we will describe here has been under continuous development since 1984, and now extends to cover nearly a hundred levels (Table 6.1)*. References to other recent CO_2 models can be found in the 'References and Further Reading' section at the end of the chapter.

Figure 6.1 shows a diagram of the vibrational energy levels and most important transitions at 15, 10, 4.3, and 2.7 μm of the principal isotope (626) of CO_2[†]. The first vibrational level of N_2 and O_2 and the bending mode of water vapour are also shown. We now develop for this model a fuller description of the input parameters, as outlined in the Introduction.

6.3.1   *Adoption of a reference atmosphere*

The reference atmosphere defines the temperature profile and the atmospheric composition for the specified conditions for which the non-LTE populations are sought. These can be some specified mean conditions such as polar midwinter or global average, or the conditions for some particular experiment whose measurements are the subject of analysis. Some examples will be given below (Sec. 6.3.6).

Very abundant molecules like N_2 and O_2, which are not themselves infrared active, but which are very important in thermalizing the energy of the vibrational levels and in exchanging quanta with CO_2 through V–V collisions, must also be

---

*We use in this book the Herzberg notation for the vibrational levels, although the nowadays widely used HITRAN notation is also given in the Tables for easy reference. Vibrational levels are represented by v, e.g., (v_1, v_2, v_3) for the level excited in the vibrational modes ν_1, ν_2, and ν_3. When only one mode is mentioned, we assume there is no excitation in the other vibrational modes, e.g., the v_3 level is equivalent to the $(0,0,v_3)$ level, and with 'the v_2 levels' we refer to the $(0,nv_2,0)$ states with n = 1, 2, 3, etc. See Appendix C for more details on the terminology used for the vibrational levels and the bands arising from them.

[†]The notation for isotopes consists of the last digit of the atomic weight of each atom, e.g. CO_2 626 stands for $O^{16}C^{12}O^{16}$, 636 for $O^{16}C^{13}O^{16}$, 628 for $O^{16}C^{12}O^{18}$, and so on.

Table 6.1   CO$_2$ vibrational states.

No.	Isotope	Level‡	Level§	Energy (cm^{-1})	No.	Isotope	Level‡	Level§	Energy (cm^{-1})
1	626	01^10	01101	667.380	39	636	12^00	20002	2595.759
2	626	02^00	10002	1285.409	40	636	04^40	04401	2643.062
3	626	02^20	02201	1335.132	41	636	12^20	12201	2700.264
4	626	10^00	10001	1388.185	42	636	20^00	20001	2750.597
5	626	03^10	11102	1932.470	43	636	01^11	01111	2920.239
6	626	03^30	03301	2003.246	44	636	02^01	10012	3527.738
7	626	11^10	11101	2076.856	45	636	02^21	02211	3580.750
8	626	00^01	00011	2349.143	46	636	10^01	10011	3632.910
9	626	04^00	20003	2548.367	47	636	03^11	11112	4147.232
10	626	04^20	12202	2585.022	48	636	03^31	03311	4194.707
11	626	12^00	20002	2671.143	49	636	11^11	11111	4287.698
12	626	04^40	04401	2671.717	50	636	04^01	20013	4748.063
13	626	12^20	12201	2760.725	51	636	04^21	12212	4770.976
14	626	20^00	20001	2797.136	52	636	04^41	04411	4832.437
15	626	01^11	01111	3004.012	53	636	12^01	20012	4887.385
16	626	02^01	10012	3612.842	54	636	12^21	12211	4983.834
17	626	02^21	02211	3659.273	55	636	20^01	20011	4991.353
18	626	10^01	10011	3714.783	56	628	01^10	01101	662.373
19	626	03^11	11112	4247.705	57	628	02^00	10002	1259.426
20	626	03^31	03311	4314.913	58	628	02^20	02201	1325.141
21	626	11^11	11111	4390.629	59	628	10^00	10001	1365.844
22	626	00^02	00021	4673.320	60	628	03^10	11102	1901.737
23	626	04^01	20013	4853.623	61	628	03^30	03301	1988.328
24	626	04^21	12212	4887.990	62	628	11^10	11101	2049.339
25	626	04^41	04411	4970.930	63	628	00^01	00011	2332.113
26	626	12^01	20012	4977.840	64	627	01^10	01101	664.729
27	626	12^21	12211	5061.780	65	627	02^00	10002	1272.287
28	626	20^01	20011	5099.600	66	627	02^20	02201	1329.843
29	636	01^10	01101	648.478	67	627	10^00	10001	1376.027
30	636	02^00	10002	1265.828	68	627	03^10	11102	1916.695
31	636	02^20	02201	1297.264	69	627	03^30	03301	1995.352
32	636	10^00	10001	1370.063	70	627	11^10	11101	2062.099
33	636	03^10	11102	1896.538	71	627	00^01	00011	2340.014
34	636	03^30	03301	1946.351	72	638	01^10	01101	643.329
35	636	11^10	11101	2037.093	73	638	00^01	00011	2265.971
36	636	00^01	00011	2283.488	74	638	01^11	01111	2897.709
37	636	04^00	20003	2507.527	75	637	01^10	01101	645.744
38	636	04^20	12202	2531.679	76	637	00^01	00011	2274.088

‡Herzberg notation. §HITRAN notation.

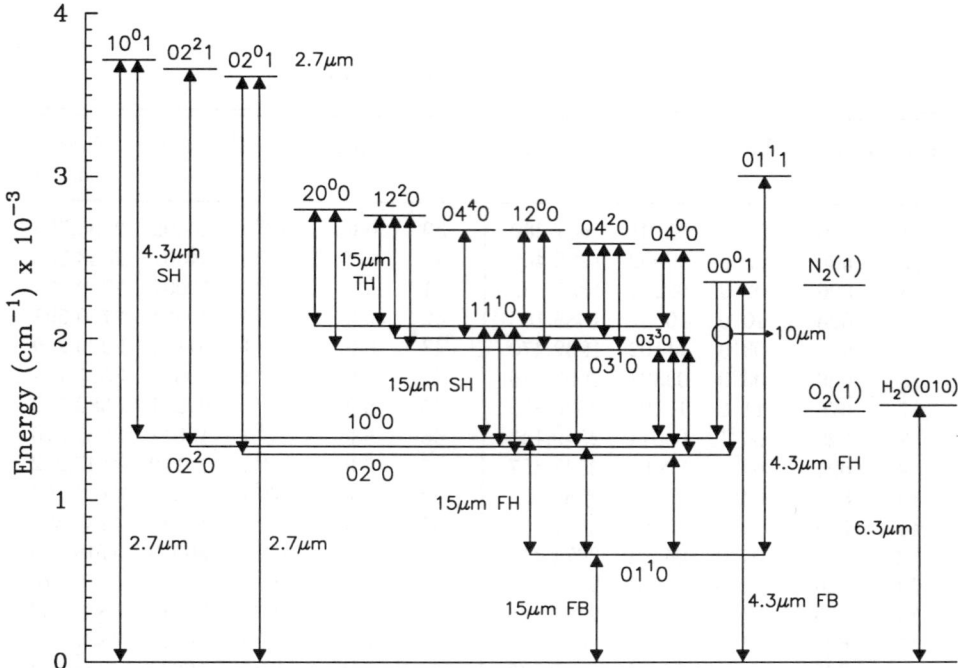

Fig. 6.1   Vibrational levels (with energy lower than 2.7 μm) and transitions for the CO_2 major isotope, and for the first vibrational levels of N_2, O_2, and H_2O.

included, of course, since otherwise the model will calculate the wrong populations for CO_2. O_2 also exerts a large influence on the populations of the H_2O and CH_4 vibrational levels (see next chapter).

The less abundant species, although they too may have levels and pathways through which significant energy transfer can occur, can be omitted if we are only interested in the CO_2 populations, for example for heating and cooling rate calculations, since the net effect on CO_2 will be small simply because collisions between the two species are relatively infrequent. Obviously, the converse is not true: if we want to calculate the radiance emitted by a trace constituent such as CO (to name an important example), we cannot do this without making sure that the exchange of quanta with a relatively highly abundant species like CO_2 is taken into account (see Sec. 7.2).

Atomic oxygen is another constituent that has to be included since it is, generally, a very efficient collision partner in exciting and de-exciting the vibrational levels, particularly above the mesopause. We also need abundances for certain excited species, specifically $O(^1D)$ and vibrationally excited $OH(v \leq 9)$, which excite some of the CO_2 vibrational levels as further discussed below. Finally, of course, we need the abundance of the ground state of CO_2 itself. The constituent abundances can be taken from compilations of reference atmospheres, the output of composition models, or from measurements. Fig. 1.4 in Chapter 1 shows typical volume mixing ratio profiles for the most important atmospheric constituents of the middle atmosphere. Some CO_2 VMR profiles are shown in Fig. 8.32.

The vertical temperature structure is an equally important input to the non-LTE model. For the regions where the deviation from LTE is small, the population of the states will depend strongly on the temperature profile (see Sec. 6.3.6). This is important when analysing non-LTE observations, since then we need a simultaneous measurement of the kinetic temperature to compute the populations of the emitting levels accurately. For theoretical studies, the usual approach is to repeat the calculation for a selection of model temperature profiles, in order to explore the dependence of the non-LTE populations on kinetic temperature.

For the non-LTE calculations described below, the abundances for N_2, O_2 and H_2O are taken from the U.S. Standard Atmosphere (see Fig. 1.4), with day and night-time atomic oxygen and ozone concentrations from Rodrigo *et al.* (1991). When a latitudinal variation of the $O(^3P)$ abundance is needed (e.g. to study the latitudinal variation of the $CO_2(0,1^1,0)$ vibrational temperature), we use the model predictions of Rees and Fuller-Rowell (1988), but fitted to the global measurements of Thomas (1990) in the 80–93 km altitude range. Other neutral compound concentrations come from Rodrigo *et al.* (1986).

The temperature profile is from the climatological reference atmosphere of the COSPAR International Reference Atmosphere, CIRA 86 (Fleming *et al.*, 1990). For the 'nominal' model, the profile for December at the equator is used, and for 'extreme' atmospheric temperature profiles, those of 80°S and 80°N are taken. The vertical structure of the atmosphere for composition, temperature and the non-LTE populations of the vibrationally excited states resulting from the model, is represented by a number of layers, typically 60 to 120 of them, each normally between 1 and 2 km thick.

6.3.2 *Boundary layers*

Although non-LTE is seldom important below the stratopause, for the strong, fundamental bands of the most abundant gases at least, sometimes the lower limit in the model needs to be extended down to the surface. This arises most frequently when treating situations where significant amounts of upwelling radiation are emitted from the troposphere, and then absorbed in the stratosphere and mesosphere. Also, for solar radiation in the weak and isotopic transitions (e.g. CO_2 at 4.3 and 2.7 μm), and when chemical reactions are considered (e.g. O_3, NO, and NO_2 bands) the downwards extension can be important.

The upper limit is set at a level where the atmosphere above is optically very thin, usually in the lower thermosphere at 120 km or higher, depending on the molecule. For the CO_2 15 and 4.3 μm bands and for NO 5.3 μm it is usually extended up to 160 or 200 km, respectively. For other less abundant species, e.g. HNO_3, 100 km or even lower is enough.

In setting up the upper boundary, we should include an approximation to allow for the atmosphere above. This is important, for example, in the computation of the absorption of solar radiation (where we should include the reduction of the solar flux above the upper boundary) and when calculating thermal cooling rates, particularly in a nearly pure CO_2 atmosphere like those of Mars and Venus. Neglecting the atmosphere above the upper boundary can lead to a large overestimation of the cooling rates in the layers close to the upper boundary.

6.3.3 *Radiative processes*

The transitions between the CO_2 vibrational levels which have an important effect on their populations are listed in Table F.1, with the band origins and strengths. The most important bands are those emitting at 15, 4.3 and 2.7 μm. The databases which give values for individual line strengths (HITRAN, GEISA or ATMOS) also list the Lorentz broadening line widths and their temperature dependence, needed in the transmission calculation, which are assumed to be of the form

$$\alpha_L(T) = \alpha_L(T_0)\,(T_0/T)^n\,,$$

where T_0 is a reference temperature, usually 296 K.

For all the bands listed in Table F.1, spontaneous emission is included, using Einstein coefficients calculated from the spectroscopic database and Eq. (3.46). Radiative transfer between layers is treated using the modified Curtis matrix method (Sec. 5.7) for all bands except those marked in the table as being very weak, for which radiative exchange is less important. The lower boundary of the model is 9 km. As we have already mentioned, tropospheric radiation can be important for exciting molecules at mesospheric altitudes by radiative transfer. Radiation from the troposphere is included by assuming a blackbody at the temperature at the lower limit of the model.

6.3.3.1 *The evaluation of the Curtis matrix*

Radiative exchange between layers is the most tedious term to evaluate in non-LTE calculations, and calculating the Curtis matrix elements takes up most of the computer time. As already noted, these elements are actually double differences of flux transmissions. They can be calculated explicitly by numerical integration, e.g., using a line-by-line code, with a summation over all of the spectral lines contributing at the frequency ν (Sec. 4.3.1) or, more quickly, with a scheme which invokes the various approximations described above, such as the following.

The quasi-line-by-line histogramming algorithm is a line-by-line scheme in the sense that integration over frequency is performed numerically, but over a reduced number of equivalent lines grouped according to their strengths. The grouping of the spectral lines of the bands into boxes is done according to a log scale in line strength, with typically five boxes per decade. In that way, all the lines in a j-th box are assumed to be equivalent to N_j lines of equal strength $S_j = \sum_i S_i/N_j$, with a Lorentz half width of

$$\alpha_{L,j}^{1/2} = \frac{\sum_i (S_i \alpha_{L,i})^{1/2}}{\sum_i S_i^{1/2}},$$

and a Doppler half width of

$$\alpha_{D,j} = \frac{\sum_i S_i \alpha_{D,i}}{\sum_i S_i}.$$

In these expressions the summation over i is extended over all the lines in the j-th box. The calculation of the equivalent width (or transmittance) for each of the intervals is then carried out by integrating over the Voigt line shape for the equivalent line represented by the box. The frequency integration over these 'equivalent' or histogrammed lines is performed numerically (with a relative precision of 10^{-4}) either on a line-by-line basis (advisable for strong bands) or using the much faster approximation of Rodgers and Williams (1974), which is more appropriate for the weaker bands.

Whether the exact or the approximate method for integrating over all lines as a function of frequency is followed, account has to be taken of the temperature dependence of the line strengths and line broadening coefficients. Line strengths depend on temperature through the vibrational and rotational partition functions and through the population of the lower state (Eq. 3.34 in Sec. 3.6.3). For hot bands, the dependence of line strength on temperature can be quite important, because the population of the lower state of the transition is very temperature sensitive, particularly if it has a non-LTE population.

The variation of the line strengths and broadening coefficients over atmospheric paths is included approximately using the Curtis-Godson approach (see Sec. 4.3.9). After histogramming, equivalent line strength and broadening coefficients are pre-computed at a set of temperatures, typically over the 100–1000 K temperature range with intervals of 10 K. The corresponding effective temperature can be obtained to useful accuracy by calculating a mass weighted value for the path. Calculating the averaged absorber amount and kinetic temperature for the equivalent homogeneous optical paths for each line strength interval, instead of for the whole band, is more accurate and allows the rotational line distribution with temperature within each band to be included more realistically.

Overlapping between lines of the same band can be included approximately by using the random band model (Sec. 4.3.6). In this, the spectrally averaged transmission for the band is calculated by means of Eq. (4.90),

$$\mathcal{T} = \exp(-W/\Delta\nu),$$

where $W = \sum_j W_j$, W_j is the equivalent width of the j-th interval of the band (considered independent of the rest of the intervals), and $\Delta\nu$ the actual width of the band in the spectrum. This gives vibrational temperatures for the $CO_2(v_2,v_3)$, $CO(1)$ and $H_2O(0,1,^10)$ states which are accurate to within 1–2 K ($\sim$1%), when compared to a line-by-line code.

Finally, the integral of the flux transmittance over zenith angle θ, or $\mu = \cos\theta$, is evaluated using a Gaussian quadrature scheme or the simple diffusion approximation, as discussed in Sec. 4.4. A four-point Gaussian quadrature has been found to give CO_2 vibrational temperature differences smaller than 0.5 K when compared to the exact integration.

6.3.3.2 *Absorption of sunlight*

Absorption of solar radiation at wavelengths of 2.0, 2.7, and 4.3 μm by CO_2 pumps up the number of molecules in the $(0,0^0,1)$ and combination $(v_1,v_2,1)$ levels, including those of the less abundant isotopes 636, 628 and 627; and produce significant heating rates by the CO_2 bands centred near 4.3 and 2.7 μm. There is also some significant pumping of $CO_2(0,2,0)$ and $CO_2(0,3,0)$ due to vibrational energy transfer from $CO_2(0,0^0,1)$. All of these need to be included in the model if it is to be valid under daylight conditions.

In the present model, solar radiation has been included for those bands falling at wavelengths shorter than or equal to 4.3 μm whose lower state energy is equal to or smaller than that of $(0,2,0)$. These bands are listed in Table F.1. Since the excited states of the minor isotopes are less abundant, the absorption of solar radiation by these levels is very small or negligible, and most of them have not been included.

The rate at which the CO_2 molecules are excited in a given level by the absorption in a given band (in units of molecules per cubic centimeter and per second) is equal to the rate at which the solar flux is reduced, given by integrating Eq. (4.22) over ν, and scaled by the energy of the band. For high solar zenith angles (χ), allowance has to be made for the sphericity of the atmosphere (Sec. 4.2.3). The rates are usually expressed per molecule in the ground state of the most abundant isotope, and caution is required with respect to the number density of the molecules actually absorbing when dealing with the hot and isotopic bands. To calculate the absorption by a hot band we need to know the non-LTE population of the lower state, which may require a first iteration of the model excluding solar absorption (see Sec. 5.4.3). For some of the lower levels, this can in fact gives a reasonably accurate result.

Finally, these calculations require knowledge of the solar flux at the near-infrared wavelengths. Common sources of this data are the paper by Thekaekara (1976) and the Kurucz model (see the 'References and Further Reading' section).

6.3.4 *Collisional processes*

The collisional processes affecting the CO_2 levels are all obviously local, i.e. they enter into the statistical equilibrium equations only for the altitudes where the collisions occur. We will consider three kinds of collisions: homogeneous, hetero-geneous, and special cases involving unstable species like atomic oxygen, and two categories of energy exchange during collisions, i.e. vibrational-translational (V–T) and vibrational-vibrational (V–V).

Homogeneous collisions are those between chemically identical molecules, al-though not necessarily the same isotope or in the same energy state. The only really important ones in the Earth's atmosphere are those involving CO_2. Among the het-erogeneous collisions, i.e. those between different molecules, the most important are those where there is a coincidental closeness in energy levels, so that the exchange of quanta occurs efficiently, for example $CO_2(0,0^0,1)$ and $N_2(1)$; $H_2O(0,1^1,0)$ and $O_2(1)$, etc. (see Fig. 6.1).

Some unstable species, like atomic oxygen in the ground and electronically ex-cited state, $O(^3P)$ and $O(^1D)$, and the vibrationally excited hydroxyl radical, are extremely efficient at activating or deactivating one or more of the vibrations of carbon dioxide, either directly, or indirectly through $N_2(1)$. They are therefore important components of any non-LTE model, in spite of their relatively small abundances. We can evaluate their effects on the populations of CO_2 using con-centrations of these species obtained from a separate (and complicated) radiative-chemical model. Suitable results have been published as plots or tables for a variety of conditions, typically as a function of latitude and solar zenith angle, and are also available as outputs of 2 and 3 dimensional chemical-dynamical models.

The collisional processes affecting the v_2 and v_3 levels are summarised in Ta-ble 6.2, together with values or estimates for their temperature-dependent rate coefficients. Note that many of these processes operate in both senses, i.e. forwards and backwards; the rate is given for the forward sense. The rate in the backward sense is calculated from the forward rate following the principle of detailed balance (see Eq. 3.54).

The thermal contribution to the populations of the $CO_2(0,v_2,0)$ states due to collisions between CO_2 and the more abundant compounds N_2, O_2 and $O(^3P)$ is represented by processes 1 and 2 in Table 6.2. Among all of the possible vibrational-translation exchanges, only those with small interconverted energy, i.e. involving a single quantum jump, need to be retained because of their relatively high prob-ability. These processes also account for the vibrational-translational relaxation of the CO_2 isotopes 636, 628 and 627 in collisions with the same species. Most of the measurements of rate constants which have been made are for the principal isotope, so these are quite well determined and available in the literature. For the minor isotopes, we have to make the reasonable assumption that they occur at the same rate as the analogous process for the principal isotope.

Process 1 drives the populations of the v_2 levels below and around the mesopause, while collisions of the $CO_2(v_2)$ levels with atomic oxygen control their population above about 100 km. The rate coefficient of process 1 is reasonably well known although that of process 2 is uncertain by around a factor 4 (see Sec. 6.3.6.4).

Table 6.2   Main collisional processes affecting the CO_2 vibrational levels.

No.	Process	Rate coefficient†	Reference
1	$CO_2^i(v_2) + M(N_2, O_2) \rightleftharpoons CO_2^i(v_2-1) + M$	$(v_2=1)$ 9.6×10^{-16} $\exp(-8.08A + 1.85B)$ $(v_2=2)$ 1.24×10^{-14} $(T/273.3)^2$ $(v_2=3)$ $(3/2)\times k_1(v_2=2)$ $(v_2=4)$ $(9/4)\times k_1(v_2=2)$	Allen et al. (1980) Taine and Lepoutre (1979) López-Puertas et al. (1998a) López-Puertas et al. (1998a)
2	$CO_2^i(v_2) + O(^3P) \rightleftharpoons CO_2^i(v_2-1) + O(^3P)$	$3.0\times10^{-12}(T/300)^{1/2}$	López-Puertas et al. (1992b)
3	$CO_2^i(v_3) + N_2 \rightleftharpoons CO_2^i(v_2=2, 3, 4) + N_2$	$2.2\times10^{-15}+1.14\times10^{-10}\exp(-76.75/T^{1/3})$	López-Puertas et al. (1986a,b)
4	$CO_2^i(v_3) + O_2 \rightleftharpoons CO_2^i(v_2=2, 3, 4) + O_2$	$2.3\times10^{-15}+1.54\times10^{-10}\exp(-76.75/T^{1/3})$	López-Puertas et al. (1986a,b)
5	$CO_2^i(v_3) + O(^3P) \rightleftharpoons CO_2^i(v_2=2, 3, 4) + O(^3P)$	2×10^{-13} $(T/300)^{1/2}$	López-Puertas et al. (1986a,b)
6	$CO_2^i(001) + CO_2^i \rightleftharpoons CO_2^i(020) + CO_2^i(010)$	3.6×10^{-13} $\exp(-1660/T + 176948/T^2)$	Lepoutre et al. (1977)
7	$CO_2^i(001) + O_2 \rightleftharpoons CO_2^i(010) + O_2(1)$	$3\times10^{-15}(1+0.02(T - 210))$	Houghton (1969)
8	$CO_2^i(v_2)+CO_2^j(v_2') \rightleftharpoons CO_2^i(v_2-1)+CO_2^j(v_2'+1)$	1.2×10^{-11}	Huddleston and Weitz (1981), Orr and Smith (1987)
9	$CO_2^i(v_1, v_2, v_3)+N_2 \rightleftharpoons CO_2^i(v_1, v_2, v_3-1)+N_2(1)$	5.0×10^{-13} $(300/T)^{1/2}$	Inoue and Tsuchiya (1975)
10	$CO_2^i(v_1, v_2, 1) + N_2 \rightleftharpoons CO_2^i(v_1', v_2', 1) + N_2$ $(2v_1+v_2=2v_1'+v_2')$	2.0×10^{-11}, 2.4×10^{-12} See text	López-Puertas and Taylor (1989)
11	$CO_2^i(v_3)+CO_2^j(v_3') \rightleftharpoons CO_2^i(v_3-1)+CO_2^j(v_3'+1)$	6.8×10^{-12} $v_3\sqrt{T}$	Moore (1973)
12	$N_2 + O(^1D) \rightarrow N_2(1) + O(^3P)$	2.4×10^{-11} $(\epsilon=0.25)$	Amimoto et al. (1979), Harris and Adams (1983)
13	$N_2 + OH^*(v \leq 9) \rightarrow N_2(1) + OH^*(v - 1)$	$1.0\times10^{-14}(v=1)$ to $4.4\times10^{-13}(v=9)$	Streit and Johnston (1976), Kumer et al. (1978)
14	$N_2(1) + O_2 \rightleftharpoons N_2 + O_2(1)$	$2.05\times10^{-20}\exp(271A - 2.32B)$	Maricq et al. (1985)
15	$N_2(1) + O(^3P) \rightleftharpoons N_2 + O(^3P)$	3.2×10^{-15} $(T/300)^{2.6}$	McNeal et al. (1974)

† Rate coefficient for the forward sense of the process in cm^3s^{-1}. $A = 10^{-4}T$; $B = 10^{-5}T^2$. T is temperature in K. i and j are different CO_2 isotopes.

A process in CO$_2$ which is of considerable importance is the transfer of the asymmetric or v_3 quanta to bending or v_2 quanta. This rearrangement occurs in collisions of CO$_2$(0,0^0,1) with N$_2$ and O$_2$ (processes 3 and 4). The greatest uncertainty here is which v_2 excited states are produced, and how efficiently. There is some experimental evidence which shows that CO$_2$ is excited in the $v_2 = 2$, 3 and 4 states, for instance. The de-activation of the asymmetric mode of CO$_2$ by O(^3P) (process 5) has proved to be even more efficient than that by N$_2$ or O$_2$, although the resulting CO$_2$ levels are even more uncertain. Given the lack of knowledge, the same partitioning as for collisions with N$_2$ and O$_2$ is normally assumed.

The model also includes the path (process 6) by which v_3 is thought to relax, during collision with a ground state molecule of CO$_2$ itself, to three v_2 quanta, as (0,2,0) and (0,1^1,0). This process is of little importance in the Earth's atmosphere but it is crucial for Mars and Venus, as we shall see in Chapter 10. Process 7 is not a significant producer of the fundamental bending mode, but is a significant loss mechanism for CO$_2$(0,0^0,1), and important also for determining the populations of O$_2$(1) and, consequently, H$_2$O(0,1^1,0) (see Sec. 7.4).

The principal V–V transfer of v_2 quanta between CO$_2$ molecules (process 8) involves one v_2 quantum but little kinetic energy exchange. For example, the de-activation of CO$_2$(0,2,0) by CO$_2$ reaches 25% of the total relaxation at altitudes below 90 km. At higher levels, deactivation by atomic oxygen tends to dominate. Process 8 also covers the sharing of v_2 quanta during collisions between CO$_2$ isotopes, although again experimental data is available only for the $v_2 = 1$ state of the most abundant isotope.

The near-resonant V–V transfer between the $(v_1,v_2,1)$ states of the four most common CO$_2$ isotopes and the first vibrational level of N$_2$ (process 9), is important in determining the populations of the former. Because of the infrared inactivity of N$_2$(1), the energy absorbed from solar radiation by CO$_2$ in the near-IR is redistributed by this collisional process among the $(v_1,v_2,1)$ states of all CO$_2$ isotopes, and it also allows excitation of these levels from other non-LTE sources such as O(^1D) and OH($v \leq 9$). This is an important source of excitation for the CO$_2$(0,0^0,1) levels of the minor isotopes. Energy is mainly absorbed from solar radiation by the most abundant CO$_2$ isotope at 4.3 and 2.7 μm, transferred to infrared inactive N$_2$(1) in V–V collisions, and then, again in similar collisions, from N$_2$(1) to the (0,0^0,1) levels of the minor CO$_2$ isotopes. The rate coefficient has been extensively studied over a wide range of temperatures in both theoretical calculations and laboratory measurements and the value is well known.

In establishing the collisional mechanisms of the high-energy solar excited levels, we have assumed that the v_3 quantum of the excited levels CO$_2(v_1,v_2,1)$ with $2v_1+v_2 \geq 3$, i.e. (2,0^0,1), (1,2^0,1), (0,4^0,1), etc. are all transferred to give CO$_2$(0,0^0,1) or N$_2$(1) before the v_1 and v_2 quanta are. This is based on the fact that all transitions involving a v_3 quantum have larger spontaneous emission coefficients than those transferring bending (v_2) and/or symmetric stretching (v_1) quanta. In addition, the V–V collisional rates for process 9, involving a v_3 quantum, are faster than those involving v_1 or v_2 (process 1). Redistribution of the v_1 and v_2 energy of combination levels of the type (v_1,v_2,v_3) also takes place in collisions with N_2 (process 10). This only occurs in allowed bands, but the exchange of energy between near-resonant high energy levels allows the emission of this energy in other hot bands, and is therefore an efficient route for the relaxation of the absorbed solar energy. For example, absorption of solar radiation in the CO_2 bands near 2.7 μm populates only the $(1,0^0,1)$ and $(0,2^0,1)$ levels (the transition from $(0,2^2,1)$ to the ground state is not allowed), but exchange between these and $(0,2^2,1)$, which is intermediate in energy between both, allows emission in the $(0,2^2,1$–$0,2^2,0)$ 4.3 μm hot band. This rate coefficient might be important for high spectral resolution measurements discriminating the contributions of the three 4.3 μm hot bands, but is of little importance for wideband measurements, since the bands are of similar strengths and closely located in frequency; hence the overall radiance emitted by all levels is essentially independent of this rate.

Due to the rapid V–V energy exchange between $N_2(1)$ and the asymmetric stretch mode of CO_2, a complete interaction scheme with $N_2(1)$ is needed in order to calculate the $CO_2(0,0^0,1)$ population. The excitation of $N_2(1)$ involves two important non-LTE processes in addition to the transfer of v_3 from $CO_2(0,0^0,1)$ and from higher energy $CO_2(v_1,v_2,1)$ levels. The first is the transfer of energy from electronically excited atomic oxygen, $O(^1D)$, and the second the transfer from OH, vibrationally excited in levels $v \leq 9$. These processes (12 and 13 in Table 6.2) can give rise to non-LTE populations of $CO_2(0,0^0,1)$ and $N_2(1)$ even when radiative transfer processes or solar pumping are unimportant. The efficiency of the electronic to vibrational $[O(^1D)$–$N_2(1)]$ conversion is not well known, because although the rate for the electronic deactivation of $O(^1D)$ by N_2 is well determined, that of the vibrational transfer from $O(^1D)$ to $N_2(1)$ is not. An overall value of 25% gives a good fit to satellite measurements of emission from $CO_2(0,0^0,1)$ at 10 μm. Since the electronic energy of $O(^1D)$ is 6.8 times that of $N_2(1)$, assuming that the vibrational energy in N_2 rapidly cascades to $N_2(v=1)$, that efficiency is equivalent to producing an average of $1.7 \times N_2(1)$ molecules in each collision.

The quenching of $N_2(1)$ is mainly due to collisions with molecular and atomic oxygen, processes 14 and 15 in Table 6.2. In process 14, part of the vibrational energy of N_2 is transferred to the O_2 molecule, this being an important excitation mechanism for O_2 (for a detailed discussion of the excitation processes of $O_2(1)$ see Sec. 7.4.2). Atomic oxygen is the more effective constituent in deactivating $N_2(1)$, as the unexpectedly large rate constant for process 15 shows (but note that the only value we have for this was determined for temperatures above 300 K). The influence of $O_2(1)$ on the population of $CO_2(0,0^0,1)$, directly through process 7, and indirectly through process 14, is relatively unimportant during daytime, but at night $O_2(1)$ may be excited by $OH^*(v \leq 9)$ in the upper mesosphere and only very weakly deactivated by $O(^3P)$, in which case the process may be non-negligible. Process 7 should be included when calculating the population of $O_2(1)$ (see Sec. 7.4.2), which in turn exerts a crucial influence on the population of $H_2O(0,1^1,0)$.

In summary, it must be emphasised that there are no available measurements of the rate coefficients for many of the important collisional processes in Table 6.2. When they are not available for minor isotopic species, we must take the same values as for the major isotope. For high-energy levels, we can estimate the rate by using the harmonic oscillator approximation in which, for example, if the rate for the collisional process $CO_2(v_2=1)+M \rightarrow CO_2+M$ is k_1, the rate coefficient for process $CO_2(v_2>1)+M \rightarrow CO_2(v_2-1)+M$ is $k_{v_2} = v_2 k_1$. This approach is usually applied to thermal collisional rates, but in the absence of other information is also used for V–V processes. For V–V processes which occur with the same transfer of quanta but with different initial and post-collisional states, the use of the same rate coefficient, independent of the excitation of the state, is probably a good estimate, since there is some experimental evidence that the exchange of a v_3 quanta between CO_2 and N_2 seems to be independent of the excitation of the ν_1 and ν_2 modes of CO_2. Finally, it needs to be kept in mind that most of these processes take place in both directions. Normally, the rates in the backward sense are less important, but in some cases they are crucial, e.g. process 2 for $v_2=1$.

6.3.5 *Solution of the multilevel system*

In the Curtis matrix method, an important aspect of the solution of the multilevel system of statistical equilibrium equations is their non-linearity, which manifests itself in two different ways. The first appears in the statistical equation of those levels affected by non-fundamental transitions and in the transmittance calculation for these bands. The second comes in when the two states following a V–V collisional process are both vibrationally excited, e.g. process 6 in Table 6.2. These two classes of non-linearity can be treated by iterating, assuming the less significant population in the non-linear terms as known initially. In the solution for the CO_2 vibrational populations, the only populations required to begin are those of the $(0,1^1,0)$ and $(0,2,0)$ levels of the different isotopes, and occasionally that of $(0,3,0)$ if the populations of the $(0,4,0)$ levels (emitting very weakly at 15 μm) are required. The initial values are the non-LTE populations that result from simplified models for $(0,1^1,0)$, $(0,2,0)$ and $(0,3,0)$, solved in that order, e.g. by first running the model for $(0,1^1,0)$, then using that for $(0,2,0)$, and then for $(0,3,0)$, with the interactions with their upper levels neglected. The convergence and stability of this approach are very good, usually requiring only two iterations to get the final values.

In practice, the complete solution of the resulting system with such a large number of equations is intractable. To obtain a practical solution, it is convenient to group the whole set of the vibrational levels and their associated bands into subsets. To make the solution faster and more accurate it is advisable when classifying the levels that all states which are strongly coupled by collisional processes are grouped together, while those collisional interactions which take place between levels belonging to different groups are much slower.

Proceeding in this way, we define three major groups. Group A includes the v_2 [$(0,1^1,0)$, $(0,2,0)$, $(0,3,0)$, $(0,4,0)$] levels of the CO_2 major isotope. The radiative transitions involved are the fundamental, first, second and third hot bands at 15 μm that originate from the v_2 levels (see Tables 6.1 and F.1, and Fig. 6.1). Group B includes the $(0,0^0,1)$ and $(0,1^1,1)$ levels of the major isotope, and the $(0,0^0,1)$ levels of the minor isotopes, in addition to $N_2(1)$. The transitions involved in Group B are the 4.3 μm fundamental, 4.3 μm first hot, and the 10 μm bands of the major isotope, and the 4.3 μm fundamental bands of the minor isotopes. Group C includes the three $2v_2+v_3$ levels [$(0,2^0,1)$, $(0,2^2,1)$ and $(1,0^0,1)$] emitting two fundamental 2.7 μm bands and the three 4.3 μm second hot bands. Group D consists of the six levels near 2.0 μm [$(0,4^0,1)$, $(0,4^2,1)$, $(1,2^0,1)$, $(0,4^4,1)$, $(1,2^2,1)$, and $(2,0^0,1)$] with three important transitions at 2.0 μm, six transitions at 2.7 μm (second hot bands, see Table F.1), and six at 4.3 μm (the fourth hot bands). The $(0,0^0,2)$ level emitting a hot 4.3 μm band, is considered as a group on its own.

In the solution of the model, the populations of all levels inside each group are calculated simultaneously by solving the system of statistical equilibrium equations for each level and the radiative transfer equation for each band. This solution, involving the coupling of the atmospheric layers, through non-local radiative transfer excitation, as well as the collisional coupling of several energy levels locally, is very time consuming and therefore not advisable for routine calculations in retrieval schemes. The weaker interactions between the different groups are then taken into account by the iterative procedure mentioned above. In each iteration, the groups of levels are solved sequentially, the order depending on whether we are considering night or daytime conditions.

For daytime, excitation takes place mostly in the high energy levels and when these relax they excite those with smaller energy. Therefore, we proceed from the groups of more energetic levels to those with less energy, i.e. from D to C, B, and finally group A. Note that, for the calculation of the photoabsorption coefficients in groups B, C, and D, we need the populations of the lower absorbing states $(0,1^1,0)$, $(0,2,0)$, and $(0,3,0)$. They are obtained in advance from simplified isolated models for each of those levels, excluding the interaction with their respective upper states, as explained above.

For night-time conditions, the main sources of excitation are collisional processes and absorption of atmospheric radiation. The excitation therefore takes place mainly through the lower energy levels, and the sequence is inverted, i.e. we proceed from the v_2 levels (group A), then group B, C, and so on. The levels in groups B, and C are of little relevance for night-time conditions and can be omitted.

A similar grouping is applied to the energy levels of the CO_2 minor isotopes. There are some particularities for these states, however. First, the coupling of a given level of a minor isotope with the corresponding level of the major isotope can be as important (or more as in the case of Mars and Venus) than the coupling with the higher states of the isotope itself. For example, the collisional coupling of any isotopic v_2 level with $626(0,1^1,0)$ can be more important than with its corresponding higher v_2 or v_3 levels. Secondly, as mentioned above, the $(0,0^0,1)$ isotopic levels are more strongly coupled with $N_2(1)$ (and to a lesser extent with $626(0,0^0,1)$) than with the high energy $(v_1,v_2,1)$ states of the isotope itself. Note that the losses of $626(0,1^1,0)$ to excite the v_2 isotopic levels, and of $N_2(1)$ and $626(0,0^0,1)$ to excite the v_3 isotopic states, have to be included in the solutions of group B of the major isotope. Because of this, some iteration between the solution of the major and minor isotopes is necessary. Some of the groups for the minor isotopes do not include all the levels used for the major isotope, particularly the highly energetic levels (see Tables 6.1 and F.1), because their populations are not large enough to contribute significantly to atmospheric emission or heating/cooling rates.

The speed and efficiency of the model can be sometimes improved by splitting the full solution of a group or subset of coupled levels into: (a) single isolated routines for each of the levels involved (including radiative transfer, and the coupling with other levels in the group iteratively); and (b) a system for all levels where they are coupled only locally by collisions, but with no exchange of radiation with other atmospheric levels. That is, to separate the full coupling of the atmospheric layers by radiative transfer from the local coupling by collisions. As a result, more iterations are needed, but, overall, it can be more efficient for certain groups. For example, it is better to solve separately for the populations of the $(0,2^0,1)$, $(0,2^2,1)$ and $(1,0^0,1)$ levels (including radiative transfer) and iterating between them, than to solve the system all at once. If, in addition, we start the iteration from the solution of the three levels coupled collisionally but with no radiative transfer (which is very quick), then the iterative procedure is considerably reduced.

6.3.6 *Non-LTE populations*

The output of the non-LTE model is generally twofold: (i) the populations of the emitting levels, and (ii) the cooling/heating rates generated by the transitions between them. The populations are used to predict the effect of non-LTE on the quantities measured by instruments, usually emission/absorption spectra or calibrated radiances. This is dealt with in detail in Chapter 8. The cooling and heating rates constitute a result by themselves, and are important for the energy budget of the atmosphere: Chapter 9 is devoted to a discussion of them.

The non-LTE population of a level v with energy E_v is usually described in terms of its vibrational temperature:

$$T_v = -\frac{E_v}{k \ln\left(\dfrac{n_v g_0}{n_0 g_v}\right)}, \tag{6.1}$$

where n_v and n_0 are the number density of level v and of the ground vibrational state, and g_v and g_0 their respective degeneracies. Sometimes T_v is defined relative to the total number density of absorbing molecules, n_a, instead of to the population in the ground vibrational level, n_0. This can be obtained by introducing Eq. (3.47) into Eq. (6.1). The vibrational temperature has the advantage, over the number density n_v, that it more clearly visualises the departure from non-LTE when compared to the kinetic temperature. However, one should be careful when comparing the radiance emitted (or cooling produced) by different bands. The latter quantities are proportional to the number density of molecules in the excited state, i.e., to its vibrational temperature and also to its energy. So, states with lower energies might produce larger radiances or cooling rates than higher ones even having smaller vibrational temperatures. In fact, the radiances from lower energy levels emitted in the fundamental bands are generally larger than those emanating in the hot bands from higher states, even though the latter usually have larger vibrational temperatures.

We describe next the vibrational temperatures of the CO_2 levels, starting with the v_2 levels emitting near $15\,\mu m$, and following with those emitting at 4.3 and $2.7\,\mu m$. The former are important for both day- and night-time, while the latter are more important during the day, when they are excited by solar pumping.

6.3.6.1 *The v_2 levels emitting near $15\,\mu m$*

Figure 6.2 shows the calculated populations for night-time conditions of the $(0,1^1,0)$, $(0,2,0)$, $(0,3,0)$ and $(0,4,0)$ levels, which give rise to the fundamental, first hot, second hot, and third hot ν_2 bands near $15\,\mu m$. The populations are displayed as the vertical profile of the vibrational temperature T_v, as described above, and the kinetic temperature is also plotted for comparison.

The variations in the vibrational temperatures for each of the various levels shown arises, of course, because it depends on the strength of its associated transition(s), as well as on the kinetic temperature structure of the atmosphere which is the same for all. The overtone and isotopic levels tend to depart from LTE at lower heights, because they are connected by weaker bands, while the ν_2 fundamental remains close to LTE up to the mesopause and above. Note also that the deviations from LTE occur in either sense, i.e., larger or smaller vibrational populations than those for LTE conditions can be found depending on the conditions.

The $(0,1^1,0)$, $(0,2,0)$, $(0,3,0)$, and $(0,4,0)$ levels of the major 626 isotope are very close to LTE all the way through the stratosphere and lower mesosphere, where thermal collisions are the dominant process. In the upper mesosphere and lower thermosphere, their populations are the result of two main excitation processes: thermal collisions and the absorption of the upwelling radiative flux. Since the more energetic levels are connected through weaker transitions, the absorption in the upper mesosphere of the upwelling radiative flux , which originates in the warmer lower mesospheric regions (the transition from optically thick to optically thin takes

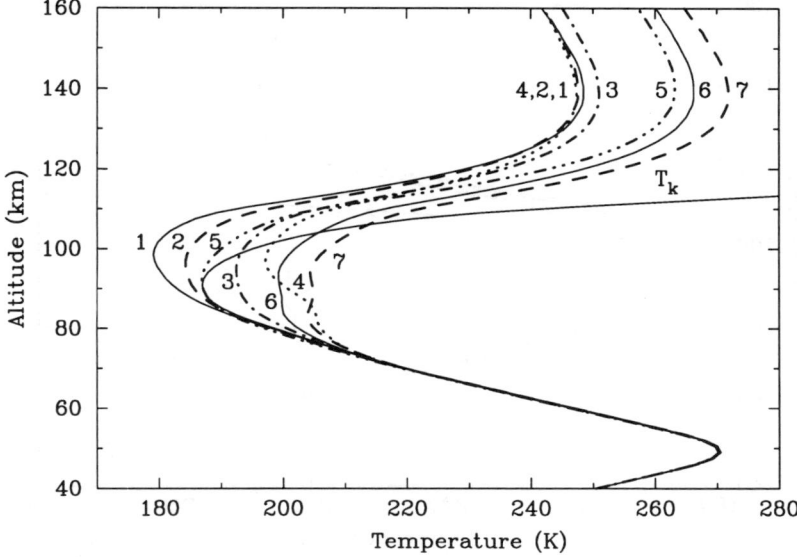

Fig. 6.2 Night-time vibrational temperatures for the CO_2 v_2 levels (emitting near $15\,\mu m$) for the U.S. Standard Atmosphere. Numbers 1 through 7 are respectively the $(0,1^1,0)$, $(0,2,0)$, $(0,3,0)$ and $(0,4,0)$ levels of the 626 isotope, and the $(0,1^1,0)$ levels of isotopes 636, 628, and 627. T_k is the kinetic temperature.

place at lower altitudes for the weaker bands), is larger in these weaker bands. This produces a larger non-LTE deviation for the more energetic v_2 levels.

In the lower thermosphere and above, thermal collisions with atomic oxygen provide the principal excitation process, and emission to space is the major loss of the vibrational excitation. In this region, the vibrational temperatures are smaller than the kinetic, because at the low pressures prevailing here the atmosphere is optically thin and collisions are much less frequent. The net result is that collisions are not fast enough to supply the emitted photons and so the populations are depleted relative to their LTE values.

The vibrational temperatures of the v_2 levels for daytime conditions (Fig. 6.3) differ from their night-time values, even when the kinetic temperature profile is the same, because they are additionally affected by the absorption of solar radiation. This takes place mostly, not in the ν_2 bands themselves where the solar radiance is quite weak, but in the near infrared, mainly at 2.7 and $4.3\,\mu m$, with subsequent internal transfer of energy into the v_2 levels. Those most affected are those with higher energy, and those close to the highly excited $4.3\,\mu m$ levels, i.e. $(0,2,0)$, $(0,3,0)$, and $(0,4,0)$. The $(0,1^1,0)$ level of 626 and the isotopic levels are very little affected. Solar absorption at $2.7\,\mu m$ directly activates the $(0,2^0,1)$ and $(1,0^0,1)$ states, which subsequently relax with the emission (or the collisional exchange) of a v_3 quantum (process 9 in Table 6.2) to the $(0,2,0)$ levels. This produces a daytime enhancement, compared to night-time conditions, in the vibrational temperatures of the $(0,2^0,0)$ and $(1,0^0,0)$ states of $\sim 5\,K$ between 80 and 105 km. The $(0,3,0)$ states are enhanced

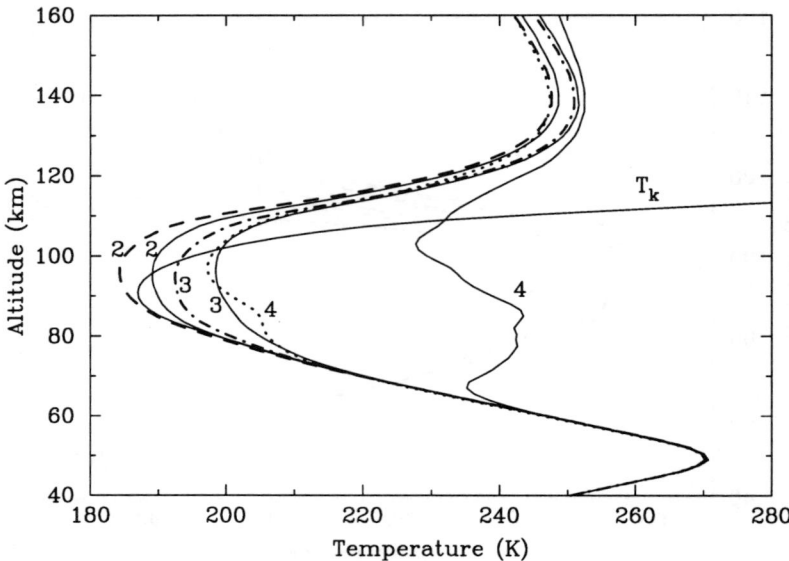

Fig. 6.3 Daytime ($\chi=60°$) vibrational temperatures for the CO_2 v_2 levels for the U.S. Standard Atmosphere. Numbers of the levels as in Fig. 6.2. Solid lines: daytime; broken lines: night-time. T_k is the kinetic temperature.

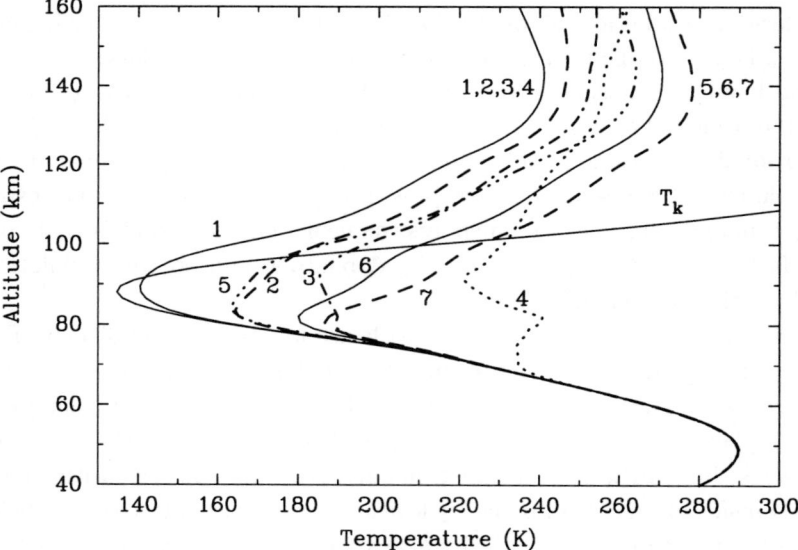

Fig. 6.4 Daytime ($\chi=60°$) vibrational temperatures for the CO_2 v_2 levels for a typical polar summer temperature profile. Numbers of the bands as in Fig. 6.2. T_k is the kinetic temperature.

by $\sim$8 K in the same region by the vibrational relaxation of the highly populated solar pumped $(0,0^0,1)$ level (processes 3, 4 and 5 in Table 6.2). Fig. 6.3 also shows that the $(0,4,0)$ levels are greatly enhanced by day, so that they have the largest T_v's

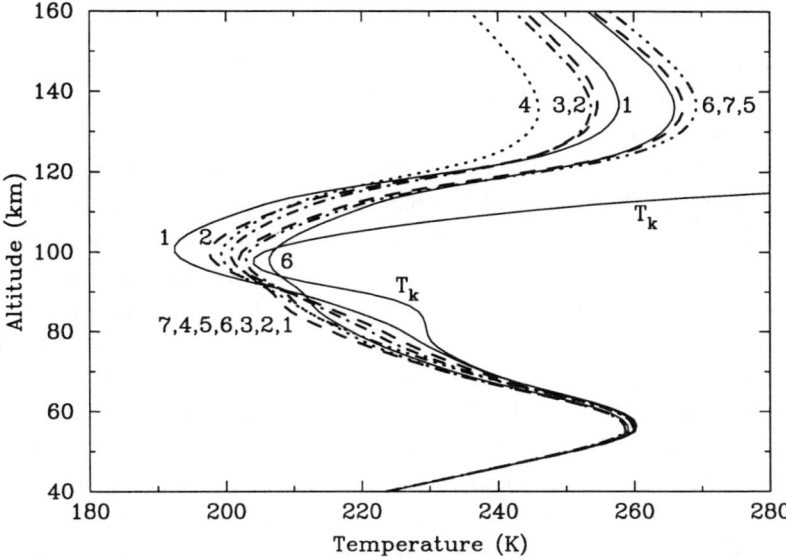

Fig. 6.5   Night-time vibrational temperatures for the CO_2 v_2 levels for a typical polar winter temperature profile. Numbers of the bands as in Fig. 6.2. T_k is the kinetic temperature.

in the mesosphere. They also deviate from LTE at lower altitudes than the other v_2 levels. As for the (0,3,0) states, these are mainly excited by collisions with the solar pumped $(0,0^0,1)$ level (processes 3, 4, and 5 in Table 6.2). At altitudes above 105 km, the day-night difference is negligible since thermal collisions with $O(^3P)$ become the main excitation mechanism.

The kinetic temperature profile influences the level populations in the mesosphere in two ways: (i) locally through thermal collisions, and (ii) remotely, from lower layers in the lower mesosphere, through the absorption of the upwelling radiative flux. This is illustrated in Figs. 6.4 and 6.5 where we show the vibrational temperatures for some of these levels for the polar summer and winter atmospheres, respectively. The warmer stratopause and colder mesopause at the summer pole make the vibrational temperatures of these levels warmer than the kinetic temperature around the mesopause (Fig. 6.4), and hence the non-LTE deviations are positive and larger in magnitude. Part of this enhancement, however, particularly for the $v_2 \geq 2$ levels, is due to the daytime conditions in the polar summer. Note again that around the mesopause the upper state populations vary inversely with the corresponding band strengths, reflecting the importance of radiative over thermal processes for populating the levels from which the weaker transitions originate. Under polar winter conditions, however, the situation is nearly reversed (Fig. 6.5). The populations are depleted relative to the Boltzmann distribution in the higher mesosphere, and only the weaker bands are close to LTE or slightly overpopulated around the mesopause.

6.3.6.2 *Global distribution of the v_2 levels*

We can extend the summer-winter comparisons of the vibrational temperatures of levels in non-LTE to make global maps of the difference between vibrational and kinetic temperature, T_v–T_k.

Figures 6.6–6.10 show T_v–T_k differences for the upper levels of the fundamental, first hot, and second hot bands of the 626 isotope and the fundamental bands of the 636 and 628 isotopes at night-time for the solstice conditions of the CIRA 1986 reference atmosphere shown in Fig. 1.2. Solar excitation in the illuminated summer hemisphere at latitudes larger than 66.56° has not been included. The model used to produce these maps covers altitudes from 40 to 120 km divided into 50 height levels 1.6 km apart, and the latitude ranges from 80°S to 80°N in 5° steps. Departures from LTE of $CO_2(0,1^1,0)$, the state responsible for the fundamental band near 15 μm, become significant (>1 K) just below the 80 km altitude level at nearly all latitudes (Fig. 6.6). From 75 to 95 km, the vibrational temperature is warmer over the cold summer pole, and colder over all of the other latitudes. Around the 93 km level, there is a region ~ 5 km in height (narrower in summer and wider in winter) where the fundamental band is again close to LTE. In this region, T_v–T_k is between –1 and –2 K near the winter pole, increasing a little over the equator, and changes from positive (at lower altitudes) to negative in the vicinity of the summer pole. This second near-LTE region occurs to some extent for the upper states of all bands and is a consequence of the temperature structure. In the summer, the kinetic temperature abruptly increases with altitude above the mesopause, and the T_v–T_k difference is then large. In wintertime this atmospheric region is more isothermal and hence T_v–T_k varies with altitude more smoothly.

At night, the populations of the upper states of weaker bands in the upper mesosphere are maintained principally by absorption of upwelling thermal radiation and, therefore, show larger deviations from LTE than the stronger fundamental band. Fig. 6.7 shows the deviation of the vibrational from the kinetic temperature for the (0,2,0) level of the main isotope which starts breaking away from LTE significantly (>1 K) at around 75 km for most latitudes. The (0,3,0) levels (Fig. 6.8), which give rise to the weaker second hot bands, start deviating from LTE at slightly lower altitudes, around 70 km, at latitudes south of 40°S in the summer hemisphere, and around 65 km at latitudes north of 40°N in the winter hemisphere. From 20°S to 20°N its deviation from LTE is small up to altitudes as high as 85 km.

Figures 6.9 and 6.10 show T_v–T_k for the $(0,1^1,0)$ level of the less abundant 636 and 628 isotopes. The former becomes non-Boltzmann at around 70 km over the equator and the tropics, slightly higher near the summer pole, and around 65 km for latitudes north of 50°N. The $(0,1^1,0)$ level of 628 generally deviates from LTE lower than 636, but T_v–T_k is less than 1 K below around 68 km at latitudes south of 40°S while the equator and tropics retain LTE up to nearly 80 km. The height of LTE departure varies from 65 km near 40°N to only 60 km near the winter pole.

A general pattern in the non-LTE departures of upper states of the hot and iso-

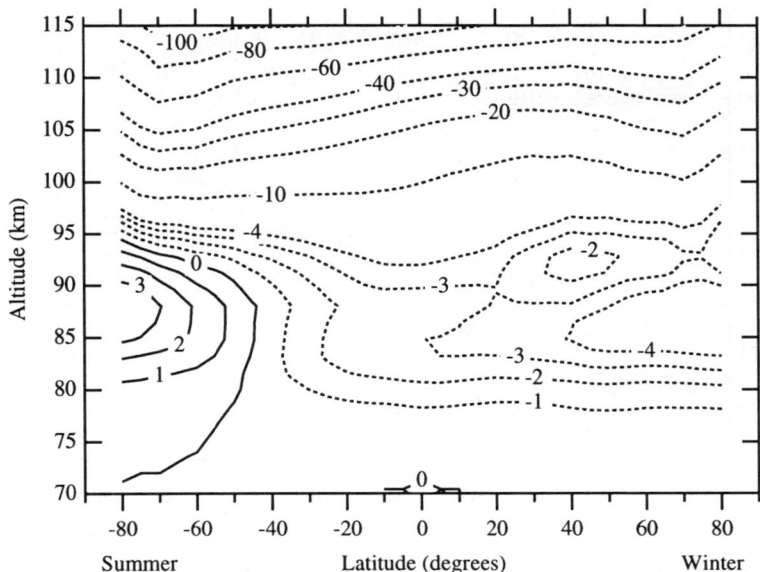

Fig. 6.6   Global distribution of $T_v - T_k$ for the $CO_2(0,1^1,0)$ level at night-time for the CIRA 86 temperature structure (see Fig. 1.2).

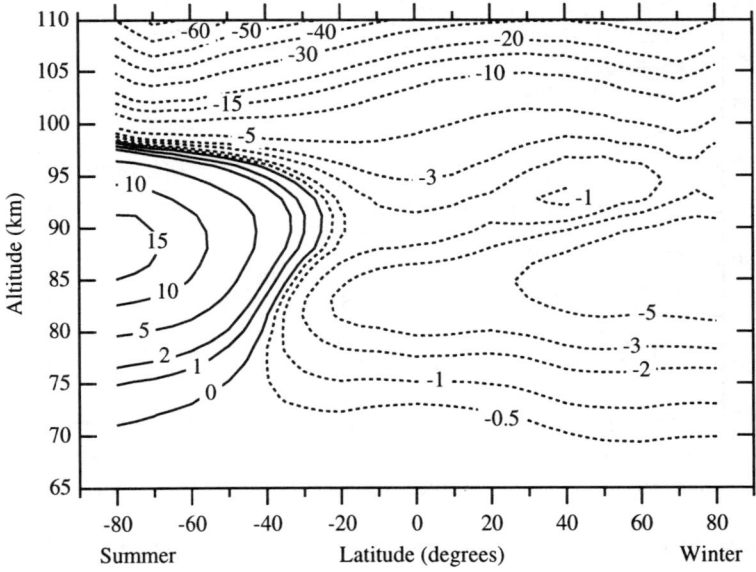

Fig. 6.7   Global distribution of $T_v - T_k$ for the $CO_2(0,2,0)$ level at night-time.

topic bands (Figs. 6.6–6.10) is the large over-population at summer latitudes around the mesopause, and the under-population at winter latitudes at about 5 km lower. The transition between the two regimes gives rise to a region where these levels are

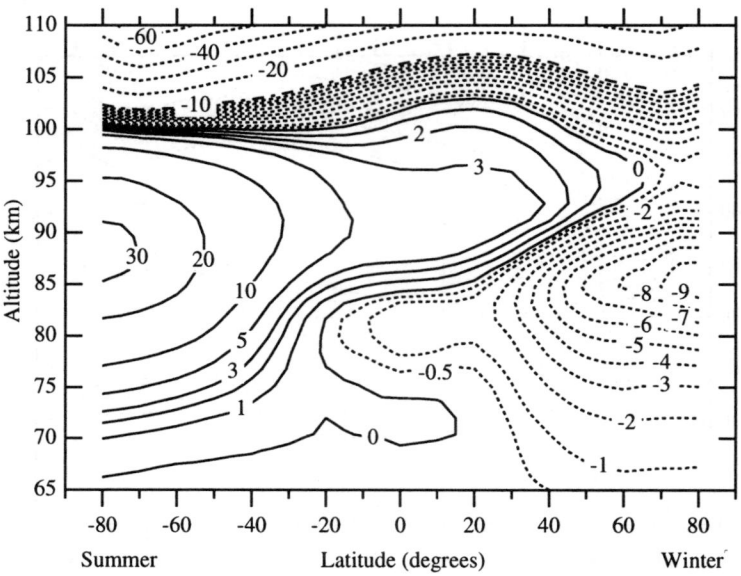

Fig. 6.8   Global distribution of $T_v - T_k$ for the $CO_2(0,3,0)$ level at night-time.

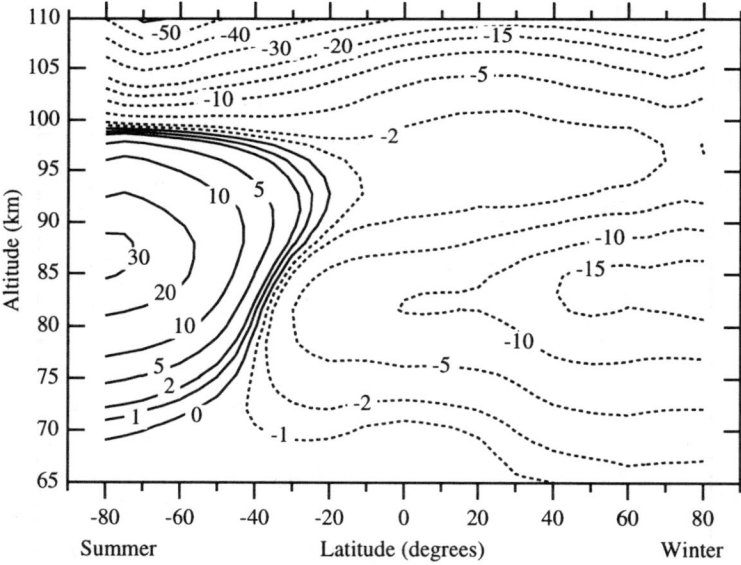

Fig. 6.9   Global distribution of $T_v - T_k$ for the CO_2 636 isotope $(0,1^1,0)$ level at night-time.

close to LTE. Thus, the $626(0,2,0)$ and $636(0,1^1,0)$ levels have vibrational temperatures very similar to the kinetic between 90 and 100 km over the winter hemisphere (Figs. 6.7 and 6.9). The corresponding region for $626(0,3,0)$ and $628(0,1^1,0)$ is narrower and confined to latitudes north of 40°N (Figs. 6.8 and 6.10).

Under daytime conditions, as explained above, we have the added complexity

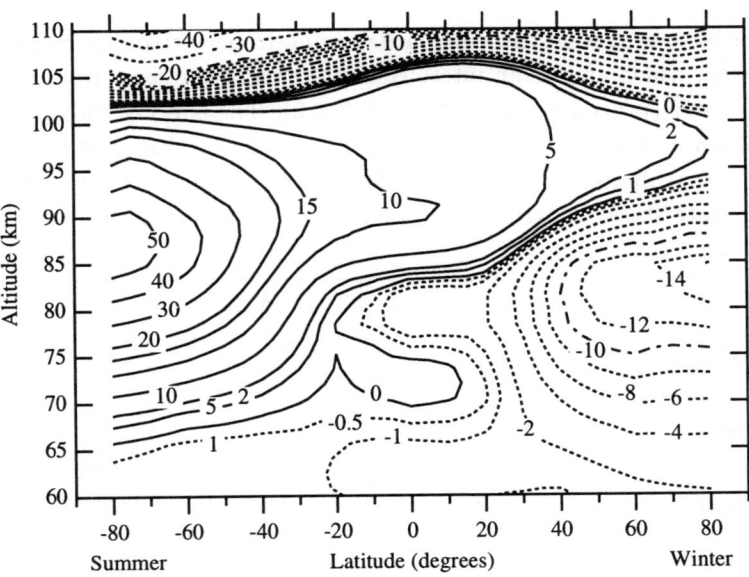

Fig. 6.10 Global distribution of $T_v - T_k$ for the CO_2 628 isotope $(0,1^1,0)$ level at night-time.

of the effect on the v_2 levels of the absorption of solar radiation in the near in-frared. The $(0,2,0)$ level is mainly populated by the relaxation of $(0,2^0,1)$, $(0,2^2,1)$ and $(1,0^0,1)$, which have been excited by the absorption of solar radiation near 2.7 μm and then radiated, or transferred to CO_2 or N_2, a v_3 quantum. The subse-quent relaxation of $(0,2,0)$ by emitting a v_2 photon then increases the population of $(0,1^1,0)$, although only very slightly. The population of $(0,3,0)$ is enhanced by the collisional relaxation of the $CO_2(0,0^0,1)$ level and, therefore, is caused mainly by the absorption of solar radiation at 4.3 μm. The deviations from LTE produced by these (and other less important processes) in the daytime population of the $(0,1^1,0)$ level are very similar to those shown above for night-time (see Fig. 6.6). There are appreciable differences only around the mesopause where the vibrational tempera-tures during daytime are generally warmer by around 1 K. The $(0,2,0)$ and $(0,3,0)$ levels increase their populations relative to night-time in the 80–110 km region (see Fig. 6.3), extending in height and to winter latitudes the region where T_v is warmer than T_k. However, the altitude for the onset of departure from LTE by day is roughly the same as at night.

6.3.6.3 *The v_3 and $2v_2+v_3$ levels emitting near 4.3 and 2.7 μm*

The upper states of the CO_2 4.3 μm and 2.7 μm bands, $(0,0^0,1)$, $(0,1^1,1)$, $(0,2^0,1)$, $(0,2^2,1)$ and $(1,0^0,1)$, remain in LTE up to about 65 km at night (see Fig. 6.11). Between 65 and 95–100 km, the absorption of upwelling radiation from the warmer lower mesosphere in the 4.3 μm bands results in T_v's larger than the kinetic temper-ature. There is also a population enhancement of about 25 K for the $(0,0^0,1)$ state of the major isotope around the mesopause, owing to near resonant collisions with the $N_2(1)$ level, which acts as an intermediary for energy transfer from the excited $OH^*(v \leq 9)$ levels (processes 9 and 13 in Table 6.2). The effect of radiative transfer is clearly illustrated with this example. Although the local excitation of $(0,0^0,1)$ by $OH^*(v \leq 9)$ is limited to the OH airglow layer (a region ~ 10 km wide around 85 km), this state shows a large enhancement up to very high altitudes above.

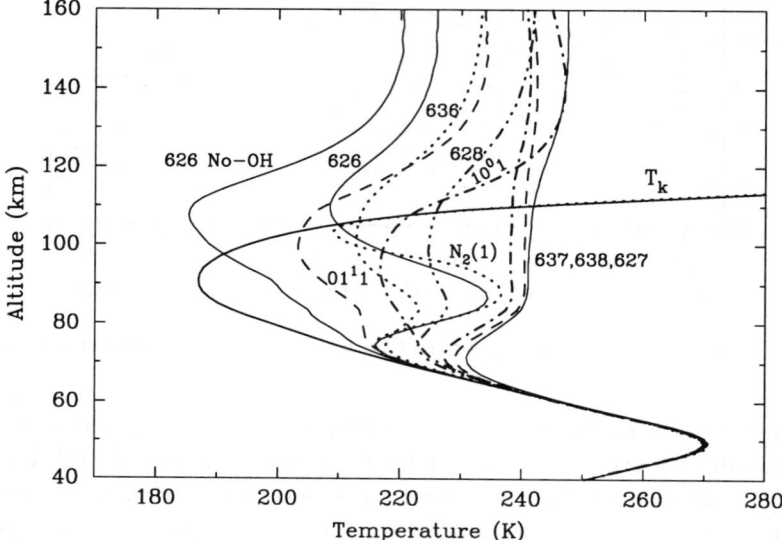

Fig. 6.11   Vibrational temperatures for the CO_2 v_3 and nv_2+v_3 levels of the CO_2 main isotope and of $N_2(1)$ for the U.S. Standard atmosphere at night. '626 No-OH' represents the population of $626(0,0^0,1)$ without OH excitation (process 13 in Table 6.2). T_k is the kinetic temperature.

Note also that $N_2(1)$ follows closely the population of $CO_2(0,0^0,1)$ up to the lower thermosphere (around 100 km), due to the strong V–V coupling between them (process 9 in Table 6.2). $N_2(1)$ couples to the kinetic temperature because of atomic oxygen collisions. Above 95 km the CO_2 vibrational temperatures are governed by the balance between the absorption of photons coming from the upper mesosphere, and photon escape to space.

The vibrational temperatures of the v_3 states at night for a temperature structure typical of winter pole conditions show features similar to those for the mid-latitude temperature profile. Their deviations from LTE in the mesosphere are smaller, however, because there is less absorption of radiation coming from the cooler stratopause. For summer conditions, however, as for the v_2 levels, the deviations from LTE around the mesopause are larger.

The day-time populations of these levels are enhanced compared to night-time by the absorption of solar radiation in the 4.3 μm and 2.7 μm bands (see Fig. 6.12). The T_v's of the v_3 levels are larger than T_k throughout the mesosphere and lower thermosphere, mainly excited by the absorption of solar radiation at 4.3 μm for the

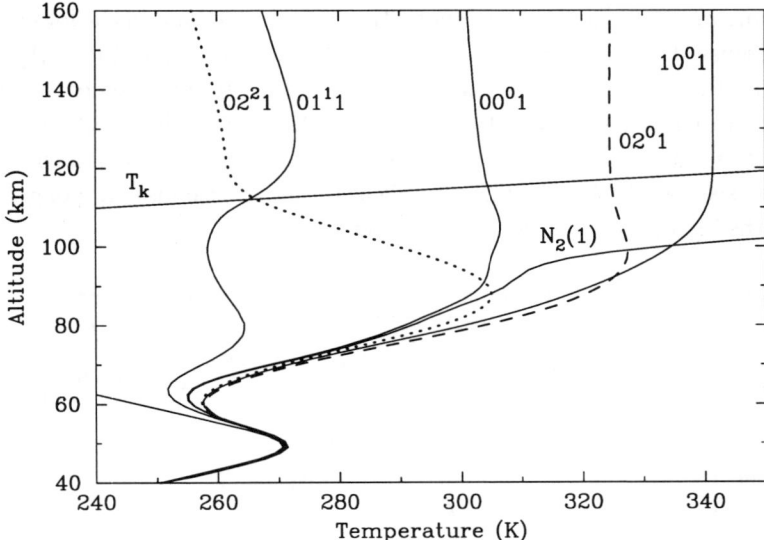

Fig. 6.12 Vibrational temperatures of the $(0,0^0,1)$, $(0,1^1,1)$, $(0,2^0,1)$, $(0,2^2,1)$ and $(1,0^0,1)$ states of the 626 isotope and of $N_2(1)$ at daytime ($\chi=60°$). T_k is the kinetic temperature.

$(0,0^0,1)$ and $(0,1^1,1)$ states, and at $2.7\,\mu m$ for the $(0,2^0,1)$, $(0,2^2,1)$ and $(1,0^0,1)$ levels. Between about the stratopause and $80\,km$, the V–V transfer of v_3 quanta among these states and $N_2(1)$ (process 9 in Table 6.2) also affects their populations. For example, $(0,0^0,1)$ is significantly populated by the absorption of solar radiation in the more weakly absorbing (and hence more deeply penetrating) $2.7\,\mu m$ spectral region, and subsequently transferred to $(0,0^0,1)$ by V–V collisions. Simultaneously, this process also provides an important relaxation of the $2.7\,\mu m$ levels. Absorption of solar radiation near $2.0\,\mu m$ is of some importance for the population of $CO_2(0,0^0,1)$ and $N_2(1)$ in the lower mesosphere, leading to an enhancement of 1–$2\,K$. Radiative transfer between atmospheric layers in the $4.3\,\mu m$ band has an important effect on the mesospheric and lower thermospheric vibrational temperatures of the $(0,0^0,1)$ levels, but for the other v_3 states this is much less important.

The transfer of electronic energy from $O(^1D)$ to $N_2(1)$ (process 12 in Table 6.2) is important for populating the latter, and in consequence $626(0,0^0,1)$, in the illuminated mesosphere and thermosphere. The inclusion of this process leads to an enhancement of the T_v of $(0,0^0,1)$ of a few kelvins (2–$6\,K$, depending on the actual $O(^1D)$ profile and altitude) from the stratopause up to the thermosphere (see Fig. 6.15). This enhancement may be even larger depending on the $O(^1D)$ profile and solar illumination conditions, and is relatively more important for twilight and low-Sun conditions. It also has some influence on the rest of the v_3 and nv_2+v_3 states through collisions with $N_2(1)$ (process 9).

If we express the daytime populations of the $(0,0^0,1)$ and $(1,0^0,0)$ states in number densities, we find that there exists a population inversion among them, i.e.

there are more molecules in the more energetic $(0,0^0,1)$ state and in the lower $(1,0^0,0)$ level. As discussed above, this is produced by the solar pumping of $(0,0^0,1)$, mainly at $4.3\,\mu m$, and gives rise to a maximum $(0,0^0,1)/(1,0^0,0)$ enhancement of 20%; or 35% if we consider also the excitation from $O(^1D)$. This produces a 'natural' laser emission in the $(0,0^0,1)–(1,0^0,0)$ $10.4\,\mu m$ band, which is more typical of the Martian atmosphere (see Sec. 10.4.3.3). The amplification for laser emission along the limb tangent paths in the Earth is, however, negligible ($\sim 10^{-5}$).

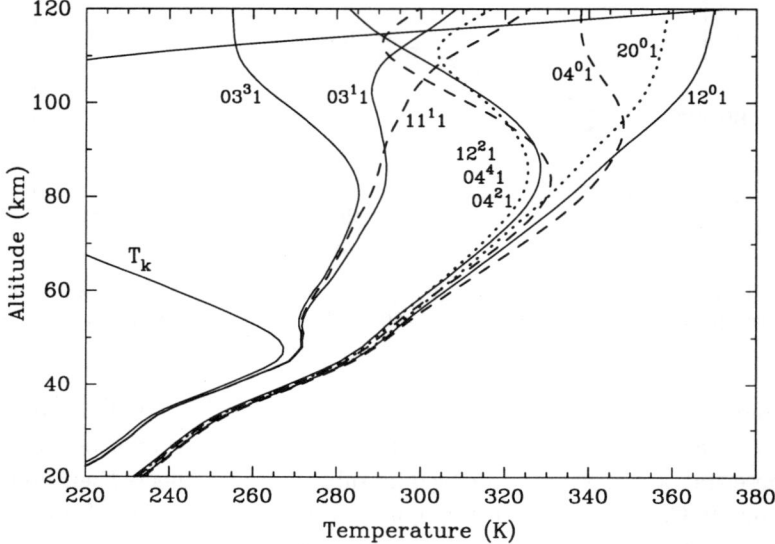

Fig. 6.13 Vibrational temperatures for the $3v_2+v_3$ $[(0,3^1,1), (0,3^3,1), (1,1^1,1)]$ levels (emitting the $4.3\,\mu m$ third hot bands) and $4v_2+v_3$ $[(0,4^0,1), (0,4^2,1), (0,4^4,1), (1,2^0,1), (1,2^2,1), (2,0^0,1)]$ states (emitting the $4.3\,\mu m$ fourth hot bands) of the 626 isotope calculated for daytime ($\chi=60°$) conditions. T_k is the kinetic temperature.

Figure 6.13 shows the vibrational temperatures of the 626 levels corresponding to the upper states of the third and fourth hot bands (4.3 TH and 4.3 FRH in Table F.1) for the U.S. Standard Atmosphere and $\chi = 60°$. Compared to the lower-energy $4.3\,\mu m$ levels, these states exhibit much larger non-LTE deviations, particularly in the lower mesosphere and in the stratosphere. They are excited by absorption of solar radiation near $2.7\,\mu m$, that is, by absorbing molecules already excited in the (01^10) level, and near $2.0\,\mu m$ by molecules in the ground state. The weaker absorption in these bands allows the solar radiation to penetrate deeper into the atmosphere, and hence to cause a larger excitation at lower altitudes, producing non-LTE populations down to the troposphere. Above about $80\,km$ the states exhibiting largest vibrational temperatures are those directly excited by absorption of solar radiation, for example, $(1,1^1,1)$ and $(0,3^1,1)$ at $2.7\,\mu m$, and $(2,0^0,1)$, $(1,2^0,1)$ and $(0,4^0,1)$ near $2.0\,\mu m$.

Figure 6.14 shows the vibrational temperatures for the corresponding levels of

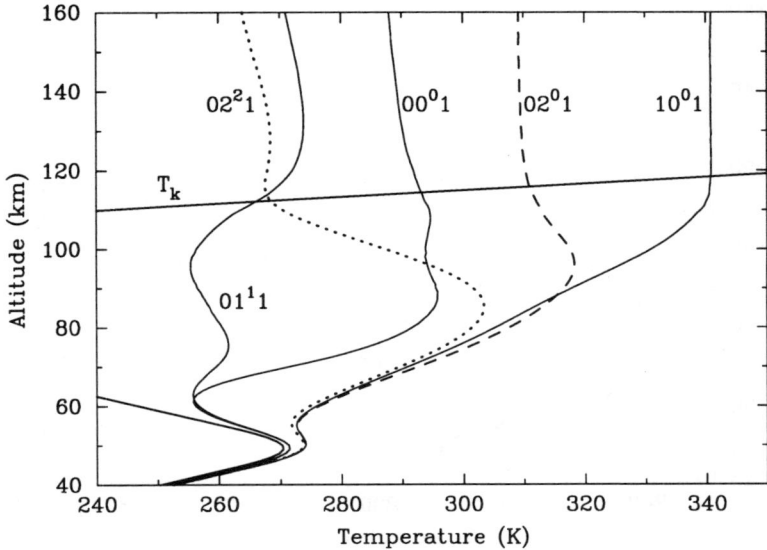

Fig. 6.14   Vibrational temperatures of the $(0,0^0,1)$, $(0,1^1,1)$, $(0,2^0,1)$, $(0,2^2,1)$, and $(1,0^0,1)$ states of the 636 isotope at daytime $(\chi = 60^\circ)$. T_k is the kinetic temperature.

the 636 isotope. Analogously to the v_3 and nv_2+v_3 626 states, the enhancement of the $636(0,0^0,1)$ and $(0,1^1,1)$ daytime vibrational temperatures compared to night is essentially due to absorption of solar radiation in the fundamental and $4.3\,\mu m$ first hot bands of this isotope, and also indirectly by absorption in the 4.3 and $2.7\,\mu m$ bands of the major isotope followed by V–V energy transfer involving collisions with $N_2(1)$. The vibrational temperature of the $(0,0^0,1)$ level is significantly smaller than that of the 626 isotope above about 80 km. Those of the $(0,2^0,1)$, $(0,2^2,1)$, and $(1,0^0,1)$ are very similar in the upper mesosphere, but significantly larger in the lower mesosphere, where they are excited by the absorption of the weaker isotopic 4.3 and $2.7\,\mu m$ bands, in which the solar radiation penetrates deeper in the atmosphere. The vibrational temperatures of the $3v_2+v_3$ and $4v_2+v_3$ states do not vary significantly compared to those of the corresponding states of the main isotope.

The populations of all these states $[v_3, nv_2+v_3,$ and $N_2(1)]$ show a significant dependence on the solar zenith angle, as illustrated for some levels of the 626 isotope in Fig. 6.15. The effect is largest for the level connected by the strong $4.3\,\mu m$ 626 fundamental band in the upper regions, and is significant at altitudes above about 55 km. For larger solar zenith angles, solar radiation is more strongly absorbed at higher altitudes, leading to smaller vibrational temperatures in the mesosphere.

The weaker first and second hot bands exhibit less dependence; in particular the second hot bands, which remain optically thin from Sun overhead conditions until close to twilight. The corresponding $2v_2+v_3$ states, excited mainly by solar radiation in the $2.7\,\mu m$ bands, only show variations with solar zenith angle below about 70 km.

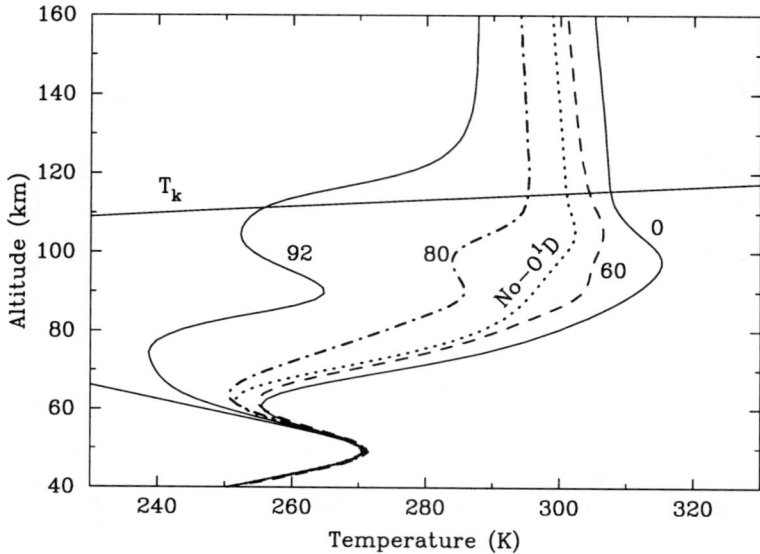

Fig. 6.15 Solar zenith angle dependence of the vibrational temperature of $626(0,0^0,1)$. The dotted line is the T_v for $\chi = 60°$ without the excitation from $O(^1D)$. T_k is the kinetic temperature.

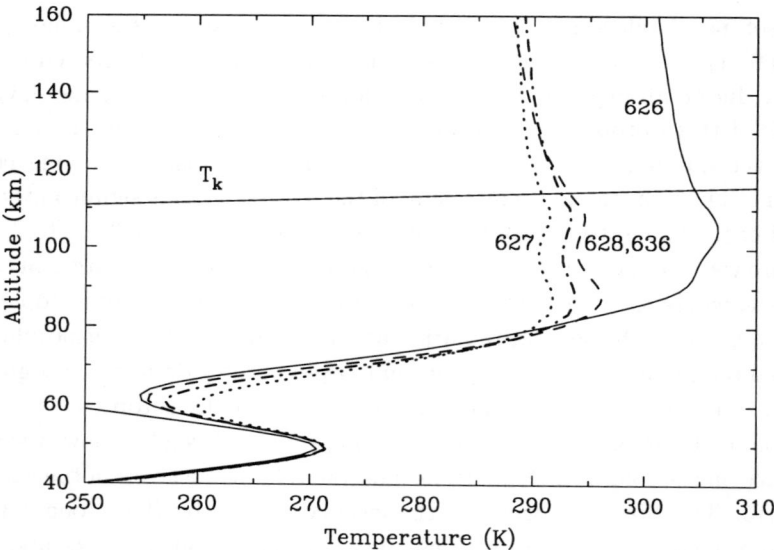

Fig. 6.16 Daytime $(\chi = 60°)$ vibrational temperatures for the $(0,0^0,1)$ states of the four CO_2 major isotopes 626, 636, 628, and 627. T_k is the kinetic temperature.

Figure 6.16 shows the vibrational temperatures of the $(0,0^0,1)$ state of the 628, and 627 isotopes, compared to those for 626 and 636. The populations of the former are essentially driven by the same processes that control the population of the 636 isotope. Collisions with $N_2(1)$ and the absorption of solar radiation in the corresponding $4.3\,\mu$m band are the major excitation processes. Like for 636, the exchange of radiation within the atmospheric layers is less important than it is for the $4.3\,\mu$m band of the major 626 isotope. Note also that the weaker the transition, the smaller is the solar excitation in the lower thermosphere. In the mesosphere, collisions with $N_2(1)$ become the dominant excitation process and all the T_v's tend to similar values.

6.3.6.4 *Uncertainties in the CO₂ populations*

The primary source of error in the computed populations of the CO_2 vibrational states is probably the rates of the processes listed in Table 6.2, as they are known today. We describe here the effects of some estimates in the uncertainties of these rates on the vibrational temperatures of the various levels. Remember that, for several of the CO_2 vibrational levels (principally those giving rise to the weak hot and isotopic bands), the rate coefficients have not been measured at all, and been assumed to be the same as those measured for the more populated levels. The scale of the uncertainties this introduces into the populations is unknown, and has been neglected in the following analysis. So, the uncertainties given should be regarded as lower limits.

The rate coefficients of processes 1, 3 and 4 are rather well determined and the effect of their uncertainties on the populations of the v_2 and v_3 levels is small. The uncertainties in the T_v of the (0,3,0) and (0,4,0) levels in the upper mesosphere are estimated as about ± 5 K and ± 10 K, respectively. The estimated uncertainties in the $(0,0^0,1)$ and $(0,1^1,1)$ levels are only ± 2 K, and the $2v_2+v_3$ levels emitting near $2.7\,\mu$m are hardly affected at all by these processes.

The v_2 levels are affected above about 90 km by the variability of atomic oxygen and the uncertainty in the collisional coefficient $k_{CO_2\text{-}O}$ (see Sec. 6.3.4). The $(0,0^0,1)$ level is affected by about ± 2 K above 80 km by the $O(^3P)$ variability and the uncertainty in the rate coefficient of process 15. The $(0,1^1,1)$ and $(0,2^2,1)$ states, populated mainly by absorption of solar radiation in hot bands arising from the v_2 levels, are also significantly affected above about 90 km. For the $(0,2^0,1)$ and $(1,0^0,1)$ states, which are solar-pumped from the ground state, the effect is very small.

The rate for the V–V exchange of v_3 quanta in process 9 has a significant influence on the populations of the $2v_2+v_3$ levels, mainly in the lower mesosphere. These levels show larger populations for a slower rate constant, while the opposite behaviour is displayed by the $(0,0^0,1)$ state, and with a smaller variation. This change is propagated to the upper regions because of radiative transfer in the ν_3 fundamental band. Since the population of $(0,0^0,1)$ is a source of excitation of the (0,2,0), (0,3,0) and (0,4,0) levels, they also show a small change in their vibrational temperatures. The change in the vibrational temperatures of the other levels does not exceed 5 K when the rate is perturbed by a factor of 1.5, except for the $2v_2+v_3$ levels in the lower mesosphere, where it can be around 10 K.

The variability of the $O(^1D)$ abundances and uncertainty in the rate coefficient and efficiency of process 12 also affect the v_3 and $2v_2+v_3$ populations. The most perturbed level is $(0,0^0,1)$, with overall changes of around $10\,K$ for a rather extreme $O(^1D)$ abundance variability. The $(0,1^1,1)$ and $(0,4,0)$ states show smaller variations than $(0,0^0,1)$, and the rest of the levels do not suffer significant changes.

The abundance of CO_2 is also rather uncertain, as we show in Sec. 8.10.3, but this has a relatively small effect on the vibrational temperature since the latter is a ratio of populations. The maximum changes of $2\,K$ occur near the polar summer mesopause in the vibrational temperatures of the upper levels of the weaker bands.

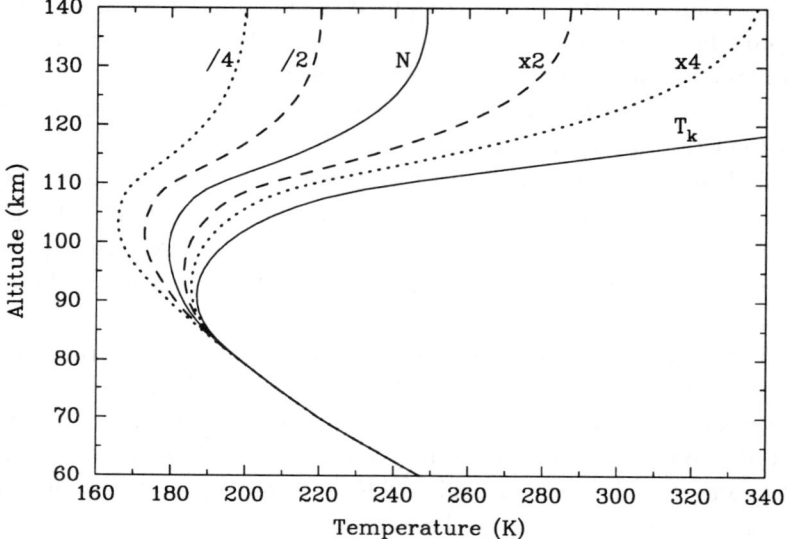

Fig. 6.17   Effect of the $k_{CO_2\text{-}O}$ rate coefficient and the $O(^3P)$ concentration on the $CO_2(0,1^1,0)$ vibrational temperature for the U.S. Standard. 'N' is the nominal T_v and the other curves are the results for the product $k_{CO_2\text{-}O}\times[O(^3P)]$ changed in the stated factors.

The least well known parameters which significantly affect the departures from LTE of the $CO_2(v_1,v_2,0)$ levels in the upper mesosphere and thermosphere are the rate coefficient of collisions of $CO_2(v_2)$ with $O(^3P)$, $k_{CO_2\text{-}O}$, and the volume mixing ratio (VMR) of atomic oxygen itself. To quantify the effect of $k_{CO_2\text{-}O}$ we estimate the current uncertainty as varying between $1.5\times10^{-12}\,cm^3s^{-1}$ from laboratory measurements to the upper values of $6.0\times10^{-12}\,cm^3s^{-1}$ derived from the analysis of rocket and satellite measurements. The atomic oxygen concentration, $[O(^3P)]$, affects T_v in the same way as the $k_{CO_2\text{-}O}$ rate and it is very variable in the upper mesosphere and thermosphere. Considering a factor of 2 in the variability of $[O(^3P)]$, we show the effects of both parameters together by multiplying and dividing the product $k_{CO_2\text{-}O}\times[O(^3P)]$ by factors of 2 and 4 (see Fig. 6.17).

The effect of using a faster rate is to drive the vibrational temperature of the v_2 levels closer to the kinetic temperature, the magnitude generally being larger for the stronger bands, particularly under summer conditions, when the temperature gradient is steeper. A variation in $k_{CO_2\text{-}O} \times [O(^3P)]$ by a factor of 2 (either in $k_{CO_2\text{-}O}$ or in $[O(^3P)]$), changes the vibrational temperature of $CO_2(0,1^1,0)$ by about 25 K at 120 km to close to 40 K at 140 km. Increasing the $k_{CO_2\text{-}O} \times [O(^3P)]$ product by a factor of 4 lead to much larger changes, about 50 K at 120 km to 85 K at 140 km. If the product is decreased by the same factor, the change in the vibrational temperature is smaller. The larger effect of $k_{CO_2\text{-}O} \times [O(^3P)]$ at higher altitudes and for larger values is a consequence of the greater importance of collisional, relative to radiative, processes in populating those levels.

In the region below 100 km, down to 80 km, where the concentration of atomic oxygen is significantly reduced, the vibrational levels are differently populated depending on the band strength and the kinetic temperature lapse rate prevailing in the mesosphere. Thus, while an increase in $k_{CO_2\text{-}O} \times [O(^3P)]$ produces an increase in the T_v of $(0,1^1,0)$ of the main isotope, it leads to a decrease in the T_v's of the minor isotopes and in those of the v_2 overtone levels. Although of different sign, the change in the T_v's of the moderate bands produced by a change in $k_{CO_2\text{-}O} \times [O(^3P)]$ is similar to that in Fig. 6.17, except for the weaker bands, which is larger.

In summary, while it is clear that a comprehensive radiative transfer scheme is required for most applications if realistic answers are to be obtained, many of the important energy transfer paths in collisions are still not well understood. In addition to more and better laboratory data, there is a great need for more experimental studies of non-LTE effects in the atmosphere and comparisons of these with models such as that presented here. We shall discuss some of these comparisons in Chapter 8.

6.4 References and Further Reading

The treatment in this chapter is based on the model described by López-Puertas *et al.* (1986a,b), Edwards *et al.* (1993), López-Puertas and Taylor (1989) and López-Puertas *et al.* (1998a), which used a modified Curtis matrix formulation. Other non-LTE models for the CO_2 levels have been: (i) the first one developed by Curtis and Goody (1956) for computing the CO_2 15 μm infrared cooling at night-time; (ii) that derived by Kuhn and London (1969) for calculating the CO_2 15 μm infrared cooling at night-time, including rotational distribution and exchange between vibrational levels; (iii) that developed at Oxford University in the late 60's and early 70's (Houghton, 1969; Williams, 1971a,b; Williams and Rodgers, 1972) extending the model to the v_3 and $2v_2+v_3$ levels emitting near 4.3 and 2.7 μm for daytime conditions; that developed at St. Petersburg University (Kutepov and Shved, 1978; Shved *et al.*, 1978) including 15, 4.3, and 2.7 μm; that developed in the 70's, mainly for the 4.3 and 2.7 μm emissions, by Kumer and co-workers (Kumer and James, 1974; Kumer, 1975, 1977a,b; Kumer *et al.*, 1978); and that developed by Dickinson (1984) as an extension of his model for the Venusian atmosphere. More recently we can cite the model developed in the 90's at Air Force Research Laboratory (Sharma and Wintersteiner, 1990; Wintersteiner *et al.*, 1992) and the revision to the model developed at St. Petersburg University including all bands up to $1.0\,\mu m$ (Shved *et al.*, 1998; Ogibalov *et al.*, 1998). Enhancements of $4.3\,\mu m$ emissions above thunderclouds have been discussed by Picard *et al.* (1997). Ogivalov and Shved (2001) have proposed a hierarchy of simplified CO_2 non-LTE models. Natural population inversion in the Earth's atmosphere has been studied by Shved and Ogivalov (2000).

Other models were also developed specifically for computing the CO_2 $15\,\mu m$ cooling rates, such as those of Wehrbein and Leovy (1982), Fels and Schwarzkopf (1981), Haus (1986), Apruzese *et al.* (1984), Zhu (1990), and Zhu *et al.* (1992). More details on these are given in the 'References and Further Reading' sections of Chapters 5 and 9.

The importance of including a complete radiative transfer treatment in calculations of the populations of the CO_2 states has been pointed out by many workers, for example Williams and Rodgers (1972), Rodgers and Williams (1974), Kutepov and Shved (1978) and Dickinson (1984), and more recently by López-Puertas *et al.* (1994). The difference between the quasi-line-by-line method of Sec. 6.3.3.1 and full line-by-line codes for the calculation of the vibrational temperatures and cooling rates is discussed by López-Puertas *et al.* (1994) for the CO_2 $15\,\mu m$ bands; for $CO(1)$ in López-Puertas *et al.* (1993); and for H_2O in López-Puertas *et al.* (1995).

The COSPAR International Reference Atmosphere 'grand mean' kinetic temperature representing global mean conditions may be found in Barnett and Chandra (1990) and Fleming *et al.* (1990). The neutral composition models used in this chapter are found in Allen *et al.* (1981) and Rodrigo *et al.* (1986), with atomic oxygen abundances from Van Hemelrijck (1981), Rees and Fuller-Rowell (1988), and Thomas (1990). The carbon dioxide volume mixing ratio profile was taken from the ISAMS measurements (López-Puertas *et al.*, 1998b). A recent review of available CO_2 measurements can be found in López-Puertas *et al.* (2000).

Exospheric solar fluxes for the calculation of the absorption of solar radiation by atmospheric molecules in the infrared and near-IR can be found in Thekaekara (1976), Kurucz (1993), and Tobiska *et al.* (2000).

A detailed review of the collisional rates can be found in López-Puertas *et al.* (1986a,b), López-Puertas and Taylor (1989), Shved *et al.* (1998), López-Puertas *et al.* (1998a) and Clarmann *et al.* (1998), and the references given in Table 6.2. General texts on collisional processes are Herzfeld and Litovitz (1959), Cottrell and McCoubrey (1961), Lambert (1977), Yardley (1980), Bransden and Joachain (1982). Good reviews, some parts out of date but interesting for the basic theory they provide, can be found in Taylor (1974) and Moore (1973), and a more recent review in Flynn *et al.* (1996). Results for the detailed studies of the sensitivity of the CO_2 level populations to uncertainties in the various model parameters are given in López-Puertas and Taylor (1989), López-Puertas *et al.* (1992a, 1998c), Clarmann *et al.* (1998), and Mertens *et al.* (2001).

Chapter 7

Non-LTE Modelling of the Earth's Atmosphere II: Other Infrared Emitters

7.1 Introduction

The previous chapter described a non-LTE radiative transfer model for the carbon dioxide vibrational levels in the Earth's atmosphere, including only those additional species like N_2 and O_2 which interact with CO_2 and hence affect its behaviour. CO_2 is probably the most important molecule in the atmosphere, because it dominates the heating and cooling rates, at least for those levels where non-LTE is an issue. CO_2 is also important because it is the species used most often for remote sounding of atmospheric temperature. Water vapour is the dominant greenhouse gas in the troposphere while NO takes over in the middle and upper thermosphere (see Fig. 9.18). O_3 is also a greenhouse gas affecting the temperature of the Earth's surface, and its infrared emissions plays an important role in cooling the region around the stratopause and mesopause.

Now we will address the question of how atmospheric species other than CO_2 behave radiatively, and how their energy levels can be incorporated into non-LTE models, proceeding molecule by molecule. However, since energy is exchanged between species as radiation and by collisional interactions, it is necessary to include such exchanges explicitly in the models whenever the amount of energy involved is significant. This modelling is important, besides the role played by of some of these molecules in the radiative budget mentioned above, because of its potential influence on their retrievals from measurements of their infrared emissions.

7.2 Carbon Monoxide, CO

CO has many fewer levels than CO_2 since it is a diatomic molecule and so has only one fundamental vibrational mode. Because it is several orders of magnitude less abundant than CO_2 (see Fig. 1.4), CO needs to be modelled only if we are interested in the emissions from CO itself; it does not play a large role in the reverse direction, for example, in the energetics of CO_2, in the way that N_2 does.

However, CO is an important molecule in the atmosphere for other reasons. As it is a major component of atmospheric pollution, there is much interest in mapping its abundance, both locally and globally. In the middle atmosphere, CO is produced by the oxidation of methane and, at higher levels, by the dissociation of CO_2. The main sink is conversion back into CO_2 by reaction with the hydroxyl radical, OH, which is derived from water vapour and ozone. Thus, a mixture of natural and anthropogenic sources and sinks contribute to a fairly complicated global distribution for CO. This is modified by dynamics, and CO is also a valuable tracer of stratospheric motions, particularly in view of its intermediate chemical lifetime of weeks to months which produces large horizontal and vertical gradients.

The global distribution of CO in the middle and upper atmosphere can be studied using remote sensing measurements from satellites. These usually measure the emission from the first vibrationally excited state, CO(1), which departs from LTE in the stratosphere and above. The band is strongly pumped by solar radiation in the mesosphere and lower thermosphere, which gives rise to a large dayside radiances at $4.7\,\mu m$ which can be measured up to 120 km or even higher. It also, of course, means that a non-LTE model is required to interpret the measurements in terms of the CO abundance. The CO(1) level is populated by collisional interactions with N_2, O_2, CO_2, and $O(^3P)$, as well as by solar pumping and radiative exchange with atmospheric layers. The populations can be calculated by coupling the CO levels to the solution of the main non-LTE model for CO_2, developed in the previous chapter. Before proceeding with this let us consider the processes which populate the CO(1) levels in more detail.

7.2.1   *Radiative processes*

The relevant radiative processes (spontaneous emission, absorption of solar radiation, and exchange of photons within the atmosphere and with the surface) can be incorporated into the Curtis matrix calculation in the same way as for CO_2. The population of CO(1) in the stratosphere and above is quite sensitive to the upwelling flux from the troposphere, and even factors such as low-level cloudiness affect the vibrational temperature at much higher levels in the mesosphere, particularly at night. The tropospheric flux is included as explained in Sec. 6.1 and using the effective temperatures derived from the data provided by the TOVS suite of instruments on board the NOAA-9 and NOAA-10 satellites (see Table 7.1).

The absorption of sunlight can be included in the same way as for CO_2, except that account must be taken of the existence of large amounts of carbon monoxide in the photosphere of the Sun. Partly because of the nature of the CO spectrum, which consists of well-spaced, individual, strong lines, the solar absorption lines affect the $4.7\,\mu m$ spectral flux incident at the Earth to an extent which has a significant effect on the CO(1) population. Measurements by the ATMOS instrument (Sec. 8.10.1) show that the absorption at the centre of a typical strong line is about 30%, and

Table 7.1   Effective temperatures for the emerging tropospheric flux in the spectral region of the CO(1–0) band near 4.7 μm.

Latitude/solar illumination	Mean temperature (K)	2σ (K)
Mid. Latitudes/Day	272.8	26.4
Mid. Latitudes/Night	272.8	26.4
Polar winter/Night	237.2	13.3
Polar summer/Day	257.9	15.4

that the line widths are about $0.1\,\text{cm}^{-1}$. We can take approximate account of this by reducing the solar flux by the same factor in our calculations, noting that this may underestimate the flux in the line wings, but that the latter are generally only important in the higher-pressure regions where LTE applies anyway.

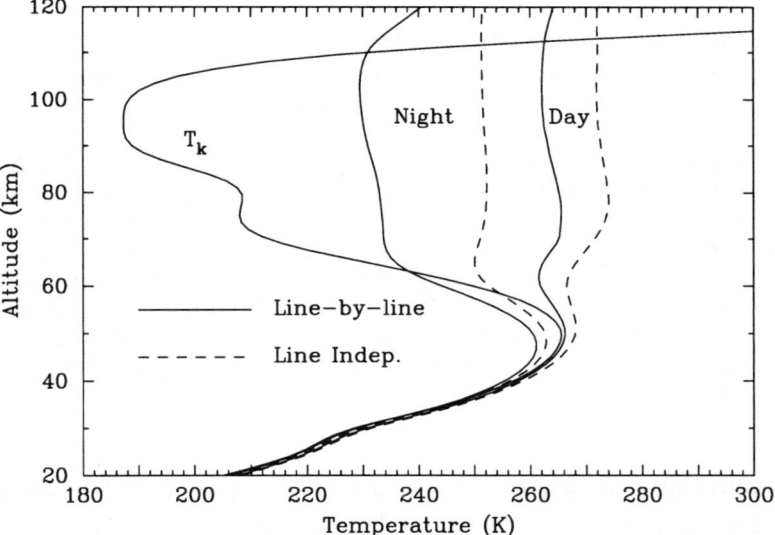

Fig. 7.1   Day and night-time vibrational temperatures for carbon monoxide, compared with the kinetic temperature profile. The dashed lines uses the quasi-line-by-line histogramming algorithm to calculate the Curtis matrix; the solid lines use the more precise GENLN2 line-by-line code.

For the calculation of the CO Curtis matrix or its transmittances, one should be careful. Because the CO(1–0) transition is weak enough to transmit radiation to space from as low as the troposphere, the flux transmittance calculations involve large gradients in temperature, pressure, and absorber amount. Under such conditions, the quasi-line-by-line histogramming algorithm described in Sec. 6.3.3.1 together with the Curtis-Godson approach might not give accurate results for the mesospheric CO(1) populations. Fig. 7.1 shows a test of the quasi-line-by-line histogramming approach (line independent) for carbon monoxide, which demonstrates that the errors, when compared to the more accurate line-by-line schemes (GENLN2 in this case), are very large (more than 20 K at night-time and to 10 K at daytime), nearly as large as the day-night difference. This shows that the approximate method is worse than useless in this case, since the error is comparable to the effect under study. A highly accurate transmission calculation is required, demonstrating that a line-by-line code, as those described in Sec. 4.3.1, is an essential tool in the inventory of the non-LTE modeller.

7.2.2 *Collisional processes*

Table 7.2 shows the important collisional processes affecting the population of the $CO(1)$ state. The collision partners are, in general, themselves in non-LTE, and it is by introducing the populations of N_2, O_2, and CO_2 to these reactions that we couple the calculation for the populations of carbon monoxide to the non-LTE model for the most abundant molecules.

Table 7.2   Principal collisional processes affecting the $CO(1)$ state.

No.	Process	E^*	Rate coefficient[†]	Ref.[‡]
1	$CO(1)+N_2 \rightleftharpoons CO+N_2(1)$	-186.4	$5.47\times10^{-15}\exp(3.82A - 5.47B)$	1
2	$CO(1)+O_2 \rightleftharpoons CO+O_2(1)$	587.1	$9.79\times10^{-17}\exp(8.02A - 2.05B)$	2
3	$CO(1)+CO_2 \rightleftharpoons CO+CO_2(00^01)$	-205.6	$7.34\times10^{-14}\exp(7.36A - 1.01B)$	3
4	$CO(1)+O(^3P) \rightleftharpoons CO+O(^3P)$	2143.3	$2.85\times10^{-14}\exp(9.50A + 1.11B)$	4

[*]Excess of energy in cm^{-1}. [†]Rate coefficient for the forward sense of the process in cm^3s^{-1}. $A = (T - 300) \times 10^{-3}$. $B = (T - 300)^2 \times 10^{-5}$. T is temperature in K. [‡]Ref. 1: Allen and Simpson (1980); 2: Doyennette *et al.* (1977); 3: Starr and Hancock (1975); 4: Lewittes *et al.* (1978).

The collisional processes with $N_2(1)$ are the most important interactions for $CO(1)$ in the atmosphere. Because of the large daytime population of $CO_2(0,0^0,1)$, collisional exchange with this level is the most likely path for excitation of CO from CO_2, but this is still more than 100 times less important than collisions with N_2, because of the much greater abundance of the latter.

Vibrational-vibrational energy transfer from $CO(1)$ to $O_2(1)$ (process 2) is strongly exothermic and the rate coefficient is not known to within two orders of magnitude. If the true value is near the high end of the range, then this exchange could have an important effect on the population of $CO(1)$; otherwise, process 1 with N_2 dominates.

The deactivation of $CO(1)$ by collision with atomic oxygen $O(^3P)$, process 4, is probably the most important of the remaining collisional exchanges which take place. Collisions with $O(^3P)$ are significant above 80 km and comparable to those with $N_2(1)$ above 100 km. Collisions with molecules such as NO, O_3 and H_2O make a negligible contribution.

7.2.3 *Non-LTE populations*

Figure 7.2 illustrates the vibrational temperatures for CO(1), at night and for different solar zenith angles during the day, calculated using the U.S. Standard Atmosphere (1976). The species is out of LTE from as low as 40 km altitude. The population is less than Boltzmann up to about 60 km and then much greater, even at night. Collisional processes dominate below about 50 km while radiative transfer controls the population of CO(1) in the mesosphere and above.

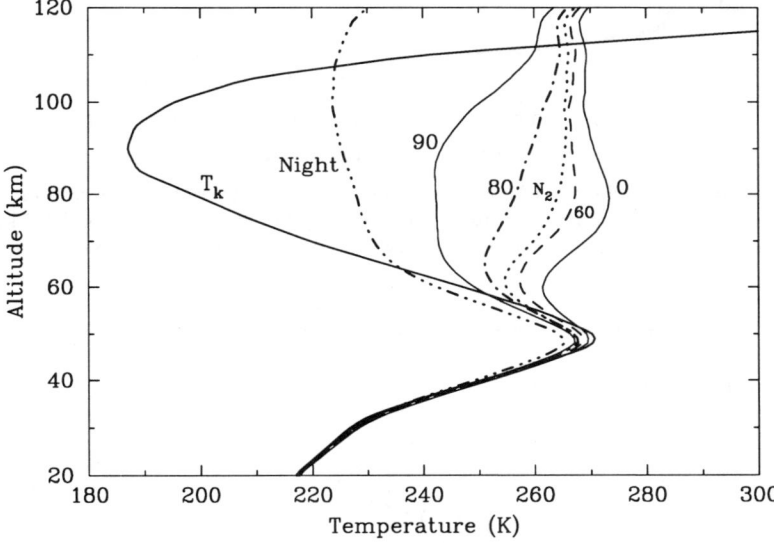

Fig. 7.2 Vibrational temperatures of CO(1) for the various solar zenith angles marked, including night-time. The daytime ($\chi=0°$) T_v when $N_2(1)$ is assumed to be in LTE (N_2) and the kinetic temperature T_k are also plotted.

During night-time, the non-LTE population of CO(1) above the stratopause is determined by the absorption of photons coming mainly from the troposphere and, to a lesser extent from other atmospheric layers (see Fig. 7.3). Note in particular the large contribution from the lower boundary: at night, over 40% of the flux absorbed near 117 km altitude originates in the troposphere. Note also the increase with altitude in the vibrational temperature above about 100 km, which is due to the absorption of photons from the troposphere, enhanced as a result of Doppler broadening of the lines at thermospheric temperatures. The deviation of the CO(1) T_v from the kinetic temperature is significant above 50 km even at night, although this depends strongly on the temperature profile. For example, for a typical polar winter atmosphere, T_v is close to T_k up to nearly 90 km, in comparison with the 60 km departure for a mid-latitude T_k profile.

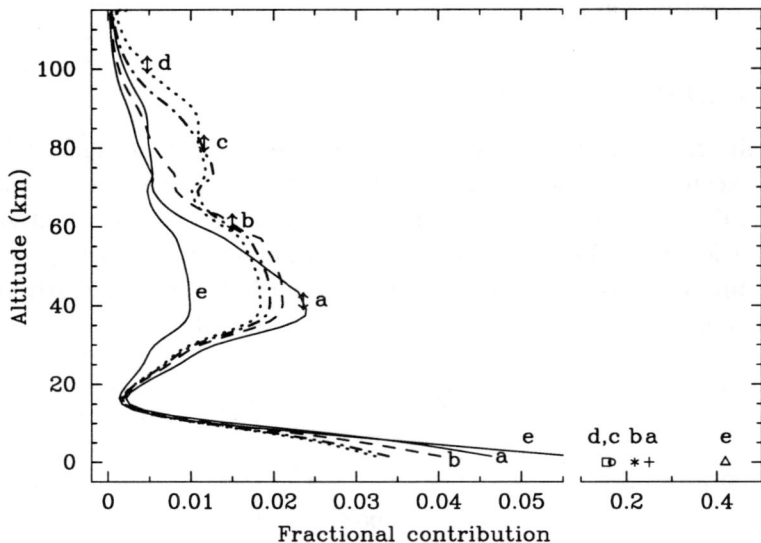

Fig. 7.3 Fractional contribution of radiative transfer between layers to the excitation of CO(1). Cases a–e correspond to excitation at observation points (↕) of 40, 60, 80, 100, and 117 km respectively. The contribution from the lower boundary is shown on an extended scale at the right.

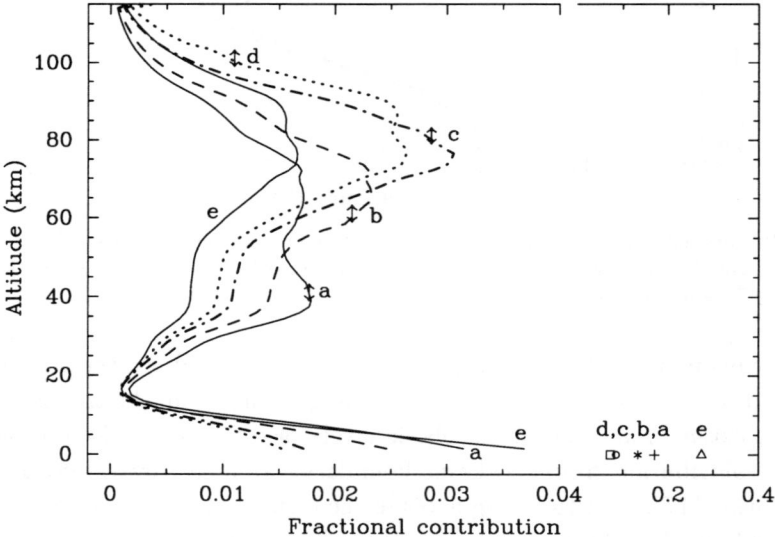

Fig. 7.4   As Fig. 7.3 but for daytime ($\chi=0°$).

The daytime non-LTE populations of CO(1) are mainly driven by absorption of solar radiation at 4.7 μm, and show a strong dependence on the solar zenith angle. Radiative transfer between atmospheric layers is also very important, as Fig. 7.4 shows. The important enhancement given to the daytime mesospheric

CO(1) population by vibration-vibration energy transfer from $N_2(1)$ can be seen by comparing the curves 'N_2' and '0' in Fig. 7.2. The $N_2(1)$ vibrational state itself is excited in V–V collisions with $CO_2(0,0^0,1)$, which in turn is excited by absorption of solar radiation at 2.7 and 4.3 μm, and by the relaxation of the electronic energy of $O(^1D)$ (Sec. 6.3.6.3).

7.2.4 Uncertainties in the CO(1) population

The most uncertain parameters affecting the daytime population of CO(1) are the population of $N_2(1)$, the variability in the abundance of $O(^1D)$, and the rate coefficient of process 1 in Table 7.2. Typical uncertainties are about (+6, –3) K at 80 km. The effect of the tropospheric flux is smaller, of the order of 1 K, while the uncertainty in the reduction in the solar flux of ±10% leads to a change of ±2 K in T_v from the top of the atmosphere down to 60 km.

At night, the parameter that most influences the population of CO(1) in the mesosphere is the tropospheric flux. A change in temperature of the tropospheric effective emitting layer by ±25 K produces a change in the night-time CO(1) T_v of about (+8, –3) K above 70 km for a mid-latitude temperature profile, and (+3, –1) K for a typical polar winter profile.

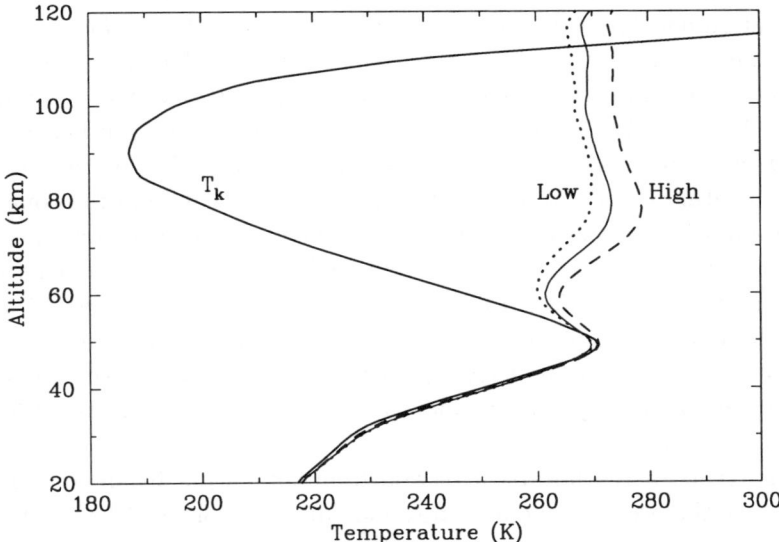

Fig. 7.5 Daytime ($\chi=0°$) vibrational temperatures of CO(1) for extreme high and low CO VMR profiles, typical of winter and summer, respectively. The T_v for an intermediate CO VMR profile, and the kinetic temperature T_k are also plotted.

As a consequence of the importance of radiative transfer, and because CO is very variable in the atmosphere, the CO abundance profile significantly affects the level of excitation, i.e., the [CO(1)]/[CO] ratio. This is illustrated in Fig. 7.5 for two rather extreme CO profiles, representing typical polar winter and polar summer profiles, which can lead to changes of (+5, –3) K in the CO(1) vibrational temperature.

Normally the changes in the vibrational temperatures of levels excited mainly from the ground state, due to changes in the number density of the latter, are very small, because the vibrational temperature is defined in terms of the ratio between the populations of the upper and lower levels. CO(1) is an exception, however, and it is important when retrieving the CO abundance from its 4.7 μm emission to account for the dependence of its vibrational temperature on the CO VMR itself.

7.3   Ozone, O_3

Although present in small amounts, ozone is an important constituent of the atmosphere, since it absorbs the ultraviolet solar radiation which is hazardous for the biosphere. It also controls the heating rate and hence the temperature structure of the stratosphere and mesosphere, and is a greenhouse gas affecting the temperature of the Earth's surface.

The ozone molecule has a dipole moment in the ground state and possesses three non-degenerate modes of vibration: a symmetric stretching mode (ν_1), a bending mode (ν_2) and an asymmetric stretching mode (ν_3). The ν_1 mode is the most energetic, at around 1103 cm^{-1}, followed by ν_3 at 1042 cm^{-1}. The fact that the two are so close in energy, together with their anharmonicity, which leads to the higher vibrational levels being progressively more closely spaced, gives rise to a number of resonances between the two, contributing to the important 9.6 μm bands. The ν_2 band emits at around 14.5 μm, but is weak, and usually masked by the CO_2 15 μm bands in the atmosphere. Another strong ozone band, that from the combination (1,0,1) level to ground, is located near 4.8 μm. There are also many hot bands rising from the higher v_1 and v_3 states which occur slightly to the long-wavelength side of the fundamental at 9.6 μm(see Table G.1). These very hot bands would not emit significantly in the atmosphere if LTE applied, but generally contribute radiances orders of magnitude above the LTE ones, which are significant in comparison with those from the fundamental bands under the non-LTE conditions found for ozone in the real atmosphere.

Measurements of the infrared emission from the 9.6 μm bands are commonly used to derive the ozone abundance in the middle atmosphere (Chapter 8). There is ample evidence (see Sec. 8.4) to show that the populations of the levels responsible for these emissions are far out of LTE in the mesosphere and above, and that the disequilibrium extends down to the upper stratosphere during daytime. Thus, non-LTE modelling is an essential part of the retrieval process for ozone, just as it is for CO.

Several such models have been developed (see references in Sec. 7.15). These differ in such key factors as: the number of vibrational levels included; assumptions concerning the nascent distribution of excited states which results following the chemical formation of O_3; the collisional relaxation processes, including coupling between the different modes of vibration, and the rate coefficients incorporated; and the treatment of radiative transfer between layers, some neglecting this altogether. They also differ in the extent to which they have been compared with observations. The model we use here includes: (i) a new collisional scheme for the relaxation of the O_3 levels; (ii) a full inter-layer radiative transfer scheme for the fundamental ν_1 and ν_3 bands; (iii) a larger number of vibrational states, 119 in all, extending up to the (0,0,7) state with a energy of 6554.29 cm^{-1}; (iv) excitation from vibrationally excited $N_2(1)$ and $O_2(1)$; and (v) absorption of solar radiation in the 4.8 μm bands.

7.3.1   *Non-LTE model*

The vibrational levels of O_3 treated in the model are listed in Table 7.3, and illustrated in Fig. 7.6. The transitions covered include the most prominent vibration-rotation bands emitting in the 4.5–17.7 μm spectral range (Table G.1).

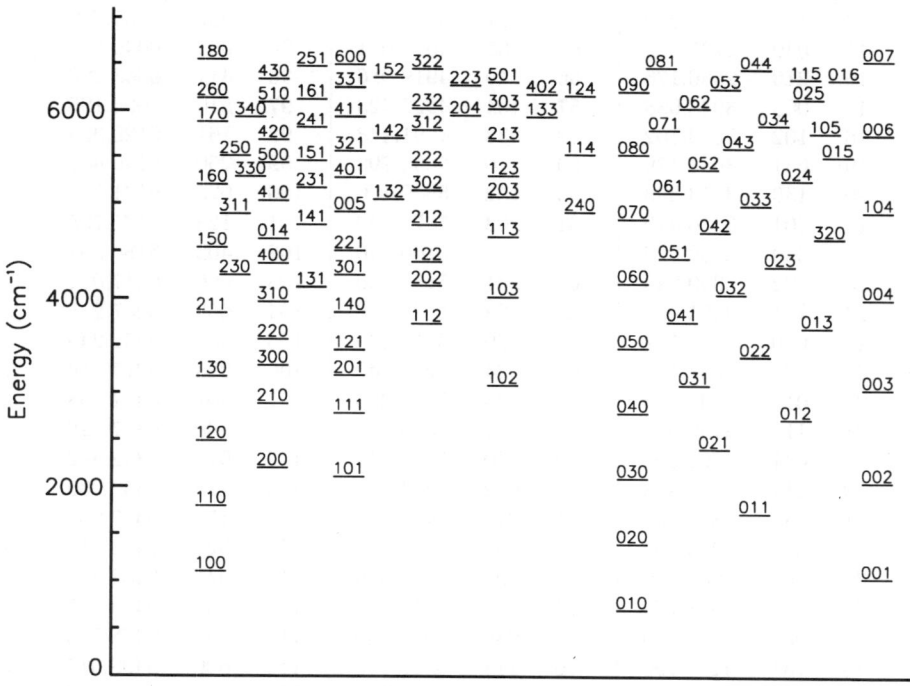

Fig. 7.6   Vibrational energy levels for O_3.

Transitions in which the total $v_1+v_2+v_3$ quantum number decreases by one or two are included, except those ν_2 mode levels with $v_2 > 2$, because no evidence has been found for them in the atmospheric spectrum. Transitions in which each

Table 7.3 O_3 vibrational levels.

No.	Level	Energy (cm^{-1})	No.	Level	Energy (cm^{-1})	No.	Level	Energy (cm^{-1})
1	010	700.931	41	023	4346.628	81	420	5702.010
2	001	1042.084	42	400	4370.074	82	213	5763.720
3	100	1103.137	43	122	4390.796	83	006	5766.800
4	020	1399.273	44	051	4432.489	84	071	5783.990
5	011	1726.522	45	221	4508.166	85	105	5801.360
6	110	1796.262	46	150	4538.166	86	312	5812.710
7	002	2057.891	47	014	4632.828	87	034	5884.870
8	030	2094.968	48	320	4643.809	88	170	5890.600
9	101	2110.784	49	113	4659.000	89	241	5919.510
10	200	2201.154	50	042	4710.313	90	133	5946.900
11	021	2407.935	51	212	4783.385	91	340	5970.890
12	120	2486.576	52	141	4783.505	92	204	5995.030
13	012	2726.107	53	070	4848.137	93	303	6030.753
14	111	2785.239	54	311	4897.013	94	232	6035.787
15	040	2787.822	55	005	4919.011	95	062	6040.821
16	210	2886.178	56	240	4919.986	96	411	6063.220
17	003	3046.088	57	104	4922.423	97	510	6096.195
18	102	3083.702	58	033	4991.448	98	161	6126.399
19	031	3086.179	59	410	5035.790	99	260	6136.467
20	130	3173.898	60	132	5040.100	100	025	6141.501
21	201	3186.410	61	203	5077.040	101	124	6171.705
22	300	3290.026	62	061	5100.140	102	402	6186.807
23	022	3390.920	63	231	5159.590	103	090	6217.011
24	121	3455.903	64	302	5171.220	104	053	6247.215
25	050	3477.727	65	160	5214.130	105	331	6252.249
26	220	3568.166	66	024	5265.440	106	223	6277.419
27	013	3698.203	67	123	5291.250	107	501	6312.658
28	112	3739.516	68	401	5308.050	108	430	6337.828
29	041	3761.094	69	330	5310.240	109	016	6342.862
30	211	3849.868	70	052	5364.970	110	115	6352.930
31	140	3857.893	71	222	5418.860	111	152	6357.964
32	310	3966.804	72	151	5438.920	112	322	6453.610
33	004	4001.259	73	500	5441.020	113	044	6458.644
34	103	4021.779	74	015	5518.830	114	081	6473.747
35	032	4052.330	75	080	5528.340	115	251	6478.780
36	131	4122.196	76	114	5540.950	116	600	6498.917
37	202	4141.199	77	250	5586.560	117	180	6549.257
38	060	4164.548	78	321	5632.700	118	214	6554.270
39	230	4246.444	79	142	5679.470	119	007	6554.290
40	301	4250.114	80	043	5697.570			

mode changes by more than 2 quanta have not been included for the same reason. Spontaneous emission and absorption of tropospheric radiation are included for all bands (see Table G.1). The exchange of radiation between atmospheric layers for the stronger ν_1, ν_2, and ν_3 fundamental bands is taken into account using the Curtis matrix method, with matrix elements evaluated as described in Sec. 6.3.3.1. Radiative transfer between the atmospheric layers for the ν_1, (2,0,0–1,0,0), and ν_3 (0,0,2–0,0,1) first hot bands has been shown to be negligible. Absorption of solar radiation in the 4.8 μm bands is included. This is important mainly for the (1,0,1–0,0,0) band, increasing its population by ~10% above 60 km.

7.3.2 *Chemical recombination*

The processes driving the non-LTE populations of the O₃ levels are listed in Table 7.4. Most of the excitation of the $O_3(v_1,v_2,v_3)$ levels occurs in the O₃ formation process itself, which involves the chemical recombination of O_2 and $O(^3P)$ in the presence of a third body (process 1 in the table). In this reaction, chemical potential energy is converted into internal energy of the ozone molecule and leads to the population of the highly excited vibrational levels, close to the dissociation limit near $8500\,cm^{-1}$ (~1.05 eV). This is the principal ozone-forming mechanism at nearly all altitudes in the atmosphere, and probably the only significant source in the stratosphere. Laboratory measurements have been used to infer the distribution of states in nascent ozone, i.e., the distribution of energy among the vibrational levels after a stable O₃ molecule is formed. The rotational levels are probably also in non-LTE in nascent ozone, but it is assumed that these are rapidly thermalized by collisions, leaving only non-thermal vibrational population.

The measurements suggest that about 50% of the energy released in the recombination reaction ends up in vibrational excitation, with perhaps a 20% uncertainty suggested by the spread of different results. They show that an appreciable fraction of the ozone is produced in the ν_1 and ν_3 vibrational modes, with quantum numbers larger than one.

One of the most commonly used nascent distributions is the so-called 'zero surprisal', which assumes that only the $O_3(0,0,v_3)$ levels result excited. The branching ratio for this distribution is given by

$$f(v) = \frac{(1 - E_v/D_e)^{1.5}}{\sum\limits_{v=1}^{N} (1 - E_v/D_e)^{1.5}}, \tag{7.1}$$

where D_e is the dissociation energy, E_v is the energy of the $O_3(0,0,v_3)$ levels, and the sum goes over all v_3 levels from $v_3 = 1$ to $N = 7$, the highest v_3 state populated after recombination. This is the only distribution measured so far and shows a bias towards the less energetic levels (see Fig. 7.7), in contrast to the theoretical expectations of some authors who predict a very hot nascent distribution following

Table 7.4   Collisional processes affecting the O_3 vibrational levels.

No.	Process		Rate coefficient, $k^\dagger$		Reference
1	$O_2 + O + M \rightarrow O_3(v_1,v_2,v_3) + M$	$M = N_2$	$5.7\times10^{-34}(T/300)^{-2.8}$		Baulch et al. (1984)
		$M = O_2$	$6.2\times10^{-34}(T/300)^{-2.0}$		Baulch et al. (1984)
		$M = O$	$2.15\times10^{-34}\exp(345/T)$		Eliasson et al. (1987)
2	$O_3(v_1,v_2,v_3) + M \rightleftharpoons O_3(v'_1,v'_2,v'_3) + M$	M	$k^\ddagger$	$A_{\Delta v_1,\Delta v_2,\Delta v_3}$	
a:	$\Delta v_1 = 1,\ \Delta v_3 = -1,\ \Delta v_2 = 0$	N_2	1.18×10^{-11}	1.60×10^{-11}	Doyennette et al. (1992)
		O_2	0.99×10^{-11}	1.34×10^{-11}	Doyennette et al. (1992)
b1:	$\Delta v_1 = 1,\ \Delta v_3 = 0,\ \Delta v_2 = -1$	N_2	4.04×10^{-14}	3.02×10^{-13}	Menard et al. (1992)
		O_2	2.95×10^{-14}	2.20×10^{-13}	Menard et al. (1992)
b2:	$\Delta v_1 = 0,\ \Delta v_3 = 1,\ \Delta v_2 = -1$	N_2	4.04×10^{-14}	2.22×10^{-13}	Menard et al. (1992)
		O_2	2.95×10^{-14}	1.62×10^{-13}	Menard et al. (1992)
c:	$\Delta v_3 = \Delta v_1 = 0,\ \Delta v_2 = 1$	N_2	3.11×10^{-14}	1.03×10^{-12}	Menard et al. (1992)
		O_2	3.42×10^{-14}	1.14×10^{-12}	Menard et al. (1992)
d:	$\Delta v_3 = 1,\ \Delta v_1 = \Delta v_2 = 0$	N_2, O_2	3.10×10^{-15}	5.68×10^{-13}	Menard et al. (1992)
e:	$\Delta v_1 = 1,\ \Delta v_2 = \Delta v_3 = 0$	N_2, O_2	3.10×10^{-15}	7.70×10^{-13}	Menard et al. (1992)
3	$O_2(1) + O_3 \rightarrow O_2 + O_3(100,001)$		10^{-16}		Parker and Ritke (1973)
4	$N_2(1) + O_3 \rightarrow N_2 + O_3(200)$		2.3×10^{-14}		Robertshaw and Smith (1980)
5	$O_2(v=2) + O_3 \rightarrow O_3(102,003) + O_2$		3×10^{-11}		Rawlins (1985)
6	$N_2(v=8\text{–}10) + O_3 \rightarrow N_2(v-1) + O_3(002,101)$				
7	$e^- + O_3 \rightarrow e^- + O_3(v)$				
8	$O_2(b^1\Sigma_g^+) + O_3 \rightarrow O_3(v) + O_2(a^1\Delta_g \text{ or } X^3\Sigma_g^-, v)$		2.2×10^{-11}		Slanger and Black (1979)
9	$O_2(A^3\Sigma_u^+) + O_2 \rightarrow O_3(v) + O$		2.9×10^{-13}		Kenner and Ogryzlo (1980)

†For the forward sense of the process in cm^3s^{-1}. T is temperature in K. ‡For processes 2, k is given for collisions with the lowest permitted v_1, v_2, and v_3 quanta at $T = 300$ K. The temperature dependence is described in the text. $\Delta v_i = v_i - v'_i$.

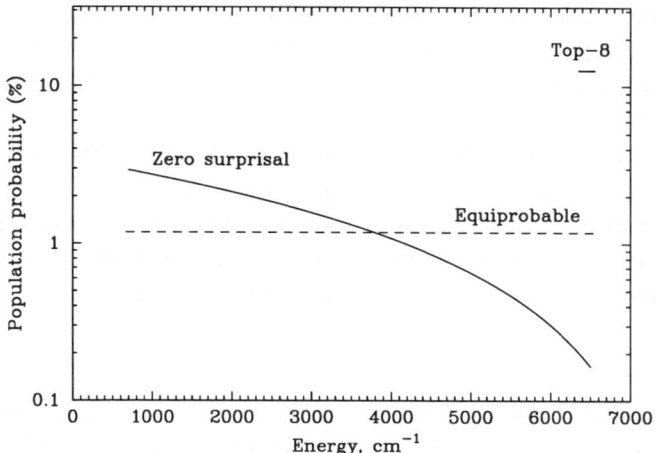

Fig. 7.7 Nascent distributions of the $O_3(v_1,v_2,v_3)$ vibrational levels. The zero surprisal extends over the 119 levels considered here.

a molecular recombination reaction. In this model we have used a similar formula for the nascent distribution but, instead of considering only the v_3 levels, it was extended to all the $O_3(v_1,v_2,v_3)$ states considered in the model.

In view of the uncertainty which exists in the nascent distribution, it is useful to study the sensitivity of the $O_3(v_1,v_2,v_3)$ populations to two extreme distributions: (i) one which favours the excitation of the lower energy levels, the zero surprisal; and (ii) one in which all the excited ozone molecules are produced in the top eight (very close in energy) states of the model (see Fig. 7.7). The results are shown in Fig. 7.11 and discussed in Sec. 7.3.6.

7.3.3 *Collisional relaxation*

It is difficult to model the relaxation of nascent $O_3(v_1,v_2,v_3)$ to lower states, because there are so few results describing the collisional processes which affect highly excited O_3. The major problems with the available data are: (i) the lack of measurements involving levels with more than one or two quanta; (ii) the fact that the measurements do not cover all atmospheric temperatures; (iii) the complete lack of measurements for collisional deactivation by atomic oxygen; and (iv) the fact that the measurements do not distinguish unambiguously between the collisional rate coefficients for the (0,0,1) and (1,0,0) levels, which are strongly coupled.

Approximations which have been used in the past include restricting the model to the asymmetric stretching (ν_3) mode and using a simple scaling formula dependent on the temperature for the relaxation rate, or a scheme based on the unimolecular reaction rate theory whereby the rate coefficient depends on the energy difference between the two states. These have obvious deficiencies, for example, the energy difference between two states is probably not the only relevant factor determining the collisional relaxation rate. A process that requires large changes in the excitation energy of each vibrational mode is probably less likely to occur than another in which the excitation energy of each mode is slightly changed or where only one mode changes.

The relaxation of the $O_3(v_1,v_2,v_3)$ states with lowest quantum numbers is assumed to take place through five major pathways. These are described below in ascending order of the energy converted from vibrational to kinetic. First we discuss the redistribution of the energy between the stretching modes (1,0,0) and (0,0,1) in thermal collisions with the major atmospheric constituents:

$$O_3(1,0,0) + M \rightleftharpoons O_3(0,0,1) + M, \qquad (7.2)$$

e.g., processes 2a in Table 7.4 with $v_1 = 1$ and $v_2 = v_3 = 0$. This rearrangement of vibrational energy is very fast (even faster than the gas kinetic rate) for the case of O_3 self-relaxation. The rearrangement of these two levels by collisions with $O(^3P)$ is negligible in comparison with the fast redistribution with N_2 and O_2.

The second pathway is the thermal relaxation of the stretching (1,0,0) and (0,0,1) levels to the lower bending (0,1,0) state through collisions with N_2 and O_2:

$$O_3(1,0,0;0,0,1) + M \rightleftharpoons O_3(0,1,0) + M, \qquad (7.3)$$

e.g., processes 2b in Table 7.4 with $v_1 + v_3 = 1$, and $v_2 = 0$.

The thermalization rate of the (0,1,0) level in collisions with N_2 and O_2 has also been included,

$$O_3(0,1,0) + M \rightleftharpoons O_3 + M, \qquad (7.4)$$

e.g., processes 2c with $v_2 = 1$, and $v_1 = v_3 = 0$ in Table 7.4.

Finally, the complete thermalization of levels (0,0,1) and (1,0,0) is represented by

$$O_3(0,0,1) + M \rightleftharpoons O_3 + M, \quad \text{and} \qquad (7.5)$$
$$O_3(1,0,0) + M \rightleftharpoons O_3 + M. \qquad (7.6)$$

These correspond to processes 2d in Table 7.4 with $v_3 = 1$ and $v_1 = v_2 = 0$, and 2e with $v_1 = 1$ and $v_2 = v_3 = 0$, respectively.

The collisional rates for the relaxation of the higher vibrational levels are not well known. In this model, the relaxation of these high energetic levels takes place through the same pathways as for the lower ones, e.g., processes 7.2–7.6, for any permitted value of v_1, v_2 and v_3 (processes 2a–e in Table 7.4). The rate coefficients for these processes use a formula that considers both the change in vibrational energy converted to kinetic energy (ΔE) and the change in the quantum numbers of the pre- and post-collision states, i.e.,

$$k(v_1, v_2, v_3 \rightarrow v_1', v_2', v_3') = (v_1 + v_2 + v_3) \, A_{\Delta v_1, \Delta v_2, \Delta v_3} \exp(-\Delta E/E_0), \qquad (7.7)$$

where E_0 is the average energy loss in a collision, estimated in $200\,\mathrm{cm}^{-1}$. This expression also generalizes the linear v-dependence of the 1D harmonic oscillator. The $A_{\Delta v_1, \Delta v_2, \Delta v_3}$ constants depend only on the change in the quantum numbers, that is on the 5 pathways 7.2–7.6, and on the collisional partner in some cases (see Table 7.4). They have been obtained from the measurements of these processes for the fundamental modes, i.e. the collisions with the lowest v_1, v_2, and v_3 quanta.

7.3.4 *Other excitation processes*

In addition to the known mechanisms described above, we have also included other less-explored collisional processes. For example, the near-resonant vibrational energy exchange between the $O_3(1,0,0)$ and $(0,0,1)$ levels and $O_2(1)$ (process 3 in Table 7.4). The V–V coupling of $O_3(2,0,0)$ with $N_2(1)$ (process 4) is also included, although this process is much slower than with $O_2(1)$. For this process we assume a square-root temperature dependence, which follows from the assumption that the probability of transition at collision is independent of temperature.

Atmospheric experiments have shown evidence for a dramatic auroral enhancement of O_3 $9.6\,\mu\mathrm{m}$ emission above $100\,\mathrm{km}$. $O_2(v)$, $N_2(v)$ and metastable O_2 (processes 5–9) may be possible precursors to auroral vibrationally excited O_3. $O_2(v=2)$ is just 5 and $43\,\mathrm{cm}^{-1}$ above $O_3(1,0,2)$ and $(0,0,3)$, respectively, and excitation of these O_3 levels in near-resonant V–V energy transfer from $O_2(v)$ has been observed in laboratory experiments. This rate is very fast and the efficiency of this process requires only that sufficient vibrational excitation in O_2 is built up during an aurora.

The vibrational states $(0,0,2)$, $(2,0,0)$ and $(1,0,1)$ can also be excited from $N_2(v=8\text{--}10)$ (process 6). This is a V–V nearly resonant process for $N_2(\Delta v=1)$, and $N_2(v=8\text{--}10)$ can be created in the vicinity of an aurora through prolonged particle bombardment. The rate for V–V transfer from $N_2(v)$ to ozone has not yet been measured.

Although some evidence for electron-related $O_3(v)$ excitation has been seen in laboratory discharges, this process has not yet been quantified. The quenching of $O_2(b^1\Sigma_g^+)$ by O_3 might result in a significant vibrational excitation of the latter (process 8). Both of these, plus the chemical reaction of $O_2(A^3\Sigma_u^+)$ with O_2 (process 9), are possible sources of excitation of O_3 in auroral conditions.

7.3.5 *Solution of the system*

With the exception of the fundamental bands originating from $(1,0,0)$ and $(0,0,1)$, the O_3 bands are optically thin even for paths through the whole atmosphere, so the exchange of radiation between layers is negligible. Similarly, the absorption of tropospheric flux ('earthshine') needs to be included only for the fundamental bands. Hence, except for the fundamental bands, the statistical equilibrium equation for level v is relatively simple (c.f. Eqs. 3.78 and 3.79) and given by

$$n_v \simeq \frac{P_{c,v} + \sum_{v'>v} \alpha_{v,v'} P_{c,v'} + P_{t,v} + P_{nt,v}}{A_v + l_{t,v}}, \qquad (7.8)$$

where n_v is the number density of level v, $P_{c,v}$ is its initial production from the recombination reaction, the sum is the production from the relaxation of the upper states (which is essentially proportional to the initial production of these levels, $P_{c,v'}$), $P_{t,v}$ and $P_{nt,v}$ are the production of thermal and non-thermal processes (3–9 in Table 7.4), A_v is the total Einstein coefficient emission from all bands originating in level v, and $l_{t,v}$ are the thermal losses. This equation only contains local terms and can be easily solved without the need to consider coupling with other altitudes.

So we have a set of local linear equations, with those for the lower levels depending on the populations of the higher. We started by solving the equations for the top level, then work down to $(1,0,0)$ and $(0,0,1)$, for which the coupled system of the statistical equilibrium and the radiative transfer equations can be solved by means of the Curtis matrix formulation.

7.3.6   *Non-LTE populations*

As for previous non-LTE models, to obtain the O_3 vibrational populations we require a model atmosphere which specifies temperature and constituent abundance profiles as function of pressure or altitude. Care has to be taken in the selection of O_3 and $O(^3P)$ abundances to make sure they are consistent, since they are not independent but linked by photochemical equilibrium in the stratosphere and mesosphere. Ozone is also unusual in that $O_3(v)$ is formed mainly from O_2 and $O(^3P)$, and not from the ground state of O_3, again requiring consistency between the O_3 and $O(^3P)$ abundances or the O_3 vibrational temperatures will be misleading. Here we have used the output from the 2D model of Garcia and Solomon (see Sec. 7.15).

The calculated populations, which again we will express as vibrational temperatures, run from the surface of the planet up to 120 km, for layers 1 km thick. In the middle atmosphere, the parameters that affect the vibrational temperatures, in order of importance, are: (i) the branching fractions for the recombination reaction; (ii) the rate coefficients for collisional deactivation of O_3 modes; (iii) the spectroscopic line strengths; (iv) the kinetic temperature structure; and (v) the O_3 and atomic oxygen volume mixing ratios.

Figure 7.8 shows the daytime and night-time vibrational temperature profiles for mid-latitude conditions for v_3 levels from $v_3 = 1$ to 7. During the day, the vibrational temperature of $O_3(0,0,1)$ starts to deviate from the kinetic temperature at around 70 km. Up to this altitude, thermal collisions are the dominant process in populating this state. Above 70 km, its population is mainly controlled by earthshine in the 9.6 μm band and to a lesser extent by the recombination reaction.

The effect of earthshine on the population of $O_3(0,0,1)$ is illustrated in Fig. 7.9

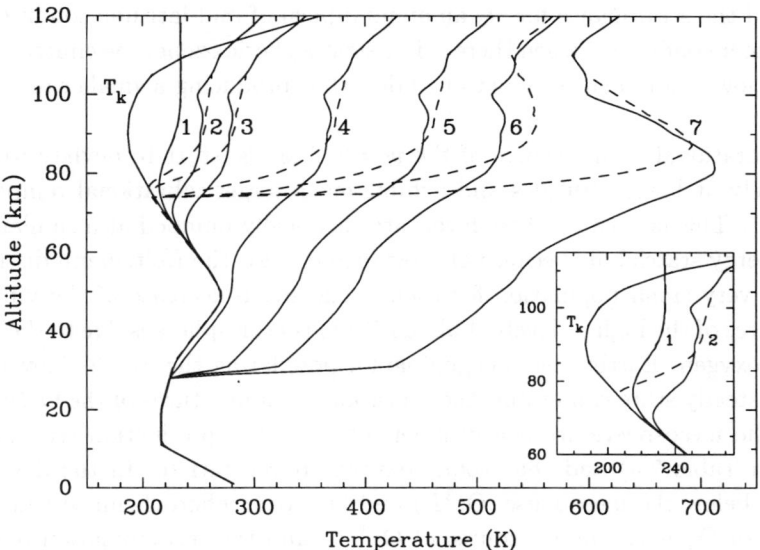

Fig. 7.8   Daytime (solid lines) and night-time (dashed lines) vibrational temperatures for the $O_3(0,0,v_3 = 1$–$7)$ levels for mid-latitude conditions. T_k is the kinetic temperature. The $(0,0,1)$ level does not present any diurnal variation.

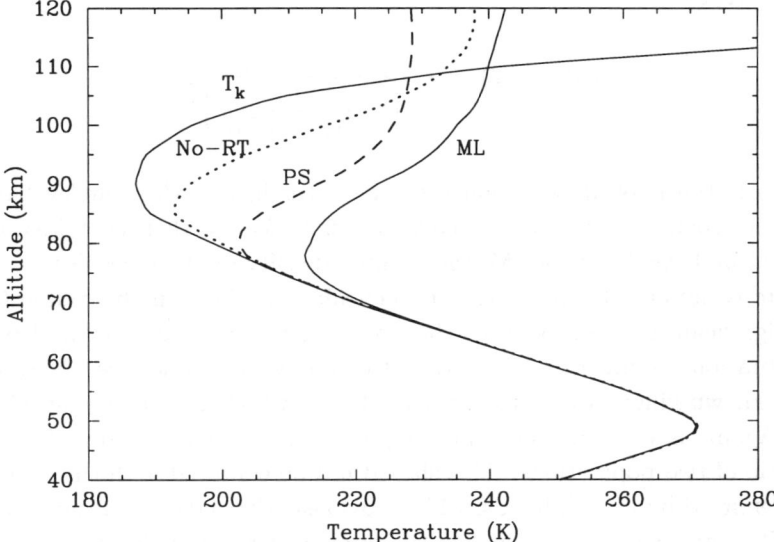

Fig. 7.9   Effect of radiative transfer ('earthshine') on the T_v of $O_3(0,0,1)$ at daytime. Solid (ML) and dotted ('No-RT') show the T_v's for mid-latitude conditions with and without radiative transfer, respectively. The case with radiative transfer assuming a smaller stratospheric O_3 profile, typical of summer conditions, dashed (PS) line, is also shown. T_k is the kinetic temperature.

(compare curves 'No RT' and 'ML'). It can be seen that it is very important above about 70 km, and depends quite significantly on the amount of stratospheric O_3. The emission absorbed in the mesosphere comes mainly from the upper stratosphere and around the stratopause for O_3 amounts typical of mid-latitude conditions. For polar summer conditions, when there is less stratospheric ozone, the emitting layer is located at lower, and hence, colder altitudes, thus producing a smaller mesospheric excitation.

The vibrational temperatures of the $v_3 = 2$–7 levels, start to deviate from LTE progressively at lower altitudes and exhibit much larger vibrational temperatures than $v_3 = 1$. This is because these levels are chemically pumped at a similar rate as the low energy states but, because of their high energy, the Boltzmann distribution predicts a very small population for them. The sharp decrease of the vibrational temperatures of the high v_3 levels at about 30 km thermosphere is due to the absence of atomic oxygen. Earthshine is negligible for populating the $v_3 = 2$–7 levels.

Under steady state conditions, the vibrational temperatures of the $O_3(v_3 = 2$–7) levels in the mesosphere are nearly independent of the production recombination rate (k_1 in Table 7.4) and the atomic oxygen concentration. In the daytime atmosphere, below the mesopause, $O(^3P)$ and O_3 are in photochemical equilibrium: photolysis of O_3 is the major source of $O(^3P)$, and the recombination of O_2 and $O(^3P)$ to form $O_3(v)$ is its major sink. Hence, considering only the chemical recombination as excitation and the thermal collisions and spontaneous radiation as the major losses, an approximate expression for the vibrational temperatures of these levels is given by

$$T_v(v_3 \geq 2) = -\frac{E_v}{k \ln\left(\frac{J_{O_3}}{A_v + k_t[M]}\right)}, \qquad (7.9)$$

where J_{O_3} is the photodissociation rate of O_3 (mainly in the Hartley band), A_v the Einstein coefficient for spontaneous emission, k_t the thermal relaxation rate (process 2d in Table 7.4), and [M] the number density of air molecules.

If there is significant atomic oxygen quenching of vibrationally excited O_3, or if there is significant reaction between the two, then there will be a weak dependence of the vibrational temperatures on the atomic oxygen abundance. However, the currently known kinetic rates suggest that this is of little importance at least up to 100 km. Again, as we show below for $NO_2(v_3)$, this happens because T_v is defined as the ratio of two populations. This should not be confused with the atmospheric radiance emitted by the O_3 bands, which obviously depends on both quantities, k_1 and $O(^3P)$, since it is proportional to the number density of the excited state (see Chapter 8).

Figure 7.8 also shows the vibrational temperatures at night-time. The $v_3 \geq 2$ levels are in LTE up to the upper mesosphere. This is due to the absence of atomic oxygen at night below about 80 km, where it recombines to form O_3 after photodissociation ceases. In the lower thermosphere the $v_3 \geq 2$ levels have similar vibrational temperatures than in the daytime. In that region, atomic oxygen is mainly controlled by dynamics, with a strong downwelling molecular diffusion, thus

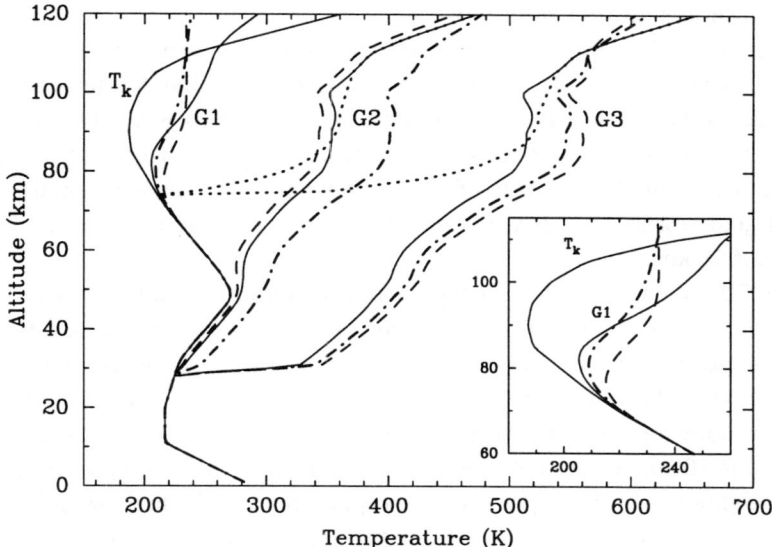

Fig. 7.10   Vibrational temperatures for 3 groups of levels close in energy. Daytime T_v's for Group 1: (0,1,0), (0,0,1) and (1,0,0) states (solid, dashed and dot-dashed lines, respectively); Group 2: (2,3,0), (3,0,1), and (0,2,3); and Group 3: (2,6,0), (0,2,5) and (1,2,4) states. Night-time T_v's for (2,3,0), and (2,6,0) are represented with dotted lines. The (0,1,0), (1,0,0), and (0,0,1) levels do not present any diurnal variation. T_k is the kinetic temperature.

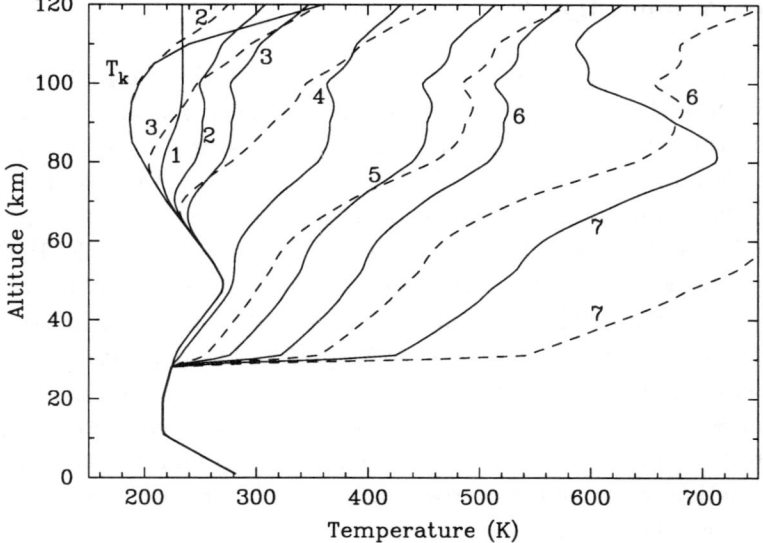

Fig. 7.11   Effects of the nascent distribution on the daytime vibrational temperatures of the $O_3(v_3 = 1-7)$ levels. Solid lines and dashed lines are calculations with the 'zero surprisal' and 'Top-8' distributions, respectively. The T_v's of (0,0,1) for the two cases are overlaid.

showing much less diurnal variation.

The vibrational temperatures of levels with excited v_1 and/or v_2 quanta (independent of whether they possess v_3 excitation) show similar features to those for the pure v_3 levels with comparable energy (Fig. 7.10). This is a consequence of the strong collisional coupling between the v_1 and v_3 states with the same quantum number and the rapid relaxation of these modes to v_2 through thermal processes. For the (1,0,0) state, radiative transfer between atmospheric layers is important above about 70 km.

Figure 7.11 shows the effect of the nascent distribution on the vibrational temperatures of the v_3 levels during daytime. The high-energy $v_3 = 6, 7$ levels have larger populations for the 'Top–8' distribution, while the lower lying v_3 levels exhibit the opposite behaviour. Intermediate levels, e.g., $v_3 = 5$, show a mixture effect, with larger vibrational temperatures for the 'Top–8' distribution in the upper region (above about 70 km) where collisions are less frequent but the opposite behaviour at lower heights where the larger pressure rapidly relaxes the high-lying levels. Note also the very low excitation of $v_3 = 2$ for the case favouring the high-energy levels ('Top–8'). The effects of this distribution on the atmospheric limb radiance is to decrease the emission in the low lying ν_3 bands and increase that of the high ν_3 transitions. Since the latter occurs at longer wavelengths, the effect of the 'Top–8' distribution is to shift the radiance to longer wavelength (see Fig. 8.21).

7.4   Water Vapour, H_2O

Water vapour is of comparable importance to CO_2 for the energy budget of the atmosphere through its rich and complex infrared spectrum, and it also plays a key role in atmospheric chemistry, for example as a source by photolysis of the important OH radical. Unlike CO_2, H_2O is of course highly variable in space and time, and so like ozone is a key objective for study by satellite-based remote sensing.

The model we describe below for the non-LTE populations of the H_2O vibrational levels solves the system of statistical equilibrium equations for the $O_2(1)$ and the six H_2O vibrational levels shown in Fig. 7.12, and the radiative transfer equations for the H_2O transitions listed in Table 7.5. The populations of the CO_2 and $N_2(1)$ levels which interact with these come from the non-LTE model for CO_2 described in Sec. 6.3. The lower boundary needs to be selected low enough, which turns out to be at or below 20 km, for the H_2O bands to fulfil the lower boundary model assumption, whereby the source function is equal to the Planck function at the local kinetic temperature.

7.4.1   *Radiative processes*

The populations of the vibrational states of water vapour are largely determined by radiative processes in the mesosphere and above; radiative exchange between

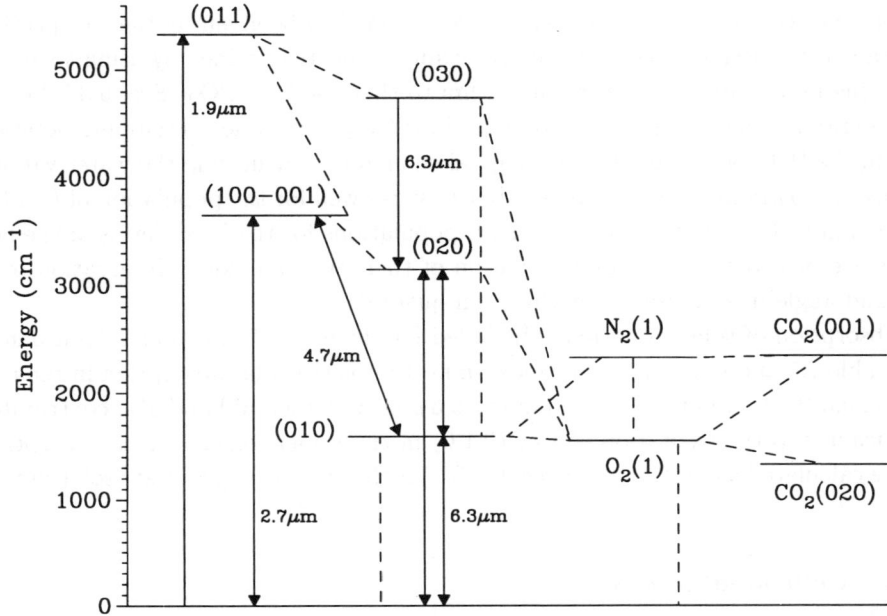

Fig. 7.12 Energy level diagram for water vapour. Solid lines represent radiative and dashed lines collisional processes.

Table 7.5   Principal H_2O infrared bands.

Band	Upper level	Lower level	$\tilde{\nu}_0$, cm^{-1}	Strength*	A, s^{-1}
1	010	000	1594.75	1.058×10^{-17}	20.34
2	020	010	1556.88	9.709×10^{-21}	41.43
3[†,‡]	030	020	1515.16	5.135×10^{-24}	39.75
4	100	010	2062.31	1.820×10^{-22}	1.34
5	001	010	2161.18	2.626×10^{-22}	2.16
6	020	000	3151.63	7.571×10^{-20}	0.57
7	100	000	3657.05	4.955×10^{-19}	5.01
8	001	000	3755.93	7.200×10^{-18}	76.80
9[†,‡]	011	010	3736.52	2.923×10^{-21}	71.75
10[†]	011	000	5331.27	8.042×10^{-19}	17.28

*Values shown are at 296 K in cm^{-1}/(molecules cm^{-2}). [†] Radiative transfer between atmospheric layers is not included for these transitions. [‡] Absorption of solar radiation is not included.

layers is particularly important for the fundamental band near 6.3 μm. In the model, radiative transfer is again calculated using a version of the Curtis matrix method for the bands shown in Table 7.5, except for the very weak transitions (0,3,0–0,2,0), (0,1,1–0,1,0), and (0,1,1–0,0,0). For these, radiative transfer between layers is neglected, and radiative losses are assumed to be given by their respective spontaneous emission rates. The flux transmittances are obtained by using a similar quasi-line-by-line histogramming algorithm to that used for CO_2 (Sec. 6.3.3.1).

The random model approximation used for CO_2 to include overlapping between lines in the H_2O bands, might not be good enough for calculating the water vapour cooling at 6.3 μm in the stratosphere, but it works well for the population of (0,1,0), which is in LTE in that region. Non-LTE populations for the lower levels of the hot bands are incorporated in the calculation of their transmissions. Integration over the solid angle uses a four-point Gaussian quadrature.

Absorption of solar radiation is included for the 6.3, 4.7, 2.7, and 1.9 μm bands (see Table 7.5 and Fig. 7.12). The most important is the solar absorption in the two bands near 2.7 μm, but absorption in the 6.3 μm fundamental band also contributes significantly to the excitation of $H_2O(0,1,0)$ in the upper regions. This absorption can be calculated as for CO_2 (see Sec. 6.3.3.2) including the appropriate solar fluxes.

7.4.2   *Collisional processes*

The main collisional processes involving the vibrationally excited states of water vapour and molecular oxygen are listed in Table 7.6. The most important one affecting the $H_2O(0,1,0)$ level is the near-resonant vibrational energy exchange with $O_2(1)$ (process 1). Its rate coefficient is still uncertain, but it is probably much more efficient than the thermal relaxation of $H_2O(0,1,0)$ by the major atmospheric constituents N_2 and O_2 (process 2). Thermal collisions of $H_2O(0,1,0)$ with $O(^3P)$ (process 3) may be important for the population of the former in the upper mesosphere and lower thermosphere, but again there are no reported experimental measurements of its rate coefficient at atmospheric temperatures. The vibrational-vibrational coupling of $H_2O(0,1,0)$ with $N_2(1)$ (process 4) is much weaker than the equivalent reaction with $O_2(1)$.

The main collisional processes that drive the population of $H_2O(0,2,0)$ are vibrational energy transfer with $O_2(1)$ and thermal quenching by N_2, O_2 and $O(^3P)$. The first of these dominates the population of $H_2O(0,2,0)$ during daytime. The H_2O (1,0,0) and (0,0,1) levels are strongly coupled by thermal collisions with the major atmospheric constituents through processes 11 and 12, which are very fast, even faster than the rate predicted by gas kinetic theory. The rearrangement of these two levels by collisions with $O(^3P)$ is negligible in comparison and can be neglected. Thermal relaxation of the (1,0,0) and (0,0,1) levels to the (0,2,0) state occurs through collisions with N_2, O_2, and $O(^3P)$ (processes 13–15, respectively).

The vibrational exchange of a v_2 quantum between $H_2O(0,1,1)$ and $O_2(1)$ (process 16) is very important for the population of the former. Thermal collisions of $O_2(1)$ with N_2 and with O_2 itself are not efficient for converting kinetic energy into vibrational excitation because of their slow relaxation rates. In spite of the low abundance of atomic oxygen in the mesosphere, thermal collisions of this

Table 7.6 Collisional processes affecting the H_2O vibrational levels.

No.	Process	E^*	Rate coefficient[†]	Reference
1	$H_2O(010) + O_2 \rightleftharpoons H_2O + O_2(1)$	38	10^{-12} to 10^{-11}	López-Puertas et al. (1995)
2	$H_2O(010) + M^{\ddagger} \rightleftharpoons H_2O + M$	1595	$4.1 \times 10^{-14}(T/300)^{1/2}$	Bass et al. (1976)
3	$H_2O(010) + O(^3P) \rightleftharpoons H_2O + O(^3P)$	1595	$1.0 \times 10^{-12}(T/300)^{1/2}$	López-Puertas et al. (1995)
4	$N_2(1) + H_2O \rightleftharpoons N_2 + H_2O(010)$	735	$1.2 \times 10^{-14}(T/300)^{1/2}$	Whitson and McNeal (1977)
5	$H_2O(020) + O_2 \rightleftharpoons H_2O(010) + O_2(1)$	1	$2 \times k_1$	López-Puertas et al. (1995)
6	$H_2O(020) + M \rightleftharpoons H_2O(010) + M$	1557	$2 \times k_2$	López-Puertas et al. (1995)
7	$H_2O(020) + O(^3P) \rightleftharpoons H_2O(010) + O(^3P)$	1557	$2 \times k_3$	López-Puertas et al. (1995)
8	$H_2O(020) + O_2 \rightleftharpoons H_2O(020) + O_2(1)$	-41	$3 \times k_1$	López-Puertas et al. (1995)
9	$H_2O(030) + M \rightleftharpoons H_2O(020) + M$	1515	$3 \times k_2$	López-Puertas et al. (1995)
10	$H_2O(030) + O(^3P) \rightleftharpoons H_2O(020) + O(^3P)$	1515	$3 \times k_3$	López-Puertas et al. (1995)
11	$H_2O(001) + N_2 \rightleftharpoons H_2O(100) + N_2$	99	$1.2 \times 10^{-11}\sqrt{T}$	López-Puertas et al. (1995)
12	$H_2O(001) + O_2 \rightleftharpoons H_2O(100) + O_2$		$1.1 \times 10^{-11}\sqrt{T}$	López-Puertas et al. (1995)
13	$H_2O(100,001) + N_2 \rightleftharpoons H_2O(020) + N_2$	505, 604	$4.6 \times 10^{-13}(T/300)^{1/2}$	Finzi et al. (1977)
14	$H_2O(100,001) + O_2 \rightleftharpoons H_2O(020) + O_2$	505, 604	$3.3 \times 10^{-13}(T/300)^{1/2}$	Finzi et al. (1977)
15	$H_2O(100,001) + O(^3P) \rightleftharpoons H_2O(020) + O(^3P)$	505, 604	$3.0 \times 10^{-12}(T/300)^{1/2}$	Zittel and Masturzo (1989)
16	$H_2O(011) + O_2 \rightleftharpoons H_2O(001) + O_2(1)$	19	Same as process 1	López-Puertas et al. (1995)
17	$H_2O(011) + N_2 \rightleftharpoons H_2O(030) + N_2$	665	Same as process 13	López-Puertas et al. (1995)
18	$H_2O(011) + O_2 \rightleftharpoons H_2O(030) + O_2$	665	Same as process 14	López-Puertas et al. (1995)
19	$H_2O(011) + O(^3P) \rightleftharpoons H_2O(030) + O(^3P)$	665	Same as process 15	López-Puertas et al. (1995)
20	$O_2(1) + M \rightleftharpoons O_2 + M$	1556	$4.2 \times 10^{-19}(T/300)^{1/2}$	Parker and Ritke (1973)
21	$O_2(1) + O(^3P) \rightleftharpoons O_2 + O(^3P)$	1556	$1.3 \times 10^{-12}(T/300)$	Breen et al. (1973)
22	$O_2(1) + CO_2 \rightleftharpoons O_2 + CO_2(020)$	221	$9.1 \times 10^{-15}\sqrt{T}\exp(-56.7/\sqrt{T})$	Bass (1973)
23	$O_3 + h\nu \rightarrow O_2(1) + O(^3P)$		$\varepsilon = 4$	Zaragoza et al. (1998)
24	$O_2 + OH^*(v \leq 9) \rightarrow O_2(1) + OH^*(v-1)$		$9.34 \times 10^{-14}\exp(0.578v)$	Dodd et al. (1991), Chalamala and Copeland (1993)

*Excess of energy in cm^{-1}. †Rate coefficient for the forward sense of the process in $cm^3 s^{-1}$. ‡$M = N_2, O_2$. T is temperature in K.

constituent with $O_2(1)$ (process 21) regulate its daytime population and, through process 1, that of $H_2O(0,1,0)$. The exchange of vibrational energy between $O_2(1)$ and $N_2(1)$ (process 14), on the other hand, is of only slight importance for the excitation of the former in the daytime mesosphere. The exchange of vibrational energy between $O_2(1)$ and the CO_2 levels can be important for the de-excitation of O_2 (process 22). Process 7 in Table 6.2 is likely to excite $O_2(1)$ during daytime because of the large excitation of $CO_2(0,0^0,1)$.

Although it has been known for some time that vibrationally excited O_2 is produced by the photolysis of O_3 in the ultraviolet Hartley band (process 23), it was not recognised until recently that it could have an effect on the excitation of $H_2O(0,1,0)$ in the daytime mesosphere. Photolysis of O_3 in the Hartley band occurs predominantly through the singlet and triplet channels:

$$O_3 + h\nu \, (175 - 310\,\text{nm}) \; \rightarrow \; O(^1D) + O_2(a^1\Delta_g), \quad \text{and} \tag{7.10}$$

$$O_3 + h\nu \, (175 - 310\,\text{nm}) \; \rightarrow \; O(^3P) + O_2(X^3\Sigma_g^-, v). \tag{7.11}$$

The yields into the singlet and triplet channels are typically 0.9 and 0.1, respectively. The triplet channel (7.11) produces $O_2(X^3\Sigma_g^-, v)$ excited in vibrational levels from about $v = 15$ up to 34 which, after vibrational relaxation and quenching, produces $O_2(1)$. The effective quantum yield for $O_2(1)$ in this triplet channel, including its 0.1 yield, is estimated to be 0.6.

The singlet channel (7.10) produces $O_2(1)$ from the relaxation of $O_2(b^1\Sigma_g^+)$, previously excited in collisions with $O(^1D)$ (process 7.23), and possibly by the electronic-to-vibrational energy redistribution of $O_2(a^1\Delta_g)$ in collisions with O_2. Concerning $O_2(1)$ production from $O(^1D)$, about 30% of $O(^1D)$ is quenched by O_2, of which 25% might give $O_2(v)$. Assuming an average excitation of two quanta from this process, this results in an efficiency of about 0.13 $O_2(1)$ molecules per ozone molecule photodissociated. The rate of production of $O_2(1)$ from $O_2(a^1\Delta_g)$ in the single channel is the most uncertain. If quenching of $O_2(a^1\Delta_g)$ by O_2 takes place with no energy loss into $O_2(X^3\Sigma_g^-)$, it can produce up to 5 $O_2(1)$ molecules. Hence, considering the two channels, we can have a quantum yield of up to nearly 6 $O_2(1)$ molecules per O_3 molecule. This mechanism produces excitation of the daytime mesospheric $H_2O(0,1,0)$ level through process 1; a quantum yield of 4 has been inferred from ISAMS H_2O observations, although all values between 2 and 6 are possible within the measurements errors.

Excitation of $O_2(1)$ from the electronically excited hydroxyl radical, $OH(v \leq 9)$ (process 24), could be important around the night-time mesopause, especially if the collisional de-excitation of $O_2(1)$ by $O(^3P)$ turn out to be much weaker than currently thought. It is difficult to test this, however, since $O_2(1)$ does not emit and the V–V collision between $H_2O(0,1,0)$ and $O_2(1)$ is negligible compared to radiative exchange at that height.

7.4.3 Non-LTE populations

The (0,1,0) vibrational level of water vapour is in LTE in the stratosphere, and departs from LTE above about 60 to 65 km, depending on the temperature structure. The altitude of LTE breakdown is approximately the same by day or night, but the extent of the departure is much greater by day. Several non-LTE processes contribute: (i) absorption of radiation emitted at lower layers; (ii) solar pumping directly at 6.3 μm, and indirectly at 2.7 μm with subsequent relaxation to $H_2O(0,2,0)$ and (0,1,0); and (iii) vibration-vibration coupling of $H_2O(0,1,0)$ with $O_2(1)$.

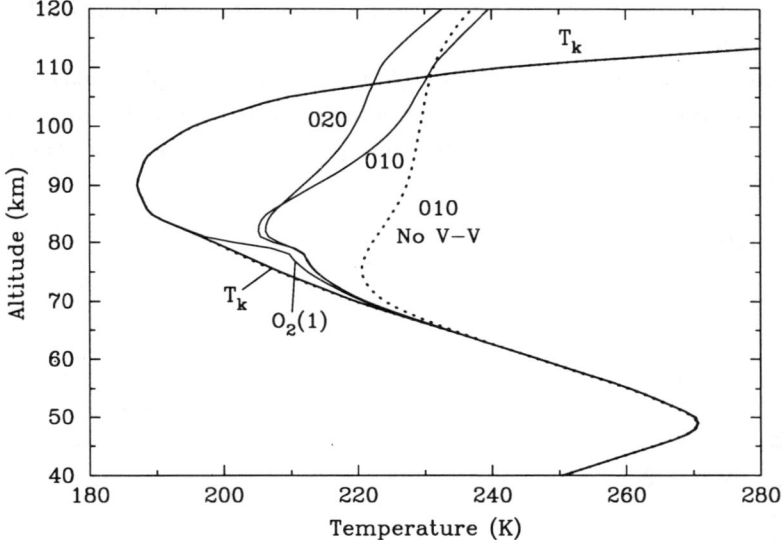

Fig. 7.13 Vibrational temperatures of $H_2O(0,1,0)$, $H_2O(0,2,0)$ and $O_2(1)$ at night-time. With vibrational-vibrational (V–V) coupling between $H_2O(0,1,0)$ and $O_2(1)$ included (solid lines) and without (dotted lines). The uncoupled T_v of $O_2(1)$ coincides near exactly with the kinetic temperature, T_k. The uncoupled T_v of (020) is not shown.

Figure 7.13 shows model vibrational temperatures of the $H_2O(0,1,0)$ and $O_2(1)$ levels at night, with and without V–V coupling between $H_2O(0,1,0)$ and $O_2(1)$. Below about 65 km, the coupled $H_2O(0,1,0)$–$O_2(1)$ system is controlled by $O_2(1)$ which, in turn, is driven by $CO_2(0,2,0)$, which is in LTE in this region (Sec. 6.3.6.1). Thermal collisions of $H_2O(0,1,0)$ with ground-state O_2, N_2, and $O(^3P)$ are also significant. As a consequence, the $H_2O(0,1,0)$ and $O_2(1)$ states have Boltzmann populations at the local kinetic temperature up to about 65 km.

Between 70 and 120 km, radiative absorption in the 6.3 μm band dominates in populating $H_2O(0,1,0)$. Up to 110 km, this induces larger populations than would exist in LTE (see Fig. 7.13). Above this altitude, thermal collisions and radiative absorption are not enough to balance the spontaneous emission, leading to vibrational temperatures much lower than the kinetic temperature.

As expected from the strong V–V coupling, radiative absorption in the water vapour band is reflected in the population of $O_2(1)$, which is greater than the LTE value in the region between about 70 to 80 km. Above 80 km, productions and losses of $O_2(1)$ are dominated by thermal collisions with $O(^3P)$, which results in an LTE population for $O_2(1)$ in the upper mesosphere and lower thermosphere.

At night, $O_2(1)$ can be produced by vibrationally excited OH in the upper mesosphere (process 24). The excitation of $O_2(1)$ is, however, offset by thermal relaxation in collisions with $O(^3P)$, which, in spite of its low mesospheric abundance, is very efficient at quenching $O_2(1)$. The rate of this relaxation (process 21) is poorly known at mesospheric temperatures. If it turns out to be much slower than the value in Table 7.6, then $O_2(1)$ population would be enhanced over LTE around 80–85 km (see Sec. 7.4.4).

The vibrational temperature of the $H_2O(0,2,0)$ level above about 65 km is mainly controlled by radiative absorption of upwelling photons and V–V processes involving $H_2O(0,1,0)$ and $O_2(1)$. This night-time population is much enhanced during daytime (see below).

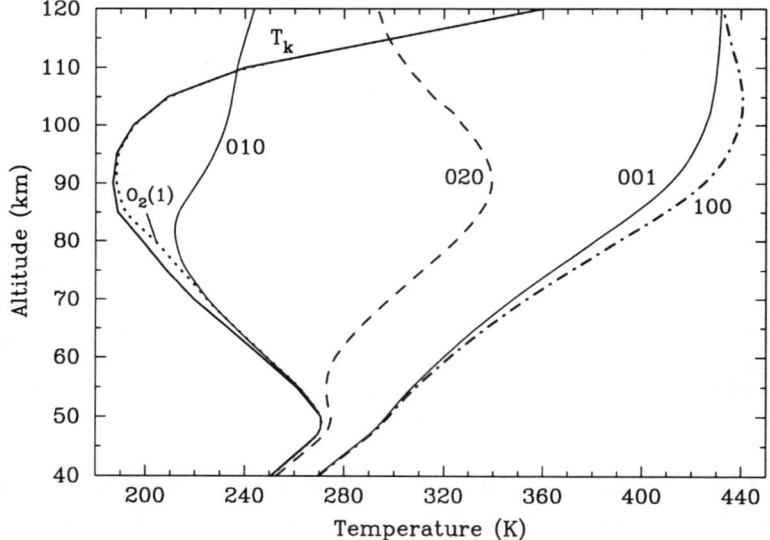

Fig. 7.14 Vibrational temperatures for the $H_2O(0,1,0)$, $H_2O(0,2,0)$, $H_2O(1,0,0)$, $H_2O(0,0,1)$, and $O_2(1)$ vibrational levels for the U.S. Standard 1976 atmosphere for daytime conditions ($\chi = 0°$).

The vibrational temperatures for the H_2O and $O_2(1)$ levels for daytime conditions ($\chi = 0°$) are shown in Fig. 7.14. The main difference from night-time is the large increase in the population of $H_2O(0,2,0)$, mainly as a consequence of the absorption of solar radiation by H_2O in the $(0,0,1$–$0,0,0)$ 2.7 μm band, and its subsequent collisional relaxation to the $(1,0,0)$ and $(0,2,0)$ states. Its vibrational temperature decreases above about 90 km because collisions are not fast enough to transfer the energy from the solar pumped $(0,0,1)$ state. The V–V relaxation of $H_2O(0,2,0)$ by O_2 is of little significance for activating the $H_2O(0,1,0)$ level but is a very important mechanism for regulating the large daytime mesospheric population of $H_2O(0,2,0)$.

The daytime populations of $H_2O(0,1,0)$ and $O_2(1)$ above about 60 km are slightly larger than at night as a result of several processes. The direct absorption of solar radiation at 6.3 μm, and the excitation of $H_2O(0,1,0)$ by absorption of solar radiation at 2.7 μm followed by V–V relaxation to $H_2O(0,1,0)$ in collisions with O_2, are both significant, as well as the absorption of upwelling terrestrial radiation at 6.3 μm, which of course also occurs at night. Direct excitation from V–V collisions with $N_2(1)$ is less important.

$O_2(1)$ has a number of daytime excitation processes, most significantly in the 60–80 km region, in addition to coupling with $H_2O(0,1,0)$. These include activation from the photolysis of O_3, and to a lesser extent vibrational transfer from $N_2(1)$. Indirect excitation by solar absorption at 1.9 μm, and especially at 2.7 μm, and its subsequent V–V transfer to $O_2(1)$, are also important, as is the excitation of $O_2(1)$ in collisions with $CO_2(0,0^0,1)$ previously excited by absorption of solar radiation by CO_2 at 4.3 and 2.7 μm. Collisional de-activation of $O_2(1)$ by $O(^3P)$ in this region acquire a larger importance than at night-time because of the larger abundance of $O(^3P)$ by day.

The effect of the various processes on the daytime population of $H_2O(0,1,0)$ is shown in Fig. 7.15, where we show the changes in its daytime vibrational temperature produced individually by each of the following processes: (i) O_3 photolysis; (ii) excitation from $CO_2(0,0^0,1)$ and $N_2(1)$; and (iii) absorption of solar radiation at 6.3 μm, and in the near-infrared, 1.9 and 2.7 μm bands (NIR). Curve '$O(^3P)$' is the $T_{v,t}(\text{day})$–$T_v(\text{night})$ difference, where $T_{v,t}(\text{day})$ is the daytime vibrational temperature obtained by neglecting the daytime excitation mechanisms, i.e., absorption of solar radiation, excitation from O_3 photolysis, and assuming the populations of $CO_2(0,0^0,1)$ and $N_2(1)$ are in LTE. The remaining difference is the change induced by the larger daytime concentration of $O(^3P)$, which produces a significant daytime decrease in the $H_2O(0,1,0)$ vibrational temperature between 65 and 85 km.

Figure 7.14 also shows the vibrational temperatures of the H_2O (0,0,1) and (1,0,0) levels, which are important in the study of atmospheric emission near 2.7 μm. The populations of these levels are principally maintained by the absorption of solar radiation by H_2O in the (0,0,1–0,0,0) band. The contribution of the (1,0,0–0,0,0) band is much smaller but the strong, near-resonant V–V coupling between (1,0,0) and (0,0,1) maintains their populations in mutual equilibrium up to the upper boundary of the model. Both levels are depleted by thermal collisions with the major atmospheric species N_2 and O_2 which relax them to the (0,2,0) state. The solar pumping is so strong that (1,0,0) and (0,0,1) have populations larger than LTE even below the stratopause. These levels are the principal source of (0,2,0) excitation during daytime which, in turn, enhances the population of $H_2O(0,1,0)$ and therefore the atmospheric emission at 6.3 μm.

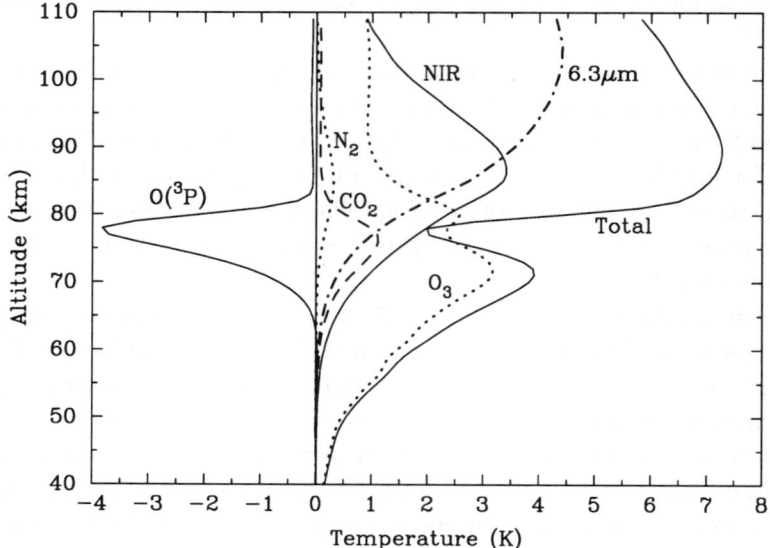

Fig. 7.15   Day-night difference in $H_2O(0,1,0)$ vibrational temperatures produced by different processes, considered individually. See text for details of the processes included.

7.4.4   *Uncertainties in the H_2O populations*

The effects of current uncertainties in the rate coefficients for the excitation of $O_2(1)$ by O_3 photolysis and by the thermal quenching of $O_2(1)$ by $O(^3P)$ on the population of $H_2O(0,1,0)$ in the mesosphere is illustrated in Fig. 7.16. An efficiency of 4 (instead of 0) for the production of $O_2(1)$ from O_3 photodissociation can give rise to a day-night enhancement of around 4 K at 70 km.

The use of very low published rate coefficients for thermal quenching of $O_2(1)$ by $O(^3P)$ (see Sec. 7.15) gives rise to a very large difference in the vibrational temperature of $H_2O(0,1,0)$ above 70 km, reaching 50 K at the daytime mesopause. Changing the rate coefficient by more reasonable values, e.g. multiplying and dividing by 2, produces variations of ± 3 K between 60 and 80 km and negligible changes in all other regions.

The population of $H_2O(0,1,0)$ above 80 km, and that of $(0,2,0)$ in the whole altitude range, change very little with solar zenith angle except for values very close to the twilight. The reason is the low abundance of water vapour in the mesosphere and upper thermosphere, which means that the solar flux at 6.3 and 2.7 μm is not substantially absorbed until very low solar elevation angles are reached. Below 80 km, the variation in $H_2O(0,1,0)$, principally caused by the photodissociation of O_3, is more sensitive to the solar zenith angle.

The analysis of the H_2O 6.3 μm measurements by ISAMS has shown that the deviation from LTE of the $H_2O(0,1,0)$ vibrational temperature depends significantly on the mesospheric temperature profile. The T_v–T_k difference in the lower mesosphere is not very large in absolute terms, only a few kelvins, but significant. For

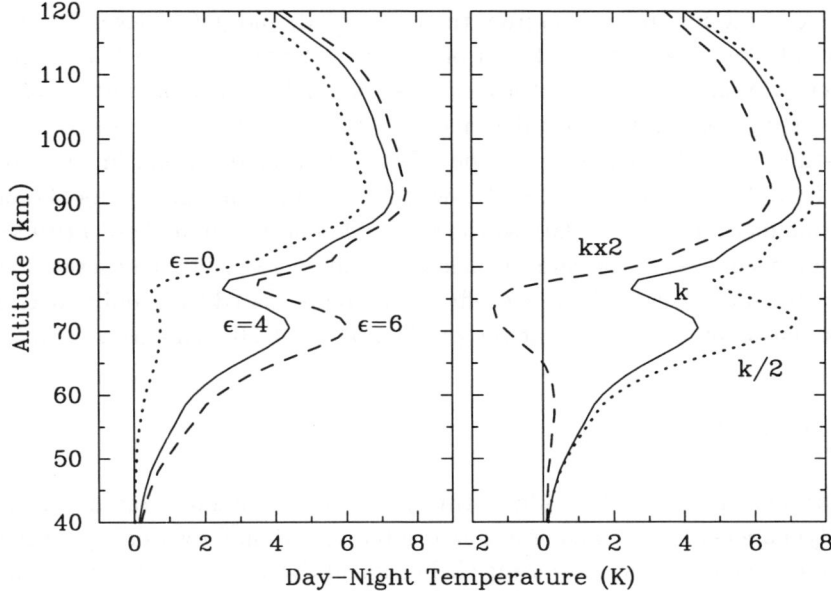

Fig. 7.16   Effects of the efficiency of O_3 photolysis, ε, and of the rate coefficient of V–T collisions of $O_2(1)$ with $O(^3P)$, k, on the day-night difference in $H_2O(0,1,0)$ vibrational temperatures.

the colder upper mesospheric temperatures, $O_2(1)$ vibrational temperatures are colder and the $H_2O(0,1,0)$ population is more decoupled from LTE. This is generally associated with warmer upper stratosphere–lower mesosphere temperatures which enhance the absorption of terrestrial upwelling radiation and hence increase $H_2O(0,1,0)$ T_v. As a result, we expect larger deviations of $H_2O(0,1,0)$ T_v's for polar summer conditions and smaller for polar winter.

Since the population of $H_2O(0,1,0)$ is dominated by radiative transfer in the upper mesosphere, it will depend on the abundance of water vapour in that region, which is not well known and probably quite variable. Daytime vibrational temperatures assuming nominal, times 2, and times 0.2 water vapour concentration profiles show differences as large as 8–10 K above about 80 km. The vibrational temperatures are larger for a drier atmosphere because the H_2O $(0,1,0–0,0,0)$ band remains optically thin down to lower and warmer mesospheric altitudes, and therefore the absorption of upwelling photons in the upper mesosphere and lower thermosphere is greater.

The vibrational temperatures of $H_2O(0,2,0)$ are very sensitive to the rate of the vibrational coupling with $O_2(1)$ (process 5), as well as to that of the thermal relaxation of $H_2O(1,0,0)$ and $(0,0,1)$ (processes 13, 14 and 15 in Table 7.6). The effect of varying the rate of process 5 between its extreme values deduced from ISAMS measurements of 2.0×10^{-12} and 6.8×10^{-12} cm^3s^{-1} on the T_v's of $H_2O(0,2,0)$ are of ±10 K in the altitude range from 50 to 100 km. The $(0,1,0)$ level, however, shows much smaller changes with only a few (2–3) kelvins in the 70–100 km region.

The rate of thermal relaxation from the H_2O (1,0,0) and (0,0,1) levels to (0,2,0) is estimated to be uncertain by a factor of 2. For the larger rate, the T_v's of (1,0,0) are smaller by about 15 K at all altitudes where non-LTE occurs and collisions are important, e.g., below 110 km for daytime conditions and between 60 and 110 km for night-time. For the (0,2,0) level, the effect of this modification in the rate is about 12 K, but only above 80 km. Below this altitude, this process is the only de-excitation mechanism of (1,0,0), so the production of (0,2,0) at these altitudes is not affected by changes in the rate coefficient, i.e., an increase in the deactivation of (1,0,0) due to an increase in the rate coefficient is compensated by a similar decrease of the population of (1,0,0), so that the production of (0,2,0) remains constant.

7.5   Methane, CH_4

The importance of CH_4 in the middle atmosphere derives mainly from its role as a tracer of dynamical features, and as a source, after oxidation, for water vapour in the stratosphere. Theoretical studies using chemical and dynamical models and a number of remote sounding experiments have carried out studies of methane (ATMOS on the ATLAS missions; CLAES, HALOE and ISAMS on UARS), and have helped for a better understanding of its spatial and temporal variability. CH_4 has also been studied in detail in the atmospheres of the giants planets and Titan, where it is the most abundant minor constituent. New instruments will investigate its sources and sinks with unprecedented resolution on the Earth and across the Solar System in the next decade. However, although non-LTE situations have been studied extensively for many atmospheric species, CH_4 has not received much attention yet because most of the observations have been made at relatively low altitudes, and no firm experimental evidence for non-LTE in the CH_4 infrared emissions has been reported. ISAMS measurements in the upper stratosphere and mesosphere might be an exception. As noted above, this situation is likely to change before long.

Table 7.7 CH_4 energy levels.

Equivalent level	Grouped levels	Energy (cm^{-1})
0001	0001	1310.76
0100	0100	1533.34
V3.3	$2v_4$, v_2+v_4, v_1, v_3, $2v_2$	2941.00
V2.3	$3v_4$, v_2+2v_4,v_1+v_4, v_3+v_4, v_2+v_3	4230.00
V1.7	$4v_4$, v_3+2v_4, $v_1+v_2+v_4$, $v_2+v_3+v_4$, $2v_3$	5566.00

Methane has four vibrational normal modes: two stretching modes, v_1 and v_3, with degeneracies of 1 and 3, and two bending modes, v_2 and v_4, which are doubly and triply degenerated, respectively. The most important transitions in the infrared are the fundamental bands originating from the (0,1,0,0) and (0,0,0,1) levels. These states are coupled with others that are pumped by solar radiation at shorter wavelengths, so it is necessary to extend the model to the near-IR bands. Three groups of CH_4 levels are significantly excited by absorption of solar radiation at 1.7, 2.3 and 3.3 μm, that at 3.3 μm being the most important for populating the v_2 and v_4 states. The levels within those groups are very close in energy and experience very rapid collisional equilibration between them. Hence, for each group, they can be considered to be in equilibrium among themselves at the local kinetic temperature and treated as a single 'equivalent' state (see Table 7.7). Fig. 7.17 shows a schematic diagram of the transitions and the vibrational levels that have been studied; the transitions are also listed in Table 7.8.

The first vibrationally excited level of molecular oxygen, $O_2(1)$, is included because it is strongly coupled by V–V processes with the $CH_4(v_2)$ and $CH_4(v_4)$ levels. For this reason, the coupling of a CH_4 non-LTE model with a H_2O model, which treat accurately the population of $O_2(1)$, is needed if we want to get accurate CH_4 non-LTE populations.

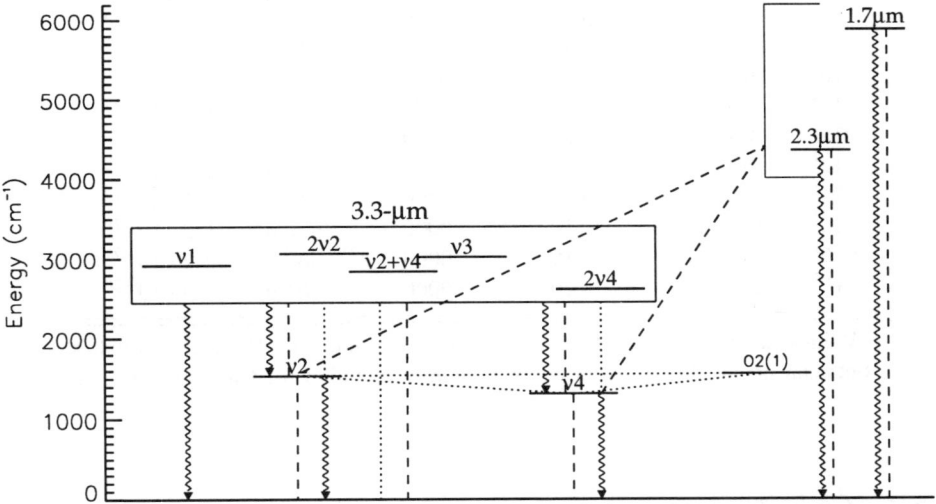

Fig. 7.17   Vibrational energy levels and the transitions (arrows) between them for the CH_4 major isotope. Dashed lines represent V–T processes; dotted: V–V processes.

7.5.1   *Radiative processes*

The most important processes are: (i) the spontaneous emission in all of the transitions in Table 7.8; (ii) the absorption of solar radiation in the 1.7, 2.3, 3.3, 6.5 and 7.6 μm bands; (iii) the absorption of tropospheric radiation in all of the bands in Table 7.8; and (iv) the exchange of photons between atmospheric layers for the stronger bands at 7.6, 6.5 and 3.3 μm.

The ν_2 and ν_4 fundamental bands are optically thin above the troposphere and the radiation absorbed in the mesosphere comes primarily from the lower region. As explained in Sec. 6.1, the tropospheric flux depends on many tropospheric parameters, such as the concentration of H_2O and the cloud coverage, and temperature. Here they have been computed with a line-by-line code for the mean tropospheric conditions at mid-latitudes. The model runs from the lower boundary located at 9 km up to 120 km.

Table 7.8   Principal CH_4 infrared bands.

Band	Label	Upper level	Lower level	$\tilde{\nu}_0$, cm^{-1}	Strength*
1	ν_4	0001	0000	1310.76	4.8×10^{-18}
2[†]	$\nu_2+\nu_4-\nu_2$	0101	0100	1296.96	5.7×10^{-21}
3[†]	$2\nu_4-\nu_4$	0002	0001	1303.50	3.6×10^{-20}
4	ν_2	0100	0000	1533.34	5.3×10^{-20}
5	$\nu_3-\nu_4$	0010	0001	1708.73	1.9×10^{-21}
6[†]	$2\nu_4$	0002	0000	2614.25	5.5×10^{-20}
7	$\nu_2+\nu_4$	0101	0000	2838.20	3.7×10^{-19}
8	ν_1	1000	0000	2916.48	1.1×10^{-21}
9	ν_3	0010	0000	3019.49	1.1×10^{-17}
10	$2\nu_2$	0200	0000	3064.39	3.4×10^{-20}
11[†]	$\nu_2+\nu_3-\nu_2$	0110	0100	3010.39	6.8×10^{-21}
12[†]	$3\nu_4$	0003	0000	3870.50	3.4×10^{-21}
13[†]	$\nu_1+\nu_4$	1001	0000	4223.46	2.5×10^{-19}
14[†]	$\nu_3+\nu_4$	0011	0000	4319.28	4.1×10^{-19}
15[†]	$\nu_2+\nu_3$	0110	0000	4540.61	6.2×10^{-20}
16[†]	$\nu_3+2\nu_4$	0012	0000	5588.00	1.2×10^{-21}
17[†]	$2\nu_3$	0020	0000	6004.99	5.9×10^{-20}

*Values shown are at 296 K in cm^{-1}/(molecules cm^{-2}). [†]Radiative transfer between atmospheric layers is not normally included for these transitions.

7.5.2   *Collisional processes*

The important collisional processes for the CH_4 levels are summarised in Table 7.9. Thermal relaxation of the CH_4 v_2 and v_4 levels occurs in collisions involving the major atmospheric constituents N_2 and O_2, processes 1 and 2 in the table. In spite of a considerable number of semi-empirical studies and experiments, our knowledge of the collisional deactivation rate for these processes is far from adequate, particularly at low temperatures.

The collisional relaxation of the $CH_4(V^i)$ levels (processes 3–5 in Table 7.9), where V^i is any of the V3.3, V2.3, or V1.7 equivalent levels, have been included with a rate coefficient obtained from a simple 'energy gap' model, $k_{V-T} = A\exp(-\Delta E/T)$,

Table 7.9 Principal collisional processes affecting the CH$_4$ states.

No.	Process	Rate coefficient, $k^\dagger$	Ref.[*]
1	CH$_4(v_2)$ + M‡ $\rightleftharpoons$ CH$_4$ + M	$\log(k) = a_0 + a_1 T + a_2 T^2$	1
2	CH$_4(v_4)$ + M $\rightleftharpoons$ CH$_4$ + M	$\log(k) = a_0 + a_1 T + a_2 T^2$	1
3	CH$_4(V^i)$ + M $\rightleftharpoons$ CH$_4(v_2)$ + M	$4\times10^{-11}\exp(-\Delta E/kT)$	1
4	CH$_4(V^i)$ + M $\rightleftharpoons$ CH$_4(v_4)$ + M	$4\times10^{-11}\exp(-\Delta E/kT)$	1
5	CH$_4(V^i)$ + M $\rightleftharpoons$ CH$_4$ + M	$4\times10^{-11}\exp(-\Delta E/kT)$	1
6	CH$_4$ + O$_2$(1) $\rightleftharpoons$ CH$_4(v_2)$ + O$_2$	5×10^{-13}	2
7	CH$_4$ + O$_2$(1) $\rightleftharpoons$ CH$_4(v_4)$ + O$_2$	5×10^{-13}	2
8	CH$_4$(V3.3) + O$_2$ $\rightleftharpoons$ CH$_4(v_2$ or $v_4)$ + O$_2$(1)	1.26×10^{-12}	2

†Rate coefficient for the forward sense of the process in cm^{3}s^{-1}. ‡M = N$_2$, O$_2$. T is temperature in K. $a_0 = -15.99$, $a_1 = 1.42\times10^{-3}$, $a_2 = 1.09\times10^{-5}$ for M = N$_2$, and $a_0 = -17.19$, $a_1 = 1.34\times10^{-2}$, $a_2 = -1.46\times10^{-5}$ for M = O$_2$. ΔE is the excess of energy of the process in cm^{-1}. V^i = V3.3, V2.3 or V1.7. [*]Ref. 1: Siddles *et al.* (1994a); 2: Avramides and Hunter (1983).

where ΔE is the energy gap between the upper and lower states and A is determined by fitting the measured collisional rate for process 1.

The populations of the $v_2 = 1$ and $v_4 = 1$ states are affected by the near-resonant V–V energy exchange with the first vibrationally excited level of O$_2$ (processes 6 and 7 in Table 7.9). Since O$_2$(1) lies very close in energy to the methane fundamental bending vibrations, rapid energy transfer occurs between these modes.

Another important V–V process affecting the populations of the CH$_4$ v_2 and v_4 states is the transfer from the equivalent level V3.3 (process 8). Vibrational-vibrational exchange from the V1.7 and V2.3 groups to these levels is much less important and can be neglected. Similarly, CH$_4$ + N$_2$(1) $\rightleftharpoons$ CH$_4(v_2$ or $v_4)$ + N$_2$, which represents the V–V exchange with N$_2$(1), is not represented in the model because it must be much slower than with O$_2$, given that the energy gap is greater.

7.5.3 *Non-LTE populations*

Figure 7.18 shows the vibrational temperatures for day and night-time, compared to the U.S. Standard Atmosphere mid-latitude kinetic temperature. The night-time vibrational temperatures remain in LTE up to about 65 km for all states considered. Below this altitude, the populations of the CH$_4$ $v_2 = 1$ and $v_4 = 1$ states are controlled by the vibrational exchange of energy with O$_2$(1). This level is coupled with H$_2$O(0,1,0), which is in turn affected by V–T and V–V energy exchange with CO$_2$(0,2,0). Hence H$_2$O(0,1,0) is in LTE and, consequently, the CH$_4(v_2)$ and CH$_4(v_4)$ states are also. Above 65 km, absorption of the upwelling flux from the troposphere in the 7.5 μm band becomes the dominant process for the population of CH$_4(v_4)$, which is larger than LTE up to 110 km.

The (0,1,0,0) or v_2 level is nearly resonant with O$_2$(1) and its population is more

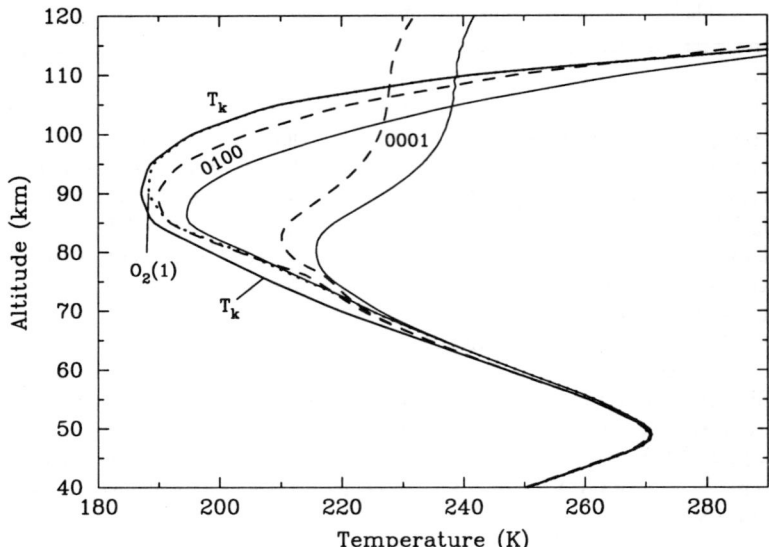

Fig. 7.18   Vibrational temperatures for the CH_4 levels for mid-latitudes at daytime (solid lines) and night-time (dashed lines). The daytime T_v of $O_2(1)$ is also shown (dotted line). T_k is the kinetic temperature.

strongly affected by V–V processes with $O_2(1)$ than that of the (0,0,0,1) state. As discussed in Sec. 7.4.3, the $O_2(1)$ population shows a small bump around 75 km at night caused by V–V transfer from $H_2O(0,1,0)$ excited by earthshine at 6.3 μm. This is reflected in the strongly coupled $CH_4(0,1,0,0)$ state, which shows a similar bump at that altitude. Above about 85–90 km the vibrational temperature of v_2 is slightly greater than T_k because of the absorption of the tropospheric flux in the ν_2 band at 6.5 μm. Vibrational-vibrational collisions with $O_2(1)$ are the major excitation process in the lower thermosphere, where spontaneous emission is the major loss.

Under polar winter conditions, the population of the (0,0,0,1) state is closer to LTE up to slightly higher altitudes, then slightly depleted relative to the Boltzmann distribution in the mesosphere. This is the result of the weaker upwelling flux from the colder troposphere. For the (0,1,0,0) state, the upwelling flux is even smaller and its population is even closer to LTE than for mid-latitudes. For the polar summer, the deviation from LTE occurs at slightly lower altitudes than for mid-latitude temperatures and it is much larger in the cold upper mesosphere.

The daytime vibrational temperatures for v_2 and v_4 at mid-latitudes are also shown in Fig. 7.18. The departures from LTE start lower than at night, and the vibrational temperatures of the (0,1,0,0) and (0,0,0,1) levels are larger than those for night-time above 85 and 75 km respectively. These effects are due to the vibrational relaxation of higher energy states (processes 3 and 4 in Table 7.9) that are strongly excited by absorption of solar radiation at 3.3 μm. The V–V transfer from the larger daytime population of $O_2(1)$ between 65 and 85 km also makes a small contribution. Solar energy absorbed in the 1.7 and 2.3 μm (see Table 7.7) is negligible in populating any of the fundamental bending modes of methane.

7.5.4 *Uncertainties in the CH$_4$ populations*

As we have seen, the populations of CH$_4$ v_2 and v_4 are very sensitive to the populations of O$_2$(1). Laboratory determinations of the O$_2$(1) + O thermalization rate, and of the excitation of O$_2$(1) from O$_3$ photolysis, are needed to reduce the current large uncertainty in the populations of these levels, as well as that of H$_2$O(0,1,0).

The uncertainties in processes 6, 7, and 8 in Table 7.9 are around 20% but these have little impact on the vibrational temperatures of the fundamental bending levels of CH$_4$ (a maximum change of 1–2 K at mesospheric altitudes). The uncertainties in the rate coefficients of the deactivation of the V3.3 equivalent level by V–T energy exchange (processes 3, 4 and 5 in Table 7.9) also have little effect on the CH$_4$ populations. For example, multiplying or dividing this coefficient rate by a factor of 3 does not change the populations of CH$_4$(v_2) and CH$_4$(v_4) significantly. The effect of neglecting the exchange between layers, on the other hand, gives rise to a large decrease in the (0,0,0,1) vibrational temperature in the upper mesosphere of about 10–20 K between 75 and 100 km.

The tropospheric flux, depending on the tropospheric conditions of temperature, cloud coverage, humidity and other constituent abundances, can change the CH$_4$(0,0,0,1) T_v in the mesosphere by as much as 10 K in the daytime and 15 K at night. The vibrational temperature of the (0,1,0,0) level is also affected, but to a smaller degree, between 2–3 K in the mesosphere.

7.6 Nitric Oxide, NO

Nitric oxide is an important trace constituent in the whole of the lower atmosphere. A knowledge of its abundance in the stratosphere is very important for understanding the odd nitrogen chemistry, especially the rapid cyclic conversions between NO and NO$_2$. Higher up, NO plays a key role in the radiative energy budget of the thermosphere.

NO has been measured using the emission from its fundamental band NO(1–0) at 5.3 μm by the ISAMS instrument (see Chapter 8) and, despite problems caused by the contribution of the large thermospheric emission to observations lower down, it has been possible to retrieve its stratospheric abundance. Other measurements of NO are summarised in Sec. 8.7. A future mission, ENVISAT/MIPAS, will record high resolution limb emission spectra enabling the retrieval of nitric oxide. The 5.3 μm emission from NO($v = 1$) is highly affected by vibrational non-LTE above the troposphere, and by rotational/spin non-LTE in the thermosphere.

7.6.1 *Radiative processes*

The spectral lines in the NO(1–0) vibration-rotation band are optically thin even for paths through the whole atmosphere. Hence the radiative field within those lines is dominated by tropospheric and solar radiation, while atmospheric NO emission does not contribute significantly. In general, less than 1% (although it can be up to 12% in polar summer conditions) of the total excitation of the NO($v = 1$) state at any height originates from NO emission. Thus, radiative transfer between atmospheric layers can be neglected, which greatly simplifies the calculation of its population. The upwelling tropospheric radiation is mainly dominated by emission from water vapour and therefore depends on temperature, humidity and cloud cover in the troposphere. Emission from the surface does not contribute significantly, since the water vapour in the atmosphere is optically thick below 5 km or so in the spectral region of the NO infrared bands.

As before, the dependence of the tropospheric flux on the tropospheric temperature can be taken into account by finding the height where the atmosphere becomes effectively opaque at the frequency of the absorbing line, $z_e(\nu_{J,J'})$. This altitude is nearly independent of the tropospheric temperature profile and depends mainly on the H_2O volume mixing ratio. The tropospheric upwelling radiation can then be expressed by the Planck function $B(T_b)$, where $T_b(\nu_{J,J'})$ is the temperature at $z_e(\nu_{J,J'})$. The values of $z_e(\nu)$ at different NO line positions varies between 4 and 12 km. The effect of clouds can be included by restricting the effective altitude range to those altitudes above the cloud top.

For a rigourous treatment, a range of effective altitudes should be used for all levels which are pumped by tropospheric radiation to account for the variability of the H_2O abundance in the troposphere. When considering only vibrational levels, however, a mean effective altitude is accurate enough for all lines. In the case of NO, we want to know the population of the rotational states as well, and so the individual contributions have to be included line by line.

7.6.2 *Collisional relaxation of the rotational and spin states*

The rotational and spin relaxation within a given vibrational state v due to collisional processes has the form

$$k_{J',S';J,S}: \quad \mathrm{NO}(v, J, S) + \mathrm{M} \rightleftharpoons \mathrm{NO}(v, J', S') + \mathrm{M}, \qquad (7.12)$$

where M is the collision partner molecule, N_2, O_2 or $O(^3P)$. The rate coefficient $k_{J',S';J,S}$ for rotational energy changes $\Delta E_{J,J'} \geq 0$ and spin energy changes $\Delta E_{S,S'} \geq 0$ can be modelled by a hybrid exponential power gap law taking into account the observed propensity rules in ΔJ and ΔS:

$$k_{J',S';J,S} = a_1 \, f_S \, f_J \left(\frac{\Delta E_{J,J'}}{B_0}\right)^{a_2} \exp\left(-a_3 \frac{\Delta E_{J,J'}}{kT}\right), \qquad (7.13)$$

with $f_S = (1 - \beta|\Delta S|)$, $f_J = \left(1 + \gamma_{\Delta S}|\Delta J|(-1)^{|\Delta J|}\right)$; where B_0 is the rotational constant and k the Boltzmann constant. The adjustable parameters a_1, a_2 and a_3, and the propensity factors β and $\gamma_{\Delta S}$ are determined by a least squares fit to spectroscopic data. Detailed balance gives the rates for energetically downward transitions.

7.6.3 *Vibrational-translational collisions and chemical production*

The collisional transitions in NO between different vibrational states has the form

$$\text{NO}(v, J, S) + \text{M} \rightleftharpoons \text{NO}(v', J', S') + \text{M}, \qquad (7.14)$$

where M can be O_2 or O. This is described by the forward rate coefficient:

$$k_{v',J',S';v,J,S} = Q(T_e^s, T_e^r, J, S)\, k_{v';v}, \qquad (7.15)$$

where $Q(T_e^s, T_e^r, J, S)$ is a normalized Boltzmann rotational/spin distribution corresponding to an effective spin temperature, T_e^s, and to an effective rotational temperature, T_e^r. $k_{v';v}$ is the total excitation rate into the vibrational state v, related by detailed balance with the quenching rates given in Table 7.10. Vibrationally reverse rates $k'_{v,J,S;v',J',S'}$ are given by detailed balance.

Table 7.10   Principal collisional processes for the NO(v, J, S) states.

No.	Process/$f(v)$	T_e^r(K)	$k^\dagger$	Refs.*
1a	$\text{NO}(1, J, S) + O_2 \rightleftharpoons \text{NO}(0, J', S') + O_2$	T	2.4×10^{-14}	1
1b	$\text{NO}(1, J, S) + O \rightleftharpoons \text{NO}(0, J', S') + O$	$0.74T + 24$	2.8×10^{-11}	2
2a	$\text{NO}(2, J, S) + O_2 \rightleftharpoons \text{NO}(1, J', S') + O_2$	T	7.4×10^{-14}	1
2b	$\text{NO}(2, J, S) + O \rightleftharpoons \text{NO}(1, J', S') + O$	$0.74T + 24$	1.3×10^{-11}	2
3	$\text{NO}(2, J, S) + O \rightleftharpoons \text{NO}(0, J', S') + O$	$0.74T + 24$	1.8×10^{-11}	2
4	$\text{NO}_2 + O \rightarrow \text{NO}(v, J, S) + O_2$	T	9.7×10^{-12}	3
	$f(v=0,1,\geq2) = 0.681, 0.222, 0.070$			4
5	$\text{NO}_2 + h\nu \rightarrow \text{NO}(v, J, S) + O$	5000		
	$f(v)$: altitude dependent			4
6	$\text{N}(^2D) + O_2 \rightarrow \text{NO}(v, J, S) + O$	5000	5.7×10^{-12}	6
	$f(v=0,1,\geq2) = 0.03, 0.05, 0.92$			5
7a	$\text{N}(^4S)_{\text{thermal}} + O_2 \rightarrow \text{NO}(v, J, S) + O$	$3000/5000^\ddagger$	$1.15\times10^{-11}\cdot$	6, 8
	$f(v=0,1,\geq2) = 0.04, 0.07, 0.89$		$\exp(-3503/T)$	
7b	$\text{N}(^4S)_{\text{super-thermal}} + O_2 \rightarrow \text{NO}(v, J, S) + O$	5000		7
	$f(v=0,1,\geq2) = 0.04, 0.07, 0.89$			8

†Rate coefficient for the forward sense of the process in cm^{3}s^{-1}. T is temperature in K.
*Refs. 1: Wysong (1994); 2: Duff and Sharma (1997), Dodd *et al.* (1999); 3: DeMore *et al.* (1997); 4: Kaye and Kumer (1987); 5: Funke and López-Puertas (2000); 6: Swaminathan *et al.* (1998); 7: Gérard *et al.* (1991); and 8: Duff *et al.* (1994). $T_e^s = T$ for collisions with O_2 and $T_e^s = 200$ K for collisions with $O(^3P)$. ‡3000 K at night-time and 5000 K at daytime.

The chemical and photochemical production terms can be described in a manner similar to the excitation rates of the vibrational relaxation (Eq. 7.15) assuming Boltzmann rotational and spin state nascent distributions:

$$k_{c;v,J,S} = Q(T_e^s, T_e^r, J, S)\, f(v)\, k_c, \qquad (7.16)$$

with an additional term $f(v)$ describing the vibrational nascent distribution, and the production rate coefficient k_c. The chemical and photochemical processes, including their rate coefficients, known currently are summarised in Table 7.10, processes 4–7. One of the remarkable aspects of NO(1), as opposite to other molecules, is the low efficiency of N_2 in vibrational-translational collisions with NO(1), in comparison with, for example, that for O_2.

The populations of the NO(v, J, S) states are calculated by solving the system of the statistical equilibrium equation for each of the levels which is expressed by:

$$n_i \sum_j \left(A_{i \to j} + l_{t;i \to j}\right) = P_{c,i} + \sum_j \left(P_{t;j \to i} + B_{j,i} \bar{L}_{0,ij}\right) n_j \qquad (7.17)$$

where n_i and n_j are the populations of the upper state i given by (v, J, S) and the lower state j given by (v', J', S'), respectively. The production rates of the state i from all other states are given: by photo-chemical and chemical processes, $P_{c,i}$; by collisional processes, $P_{t;j \to i}$; and by radiative processes, $B_{j,i} \bar{L}_{0,ij}$ (including solar and tropospheric radiances). The losses are given by radiative spontaneous emission, $A_{i \to j}$, and by collisions, $l_{t;i \to j}$, as described in the sections above. Since radiative transfer between layers is negligible, the statistical equilibrium equations for the NO(v, J, S) states can be solved locally, i.e. they are not coupled at different altitudes. It should be noted that the additional closing equation setting the sum of all excited and ground states equal to the total number of NO molecules is required.

As will be explained below, NO_2 photolysis can excite the vibrations of NO and this depends on the NO/NO_2 abundance ratio. Hence, as the role played by O(^3P) and O_3 concentrations for the O_3 non-LTE populations, it is important that the NO and NO_2 concentrations included in the model are mutually consistent.

7.6.4   *Non-LTE populations*

Figure 7.19 shows the populations of levels NO($v = 1,2$), in terms of the vibrational temperatures $T_v(v)$, as calculated for mid-latitude day conditions with the processes described above. Departure from LTE in the illuminated stratosphere is caused mainly by the photolysis of NO_2; the $NO_2 + O$ reaction has a relatively minor effect. Vibrational temperatures are about 3–7 K above the kinetic temperature for the NO($v = 1$) state and about 80–100 K above the kinetic temperature for the NO($v = 2$) level. The higher values of $T_v(2)$ in comparison to $T_v(1)$ arise because NO(2) and NO(1) both have comparable absolute excitation rates from chemical production, but the higher-energy NO(2) state has a much lower LTE number den-

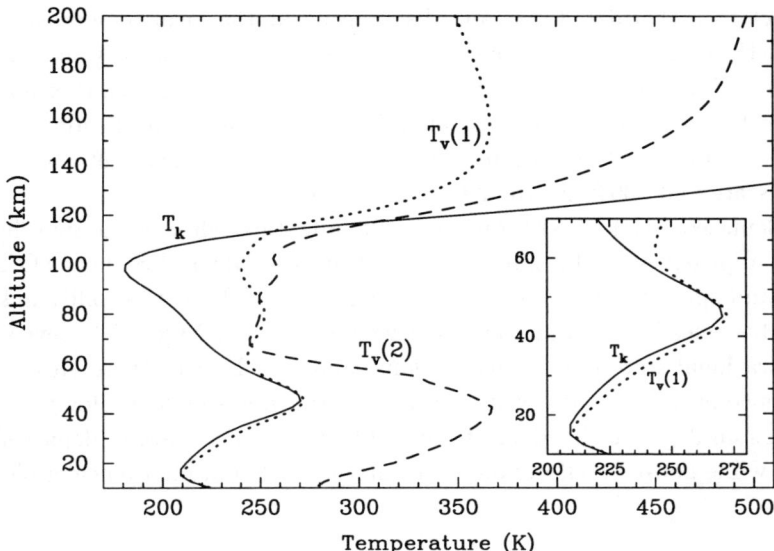

Fig. 7.19   NO vibrational temperatures $T_v(v)$ for mid-latitude daytime conditions.

sity. Thus, non-LTE enhancement, relative to LTE, is much larger for NO(2). This is similar to the case of the $O_3(v_3 \geq 2)$ and $NO_2(v_3 \geq 2)$ states (see Secs. 6.3.6, 7.3.6 and 7.7.2). At night, vibrational LTE is maintained up to 50–60 km.

The impact of NO_2 photolysis on the population of $NO(v)$ depends on the NO/NO_2 abundance ratio: daytime vibrational temperatures increase with decreasing NO and increasing NO_2 abundances. Due to the diurnal $NO \leftrightarrow NO_2$ conversion, the maximum values of $T_v(1)$ at sunrise increase to about 15–20 K above T_k.

In the mesosphere (60–90 km), the vibrational populations of $NO(v)$ are mainly affected by radiative processes (absorption of solar and tropospheric radiation) and hence depend on the tropospheric temperature profile, cloud coverage, and humidity. The typical variability of these parameters is enough to induce changes from 10 to 20 K in the mesospheric vibrational temperature. Excitation due to absorption of sunlight leads to a mid-latitude increase of the daytime vibrational temperatures of about 15 K above the night-time values. De-excitation by collisional quenching with O_2 and O is much less important than spontaneous emission.

$NO(v = 1, 2)$ excitation above 110 km is mainly caused by collisions with atomic oxygen and in the reaction $N + O_2 \rightarrow NO(v) + O$, while spontaneous emission is again the main de-excitation process. The reaction of $N(^2D)$ atoms with O_2 (process 6) is the dominant source of thermospheric NO below 130 km during daytime, but vanishes at night due to the negligible night-time $N(^2D)$ concentrations. Above approximately 130 km in the daytime and in the whole thermosphere at night, the main source of NO is the reaction of $N(^4S)$ atoms with O_2 which has an energy barrier of ~ 0.3 eV. In the daytime thermosphere, the $N(^4S)$ kinetic energy distribution is non-thermal, i.e., there is a portion above ~ 0.3 eV which is increased with respect to a thermal Boltzmann distribution at the local temperature. The reaction of those 'superthermal' $N(^4S)$ atoms with O_2 significantly increases the thermospheric NO production during the daytime. In spite of this, thermospheric NO(1) vibrational temperatures are generally lower than T_k since radiative losses dominate over collisional excitation. Maximum values of T_v are about 360 K for the NO($v=1$) state and 500 K for the NO($v=2$) state.

The calculated NO(v, J, S) rotational and spin states show a thermal distribution below approximately 110 km. Above 110 km, the NO(1, J, S) and NO(2, J, S) rotational and spin distributions start to depart from LTE, as the collisional relaxation time becomes longer than the radiative lifetime. The NO(0, J, S) levels in the ground vibrational state remain in rotational and spin LTE at these altitudes due to the absence of radiative de-excitation by spontaneous emission. However, above 160 km the spin degree of freedom of the NO(0, J, S) states start to depart slightly from LTE, while non-LTE effects on the rotational distribution are negligible below 200 km.

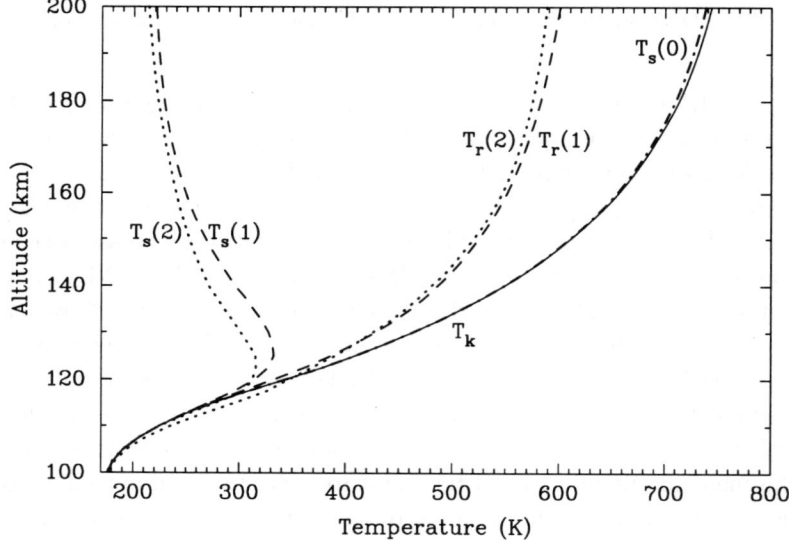

Fig. 7.20   NO rotational temperatures of the subthermal part, $T_r(v)$, and spin temperatures $T_s(v)$ for mid-latitude night-time conditions. Rotational and spin temperatures are shown for NO($v=1$) (dash) and NO($v=2$) (dotted). The spin temperatures for NO($v=0$) (dot-dash) are also shown. A spin propensity factor $\beta = 0.9$ was used. The kinetic temperature, T_k, is also plotted.

The non-LTE rotational and spin state distributions of NO(v, J, S) with $v=1$, 2 can be described by a subthermal part arising from the intervibrational NO + O collisional production, and a superthermal part arising from chemical production by $N(^4S, ^2D) + O_2$. Both parts of the distribution have been observed in CIRRIS-1A measurements. The calculated rotational and spin distributions of NO(v, J, S) with $v=1,2$ can be described by the sum of two Boltzmann terms, characterised by the corresponding rotational temperatures T_r^{low} and T_r^{high}. Since both spin-orbit states show the same rotational distribution, the ratio of the two can be expressed by a spin temperature, T_s.

Subthermal rotational temperatures T_r^{low} and spin temperatures T_s above 100 km are shown in Fig. 7.20 for mid-latitude night-time conditions. Rotational and spin states in NO($v = 1,2$) start to depart from LTE at about 120 km. At 200 km, T_r^{low} and T_s^{low} are very close to the effective rotational and spin production temperatures, T_r^e and T_s^e, respectively, of the vibrational NO + O relaxation. This suggests that at high altitudes collisional NO + O excitation is the main production process of subthermal NO($v = 1,2$). The calculated profiles of T_r^{low} and T_s are in very good agreement with the experimental results from CIRRIS-1A.

The chemical production of NO from N(^4S) + O$_2$, which gives rise to the superthermal emission, is highly variable and still needs to be investigated in more detail. Night-time NO was modelled as being produced only by N(^4S) + O$_2$. The resulting values for T_r^{high} are 3000 K at night and 5000 K by day. For the NO($2, J, S$) states, the ratio $n_{\text{super}}/n_{\text{sub}}$ of the calculated super- and subthermal parts of the distribution function is about 0.4 at 140 km for daytime, increasing up to 2.5 at 200 km. At night-time the ratio is about 50–70% of the daytime values. The $n_{\text{super}}/n_{\text{sub}}$ ratio for NO($1, J, S$) is 0.08 at 200 km and less than 0.01 below 150 km. At night-time the ratio is reduced by another factor of two. These calculations are consistent with the reported measurements (see, e.g., Armstrong *et al.*, 1994, and the references

7.7 Nitrogen Dioxide, NO$_2$

Nitrogen dioxide is another species which is measured by satellite-borne instruments, generally using the infrared emission near 6.2 μm originating from its v_3 bands. Non-LTE effects in these levels could be important for the retrieval of NO$_2$ abundance profiles, and also for H$_2$O, since the NO$_2$ v_3 bands overlap with the H$_2$O 6.3 μm fundamental. There is as yet no conclusive evidence for non-LTE in these bands. However, an airglow layer has been observed at visible wavelengths in the 35–55 km altitude region due to the chemiluminescence reaction NO + O$_3$ → NO$_2^*$ + O$_2$, which is also thought to be the responsible for the non-LTE excitation of the NO$_2$ in the infrared.

The model used here computes the vibrational temperatures for the $v_3 = 1$ to 7 states of NO$_2$ (see Table 7.11). The transitions that take place between them are listed in Table 7.12. A diagram of the most important (i.e., the less energetic) levels of NO$_2$ is shown in Fig. 7.21. So far no non-LTE modelling or experimental evidence has been found for other states and transitions in the atmosphere. The extension of the model to other NO$_2$ states and transitions, in a similar way as for O$_3$, is straightforward.

The excitation and collisional processes affecting the NO$_2$(v_3) levels are listed in Table 7.13. The main mechanisms thought to cause non-LTE in NO$_2$ are flu-

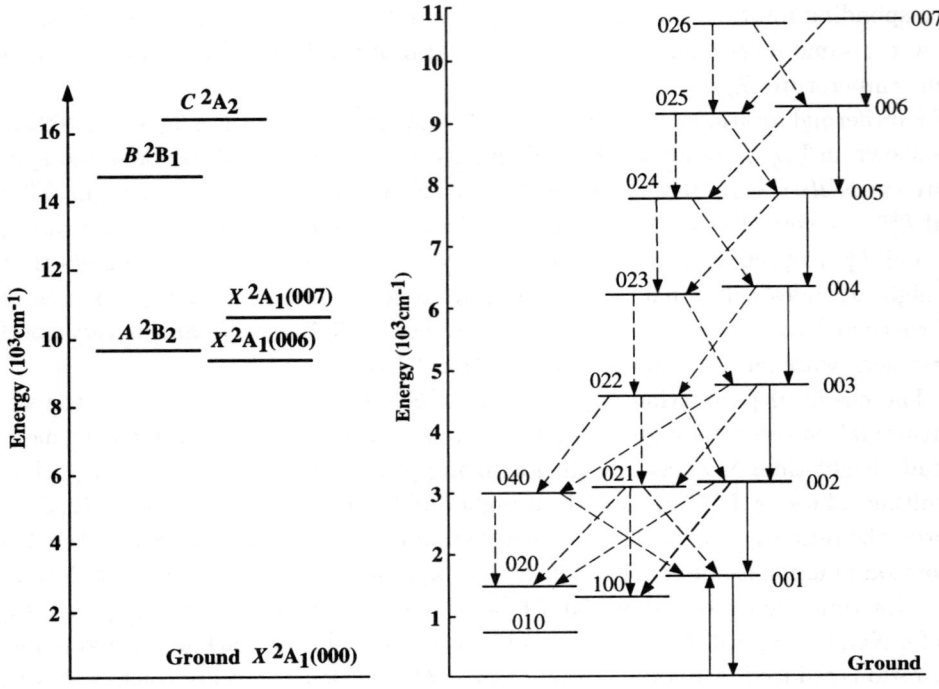

Fig. 7.21 Diagram of electronic (left) and vibrational (right) states of NO_2. The stronger vibrational transitions in the 1400–$1650\,cm^{-1}$ region are also shown. Note that only the $(0,0,v_3 - 0,0,v_3-1)$ transitions (solid arrows) have been considered in the model described. Spectroscopic data for these bands have been supplied by J.M. Flaud.

Table 7.11 NO_2 v_3 energy levels.

No.	Level	Energy (cm^{-1})
1	001	1616.85
2	002	3201.45
3	003	4754.21
4	004	6275.98
5	005	7766.28
6	006	9226.23
7	007	10659.32

orescence after absorption of sunlight in the visible and near-IR ($400 \leq \lambda \leq 800\,nm$) by NO_2 itself, and the chemiluminescence arising from the reaction of NO with O_3. The model also incorporates excitation and de-excitation by the absorption of solar and terrestrial (tropospheric) radiation, the exchange of photons between atmospheric layers in the $(0,0,1$–$0,0,0)$ fundamental band, and spontaneous emission in the v_3 fundamental and hot bands with $\Delta v_3 = 1$. Radiative transfer in the

Table 7.12   NO$_2$ ν_3 infrared bands.

No.	Upper level	Lower level	$\tilde{\nu}_0$, cm^{-1}	Strength*	A, s^{-1}
1	001	000	1616.849	5.688×10^{-17}	115.05
2	002	001	1584.600	4.350×10^{-20}	218.69
3	003	002	1552.760	2.960×10^{-23}	316.11
4	004	003	1521.771	2.100×10^{-26}	408.23
5	005	004	1490.300	1.580×10^{-29}	480.19
6	006	005	1459.950	1.320×10^{-32}	538.59
7	007	006	1433.090	1.260×10^{-35}	597.94

*Values shown are at 296 K in cm^{-1}/(molecules cm^{-2}).

fundamental band has been treated with the Curtis matrix method as before, using spectral line parameters from the HITRAN database.

The NO$_2$ bands tend to be optically thin over most of the atmosphere, even for limb viewing. As a result, the population of NO$_2$(0,0,1) in the mesosphere at night is, like that of NO(1) and CO(1), sensitive to the tropospheric conditions. The characteristic effective temperature of the upward tropospheric flux is close to the kinetic temperature at the $\sim$3 km layer for a cloud-free atmosphere. The origin of the outgoing atmospheric flux in this spectral region from such a low altitude emphasises the weakness of these bands. The lower boundary of the model was set up at 3 km and included the tropospheric flux as a blackbody at the temperature of this layer. From there it runs up to 120 km with a vertical grid of 1.5 km.

7.7.1   *Excitation and relaxation processes*

Fluorescence of NO$_2$

NO$_2$ has a complex absorption spectrum ranging from the near UV to the near-infrared with no obvious periodic or regular shape. At wavelengths smaller than $\sim$395 nm, absorption leads to the photodissociation of the molecule. At longer wavelengths, the molecule is excited from the $^2A_1$ ground state to the $^2B_1$ and $^2B_2$ electronic states. Fluorescence following absorption of radiation at wavelengths larger than the photodissociation limit can be observed in the visible, near- and mid-infrared.

The visible and near infrared fluorescence has an *anomalous* lifetime in the sense that the radiative lifetime of the observed fluorescence is significantly larger (by a factor of approximately 100) than the estimated spontaneous emission rate based on the integrated absorption coefficient measurements, even in the absence of collisions. This is explained by the interaction between the $^2B_1$ and $^2B_2$ electronic states, one of which, $^2B_2$, is not coupled radiatively to the ground state, and the mixing of vibrational levels in the ground electronic state $^2A_1$ with those of the excited $^2B_2$ state (see Fig. 7.21). The relaxation of the highly excited vibrational levels in the

$^2A_1$ ground state, by emission and by collisions with N_2 and O_2, are thought to be an important source of excitation of the $NO_2(v_3 \leq 7)$ levels in the daytime.

The present model follows the mechanism suggested by laboratory studies in which NO_2 molecules were excited by the absorption of visible light, simulating the absorption of solar radiation. In this process, as explained above, a mixture of vibrational levels in the ground electronic state $^2A_1$ with those of the excited $^2B_2/^2B_1$ states, which we denote by $NO_2(^2B_2/^2A_1, v^*)$ or by NO_2^* as in Table 7.13, is produced. Since there is a transition moment for electronic transitions to the ground state, radiative decay in transitions from these 'admixture' states represents a loss of the electronic excitation (process 2 in Table 7.13). Collisional relaxation (process 3) of these admixture states occurs very quickly, at nearly the gas kinetic rate, and is nearly independent of the collision partner. This process depopulates the admixture $(^2B_2/^2A_1, v^*)$ states and is also a direct source of the $(^2A_1, v_3 \leq 7)$ levels, albeit with poorly known quantum efficiency. An efficiency of 0.13 (as suggested by the analysis of LIMS measurements) has been considered here, and that all molecules result excited in state $NO_2(v_3 = 7)$. The thermal relaxation rate of $(^2B_2/^2A_1, v^*)$ decreases markedly for levels with energies below the $^2B_2$ origin, where it is typically in the range of ~ 0.01 times the gas kinetic rate or smaller. Thus, the admixture high-energy $(^2B_2/^2A_1, v^*)$ states suffer both radiative and collisional relaxation rates which are much faster than the corresponding processes for the $(^2A_1, v_3 \leq 7)$ states.

Table 7.13   Principal collisional processes affecting the NO_2 states.

No.	Process	Rate coefficient[†]	Refs.[‡]
1	$NO_2 + h\nu(400\text{--}800\,\text{nm}) \rightarrow NO_2^{*,\diamond}$	See text	1
2	$NO_2^* \rightarrow NO_2 + h\nu\ (\lambda \leq 400\text{--}800\,\text{nm})$	1.25×10^4	2
3	$NO_2^* + M \rightarrow NO_2[^2A_1(v_3 \leq 7)] + M$	1.2×10^{-10}	3
4	$NO + O_3 \rightarrow NO_2[^2A_1(v_3 \leq 7)] + O_2$	$2.0 \times 10^{-12} \exp(-1400/T)$	4
5	$NO + O + M \rightarrow NO_2^* + M$	See text	4
6	$NO_2(v_3) + N_2 \rightleftharpoons NO_2(v_3-1) + N_2$	$3.4\ v_3 \times 10^{-14}$	5, 6
7	$NO_2(v_3) + O_2 \rightleftharpoons NO_2(v_3-1) + O_2(1)$	1.2×10^{-13}	7

$^\diamond NO_2^*$ represents NO_2 excited in the admixture electronic/high vibrationally excited states $(^2B_2/^2A_1, v^*)$. $M = N_2$, O_2. [†]Rate coefficient for the forward sense of the process in cm^3s^{-1}. T is temperature in K. [‡]Refs. 1: Garcia and Solomon (1994); 2: Patten *et al.* (1990); 3: Donnelly *et al.* (1979); 4: DeMore *et al.* (1997); 5: Golde and Kaufman (1974); 6: Schwartz *et al.* (1952); and 7: López-Puertas (1997).

The inclusion of fluorescence as an excitation mechanism requires to compute the absorption of solar radiation by NO_2 at visible wavelengths. The calculation of the photoabsorption coefficients in the visible when considering single scattering are carried out in a similar way as in the near- and mid-infrared (see Sec. 6.3.3.2). The single scattering approach is generally accurate for most conditions, except for solar zenith angles larger than $90°$ in the atmospheric region below about 15 km.

Under those circumstances, multiple scattering has to be considered; see Sec. 7.15 for more details on this. The resulting photoabsorption rates are nearly constant with altitude with values of 3.4×10^{-2} and 4.0×10^{-2} photons s^{-1} for mid-latitudes and polar conditions, respectively.

Chemiluminescence from NO$_2$

The chemiluminescence from NO$_2$* formed in the NO + O$_3$ reaction (process 4 in Table 7.13) consists of a continuum which extends from the visible to about 3 μm, together with infrared bands at 3.7 and 6.3 μm. There is enough energy released in this reaction to excite NO$_2$ up to the $^2B_2$ electronic state, and to excite the ground electronic state $^2A_1$ up to $v_3 = 7, 8$ (see Fig. 7.21). The question is, which is the resulting electronic state, and is it vibrationally excited and in which level(s)? Experimental results, both in the laboratory and from satellite measurements lead a nascent distribution which favours the upper v_3 states and leading, after including the collisional and radiative relaxations, to similar number densities for all v_3 states. An efficiency of one was assumed for this process. For the minimum excitation, probably an efficiency of 0.1 is low enough.

The three-body recombination process whereby NO and O produce NO$_2$ comprises a pressure independent radiative recombination as well as a pressure dependent chemiluminescence (process 5 in Table 7.13). The resulting NO$_2$ molecule is excited in its electronic states with subsequent emission in the visible. Given the mixing between the $^2B_1/^2B_2$ electronic states with the vibrational states of the ground electronic state $^2A_1$ (as described in previous section) this process is also an excitation mechanism of the $(^2A_1, v_3 \leq 7)$ vibrational states, and it may be assumed that the electronic states produced in this reaction suffer the same de-excitation mechanisms as those produced by the absorption of sunlight in the visible and near-IR (processes 2 and 3 in Table 7.13). The efficiency of this recombination process for producing NO$_2(^2A_1, v_3)$ vibrational states is thought to be much weaker than for the NO + O$_3$ reaction, so it is only of importance when the O(^3P) abundances are significant, i.e., above about 80 km.

7.7.2   Non-LTE populations

Figure 7.22 shows the vibrational temperatures of the NO$_2(v_3 = 1–7)$ states computed with the non-LTE model described above for mid-latitude daytime conditions. The (0,0,1) state begins to move into non-LTE just above the stratopause and exhibits a large non-LTE enhancement near the mesopause. This is mainly excited by the collisional and radiative relaxation of the highly excited v_3 states (processes 6 and 7), which are themselves excited mainly by chemiluminescence and the absorption of solar radiation in the visible. Absorption of tropospheric radiation for this states is important only in the mesosphere under night-time conditions.

The $v_3 \geq 2$ levels show huge vibrational temperatures in the mesosphere, extend-

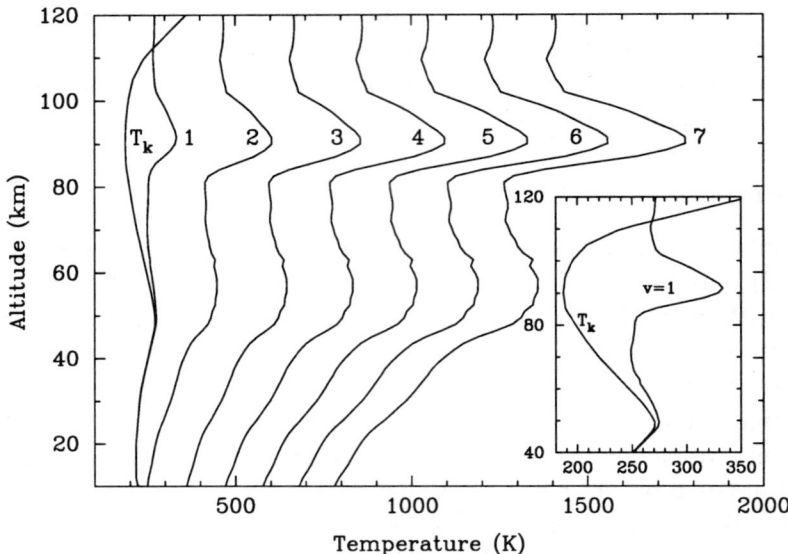

Fig. 7.22   Daytime vibrational temperatures of the $NO_2(v_3 = 1–7)$ states for the U.S. Standard atmosphere. The curves from left to right corresponds to the levels $v_3 = 1$ to $v_3 = 7$, respectively. The inset shows the T_v of the (0,0,1) level in a larger scale.

ing down into the stratosphere. The effect can be as large as $1000\,K$ and is similar to that shown by the O_3 levels (see Sec. 7.3.6). These high vibrational temperatures do not necessarily means that its contribution to the non-LTE radiance is larger than less energetic states with smaller T_v's since the radiance is proportional the number density of the excited state (Sec. 8.2.2), which is proportional to both its T_v and to its E_v. Actually, the non-LTE radiance contributions of these hot bands are usually of similar magnitude (see, e.g., Plate 11). The excitation of these levels is mainly by $NO + O_3$ chemiluminescence and the absorption of solar radiation in the visible and near-IR, the latter dominating in the middle mesosphere. The $NO + O(^3P)$ chemiluminescence is only of importance in the very upper mesosphere and lower thermosphere, where the abundance of $O(^3P)$ is larger.

The vibrational temperatures for mid-latitude night-time conditions are much smaller (see Fig. 7.23), reflecting the lack of excitation by absorption of solar radiation in the visible and near-IR; and the different abundances of NO, O_3 and NO_2. The (0,0,1) state is very close to LTE up to about 60 km, some 10 km higher than by day, and above that level exhibits a population larger than in LTE, caused mainly by the absorption of the tropospheric flux. There is essentially no nitric oxide below about 50–60 km at night, so the $NO_2(v_3 \geq 2)$ levels are in LTE in the stratosphere.

Like O_3, NO_2 is formed in a vibrationally excited state, so the vibrational temperatures are not independent of the NO_2 abundance. From the definition of T_v, Eq. (6.1), and assuming that process 4 is the major source of excitation of $NO_2(v_3)$, we find that their vibrational temperatures are proportional to $\{[NO] \times [O_3]\}/[NO_2]$. NO abundances are very similar for day and night-time above ∼60 km, while the

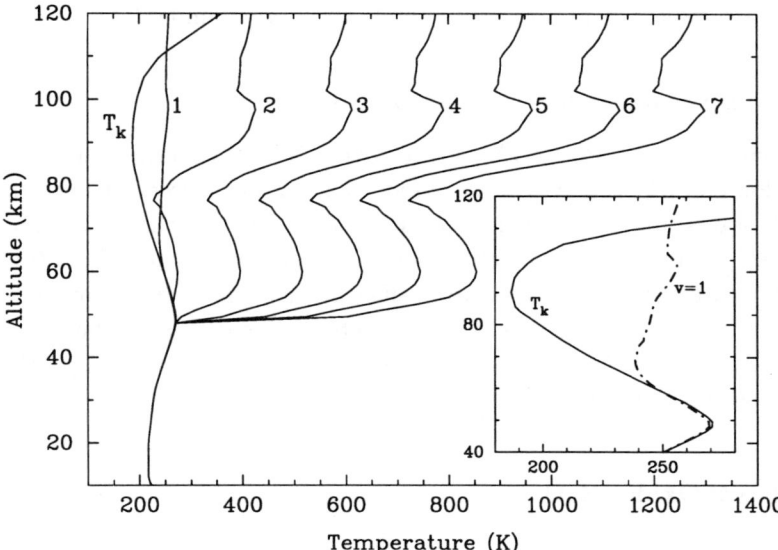

Fig. 7.23 Night-time vibrational temperatures of the $NO_2(v_3 = 1-7)$ states for the U.S. Standard atmosphere. The curves from left to right corresponds to the levels $v_3 = 1$ to $v_3 = 7$, respectively. The inset shows the T_v of the (0,0,1) level in a larger scale.

O_3 abundance is larger by about a factor of 10 in that region at night, and that of NO_2 is larger by at least 2 orders of magnitude. The overall effect leads to smaller vibrational temperatures at night. Despite this, the non-LTE limb radiances from the ν_3 hot bands at night, which are proportional to $[NO] \times [O_3]$, are larger than during daytime above about 60 km. This illustrates that in the case of the NO_2 hot bands, as for O_3 hot bands, the non-LTE effects cannot be derived uniquely from an inspection of the vibrational temperatures of their upper states. This actually happens when the excitation rates for all, low- and high-energy, levels are similar, as occurs in molecule formation.

The vibrational temperature of $NO_2(0,0,1)$ in the mesosphere depends significantly on the kinetic temperature profile. For example, for polar winter night-time conditions, this state is in LTE only up to about 50 km. Between this altitude and $\sim$75 km this exhibits a population significantly smaller than in LTE because absorption of tropospheric radiation plus excitation in thermal collisions are not fast enough to compensate for the spontaneous emission. In comparison with night-time mid-latitudes, this smaller population is caused by the colder troposphere, which produces a weaker tropospheric flux and a smaller absorption of radiation in the mesosphere. The region of non-LTE enhancement, which normally occurs in the upper mesosphere, is limited to a narrow interval between $\sim$75 and 95 km for polar winter conditions.

The populations of $NO_2(v_3 \geq 2)$, on the other hand, are more sensitive to the variability of NO and O_3 abundances than to the kinetic temperature. Hence larger populations of the $NO_2(v_3 \geq 2)$ states are expected in the polar winter upper mesosphere, for example, where both NO and O_3 are more abundant.

7.7.3   Uncertainty in the NO_2 populations

The populations of the $NO_2(v_3)$ levels are obviously very uncertain since no definite evidence for non-LTE has yet been found. The major sources of uncertainty for the $v_3 \geq 2$ states are the efficiencies of solar absorption in the visible and of the $NO + O_3$ chemiluminescence reaction, which could lie anywhere from 0.1 to 1 in each case. Also, the measured rate coefficients for the thermal and vibrational relaxation of these levels span factors from 10 to 25, respectively. These ranges lead to uncertainties in the vibrational temperatures of the order of 50 K for the (0,0,1) level to hundreds of kelvins for the high v_3 states.

7.8   Nitrous Oxide, N_2O

Nitrous oxide is the main precursor of active nitrogen in the middle atmosphere. It has both natural and man-made sources at the surface, and is carried up into the stratosphere, where dissociation by solar UV leads to the formation of chemically active species like NO_2 and NO.

The N_2O molecule has three modes of vibration. The influence of non-LTE on the populations of the vibrational levels has not been studied in as much detail as other atmospheric gases, and there is no experimental evidence yet for non-LTE in the atmosphere itself. So far, only the theoretical model of Shved and Gusev (1997) has been published. This model uses the 'optically thin' approximation, which consists of a blackbody located at the bottom of the atmosphere which radiates at the base temperature to the mesosphere without being absorbed significantly by the atmosphere in between. This approach overestimates the radiative excitation in the mesosphere, and gives upper limits for the non-LTE populations expected there.

The model we use here computes the vibrational populations for 41 levels (see Table 7.14 and Fig. 7.24). Four digits are used to identify the states, $(v_1, v_2{}^l, v_3)$, corresponding to the vibrational quantum numbers of the v_1-symmetric stretching mode, v_2-bending mode, angular momentum, and v_3-asymmetric stretching mode, respectively. The $O_2(1)$ and $N_2(1)$ populations previously computed in the CO_2 and H_2O non-LTE models are used here to obtain the interactions with the N_2O states. The model includes the usual radiative processes, e.g., radiative transfer between atmospheric layers, absorption of tropospheric and solar radiation and spontaneous emission, for the strongest N_2O fundamental bands, listed in Table 7.15.

7.8.1   Collisional processes

The states with the same quantum numbers for each mode but different angular momentum l [e.g., pairs of states $(0,2^0,0)$ and $(0,2^2,0)$; $(0,3^1,0)$ and $(0,3^3,0)$; etc.] are

Table 7.14   N$_2$O energy states.

No.	State	Energy (cm^{-1})	No.	State	Energy (cm^{-1})
1	01^10	588.768	22	06^00	3466.600
2	02^00	1168.132	23	06^20	3474.450
3	02^20	1177.745	24	14^00	3620.941
4	10^00	1284.903	25	14^20	3631.590
5	03^10	1749.065	26	22^00	3748.252
6	03^30	1766.913	27	22^20	3766.052
7	11^10	1880.266	28	30^00	3836.373
8	00^01	2223.757	29	03^11	3931.247
9	04^00	2322.573	30	03^31	3948.285
10	04^20	2331.122	31	11^11	4061.979
11	12^00	2461.997	32	00^02	4417.379
12	12^20	2474.799	33	23^10	4335.798
13	20^00	2563.339	34	12^01	4630.164
14	01^11	2798.292	35	12^21	4642.458
15	05^10	2897.813	36	20^01	4730.828
16	13^10	3046.213	37	01^12	4977.695
17	13^30	3068.721	38	32^00	5026.340
18	21^10	3165.854	39	40^00	5105.650
19	02^01	3363.978	40	21^11	5319.176
20	02^21	3373.141	41	31^10	5346.806
21	10^01	3460.821			

Table 7.15   N$_2$O infrared bands.

No.	Upper level	Lower level	$\tilde{\nu}_0$, cm^{-1}	Strength[*]
1	01^10	00^00	588.768	9.857×10^{-19}
2	02^00	00^00	1168.132	2.877×10^{-19}
3	10^00	00^00	1284.903	8.248×10^{-18}
4	00^01	00^00	2223.757	5.005×10^{-17}
5	20^00	00^00	2563.339	1.194×10^{-18}
6	10^01	00^00	3480.821	1.732×10^{-18}
7	00^02	00^00	4417.379	6.074×10^{-20}

[*]Values shown are at 296 K in cm^{-1}/(molecules cm^{-2}).

completely coupled by intra-molecular V–V energy processes, and were considered as an 'equivalent' state. This approximation is justified because the energy gap between these states is very small (9–18 cm^{-1}).

The six collisional processes included for the N$_2$O levels are listed in Table 7.16. The first represents V–T collisions for the bending mode. We assumed that these vibrational transitions obey the selection rule $l - l' = 1, -1$. Re-distribution or relaxation of N$_2$O$(0,0^0,1)$ by collisions with N$_2$ most likely occurs through pathways

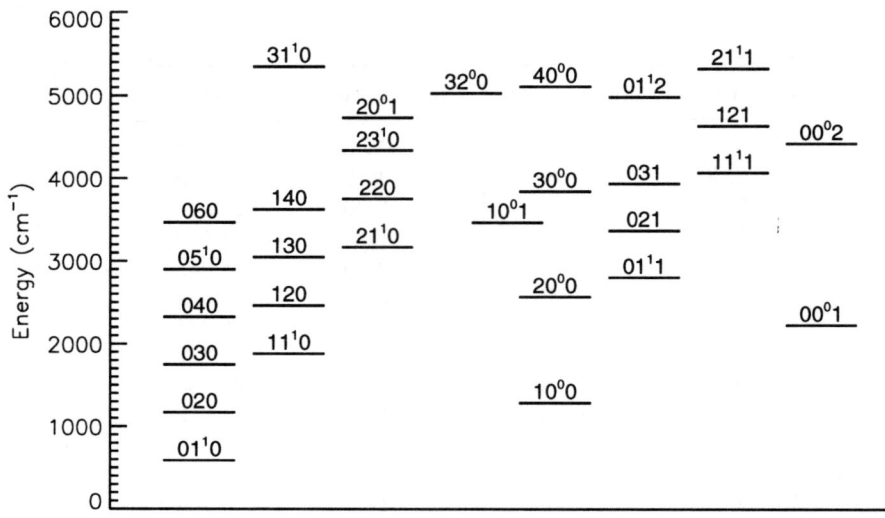

Fig. 7.24   Vibrational energy levels of the N_2O major isotope. The labels without the angular momentum digit represent the corresponding pair of levels with $l = 0, 2$, i.e., 020 represents the $(0,2^0,0)$ and $(0,2^2,0)$ states.

Table 7.16   Principal collisional processes affecting the N_2O states.

No.	Process	$k^\dagger$	Refs.[*]
1	$N_2O(v_1,v_2,v_3) + M \rightleftharpoons N_2O(v_1,v_2-1,v_3) + M$	$8.7\,v_2 \times 10^{-15}$	1, 2, 3
2a	$N_2O(00^01) + N_2 \rightleftharpoons N_2O(11^10) + N_2$	4.3×10^{-15}	1, 4, 5
2b	$N_2O(00^01) + N_2 \rightleftharpoons N_2O(03^10, 03^30) + N_2$	4.3×10^{-15}	1, 4, 5
3	$N_2O(v_1,v_2,v_3)+O_2 \rightleftharpoons N_2O(v_1,v_2+1,v_3-1)+O_2(1)$	7.0×10^{-15}	4
4	$N_2O(v_1,v_2,v_3) + N_2 \rightleftharpoons N_2O(v_1,v_2,v_3-1) + N_2(1)$	1.1×10^{-13}	6
5	$N_2O(v_1,v_2,v_3) + O_2 \rightleftharpoons N_2O(v_1-1,v_2,v_3) + O_2(1)$	1.3×10^{-15}	1, 2
6	$N_2O(v_1,v_2,v_3) + M \rightleftharpoons N_2O(v_1-1,v_2+2,v_3) + M$	1.0×10^{-13}	7

†Rate coefficient for the forward sense of the process in cm^3s^{-1} at 200 K. $M = N_2$, O_2. [*]Ref. 1: Shved and Gusev (1997); 2: Zuev (1989); 3: Schwartz *et al.* (1952); 4: Siddles *et al.* (1994b); 5: Zuev (1985); 6: Gueguen *et al.* (1975); 7: Kung (1975).

2a and 2b, and the relaxation of a v_3 quantum into a v_2 quantum by collision with O_2 through the near-resonant V–V energy exchange (process 3). Inter-molecular V–V energy exchange between the ν_3 mode and $N_2(1)$ is included by process 4, and between ν_1 and $O_2(1)$ by process 5. Process 6 represents the intra-molecular energy exchange between the ν_1 and ν_2 modes. The vibrational energy that is converted to (or taken from) kinetic energy in this process is in the 88–132 cm^{-1} range. The probability for the conversion of a v_1 quantum into two v_2 quanta is rather high.

7.8.2 *Non-LTE populations*

The daytime vibrational temperatures for the levels originating the strongest emissions are shown in Fig. 7.25. The lower energy levels (0,2,0), $(1,0^0,0)$, $(1,1^1,0)$, and (1,2,0) depart from LTE at about 75 km, having populations larger than in LTE. This departure is caused by the absorption of radiation from the lower levels and from the Sun. For the higher-energy levels, the absorption of solar radiation is of relatively greater importance and causes the $(0,0^0,1)$, $(0,1^1,1)$, and (0,2,1) levels to depart from LTE at about 60 km. The $(1,0^0,1)$ state is in non-LTE all the way down to the troposphere.

The vibrational temperatures for mid-latitude night-time conditions are shown in Fig. 7.26. Because of the lack of solar pumping, the departure from LTE is generally smaller and occurs at higher altitudes than for daytime. The levels $(1,1^1,0)$ and (1,2,0) depart from LTE at about 90 km, (0,2,0) at about 80 km, $(1,0^0,0)$ at about 75 km, while the other levels, with the exception of $(0,1^1,1)$, all depart at about 70 km. The latter is in non-LTE even at lower altitudes.

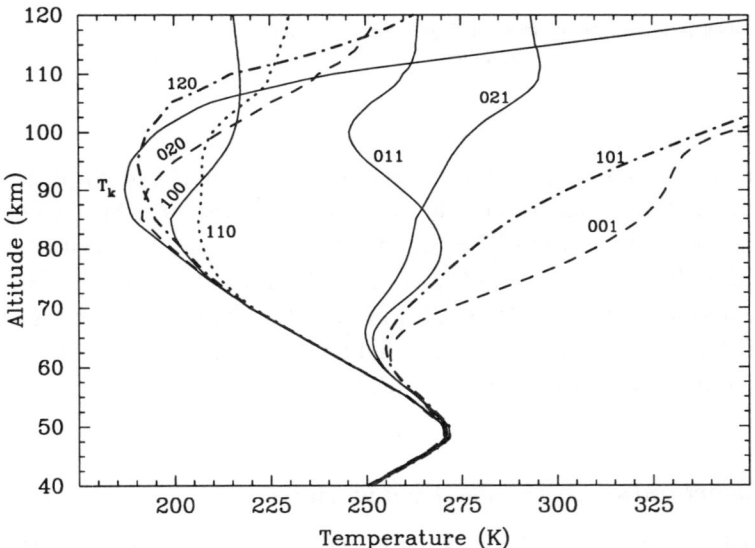

Fig. 7.25 Vibrational temperatures for N_2O levels for mid-latitudes, daytime. The angular momentum l has been omitted in the labels. T_k is the kinetic temperature.

The deviations from LTE for a typical polar winter temperature structure (cold stratopause and warm mesopause) are smaller than for night-time mid-latitudes. All vibrational levels slightly depart from LTE between 50 and 70 km. Above this altitude the non-LTE departures are significant, although smaller than for the mid-latitude night-time case.

The daytime vibrational temperatures for a typical polar summer temperature structure are similar to those for daytime mid-latitude conditions. However, because

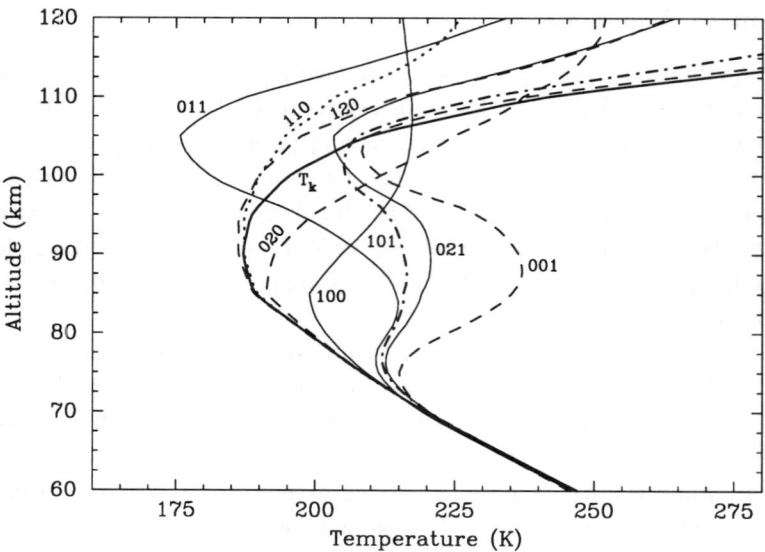

Fig. 7.26   Vibrational temperatures for N_2O levels for mid-latitudes, night-time. The angular momentum l has been omitted in the labels. T_k is the kinetic temperature.

of the warmer stratopause, the absorption of earthshine in the mesosphere is larger, and local thermal collisions in the cold mesopause are weaker, which lead the levels to have larger departures from LTE around the mesopause. The altitude of departure from LTE for the $(0,2,0)$, $(1,0^0,0)$, $(1,1^1,0)$, and $(1,2,0)$ levels is very similar to that for mid-latitudes. The non-LTE breakdown altitude for the higher-energy levels, $(0,0^0,1)$, $(0,1^1,1)$, and $(0,2,1)$, are slightly higher than for mid-latitude, about 60 km, because of the warmer lower mesosphere.

7.8.3   *Uncertainties in the N_2O populations*

From the point of view of model parameters, the major uncertainties in the populations of the N_2O levels are caused by the lack of accurate collisional rate coefficients. Assuming an uncertainty in them of a factor 2, and a 10% in the tropospheric flux, we find that the altitude of LTE departure is basically unaltered; and that the error in the vibrational temperature of the $(0,2,0)$, $(1,0^0,0)$, and $(1,1^1,0)$ levels is about 10–15 K. The high-energetic levels, which are solar pumped in the daytime, show similar uncertainties in the upper mesosphere and lower thermosphere for daytime, but have small changes for night-time conditions.

Nevertheless, we do not have yet any evidence of non-LTE in N_2O, and the departures from LTE shown in Figs. 7.25 and 7.26 need still to be confirmed experimentally.

7.9 Nitric Acid, HNO₃

Nitric acid is the main reservoir for active nitrogen in the stratosphere. Its abundance is largest in the polar night, where it also participates in the formation of polar stratospheric clouds. These are important agents in the depletion of stratospheric ozone.

Non-LTE in HNO_3 has been very little studied, because the species is mostly confined to the stratosphere and below, where thermal collisions are relatively frequent and LTE is expected to prevail. No experimental evidence for non-LTE in the HNO_3 infrared emissions from the atmosphere has been reported. Nevertheless, a model has been developed for 8 different vibrational levels of the major isotope of HNO_3: v_9, v_5, $2v_9$, $3v_9$, v_4, v_3, v_5+v_9, and v_2, which are shown in Fig. 7.27.

Table 7.17 HNO₃ energy levels.

Equivalent level	Grouped levels	Energy (cm^{-1})
V9	v_9	458.230
V5	v_5, $2v_9$	879.110
V3	v_3, v_4, v_5+v_9, $3v_9$	1326.185
V2	v_2	1709.568

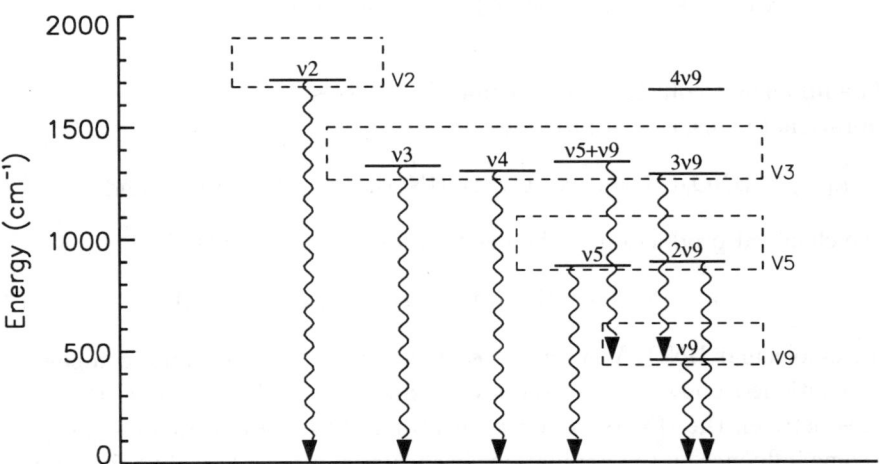

Fig. 7.27 Schematic diagram of the vibrational levels and transitions of HNO₃.

There is strong coupling between the v_5 and $2v_9$ levels of HNO_3 that emit at 11 μm, as well as between v_3 and v_4, which originate bands at 7.5 μm. Important coupling also exists between the $3v_9$ and the v_3 and v_4 levels. This allows us to group all of the HNO_3 vibrational levels that generate the relevant transitions at 11 μm and 7.5 μm into four equivalent levels, by assuming that the energy gained by any of them within an equivalent state, through any radiative or collisional processes, is quickly redistributed among all of them. This is a particularly appropriate assumption when there are major uncertainties in the magnitude of the major energy transfer processes, as is the case for HNO_3. The equivalent levels are listed in Table 7.17 and shown in the schematic diagram of Fig. 7.27.

The principal transitions between the vibrational levels are listed in Table 7.18 and shown in Fig. 7.27. The important radiative processes are: spontaneous emission; absorption of solar flux in every band with $\lambda < 12\,\mu m$; exchange between atmospheric layers; and absorption from the troposphere. The exchange between layers is small, even for the stronger ν_2, ν_5, and $2\nu_9$ transitions.

Table 7.18   Principal HNO_3 infrared bands.

Band	Upper level	Lower level	$\tilde{\nu}_0$, cm^{-1}	Strength*
1	ν_9	ground	458.23	1.082×10^{-17}
2	ν_5	ground	879.11	1.261×10^{-17}
3	$2\nu_9$	ground	895.50	9.843×10^{-18}
4	$3\nu_9$	ν_9	1288.90	2.030×10^{-18}
5	ν_4	ground	1303.06	1.245×10^{-17}
6	ν_3	ground	1326.19	2.469×10^{-17}
7	$\nu_5 + \nu_9$	ν_9	1343.65	1.216×10^{-18}
8	ν_2	ground	1709.57	4.381×10^{-17}

*Values shown are at 296 K in cm^{-1}/(molecules cm^{-2}).

The important collisional and chemical processes are:
(i) vibrational-translational processes

$$k_{V-T}: \ HNO_3(v) + M(N_2, O_2) \rightleftharpoons HNO_3(v') + M(N_2, O_2), \quad \text{and} \qquad (7.18)$$

(ii) the chemical production in the reaction formation of HNO_3

$$k_c: \ NO_2 + OH + M(N_2, O_2) \rightleftharpoons HNO_3(v) + M. \qquad (7.19)$$

Intramolecular HNO_3 V–V processes have very little importance compared with those mentioned above, due to the low mixing ratio of HNO_3. Also the V–V exchanges between the HNO_3 ν_2 and ν_3 levels and $O_2(1)$ are negligible. They might have some influence on the populations in the upper mesosphere, but the abundance of HNO_3 is so low in those regions that it is irrelevant.

In the formation of HNO_3, reaction (7.19), it is reasonable to expect that some of the molecules will be vibrationally excited and will subsequently relax to lower levels. However, there is no information either about this nascent distribution, or about relaxation from high excited states to lower ones. If $f(v)$ is the branching ratio, the chemical production of level v is then given by $P_c(v) = f(v)\,k_c\,[NO_2]\,[OH]\,[M]$. Even when a maximum efficiency $f(v) = 1$ is assumed for all levels, this excitation source is of very little importance compared with V–T processes.

Most of the HNO₃ bands are optically thin and then radiative transfer is of little importance for populating the vibrational levels. Exchange of radiation between the atmospheric layers has been included for the stronger fundamental bands arising from the v_2, v_5 and v_9 levels; whose populations were obtained by means of the Curtis matrix method. For the remaining levels, their populations were derived from the solution of their statistical equilibrium equations, which have the form:

$$n_v \simeq \frac{B_{v',v}\,\bar{L}_{0,\Delta\nu}\,n_{v'} + P_c + P_t}{A + l_t}, \qquad (7.20)$$

where n_v is the number density of level v, $n_{v'}$ is the number density of the lower level of the v–v' transition, $B_{v',v}\bar{L}_{0,\Delta\nu}n_{v'}$ accounts for the absorption of the tropospheric and solar radiation, P_c is its chemical production, P_t is the production of thermal processes, A is the total Einstein coefficient emission for all bands originating in level v, and l_t are the thermal losses.

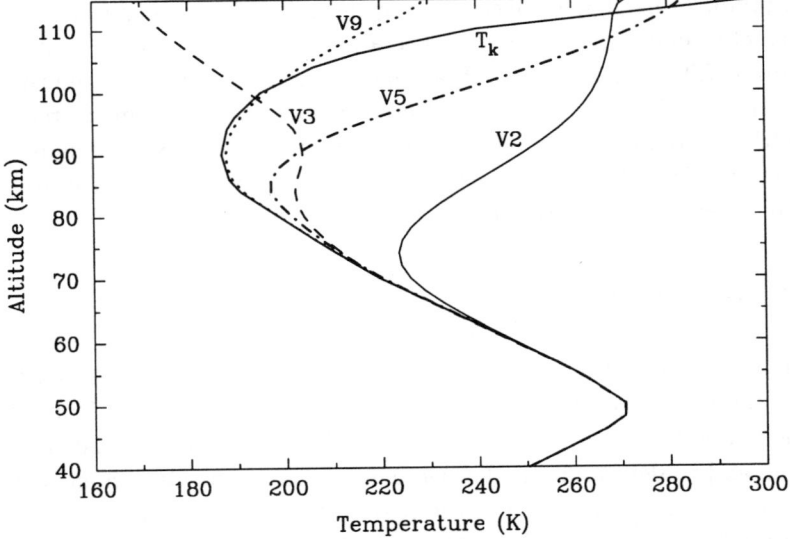

Fig. 7.28   Vibrational temperatures of the four HNO₃ equivalent levels for mid-latitudes, daytime. T_k is the kinetic temperature.

Figure 7.28 shows the populations of the four equivalent levels (Table 7.17) for daytime mid-latitude conditions. The departure from LTE occurs around 60 km or above, and very similar situations are found for all of the extreme (i.e., polar winter or polar summer) atmospheric temperature structures.

Chemical production of HNO₃ and direct solar pumping have a very small effect on the vibrational temperatures of all levels at every height. The production is completely dominated by collisional processes and, among these, V–T processes are much more important than V–V ones. Radiation can only compete with the V–T processes in the uppermost layers of the mesosphere, where the absorption of tropospheric radiation is responsible for the population enhancements observed at these heights. A larger difference between kinetic and vibrational temperatures occurs in the polar summer, due to the low temperatures near the summer mesopause. The opposite occurs in the polar winter.

Current uncertainties in the band strengths of nitric acid vary from 15% for the ν_3-ground, ν_4-ground and $3\nu_9-\nu_9$ bands, to 10% for all the other bands, while the line width coefficients may be uncertain by about 25%. A change of $\pm 15\%$ in all the band strengths produces a very small effect on the altitude of LTE breakdown for all levels and on the vibrational temperatures of most of them in the non-LTE region. Only at very high altitudes do the vibrational temperatures of the V2 and V5 levels change by as much as $5\,\mathrm{K}$ as a result of the uncertainty in the band strengths.

A decrease of an order of magnitude in the V–T rate coefficient significantly lowers the altitude where LTE breaks down, e.g., from 70 to $50\,\mathrm{km}$ for the V2 level, because when this rate is smaller the radiative processes gain relatively in importance. A change of two orders of magnitude in the V–V rate coefficients, however, produces negligible effects at every height and for every level.

The kinetic temperature structure has a negligible effect on the altitude of departure from LTE for the HNO_3 vibrational levels, although the magnitude of the deviation from LTE is significantly affected by the choice of kinetic temperature profile. Other parameters, like the surface pressure or the HNO_3 abundance, have little effect since all the bands considered are optically thin and the exchange of photons between different layers is of little relevance.

7.10   Hydroxyl Radical, OH

Vibrationally excited $OH(v \le 9)$ is responsible for one of the most characteristic feature of the nightglow: the OH emission layer centred around $85\,\mathrm{km}$. This has been widely used to study dynamics (in particular gravity waves) and chemistry in this region, and to determine the kinetic temperature. The OH emissions also cool the upper mesosphere, since part of the chemical energy converted into vibrational energy through reaction (7.21) is readily emitted to space by $OH(v \le 9)$ in the optically thin $\Delta v = 1$ and in the more energetic $\Delta v = 2$ Meinel bands.

The hydroxyl radical is excited by the Bates–Nicolet reaction:

$$O_3 + H \rightarrow O_2 + OH(v \le 9), \tag{7.21}$$

which results in vibrational populations much higher than those corresponding to a Boltzmann distribution at mesospheric temperatures. Non-LTE emission from high rotational states in low vibrational levels has also been observed recently (see Fig. 8.30). Moreover, pure rotational transitions from highly non-LTE excited rotational states in the OH ground vibrational state have been observed in the 400–$1000\,\mathrm{cm}^{-1}$ spectral region. It is believed that OH is preferentially formed in high vibrational, low rotational states and then rapidly relaxes to low vibrational, high rotational states by near-resonant collisional processes. The emission from these rotational levels, particularly those with high J, will then correspond to an OH excitation temperature much higher than the local kinetic temperature.

The vibrationally excited hydroxyl radical is deactivated by spontaneous radiation, in the $\Delta v = 1$ and mainly in the $\Delta v = 2$ bands, and quenched in the collisional process

$$\text{OH}(v \leq 9) + \text{M}(\text{N}_2, \text{O}_2, \text{O}(^3P)) \rightarrow \text{OH}(v' < v) + \text{M}. \qquad (7.22)$$

Radiative transfer does not play any role since both the $\Delta v = 1$ and $\Delta v = 2$ transitions are very optically thin everywhere in the atmosphere, and tropospheric and solar radiation gives a negligible contribution in comparison to the chemical production of reaction (7.21).

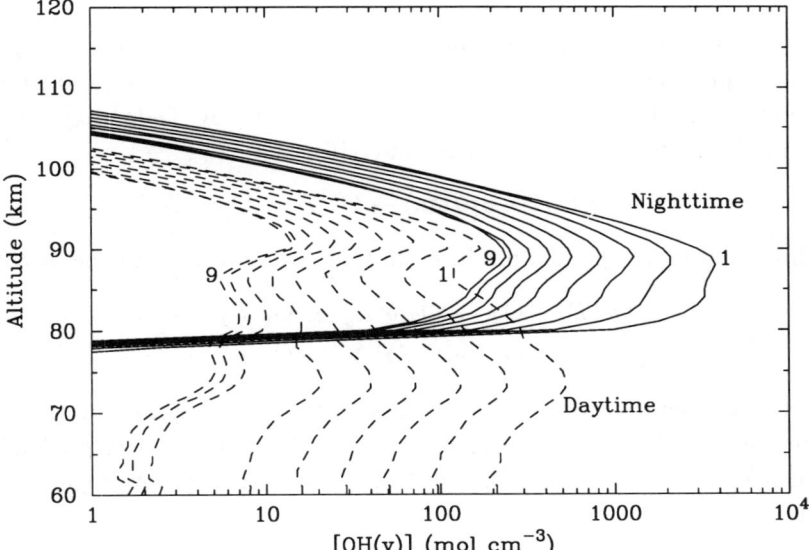

Fig. 7.29   Number densities of the vibrationally excited $v = 1$–9 (right to left) hydroxyl radical for night-time (solid) and daytime (dash) conditions. After López-Moreno *et al.* (1987).

Since the number densities of the OH(v) states are very large in comparison to that expected from an LTE situation (and actually comparable to the number density in the ground vibrational state), their populations are not represented by their vibrational temperatures but by their absolute number densities. Fig. 7.29 shows typical OH(v) number densities for vibrational levels $v = 1$–9 for night- and daytime conditions, as predicted by the model of López-Moreno *et al.* (1987) for mid-latitude conditions. Note the larger population at night, corresponding to the larger O_3 night-time abundance. The considerable abundance of the high vibrational levels (e.g., $v = 7$–9) is also noticeable in comparison with the ground state, even if they are between 20 and 25 thousand wavenumbers higher. This gives a measure of the very high non-LTE populations of these levels. As a further example, the vibrational temperature corresponding to the population of $OH(v=8)$ at 85 km is ~ 5500 K.

The contribution of these $OH(v \le 9)$ levels to the atmospheric limb emission at a tangent height of 60 km is shown in Plate 10 and discussed in Sec. 8.11. These emissions are due to the $\Delta v = 2$ bands in the near-IR, ranging from around 1 to 2.3 μm, and to the $\Delta v = 1$ bands between the near- and mid-IR (2.6–5.7 μm).

Emission from excited OH is believed to be responsible for the unusually high night-time radiance observed by the SPIRE experiment in the region of the CO_2 4.3 μm fundamental band, since around 80% of the OH(9–8) band and 10% of the OH(8–7) band emission falls between 4.17 and 4.48 μm. OH emission from those bands has been recently observed by ISAMS in the CO channel centred near 4.7 μm.

7.11   Molecular Oxygen Atmospheric Infrared Bands

The bands of O_2 at 1.27 and 1.58 μm in the near-IR originate from the transition of the electronic state $O_2(a^1\Delta_g)$ down to the ground and first vibrational levels of the ground electronic state $O_2(X^3\Sigma_g^-, v = 0,1)$, respectively. $O_2(a^1\Delta_g)$ is produced directly by O_3 photolysis through reaction (7.10), i.e.,

$$O_3 + h\nu \, (175 - 310\,\text{nm}) \rightarrow O(^1D) + O_2(a^1\Delta_g),$$

and indirectly from the $O(^1D)$ produced in that reaction through the mechanism:

$$O(^1D) + O_2 \rightarrow O_2(b^1\Sigma_g^+) + O(^3P), \quad \text{and} \tag{7.23}$$

$$O_2(b^1\Sigma_g^+) + O_2 \rightarrow O_2(a^1\Delta_g) + O_2. \tag{7.24}$$

Additional excitation of $O_2(a^1\Delta_g)$ takes place in the daytime by the resonant scattering of sunlight in the atmospheric bands. This takes place, predominantly, in the $O_2(X^3\Sigma_g^-, v=0) \rightarrow O_2(b^1\Sigma_g^+, v=0)$ band near 762 nm. At night, its major source is

$$O + O + M \rightarrow O_2(a^1\Delta_g) + M. \tag{7.25}$$

The $O_2(a^1\Delta_g)$ state is deactivated by spontaneous emission in the 1.27 and 1.58 μm bands, $O_2(a^1\Delta_g) \rightarrow O_2(X^3\Sigma_g^-, v=0)$ and $O_2(a^1\Delta_g) \rightarrow O_2(X^3\Sigma_g^-, v=1)$, respectively, and by deactivation in collisional processes mainly with O_2 itself.

The emissions from these O_2 states, whose upper state populations in the stratosphere, mesosphere and lower thermosphere exceed by far the population expected from LTE, have been extensively studied in the literature cited in Sec. 7.15.

7.12   Hydrogen Chloride, HCl, and Hydrogen Fluoride, HF

HCl has its fundamental vibration-rotation (1–0) band centred near 3.45 μm. Absorption of sunlight in this band and in the (2–0) band near 1.76 μm pumps the population of HCl(1) in the daytime stratosphere and mesosphere to a population much larger than that of LTE. This increases the HCl stratospheric limb radiance by a factor of 2 relative to LTE. The non-LTE contribution to the limb radiance is significant at altitudes as low as 20 km and becomes greater than the LTE component above 40 km.

The global distribution of atmospheric HF is of interest mainly because of its role as the reservoir for fluorine in the stratosphere. The HF molecule has its fundamental band near 2.52 μm. Direct absorption of solar radiation at this wavelength is predicted to increase the population of HF(1) significantly in the daytime stratosphere and mesosphere. Other, smaller, non-LTE excitation mechanisms are the absorption of backscattered radiation at 2.52 μm, and the transfer of the electronic energy of $O_2(a^1\Delta_g)$ produced in the O_3 photolysis, reaction (7.10), and also from the electronically excited $O_2(b^1\Sigma_g^+)$ produced after collisions of $O(^1D)$ with O_2 (reaction 7.23), e.g.,

$$O_2(a^1\Delta_g) + HF \rightarrow O_2 + HF(v \leq 2), \quad \text{and} \tag{7.26}$$

$$O_2(b^1\Sigma_g^+) + HF \rightarrow O_2(a^1\Delta_g) + HF(v \leq 1). \tag{7.27}$$

Another source of less importance is the absorption of solar and backscattered radiation near 1.29 μm in the HF(2–0) band. This, and the electronic transfer from $O_2(a^1\Delta_g)$ are also responsible for the excitation of the HF(2) level to very large non-LTE populations, and significantly enhanced emissions are expected from both the fundamental and the first hot bands of HF near 2.5 μm. However, evidence for this has not yet been reported.

7.13 NO^+

NO^+ ions are a major constituent in the E and lower F-regions of the ionosphere. They play an important role in the ion and neutral chemistry of the thermosphere, and have been found recently to contribute significantly to the cooling there. NO^+ is the most abundant positive ion in the daytime lower and middle thermosphere and, through dissociative recombination, it is the dominant source of $N(^2D)$ in this region which, in turn, is the major source of NO. NO^+ is produced in fast exothermic ion-molecule reactions, which are most relevant under auroral conditions. It has vibrational bands at wavelengths very close to the CO_2 4.3 μm bands; for this reason, and because of the low density of NO^+, their emissions have been difficult to observe, but they are relevant to any analysis of CO_2 measurements in the lower thermosphere.

There is experimental evidence for both vibrational and rotational non-LTE emissions from NO^+ in the thermosphere, although the excitation mechanisms are less clear. Several have been proposed:

$$N_2^+ + O \rightarrow NO^+(v, J) + N(^2D), \tag{7.28}$$

$$N_2^+ + O \rightarrow NO^+(v, J) + N(^4S), \tag{7.29}$$

$$N^+ + O_2 \rightarrow NO^+(v, J) + O(^1S, ^1D, ^3P), \tag{7.30}$$

$$O^+ + N_2 \rightarrow NO^+(v, J) + N, \quad \text{and} \tag{7.31}$$

$$N(^2P) + O \rightarrow NO^+(v, J) + e. \tag{7.32}$$

The analysis of data from the rocket-borne sensor which first detected the NO^+ 4.3 μm emission, and of the CIRRIS-1A measurements, has suggested that the non-LTE vibrational populations were caused by reaction (7.28). A more recent analysis has shown that the strong rotational non-LTE emissions observed by CIRRIS-1A is not consistent with the predictions from that reaction but by that produced in reaction (7.31).

7.14   Atomic Oxygen, $O(^3P)$, at 63 μm

The atomic oxygen emissions at 63 and 147 μm in the far-IR arise from the fine structure in the ground electronic state caused by magnetic dipole transitions (see Fig. 7.30). The 63 μm emission contributes significantly to the cooling of the thermosphere, most importantly above about 200 km, but also in the region down to 120 km, particularly at night. The emission at 147 μm is more than an order of magnitude weaker.

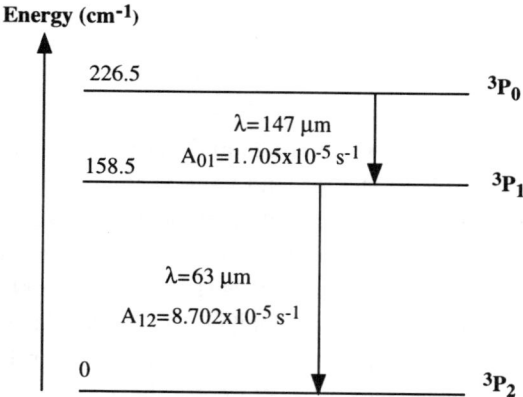

Fig. 7.30   Diagram of the atomic oxygen magnetic dipole transitions.

The 63 μm line is an example of a transition where the energy gap is so small that relatively few collisions are necessary to keep it in LTE. This, together with its long radiative lifetime ($1/A_{12} = 3.1$ hr), make it likely to be in LTE up to the very high thermosphere. We will discuss the cooling rates produced by the 63 μm line further in Sec. 9.6.

7.15   References and Further Reading

A detailed study of the population of the CO(1) level, and its variations with atmospheric parameters, may be found in López-Puertas *et al.* (1993) and Clarmann *et al.* (1998). The effects of vibrational and rotational non-LTE on limb radiance in the CO 2.3 and 4.7 μm bands have been studied by Kutepov *et al.* (1997). Other studies of CO are those of Oelhaf and Fischer (1989) and Winick *et al.* (1991). The CO abundances derived from the ISAMS measurements of its emission near 4.6 μm have been reported by López-Valverde *et al.* (1993, 1996), and Allen *et al.* (2000).

Exospheric solar fluxes for the calculation of the absorption of solar radiation by atmospheric molecules in the infrared and near-IR can be found in Thekaekara (1976), Kurucz (1993), and Tobiska *et al.* (2000).

The concentrations for O_3, $O(^3P)$, N_2O, NO_2, and NO from the Garcia and Solomon model used in Secs. 7.3, 7.7, and 7.8 can be found in Garcia and Solomon (1983, 1994) and Garcia *et al.* (1992).

The O_3 model described in the text is based on that developed by Martín-Torres and López-Puertas (2002). Several other non-LTE models for studying the O_3 infrared emissions have been developed, e.g., Ogawa (1976), Yamamoto (1977), Gordiets *et al.* (1978), Rawlins (1985), Solomon *et al.* (1986), Mlynczak and Drayson (1990a,b), Fichet *et al.* (1992), Manuilova and Shved (1992) [lately improved by Manuilova *et al.* (1998)], Pemberton (1993) [later revised with the analysis of ISAMS measurements by Koutoulaki (1998)], Rawlins *et al.* (1993), and Edwards *et al.* (1994). Spectral data for these bands may be found in the HITRAN'96 database and in Adler-Golden *et al.* (1985, 1990), Flaud (1990), Clarmann *et al.* (1998), and Tyuterev *et al.* (1999).

Laboratory work on the vibrational excitation of ozone is reported by Hochanadel *et al.* (1968), Bevan and Johnson (1973), Simons *et al.* (1973), von Rosenberg and Trainor (1973, 1974, 1975), Rosen and Cool (1973, 1975), Hui *et al.* (1975), Kleindienst and Bair (1977), Adler-Golden and Steinfeld (1980), McDade *et al.* (1980), Rawlins *et al.* (1981), Adler-Golden *et al.* (1982); Joens *et al.* (1982), Joens (1986), Locker *et al.* (1987), Steinfeld *et al.* (1987), Menard-Bourcin *et al.* (1990), Shi and Barker (1990), Doyennette *et al.* (1992), Menard *et al.* (1992), and Upschutte *et al.* (1994). The 'zero surprisal' nascent distribution of ozone is discussed by Rawlins and Armstrong (1987) and Rawlins *et al.* (1987). The collisional scheme in the model described here is based on analyses of the data obtained in 1986 by the SPIRIT-1 rocket experiment by Adler-Golden and Smith (1990), and those obtained in the laboratory experiments by Rosen and Cool (1973, 1975). See also the SPIRE rocket observations of ozone reported by Green *et al.* (1986). Non-LTE corrections to the model used for the retrieval of ozone from the LIMS channel at 9.6 μm are described by Solomon *et al.* (1986).

The populations of the H_2O levels described in the text are discussed in detailed by López-Puertas *et al.* (1995). A review of the collisional processes and rate coefficients for H_2O may be found in López-Puertas *et al.* (1995) and Zaragoza *et al.* (1998). Non-LTE models of the population of the H_2O (0,1,0) and (0,2,0) levels have been reported by Manuilova and Shved (1985) and López-Puertas *et al.* (1986b). Other non-LTE H_2O calculations are shown in Kerridge and Remsberg (1989). Estimations about the likely uncertainties in H_2O population levels may be found in Clarmann *et al.* (1998).

For evidence on non-LTE effects in the LIMS water vapour channel, see Kerridge and Remsberg (1989). Further implications of this can be found in Toumi *et al.* (1991). SAMS limb daytime $2.7\,\mu$m radiances were used to deduce $[H_2O]$ in the 35–100 km region by Drummond and Mutlow (1981), and non-LTE effects on the ISAMS water vapour channels have been studied by López-Puertas *et al.* (1995), Goss-Custard *et al.* (1996), and Zaragoza *et al.* (1998). Evidence of non-LTE effects in the H_2O fundamental and hot bands have also been found recently from the CIRRIS-1A experiment by Zhou *et al.* (1999) and from the CRISTA experiment by Edwards *et al.* (2000).

The methane model described here is based on the work of Martín-Torres *et al.* (1998) and Shved and Gusev (1997), the only two models so far developed for CH_4. Estimates of the likely uncertainties in the CH_4 populations may be found in Clarmann *et al.* (1998). SAMS and ISAMS remote sensing measurements of CH_4 are discussed by Jones and Pyle (1984) and Taylor and Dudhia (1987). In both analysis LTE was assumed for the CH_4 and N_2O bands.

The non-LTE model for nitric oxide presented here is based on the work by Funke and López-Puertas (2000). A non-LTE model for the NO(1) population was developed by Caledonia and Kennealy (1982). For a discussion of the importance of photochemistry for non-LTE in the stratosphere, see Kaye and Kumer (1987) and Sharma *et al.* (1993). SPIRE day and night-time limb radiances at 50–200 km are used to derive the rotational temperature (110–150 km), cooling rates, and the thermal collision rate with $O(^3P)$ by Zachor *et al.* (1985), and ISAMS NO results are presented by Ballard *et al.* (1993). For the effect of NO on the energy balance, see Kockarts (1980). Electron impact excitation of vibrational states of NO for an IBC II aurora has been simulated by Cartwright *et al.* (2000).

The evidence for chemically-pumped hot band emission from high v_3 vibrational levels of NO_2 in LIMS $6.9\,\mu$m measurements has been discussed by Kerridge and Remsberg (1989). A recent non-LTE model for the populations of the v_3 levels of NO_2 may be found in López-Puertas (1997), including a sensitivity study. For the calculation of the photoabsorption coefficients of NO_2 in the visible see Solomon *et al.* (1987) and the references therein.

The non-LTE populations of the N_2O vibrational levels described in the text are based on an unpublished work by Martín-Torres and López-Puertas. SAMS and ISAMS remote sensing measurements of N_2O are discussed by Jones and Pyle (1984) and Taylor and Dudhia (1987).

The HNO_3 model described here is based on that reported by Martín-Torres *et al.* (1998). Estimates of the likely uncertainties in the HNO_3 population levels may be found in Clarmann *et al.* (1998).

The $OH(v \leq 9)$ discussion in the text is based on the model and rocket experiment by López-Moreno *et al.* (1987). A good review of the $OH(v \leq 9)$ populations and measurements was compiled by López-González (1990); see also Makhlouf *et al.* (1995). Evidence for non-LTE in the rotational levels of atmospheric OH is presented by Smith *et al.* (1992), Pendleton *et al.* (1993), and Cosby *et al.* (1999).

A good review of the volume emission rate of the O_2 infrared bands was compiled by López-González (1990) (see also López-González *et al.*, 1992a,b). The analysis of the $1.58\,\mu m$ emission measured by SPIRE can be found in Winick *et al.* (1985).

Daytime non-LTE in hydrogen chloride and hydrogen fluoride is discussed by Kumer and James (1982), Kumer *et al.* (1989), and Kaye (1989).

Tentative observations of NO^+ were first reported by Picard *et al.* (1987), and Winick *et al.* (1987). Picard *et al.* (1992) and Winick *et al.* (1992) discuss CIRRIS-1A measurements and the role of NO^+ in cooling the region above $140\,km$ in the daytime. More recent experimental and modelling results from CIRRIS-1A have been reported by Duff and Smith (2000), and Smith *et al.* (2000).

Bates (1951) proposed that the $63\,\mu m$ atomic oxygen emission is an important mechanism for cooling the upper atmosphere, above $200\,km$. Gordiets *et al.* (1982) compute $O(^3P)$ $63\,\mu m$ cooling rates for day and night-time and compare them with those by CO_2 at $15\,\mu m$ and NO at $5.3\,\mu m$. Non-LTE versus LTE in $O(^3P)$ fine structure levels is discussed by Kockarts and Peetermans (1970), Grossmann and Offermann (1978), Iwagami and Ogawa (1982), Sharma *et al.* (1994), and Grossmann and Vollmann (1997). The feasibility of retrieving $O(^3P)$ abundances and temperature in the thermosphere by using the 63 and $147\,\mu m$ emissions are discussed by Zachor and Sharma (1989) and by Sharma *et al.* (1990).

Chapter 8

Remote Sensing of the Non-LTE Atmosphere

8.1 Introduction

Remote sensing of the atmosphere is an increasingly important tool, because it can be used from satellites to obtain global coverage of the Earth, while observing diurnal, seasonal and long-term changes in time. Temperature sounding from satellites is used to obtain information for weather forecasting purposes, and also to study the dynamics of the middle atmosphere, for example. Composition measurements are also coming to the fore as spectroscopic instruments improve. Two common examples of this are monitoring the stratospheric ozone layer and measuring the humidity in the troposphere and above. It is also becoming possible to map the origins and destinations of various kinds of pollution by increasingly sensitive remote sensing techniques.

The relevance of non-LTE theory to remote sensing can be looked at in two ways. The first case occurs when one does not know the nature nor the degree of the non-LTE processes that affect the observed quantity, usually thermally-emitted radiance or atmospheric transmittance; and the second when the effect of non-LTE processes on the observed quantity is reasonably well understood. In the first case, one can plan measurements of bands known to be in non-LTE and use these to study the process in detail, and to validate theoretical models such as those described in earlier chapters. In the second case, if the measurements are being used to determine temperature or composition, the presence of non-LTE effects (even if the underlying theory is well understood) will at best increase the complexity of the interpretation of the data, and at worst add an extra source of error and uncertainty. Sometimes non-LTE can help the investigator by increasing the daytime signal at high altitudes, especially when there is significant solar pumping, as is the case for CO at $4.6\,\mu$m and CO_2 at $4.3\,\mu$m, for example. In cases like these, where the non-LTE effect is large, it has to be included in the retrieval scheme. On the other hand, if the non-LTE effect is relatively small, as it is for H_2O at $6.3\,\mu$m and O_3 at $9.6\,\mu$m in the lower mesosphere, it can be sufficient to obtain results assuming LTE, and then to apply a correction, rather than facing the complexity of incorporating non-LTE into the retrieval process itself. However, this distinction might disappear soon, as the speed of computers is now fast enough to directly account for non-LTE effects in the retrieval schemes regardless of their magnitude.

Some sets of measurements, particularly if they are combined with simultaneous, co-located observations taken by a technique which is not affected by non-LTE, can be used in both senses: first, for understanding the non-LTE effect and then, once it is understood, for determining atmospheric temperature and composition. This is rare, however; very few experiments have been designed specifically to understand non-LTE, so our knowledge of non-LTE has been mostly gained from measurements made for other purposes.

In general, we proceed as follows. The kinetic temperature and the abundances of the species being studied are assumed either from climatology, from a co-located in-situ measurement or from previous iteration in the case of a retrieval process, and these are used as inputs to a model which calculates the non-LTE populations for the different molecules, as described in the preceding chapters. Alternatively, data from the same channel but under night-time conditions can be used to analyse daytime observations of solar-pumped phenomena if the species is known to have little diurnal change in abundance and temperature is measured simultaneously at night and daytime. An example of this approach, the measurement of H_2O in the mesosphere by ISAMS, is discussed in Sec. 8.5. Sometimes the species abundance can be obtained from another instrument which is less susceptible to non-LTE, as in the case of the ozone observations by ISAMS where measurements of ozone from the Microwave Limb Sounder on the same satellite were used, as discussed in Sec. 8.4. Errors tend to multiply rapidly in such multi-instrument comparisons, however.

After computing the non-LTE populations, the next step in the analysis is to estimate the measured radiances with a forward radiative transfer model and combine them with the known instrument characteristics, especially observing geometry, filter response, and detector sensitivity. This produces synthetic observations which can be compared with the real measurements, and an iteration performed in which the non-LTE models are modified by adjusting parameters and/or invoking new non-LTE processes until the predictions fit the measurements. The conclusions reached about the atmospheric properties affecting the signal and about the non-LTE processes themselves may not be unique, depending on how good the non-LTE model is to begin with, and how much information is available in the measurements. Sometimes the remaining uncertainties are large. At the present time, the priority is to develop complete and comprehensive non-LTE retrieval schemes, i.e., a merging of the non-LTE models described in Chapters 6 and 7 with the existing standard retrieval schemes which were developed for LTE. In this way we obtain a tool in which geophysical parameters (temperature and gas abundances) and non-LTE parameters (collisional coefficient rates, energy transfer efficiencies, etc.) can all be retrieved from the measurements of radiance, provided these are sufficiently comprehensive.

In this chapter, we shall look at some real observations which provide evidence of non-LTE emissions in the atmosphere, and use them to illustrate the remote sensing of atmospheric quantities under non-LTE conditions. An example of a study which revealed basic details of the non-LTE processes themselves is that of the emission from the 4.3 μm band of CO_2 made by the Stratospheric and Mesospheric Sounder (SAMS) on the Nimbus 7 satellite from 1978 to 1982 and more recently by the ISAMS instrument on UARS (Secs. 8.3.4 and 8.3.5). Another typical example of this is the H_2O 6.3 μm emission measured by ISAMS (Sec. 8.5). Examples where non-LTE models are used as part of the retrieval process are the global mapping of carbon monoxide and nitric oxide by ISAMS (Secs. 8.6 and 8.7).

Chapters 6 and 7 discussed some non-LTE situations caused by chemical processes, including that giving rise to the excitation of $CO_2(v_3)$ by $OH(v \leq 9)$ through $N_2(1)$ as an intermediary, and the production of $O_3(v_1,v_2,v_3)$ and $NO_2(v_3)$ in the stratosphere and mesosphere. Some satellite observations of this behaviour have been reported, and are discussed in Secs. 8.4 and 8.8.

Usually, our focus is on non-LTE populations of the vibrational levels, and rotational LTE is assumed at all levels of interest in the atmosphere. There is, however, some experimental evidence for non-LTE populations in rotational states, which is discussed in Sec. 8.9.

The measurements which have been used to study non-LTE have generally been of atmospheric *emission*. There are, however, a few important examples of the use of solar occultation spectroscopy to obtain simultaneous vibrational and kinetic temperatures from atmospheric *transmission* spectra. One such is the Atmospheric Trace Molecule Spectroscopy (ATMOS) spectrometer which first flew on Spacelab 3 in 1986. The effects of non-LTE on absorption spectra are much less than for emission (see Sec. 8.2.1), and are generally very small for atmospheric species, except in certain cases like thermospheric CO_2. For this case, using fundamental and hot bands, it is possible to determine the kinetic temperature, the concentration of CO_2 and the non-LTE population of $CO_2(0,1^1,0)$, all simultaneously from a single set of high resolution transmission measurements. As a result, measurements in absorption with high spectral resolution offer one of the best techniques for testing CO_2 non-LTE models in the upper mesosphere and lower thermosphere. This is in marked contrast to the analysis of emission spectra, where, as noted above, a simultaneous measurement of the kinetic temperature and the abundance of the species at work is rarely available. The absorption measurements are described in detail in Sec. 8.10.

The latest advances in emission measurements have allowed, however, the use of very high spectral resolution, thus giving them one of the advantages that was previously possible only when working in transmission using the Sun as a source. The benefits of this progress for non-LTE studies are discussed towards the end of the chapter in Sec. 8.11.

Most experimental studies of non-LTE have involved the measurement of the atmosphere at the Earth's *limb* (see Fig. 8.1), and only a few use *nadir* observations. The potential effects of non-LTE on the nadir radiances are discussed in Sec. 8.12.

Finally we outline the non-LTE retrieval schemes in Sec. 8.13. These are algorithms which systematically 'invert' measurements of non-LTE emission profiles to obtain atmospheric quantities like temperature and composition, and non-LTE parameters such as kinetic collisional rates and energy transfer efficiencies.

8.2   The Analysis of Emission Measurements

8.2.1   *Limb observations*

The radiance measured by an instrument is the integral over frequency of the product of the radiance emitted by the atmosphere at frequency ν with the instrumental response function $F(\nu)$. The response function is non-zero over a finite frequency range $\Delta\nu$, called the spectral width of the channel. Thus,

$$L(x_{\mathrm{obs}}) = \int_{\Delta\nu} L_\nu(x_{\mathrm{obs}})\, F(\nu)\, \mathrm{d}\nu,$$

where $L_\nu(x_{\mathrm{obs}})$ is the monochromatic radiance at a given observation point x_{obs}. For the case of limb viewing (see Fig. 8.1), the monochromatic radiance is calculated from (see the formal solution, Eq. 3.20)

$$L_\nu(x_{\mathrm{obs}}) = \int_{x_s}^{x_{\mathrm{obs}}} J_\nu(x)\, \frac{\mathrm{d}\mathcal{T}_\nu(x)}{\mathrm{d}x}\, \mathrm{d}x, \tag{8.1}$$

where x is position along the limb path line-of-sight and the integration is performed from the furthest extent of the atmosphere at the limb point, x_s, passing through the tangent height of the observation point, z, to the observation point at x_{obs}.

The source function $J_\nu(x)$ is given by Eq. (3.35), and the monochromatic transmittance between x and x_{obs} is defined, according to Eqs. (4.11) and (4.59), by

$$\mathcal{T}_\nu(x, x_{\mathrm{obs}}) = \exp\left(-\int_x^{x_{\mathrm{obs}}} k_\nu(x')\, n_a(x')\, \mathrm{d}x'\right), \tag{8.2}$$

where k_ν is the molecular absorption coefficient, and n_a is the number density of absorbing molecules.

Equation (8.1) gives the measured radiance when there is only one band contributing. In the case where there are contributions from several bands, the total radiance is given by a similar expression with the form:

$$L_\nu(x_{\mathrm{obs}}) = \int_{x_s}^{x_{\mathrm{obs}}} \sum_i J_{\nu,i}(x)\, \frac{\mathrm{d}\mathcal{T}_{\nu,i}(x)}{\mathrm{d}x} \prod_{j\neq i} \mathcal{T}_{\nu,j}(x)\, \mathrm{d}x, \tag{8.3}$$

where the sum is over all bands contributing at frequency ν, and the product $\prod_{j\neq i} \mathcal{T}_{\nu,j}(x)$ is over all of the bands which absorb part of the emission from the i-th band.

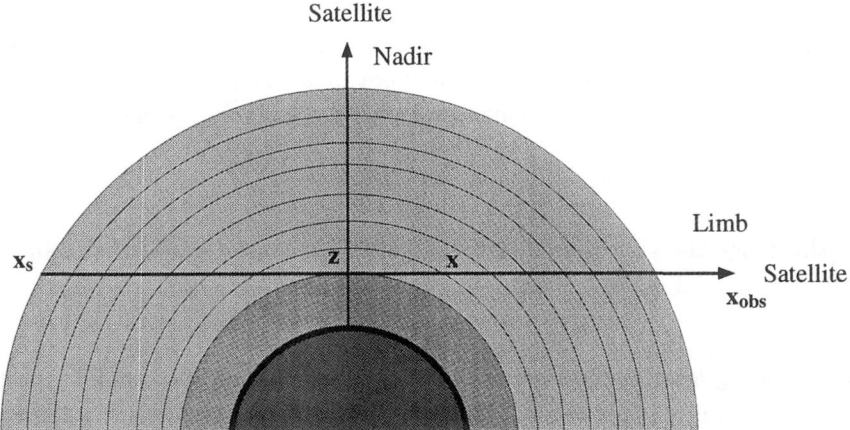

Fig. 8.1   Geometry for calculating the limb and nadir radiances.

The effects of non-LTE on the limb radiance of a transition between two states, (2–1), are principally due to the departure of its source function from the Planck function. We derived the relation between the two functions in Chapter 3, where it is given by Eq. (3.36), i.e.,

$$J_\nu = B_\nu \, \frac{n_2}{\bar{n}_2} \, \frac{\bar{k}_\nu}{k_\nu}. \tag{8.4}$$

Following the notation in Chapter 3, the overbar signifies a quantity in LTE. The principal effect comes from the factor $n_2/\bar{n}_2$, which accounts for the deviation of the population of the upper state of the transition, n_2, from its LTE value, $\bar{n}_2$.

The second factor in this equation is the ratio of the absorption coefficient to its LTE value, which also affects the limb radiance through the transmission function (Eq. 8.2). From Eqs. (3.34) and (3.37), this factor is given by

$$\frac{k_\nu(x)}{\bar{k}_\nu(x)} = \frac{n_1}{\bar{n}_1} \frac{\left[1 - \dfrac{g_1 n_2}{g_2 n_1}\right]}{\left[1 - \exp\left(-\dfrac{h\nu_0}{kT}\right)\right]} = r_1 \left[\frac{1 - \Gamma r_2/r_1}{1 - \Gamma}\right] \tag{8.5}$$

where we have assumed that $\nu = \nu_0$ in Eq. (3.34), since $E_2 - E_1 = h\nu_0$, and that the line shape profile f_a is the same for LTE and for non-LTE conditions. Also we have made use of the population ratio

$$r_v = n_v/\bar{n}_v, \tag{8.6}$$

and the Boltzmann factor, Γ,

$$\Gamma = \frac{g_1 \bar{n}_2}{g_2 \bar{n}_1} = \exp\left(-\frac{h\nu_0}{kT}\right). \tag{8.7}$$

The limb radiance calculation usually starts with the vibrational temperatures, obtained from the non-LTE models as described in Chapters 6 and 7. The population ratio (Eq. 8.6) is calculated from the vibrational temperature (Eq. 6.1) for any (except the ground state) vibrational level v with energy E_v, using

$$r_v = \frac{n_0}{\bar{n}_0} \exp\left[-\frac{E_v}{k}\left(\frac{1}{T_v} - \frac{1}{T}\right)\right]. \tag{8.8}$$

At this stage the vibrational partition sum Q_{vib} can be introduced using

$$n_0 = \frac{n_a \, f_{\mathrm{iso}}}{Q_{\mathrm{vib}}}, \tag{8.9}$$

where n_0 and n_a are the number density of the molecules in the ground vibrational level and the total number of absorbing molecules, respectively, to give the ratio of populations between the general (non-LTE) and LTE cases for the state v as

$$r_v = \frac{\bar{Q}_{\mathrm{vib}}}{Q_{\mathrm{vib}}} \exp\left[-\frac{E_v}{k}\left(\frac{1}{T_v} - \frac{1}{T}\right)\right]. \tag{8.10}$$

Q_{vib} can be evaluated from

$$Q_{\mathrm{vib}} = \sum_v^{v_{\mathrm{max}}} g_v \exp\left(-\frac{E_v}{kT_v}\right), \tag{8.11}$$

where v_{max} is the state at which the summation is truncated, determined in practice by the highest state for which populations are calculated by the non-LTE model.

Non-LTE affects the absorption coefficient in three ways: (i) through the population of the lower level of the transition; (ii) through the induced emission factor, affecting the population of the upper level; and (iii) through the vibrational partition sum, which involves the vibrational temperatures of the lower energy levels of the species involved.

The first of these only applies to hot bands. It usually does not have a major effect on the limb radiance, since non-LTE populations are usually important at high altitudes where the hot bands, in particular, are optically thin. Under these circumstances, the derivative of the transmission in Eq. (8.1) is proportional to n_1, which also appears in the denominator of the source function, and hence cancels out. Thus the limb radiance is essentially independent of whether we include the LTE or the more correct non-LTE population of the lower level of the transition.

The second effect, due to the induced emission factor in Eq. (8.5), is also usually small, and is significant only for low-energy transitions at high temperatures, and when population inversions occur.

The third effect, due to the vibrational partition function, entering through the ratio $n_0/\bar{n}_0$, can be important for some species when the population of the lower energy levels is significantly altered by non-LTE. This is not often the case, as we have already seen several times, since the lower-energy levels, which contribute most to the vibrational partition sum, are more easily thermalized and depart less from

LTE. In the low-energy ν_2 mode of CO_2 at $15\,\mu$m, however, the difference between the vibrational partition function in LTE versus non-LTE is $\sim$10% at $120\,$km and $\sim$40% at $160\,$km. This affects the emission from all CO_2 bands in the thermosphere, including those at 4.3 and $2.7\,\mu$m. For most other atmospheric species the non-LTE effects manifested through the vibrational partition sum are much less.

8.2.2 The limb radiance under optically thin conditions

We have seen that non-LTE processes become most important in the upper atmosphere, where many lines in the absorption bands are optically thin. We now examine this common situation by considering the contribution of a single spectral line to the limb radiance under these conditions. The results can be extended to complete vibrational bands by assuming that lines may be treated as independent under conditions where they are narrow compared to their mean spacing. The monochromatic transmittance from Eq. (8.2) may then be written as

$$\mathcal{T}_\nu(x, x_{\text{obs}}) = 1 - \int_x^{x_{\text{obs}}} k_\nu(x')\, n_a(x')\, dx', \tag{8.12}$$

such that

$$\frac{d\mathcal{T}_\nu(x)}{dx} = k_\nu(x)\, n_a(x). \tag{8.13}$$

Replacing this in Eq. (8.1) we have

$$L_\nu(x_{\text{obs}}) = \int_{x_s}^{x_{\text{obs}}} J_\nu(x)\, k_\nu(x)\, n_a(x)\, dx. \tag{8.14}$$

Substituting for $J_\nu(x)$ from Eq. (8.4) and using Eq. (8.10) gives

$$L_\nu(x_{\text{obs}}) = \int_{x_s}^{x_{\text{obs}}} B_\nu(x) \exp\left[-\frac{E_2}{k}\left(\frac{1}{T_2} - \frac{1}{T}\right)\right] \frac{\bar{Q}_{\text{vib}}}{Q_{\text{vib}}} \bar{k}_\nu(x)\, n_a(x)\, dx. \tag{8.15}$$

This expression shows that under optically thin conditions, the limb radiance depends mainly on the vibrational temperature of the upper level of the transition, T_2 (defined relative to the ground state population). It does not depend directly on the vibrational temperature of the lower level of the transition, T_1, except for a weak dependence through Q_{vib} in Eq. (8.15) and through Q_{vib} when calculating the vibrational temperature T_2.

An alternative expression to Eq. (8.15) for the limb radiance, using in Eq. (8.14) the expression for $J_\nu(x)$ from Eq. (8.4), the Planck function, and the relationships between $\bar{k}_\nu$ and B_{12} (Eq. 3.37), between the Einstein coefficients B_{21} and A_{21} (Eq. 3.30), and between B_{21} and B_{12} (Eq. 3.31), is

$$L_\nu(x_{\text{obs}}) = \frac{h\nu}{4\pi} A_{21} \int_{x_s}^{x_{\text{obs}}} q_{r,v}\, f(\nu)\, n_2\, dx, \tag{8.16}$$

where $q_{r,v}$ is the normalized factor for the rotational states distribution in the vibrational level v and we have considered also the normalized shape of a vibration-rotation line $f(\nu)$. Using the definition of the vibrational temperature and the relation between n_0 and Q_{vib}, it can be expressed by

$$L_\nu(x_{\mathrm{obs}}) = \frac{h\nu}{4\pi} A_{21}\, g_2 \int_{x_s}^{x_{\mathrm{obs}}} \frac{n_a}{Q_{\mathrm{vib}}}\, q_{r,v}\, f(\nu)\, \exp\left(-\frac{E_2}{kT_2}\right)\, \mathrm{d}x. \qquad (8.17)$$

These expressions allow approximate calculations of the limb radiance to be made for straightforward non-LTE situations, such as those described in Secs. 3.7 and 5.2, by using the approximate non-LTE populations n_2 described in those sections (e.g., Eqs. 3.88, 3.89, and 3.90). Those equations can also be integrated over the frequency, e.g., over the spectral width of the instrument channel or over the bandwidth, and over the solid angle subtended by the atmosphere as viewed with the instrument field-of-view.

8.2.3   *Summary of observations*

Here we summarise the various experiments which have provided the most important measurements of atmospheric infrared emissions containing evidence for non-LTE. The earlier ones are rocket-borne instruments, which provided, for a given flight, only a few vertical profiles. These are gradually being superseded by satellite instruments, which offer repeated measurements and global coverage. The experiments are generally denoted by acronyms (see Sec. A.2) and are characterised by a set of observing wavelengths or channels, which can become confusing if not defined at the outset. Table 8.1 and the summary below provide these definitions and outline the principal characteristics of each project for ease of reference.

SPIRE (Spectral Infrared Rocket Experiment) was one of the most successful rocket experiments, covering the emission features of a large range of constituents (Fig. 8.2). A continuously–variable filter covered the range 1.4–$16.5\,\mu$m with spectral resolution of $\Delta\lambda = 0.03\,\lambda$, making day and night-time limb (quasi-horizontal) observations from the troposphere up to $200\,$km. Earlier launches in the ICECAP (Infrared Chemistry Experiments–Co-ordinated Auroral Program), used a cryogenic spectrometer to observe from 1.6 to $23\,\mu$m at the same spectral resolution, making zenith observations under auroral conditions.

HIRIS (High–Resolution Interferometer Spectrometer) and FWI (Field–Widened Interferometer) both achieved higher spectral resolution than SPIRE or ICECAP by employing Michelson interferometers to cover the 4–$22\,\mu$m and 2.0–$7.5\,\mu$m spectral regions, respectively, at $\Delta\tilde{\nu} = 2\,$cm^{-1}. Both observed during auroral conditions from the upper mesosphere to $125\,$km.

EBC (Energy Budget Campaign) was an international campaign involving ground-based, balloon- and rocket-borne experiments carried out in the high latitude winter night with the aim of studying the energy production and loss processes of the mesosphere and lower thermosphere. As part of the EBC campaign there were rocket

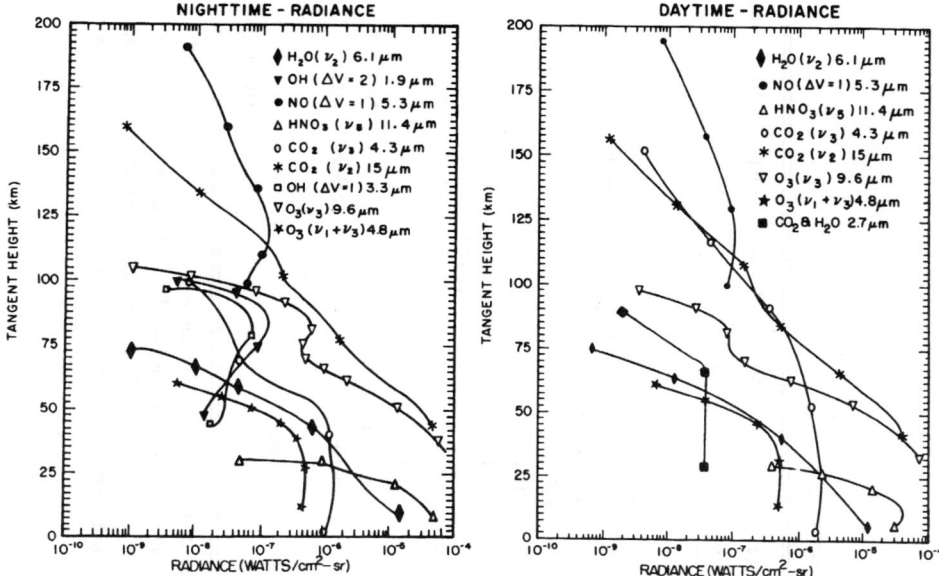

Fig. 8.2 Integrated band radiance of several atmospheric radiators in the middle and upper atmosphere as measured by the SPIRE experiment under night-time and daytime conditions.

launches with a helium-cooled circular variable filter spectrometer, covering the range from 4.7 to 23.6 μm, and a cryogenic infrared grating spectrometers measuring in the interval from 2.5 to 100 μm. Both instruments measured the major infrared emissions from the mesosphere and lower thermosphere in the zenith (see Table 8.1 for the list of emissions measured). Simultaneously EBC measured temperature and atomic oxygen using different techniques, some of them unaffected by non-LTE, which are useful for understanding the non-LTE emissions.

The Middle Atmosphere Program/Winter in Northern Europe (MAP/WINE) campaign was carried out in 1983/84 with the principal aim of studying the dynamical variability of the middle atmosphere. As part of this campaign the infrared grating spectrometer mentioned above was also launched on a rocket measuring zenith spectra of the mid-infrared vibration-rotation bands of CO_2 at 15 μm, O_3 at 9.6 μm, and H_2O at 6.3 μm.

Two early satellite experiments were more or less contemporaneous with these rocket campaigns. LIMS (Limb Infrared Monitor of the Stratosphere) and SAMS (Stratospheric and Mesospheric Sounder) were both multi-channel infrared radiometers which were launched together into a sun-synchronous, near-polar orbit on the Nimbus 7 satellite in October 1978. LIMS used filters and was cryogenically cooled for sensitivity, while SAMS used the pressure-modulator technique (discussed in more detail below) for high spectral discrimination. Both observed the atmosphere at the limb, day and night, SAMS until spring 1984, measuring thermal radiation, fluorescence, and resonance-scattered sunlight from the stratosphere up to the lower thermosphere, approximately 10 to 80 km, covering latitudes from 50°S to 70°N.

Table 8.1   Measurements of non-LTE emissions.

Molecule/ Band	CO$_2$ 15 μm	CO$_2$ 10 μm	CO$_2$ 4.3 μm	CO$_2$ 2.7 μm	O$_3$ 9.6 μm	CO 4.7 μm	H$_2$O 6.3 μm	H$_2$O 2.7 μm	NO 5.3 μm	NO 2.8 μm	NO$_2$	OH 1-3 μm	NO$^+$ 4.3 μm
ICECAP	60-150	–	70-120	–	40-100	–	50-75	–	70-140	70-130	–	–	–
HIRIS	70-120	–	80-105	–	70-110	–	–	–	70-125	–	–	–	–
SPIRE	50-160	–	10-150	25-90	30-105	–	10-75	25-90	100-200	–	–	40-110	–
FWI	–	–	85-140	–	–	90-95	–	–	85-140	–	–	✓	109
SPIRIT I	125-200	–	–	–	67-105	–	–	–	120-170	–	–	–	–
EBC	70-150	–	–	–	70-105	–	–	–	70-185	–	–	–	–
MAP/WINE	55-130	–	–	–	53-95	–	54-80	–	80-150	–	–	–	–
LIMS	–	–	–	–	50-70	–	20-70	–	–	–	20-45?	–	–
SAMS	–	–	30-110	–	–	30-70	✓	40-95	–	–	–	–	–
SISSI	60-140	✓	60-120	–	60-100	–	–	–	95-185	–	–	✓	–
CIRRIS-1A	80-170	✓	✓	–	60-100	70-150	–	–	100-170	✓	–	✓	100-215
ISAMS	30-90	–	50-120	–	30-70	30-90	30-70	–	30-150	–	–	80-90	120?
CLAES	–	20-60	–	–	–	–	✓	–	–	–	–	–	–
CRISTA	40-150	✓	15-120	–	15-90	✓	15-80	–	90-180	–	–	–	–
SPIRIT III	65-130	–	0-120	–	✓	–	✓	–	–	–	–	–	100-120

See the text and Sec. A.2 for a definition of the experiment acronyms in the first column. Only the measurements made by each experiment which are non-LTE over at least part of the altitude range of the observations are included. The altitude intervals (km) indicate the range of measurements, not the range where the emissions are in non-LTE. A check mark (✓) means either the non-LTE emission has been measured but not analysed or the altitude range has not been published. The question marks denote doubtful non-LTE detections.

The SISSI (Spectroscopic Infrared Structure Signatures Investigation) payload contains an infrared grating spectrometer similar to that flown in the EBC and MAP/WINE campaigns mentioned above. The payload was launched on a rocket during the DYANA (Dynamics Adapted Network for the Atmosphere) campaign on 6 March 1990 and in three more subsequent flights in the summer of 1990 and spring of 1991. As in the previous campaigns, the infrared spectrometer measured the major infrared emissions of the middle and upper atmosphere: CO_2 at 15 and $4.3\,\mu m$, O_3 at $9.6\,\mu m$, and NO at $5.3\,\mu m$.

Infrared spectrally-resolved measurements were made from space by the Cryogenic Infrared Radiance Instrumentation for Shuttle (CIRRIS-1A) which operated on the space shuttle Discovery from April 28th to May 6th 1991 at an orbital altitude of $\sim$260 km. Among the experiments conducted was the measurement of the infrared emission from the atmosphere in the 2.5–25 μm interval with filter radiometers and a Michelson interferometer with a resolution of $\sim 1\,cm^{-1}$. The data were used to reveal the evidence of non-LTE effects in both vibrational and rotational states in many atmospheric species which include CO_2 at 15 and $4.3\,\mu m$, O_3 near $9.6\,\mu m$, mesospheric water vapour near $6.3\,\mu m$, CO near $4.6\,\mu m$, NO at $5.3\,\mu m$, NO^+ near $4.3\,\mu m$, and OH in pure rotational and vibrational emissions. This small set of spectral data uniquely provides the spectra needed to identify the enhanced $H_2O(0,2,0-0,1,0)$ transitions of daytime emissions and the rotational non-LTE in OH, NO and NO^+.

Other recent satellite instruments include the Cryogenic Limb Array Etalon Spectrometer (CLAES), and the Improved Stratospheric and Mesospheric Sounder (ISAMS), which flew on the Upper Atmosphere Research Satellite (UARS) launched in 1991. As earlier with SAMS, the analysis of those measurements to retrieve species abundances and kinetic temperature needs to invoke non-LTE models. The most interesting cases are carbon dioxide, carbon monoxide, ozone, water vapour, and nitric oxide, measurements of all of which show day–night differences in radiance emission so pronounced that they can only be due to non-LTE effects. The CLAES and ISAMS non-LTE emission observations were taken simultaneously with measurements of temperature and species abundances by other instruments, in particular MLS (Microwave Limb Sounder) and WINDII (Wide-Angle Doppler Imaging Interferometer) which use techniques that are not prone to non-LTE effects.

The CRISTA (Cryogenic Infrared Spectrometers and Telescopes for the Atmosphere) instrument is a high-sensitivity limb-scanning experiment measuring the atmospheric radiation in the 4–71 μm interval, which was first flown on Shuttle flight STS66 in November 1994. The SPIRIT III (SPectral InfraRed Interferometric Telescope) high-spatial resolution radiometer array and interferometer/spectrometer flew onboard the MSX (Midcourse Sensor Experiment) satellite and obtained > 200 data sets of limb and nadir data between April 1996 and February 1997 in the 2.5–28 μm spectral region.

The wide range of measurements by these instruments (see Table 8.1) has helped considerably to understand atmospheric non-LTE processes, as will be described below, and several new experiments will soon add to the pool of knowledge. These include the Michelson Interferometer for Passive Atmospheric Sounding (MIPAS) Instrument on the European Environmental Satellite (ENVISAT). The high spectral resolution, wide spectral range and high sensitivity of MIPAS make it very valuable for studying non-LTE emissions. In addition, the SABER (Sounding of the Atmosphere using Broadband Emission Radiometry) instrument on the NASA TIMED mission and the HIRDLS (High Resolution Dynamics Limb Sounder) planned for the NASA EOS Aura mission are both wide-band radiometers capable of measuring the strongest non-LTE emissions up to the lower thermosphere.

8.3   Observations of Carbon Dioxide in Emission

The relatively high abundance and strong infrared bands of carbon dioxide provide a significant signal up to high altitudes and ensure the importance of the molecule in the atmospheric energy budget. Its mixing ratio is well known and nearly constant, below the mesosphere at least, and this additional feature makes it convenient for atmospheric temperature sounding. Consequently, many experiments have measured CO_2 emissions, some routinely, including instruments on operational weather satellites. The near- and mid-infrared spectrum is rich in observable CO_2 bands; we will consider each major spectral region in turn, beginning with that most commonly used for temperature sounding.

8.3.1   *Observations of the CO_2 15 µm emission*

Most of the recent evidence suggests that the 15 µm fundamental band is close to LTE up to at least 100 km (see Sec. 8.10.2). However, EBC rocket measurements of the zenith night-time radiance at 70–150 km, made in the mid-1980s, were used to retrieve vibrational temperatures up to 90 km, and these were reported as showing significant departures of up to 25 K from the kinetic temperature between 70 and 85 km. Similar rocket observations as part of the MAP/WINE project measured the zenith night-time radiance at 55–130 km and, in this case, the derived vibrational temperature at 54–90 km was much closer to the kinetic temperature. A similar conclusion have been drawn from the analysis of the measurements of the SPIRE rocket experiment (see Fig. 8.3). They showed that the transfer rate of vibrational energy of $CO_2(0,1^1,0)$ to and from the translational energy of atomic oxygen, k_{CO_2-O}, is much faster than had been previously thought. This leads the $CO_2(0,1^1,0)$ level to have a nearly LTE population up to altitudes around 100 km where non-LTE models had, using the old rate, predicted a large deviation from LTE at altitudes above ∼80 km. Without the fast atomic-oxygen rate, the prominent ledge seen in the SPIRE data of Fig. 8.3 between 100 and 110 km would also be absent. The rotational temperature from the CO_2 emissions at ∼110–160 km

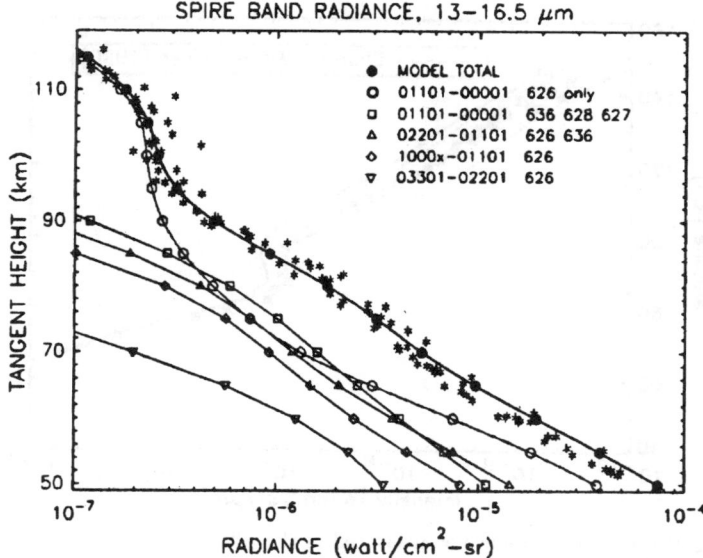

Fig. 8.3 SPIRE limb radiance in the 13–16.5 μm band as a function of tangent height (asterisks). Contributions of different CO_2 bands near 15 μm to the total limb radiance are also shown. Notice the significant contribution of the weak bands for tangent heights below 90–100 km. Above 110 km, the 626 fundamental band is almost entirely responsible for the observed radiance.

could also be inferred from SPIRE and SPIRIT I measurements of high J lines. The results showed that rotational equilibrium in the CO_2 15 μm levels applies up to $\sim$150 km, in line with theoretical expectations.

Figure 8.4 shows the sensitivity of the CO_2 15 μm radiance as measured by the SISSI experiment to the atmospheric and non-LTE model parameters. The solid line is a calculation with the $k_{CO_2\text{-}O}$ rate coefficient from Sharma and Wintersteiner (1990), close to 6×10^{-12} cm^{3}s^{-1}. The calculation using the $k_{CO_2\text{-}O}$ rate of Shved *et al.* (1990) of 1.5×10^{-12} cm^{3}s^{-1} measured in the laboratory at 300 K (not shown in Fig. 8.4) nearly lies on top of the SISSI measurements. The dotted lines shows the uncertainty in the radiance due to the uncertainties in the CO_2 number density ($\pm30\%$), atomic oxygen number density ($\pm30\%$), and kinetic temperature ($\pm10\%$). Thus, while SPIRE measurements seem to suggest a $k_{CO_2\text{-}O}$ rate close to 6×10^{-12} cm^{3}s^{-1}, SISSI observations fit better with a rate about 4 times slower. The uncertainties in other atmospheric parameters as the CO_2 and O(^3P) abundances and kinetic temperature, could possibly explain part of this discrepancy. The ATMOS measurements, in which the kinetic temperature and the CO_2 abundance were simultaneously measured (see Sec. 8.10.1) suggest a value between 3 and 6×10^{-12} cm^{3}s^{-1}, depending on the assumed O(^3P) density. The uncertainty in this crucial rate is still quite large, as is its temperature dependence.

An analysis of the measurements of the CO_2 15 μm radiance by the ISAMS channel and the kinetic temperature from WINDII taken at northern latitudes around the mesopause clearly shows evidence of non-LTE emissions in the $CO_2(v_2{=}1)$ levels

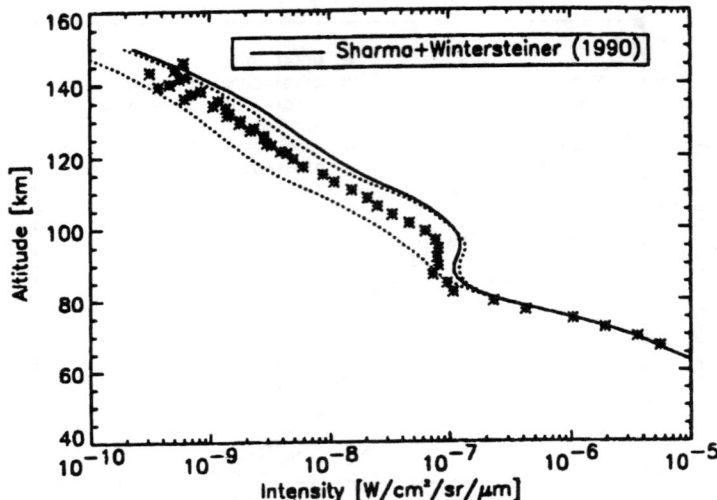

Fig. 8.4 CO_2 15 μm zenith radiance as measured in the SISSI-3 experiment (stars) and several calculations. Solid line: with the $k_{CO_2\text{-}O}$ rate coefficient from Sharma and Wintersteiner (1990); dotted lines shows the uncertainty in the radiance due to the uncertainties in the CO_2 number density ($\pm$30%), atomic oxygen number density ($\pm$30%), and kinetic temperature ($\pm$10%).

of the minor isotopes 636, 628, and 627 in the 70–90 km region.

The procedure outlined in the introduction to this chapter was followed in this analysis. First, the radiances that ISAMS would have measured if the CO_2 emitting levels were in LTE at the kinetic temperatures measured by WINDII were computed and compared with measurements. These LTE calculations clearly underestimate the observed ISAMS radiances over most of the altitude range, particularly below 85 km (see Fig. 8.5). Then, vibrational temperatures for the levels emitting in the ISAMS channel were computed with a non-LTE model similar to that described in Chapter 6, using the kinetic temperature measured by WINDII (see Fig. 8.6).

In order to find out for which levels we have to compute non-LTE populations it was necessary to calculate individual radiance contributions of the candidate bands. This calculation also tells us, once the measurements have been explained, which non-LTE populations are actually being verified. Fig. 8.7 shows the contributions of the different groups of bands for a measured profile. It can be seen that, for most of the 70–90 km region, the main contribution (60–80%) comes from the 15 μm fundamental bands of the minor isotopes 636, 628 and 627. Note the small contribution to this channel of the CO_2 fundamental band, whose originating level ($v_2{=}1$) is quite close to LTE in this region because of the efficient thermalization with $O(^3P)$ (see Sec. 8.10.1).

Once we have the vibrational temperatures, non-LTE forward radiances as measured by ISAMS were computed and compared again with the measurements. The agreement of the non-LTE radiances with ISAMS measurements is not perfect but it is much better than for the LTE radiances at most altitudes (see Fig. 8.5). One

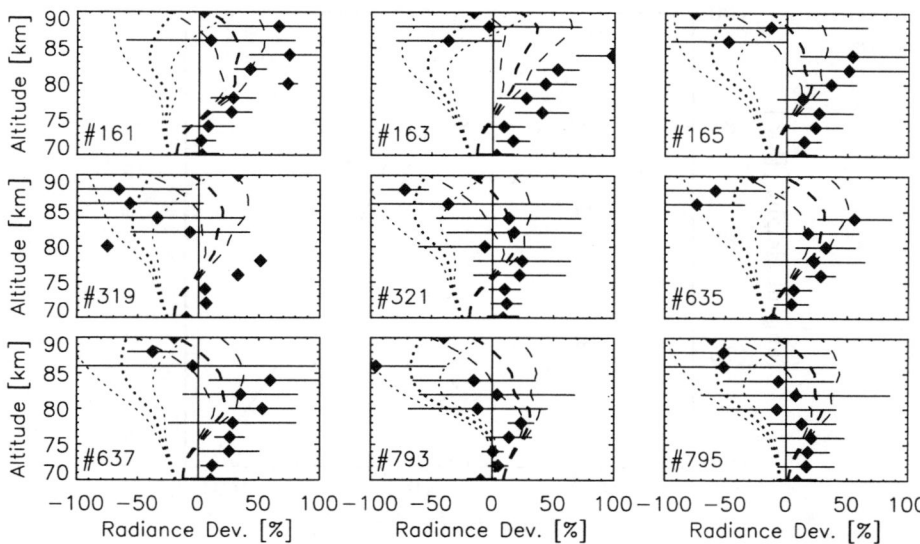

Fig. 8.5   Comparison of simulated and measured ISAMS radiances near 15 μm, shown in percentage deviation from the subarctic summer LTE case, for several WINDII/ISAMS coincident measurements. The dotted and dashed lines are the LTE and non-LTE radiances, and their error limits, corresponding to the kinetic temperatures measured by WINDII; the diamonds with bars are the corresponding ISAMS measurements with their standard deviations.

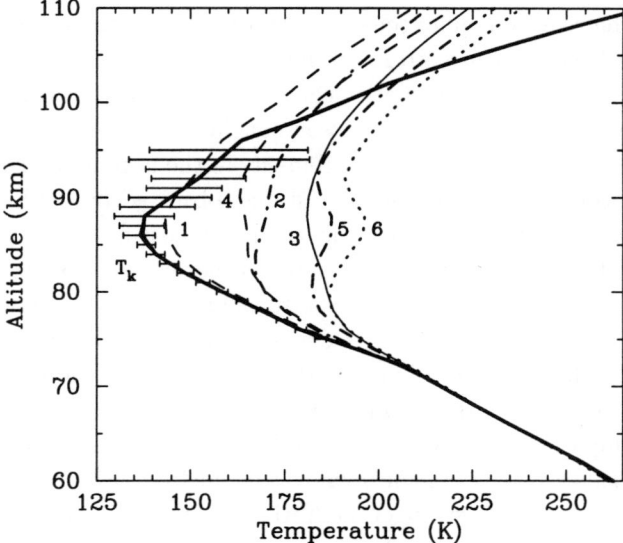

Fig. 8.6   Vibrational temperatures for the $CO_2(v_2)$ levels for a measured WINDII kinetic temperature profile. Curves 1, 2 and 3 denote the $(0,1^1,0)$, $(0,2,0)$, and $(0,3,0)$ levels of the 626 isotope, respectively; 4, 5 and 6, represent the $(0,1^1,0)$ levels of the 636, 628 and 627 isotopes, respectively. See Table 6.1.

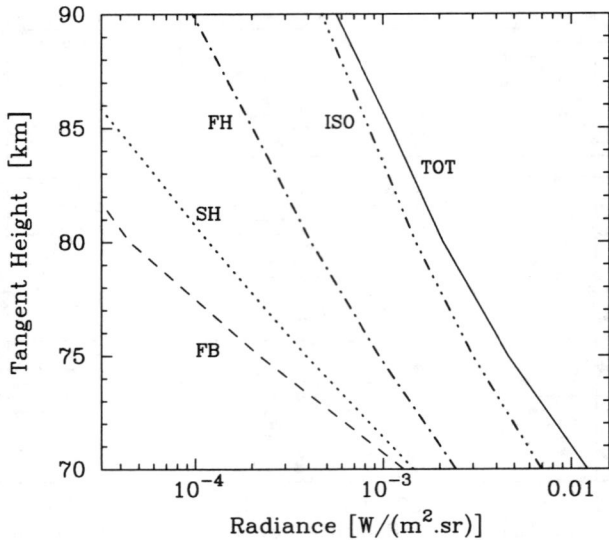

Fig. 8.7   The CO_2 ν_2 band contributions to the radiances measured by ISAMS channel 30W for a WINDII kinetic temperature profile. The contributions are shown for the 626 fundamental band (FB), the 626 first hot bands (FH), the 626 second hot bands (SH), the fundamental bands of the minor 636, 628 and 627 isotopes (ISO), and the sum of all bands (TOT). See Table F.1.

alternative explanation to non-LTE could be that the CO_2 mixing ratio is under-estimated, but the required enhancement is well beyond currently accepted values. Also, the WINDII temperature errors produce a change in the computed radiances (see Fig. 8.5), but this is much smaller than the difference between LTE and the ISAMS measured radiances. Hence, it was concluded that such large differences can only be explained by non-LTE enhancements in the populations of the $CO_2(v_2)$ levels (including the overtone and isotopic states) observed by ISAMS.

The non-LTE models also predict that the bands arising from these levels give rise to a net radiative heating around the summer mesopause (see Sec. 9.2.1.1). The measurements can therefore be considered also as indirect evidence of such heating.

On April 29–30 1991 the CIRRIS-1A experiment observed the CO_2 15 μm emission globally for different latitudinal, diurnal and geomagnetic conditions in the altitude range from 80 to 170 km. The measurements showed an enhancement by a factor 2–5 at high latitude (where a class III aurora had developed) relative to mid and low latitudes. This enhancement occurred from 100 to 160 km, and was thought to be due, not to a stronger excitation of the $CO_2(v_2)$ levels, nor to the atomic oxygen abundance which is known to diminish during auroral activity, but to an increase in the CO_2 concentration at high latitudes, caused by upwelling forced by Joule heating at 120–130 km. A large enhancement in the CO_2 15 μm emission was observed at 140–160 km between 04:00 and 06:30 local time, again probably due to increased CO_2 upwelling and temperature changes resulting from semidiurnal tidal effects.

8.3.2 *Observations of the CO_2 10 µm emission*

The Cryogenic Limb Array Etalon Spectrometer (CLAES) scanned the spectral range from 921–925 cm^{-1} with moderate spectral resolution (0.25 cm^{-1}) to make stratospheric measurements of CF_2Cl_2 and aerosol. However, in the upper stratosphere and above, some lines of CO_2 and O_3 dominate the signal in this spectral region, enabling measurements of those species in the 35–60 km altitude range.

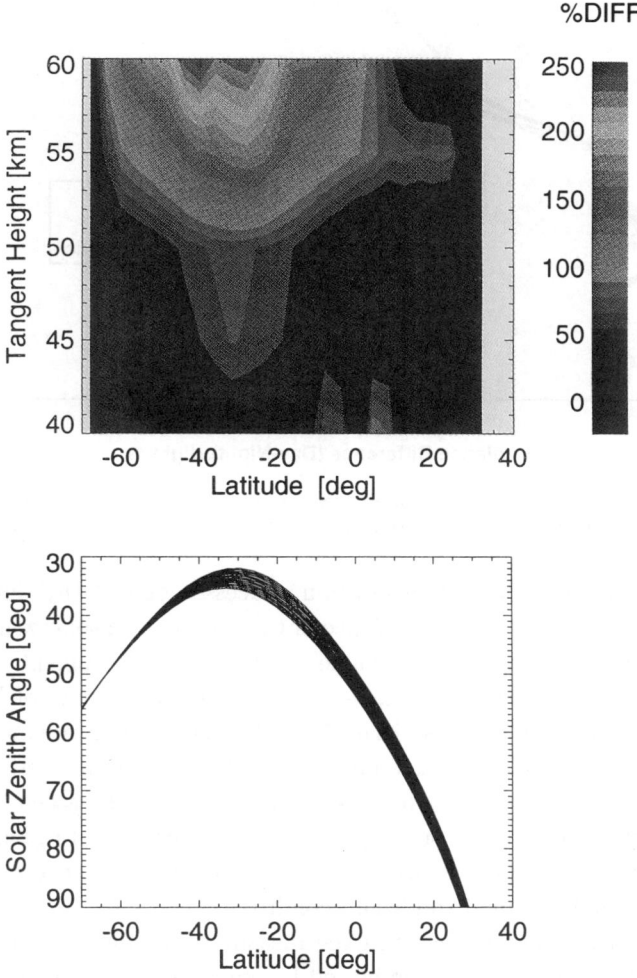

Fig. 8.8   Top panel (see also Plate 1): Diurnal difference in CLAES 10 µm radiance as a percentage of the night-time value [(day-night)/night%]. Day 92-02-12. Bottom panel: Corresponding daytime solar zenith angle variation.

The individual contributions of the 626 and 636 isotopes of CO_2, and two hot bands of O_3 ($\Delta v_3 = 1$ from $v_3 = 5$ and 4), can be discriminated in the data. This is consistent with non-LTE model calculations of the populations of the CO_2 626 and 636 $(0,0^0,1)$ states, particularly in the lower mesosphere and upper stratosphere.

The CLAES measurements also showed further experimental evidence for the excitation of the CO_2 ν_3 mode by energy transfer from electronically excited $O(^1D)$ through $N_2(1)$.

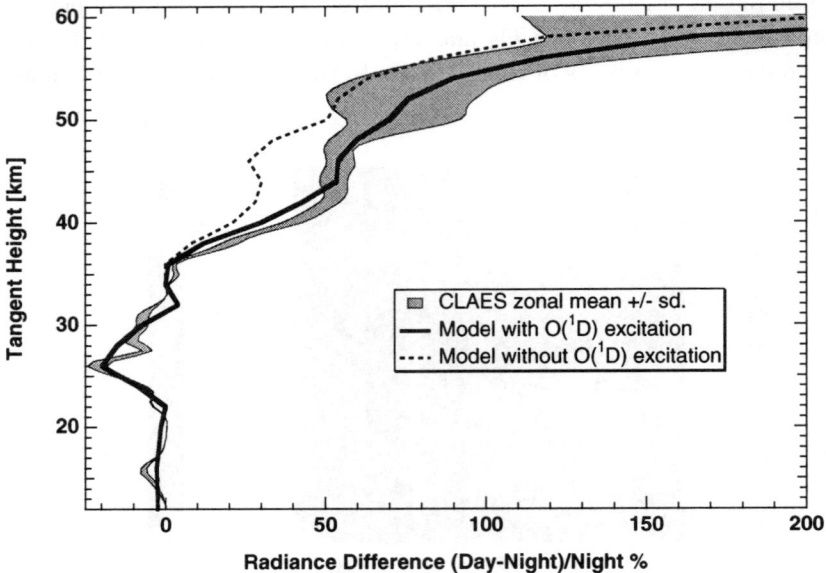

Fig. 8.9   Day-night CLAES radiance differences (mean±standard deviation) near $10\,\mu m$ as a function of tangent height for equatorial latitudes on 92-01-12. Model predictions are also shown.

Because the major non-LTE excitation processes of $CO_2(v_3)$ levels are solar driven, we expect to observe a dependence of the radiance on solar zenith angle (χ). During a UARS orbit, solar zenith angle at the observed tangent point changes with latitude along the orbital path, as shown in the lower part of Fig. 8.8. The upper panel in this figure (see also Plate 1) shows the diurnal radiance change as a function of latitude and altitude as a percentage of the night-time value. Where the difference approaches zero, the radiances measured during the ascending (day) and descending (night) parts of the orbit are equal. The contours of radiance difference follow the solar zenith angle variation, and the radiance difference is greater when the solar zenith angle is small (high Sun conditions).

Model calculations agree well with the measured day-night radiance difference profiles (Fig. 8.9). The signal is dominated by emission from the CO_2 626 $(0,0^0,1–1,0^0,0)$ band above 40 km, and the diurnal variation is due to a combination of the direct effect of solar incidence angle, and the indirect effect of the concentration of $O(^1D)$. The latter can be seen in Fig. 8.10, where the radiance difference as a function of solar zenith angle for a representative tangent height of 57 km is shown. The data shows a sharp decrease across the terminator, which is reproduced in the model calculations only when the effect of atomic oxygen is included, using the variation of the $O(^1D)$ concentration taken from a 3D chemical transport model.

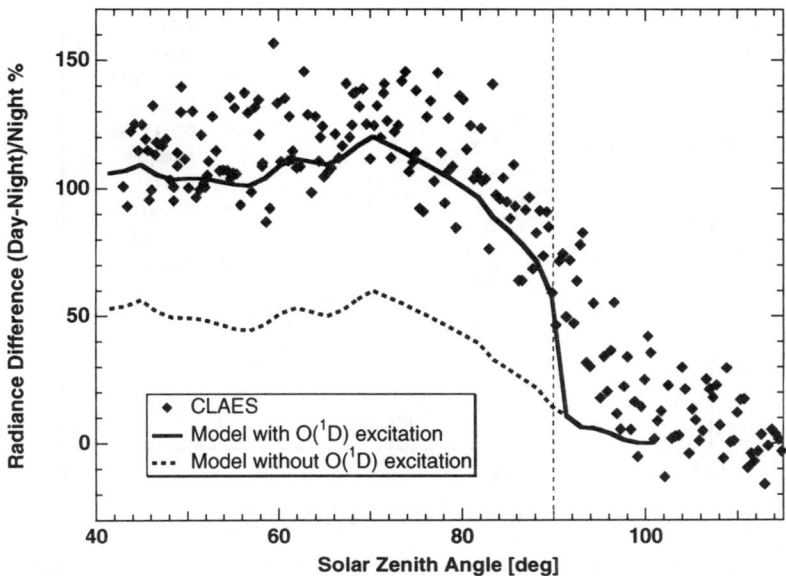

Fig. 8.10   CLAES radiance measurements near $10\,\mu m$ taken over the latitude range 20°S to 34°S on 92–01–12, at a limb view tangent height of 57 km. The diurnal variation is shown as a percentage of the nominal night-time radiance of $4.82\times10^{-6}\,\mathrm{W}/(\mathrm{m^2\,sr\,cm^{-1}})$.

Since $O(^1D)$ is a photolysis product of O_3 and O_2, its concentration profile is also strongly dependent on solar zenith angle.

The limb radiance at a tangent height of 57 km is also quite sensitive to any variation in the kinetic temperature. The data in Fig. 8.10 varies over 15 degrees of latitude, and the small increase in radiance between $\chi = 40$ and 80° (20 to 35°S) is a result of a slightly warmer lower mesosphere at the higher latitudes. The reason for the more gradual drop in radiance difference across the terminator in the data, as compared to the model, has a less obvious cause. It may be a limitation of the model, which calculates the vibrational temperatures for the solar zenith angle at the tangent point, whereas in fact the solar illumination may change significantly along the instrumental line of sight, particularly where the lines are optically thick.

8.3.3 CO_2 $4.3\,\mu m$ emission observed by rockets

Figure 8.11 shows a radiance profile for the emission from the CO_2 bands during the daytime, as measured during the SPIRE rocket campaign. Note the good agreement with non-LTE model calculations for the appropriate illumination conditions. Again, this gives us confidence to dissect the signal, using the model, to see the relative contributions of the various bands. Notice the large contribution of the weak bands originating from the $(v_1,v_2,1)$ levels (bands 4, 5, and 6 in Fig. 8.13) below 100 km. Above 110 km, the 626 fundamental band is almost entirely responsible for the observed radiance.

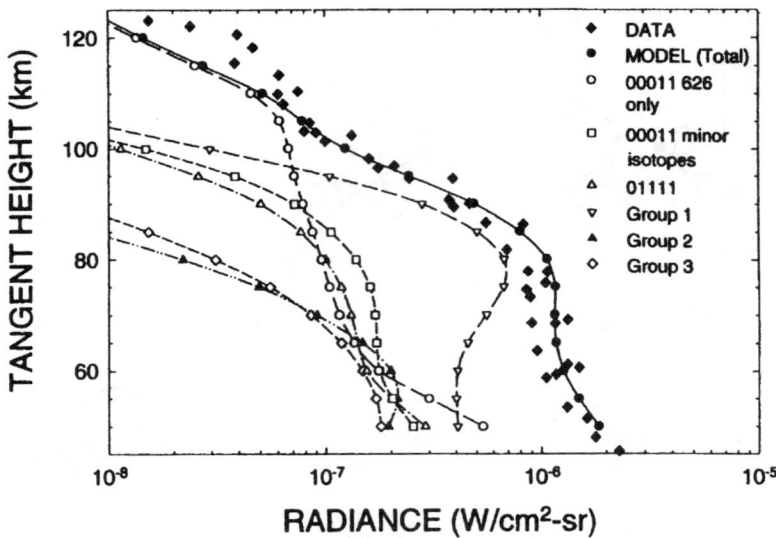

Fig. 8.11 Comparison of daytime SPIRE radiances in the band 4.12–4.49 μm with daytime model calculations. The solar zenith angles, χ, of all the scans shown vary from 78° to 84° for a tangent height of 80 km. Contributions of different CO_2 bands near 4.3 μm to the total limb radiance are also shown.

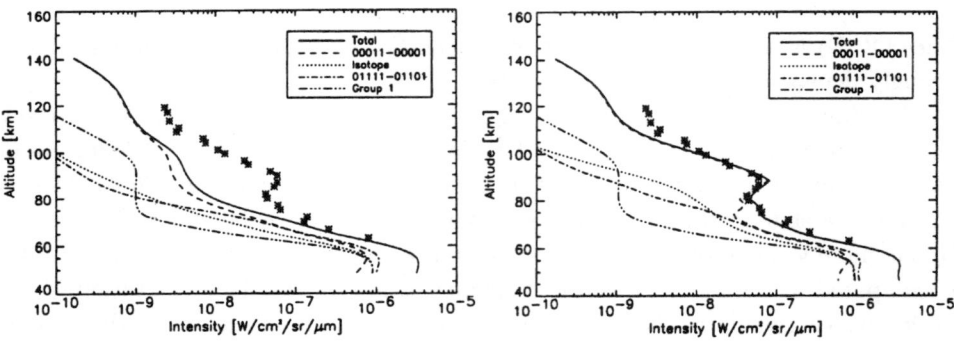

Fig. 8.12 SISSI–3 CO_2 4.3 μm measured and modelled zenith radiance of the spectral peak of the CO_2 band at 4.26 μm for twilight conditions ($\chi = 94°$). Spectral resolution is 0.04 μm. Stars are the measurements and lines the calculations. Left: Calculations including only solar excitation. Right: Calculations including also the excitation of $N_2(1)$ from $O(^1D)$.

Figure 8.12 shows zenith radiances at the peak of the CO_2 4.3 μm band under twilight conditions ($\chi = 94°$), as measured by the SISSI–3 experiment with a resolving power of $\lambda/\Delta\lambda = 100$. From comparison of the left and right frames in the figure, the importance of the $O(^1D)$ excitation of $N_2(1)$ (process 12 in Table 6.2) and its subsequent excitation of CO_2 near 4.3 μm is evident. This figure also shows that the fundamental band of the main isotope is the major contributor above 70 km under twilight conditions. The measured radiance above about 110 km is not explained even when the $O(^1D)$ excitation mechanism is invoked. Similar deficits in explaining the 4.3 μm radiances have also been found in the analysis of the SPIRE measurements (see Fig. 8.11) and of the ISAMS measurements (Figs. 8.18, 8.19). This could be due to a an error in the CO_2 number density, which is not well understood at these heights, or it might be that another non-LTE radiator, e.g., $NO^+(v)$.

8.3.4 *CO_2 4.3 μm emission observed by Nimbus 7 SAMS*

The Stratospheric and Mesospheric Sounder (SAMS) measured temperature and composition profiles of several gases in the altitude range of 15–80 km by detecting thermal emission, fluorescence and resonantly scattered sunlight. It also had a channel designed to study non-LTE effects in carbon dioxide, since absorption and emission by CO_2 is the main factor controlling the energy budget of the mesosphere, and non-LTE processes play an important part. This channel observed the CO_2 limb infrared emission at 4.3 μm for over four years, for both day and night conditions, providing measurements of the vibrational temperatures of CO_2 excited in the ν_3 mode over an extensive altitude range (30–100 km). This near-global survey of CO_2 4.3 μm atmospheric radiance is still one of the most comprehensive data sets of its kind obtained to date. It provides an exceptional opportunity to test non-LTE models and to study in detail and under different atmospheric situations the effects of non-LTE on the vibrational levels of carbon dioxide.

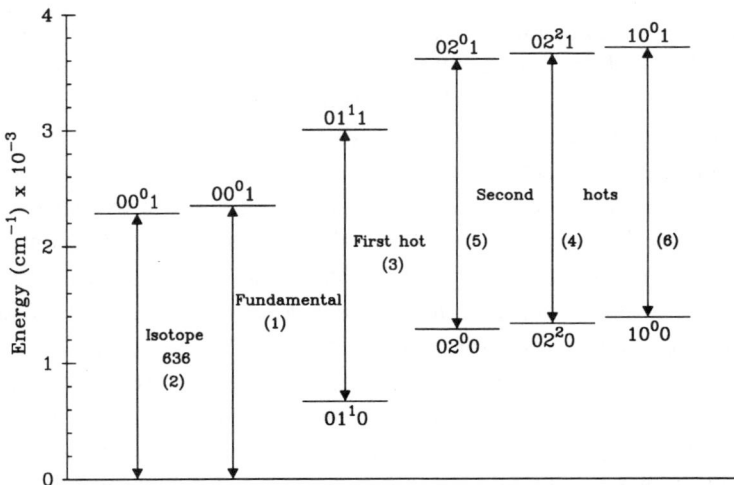

Fig. 8.13 Energy levels and transitions associated with the main CO_2 4.3 μm bands contributing to SAMS radiances.

SAMS measurements contain information about the population of the $(0,0^0,1)$, $(0,1^1,1)$, $(0,2^0,1)$, $(0,2^2,1)$, and $(1,0^0,1)$ vibrational levels of the most abundant 626 isotope, and the $(0,0^0,1)$ level of the 636 isotope, and their diurnal, latitudinal, and seasonal variations (Fig. 8.13). From the data we can deduce which processes dominate the population of these levels under different conditions.

SAMS used the pressure modulation technique to select the emission of a specific gas. The basic principles of this technique are quite simple: a pressure modulated cell (PMC) filled with CO_2 act as an optical filter. The varying absorption in the spectral lines of the gas in the PMC 'labels' the emission from the atmosphere, which of course occurs in the same spectral lines. At the typical conditions of pressure, temperature and illumination, lines in the PMC are in LTE. The original radiance of the atmospheric emission can be worked out from the component of the signal at the detector which is at the frequency of the pressure modulator. Allowance must be made for the latitude-dependent speed of the Earth's surface and atmosphere relative to the spacecraft, which has a small effect on the radiances by Doppler-shifting the emission lines. For this reason, PMC radiometers must view sideways to the orbit to minimize the Doppler shift.

SAMS obtained the vertical distribution of kinetic temperature by measuring the CO_2 15 μm emission, thus providing simultaneous measurements of temperature and 4.3 μm radiance profiles. Broad spectral selection of the 4.3 μm band used a 3.6 to 5.9 μm (2800–1700 cm^{-1}) filter. Within these limits lies not only the CO_2 fundamental band, but also a large number of isotopic and hot bands of the same molecule. The use of a PMC gives high effective spectral resolution, of the order of 0.1 cm^{-1}, but lacks the ability to discriminate between vibrational bands arising from the same gas which fall within the overall spectral band. Hence, as for the rocket experiments discussed above, the PMC signal originates in the CO_2 fundamental, isotopic and hot bands simultaneously, but in different proportions due to the use of the modulator as a filter. This means that SAMS and, for example, SPIRE radiance profiles are not directly comparable.

With the large amount of data generated by a satellite instrument, it is convenient to average the measurements zonally over latitude and over the seasons. This smoothes out instrument noise and daily fluctuations, and provides a seasonal data set which is readily compared to model calculations. Annual fluctuations are also reduced by averaging over the four and a half years of measurement, but in fact, if the radiances for each month of the different years are compared, the difference is found to be less than the measurement noise.

The data are regularly distributed amongst the seasons except for autumn, when the instrument was observing the atmosphere on a significantly smaller number of days, and they are approximately uniformly distributed over each season, except for autumn, when they were taken early in that period. The Sun-synchronous orbit of the Nimbus 7 satellite allowed SAMS to observe the atmosphere at a given tangent point at approximately the same local time for the whole period of observation.

The SAMS CO_2 4.3 μm radiances exhibit large differences between daytime and night-time which are always present. An example is shown in Fig. 8.14 for latitudes 10°S–30°S in winter. Because the instrument use an internal blackbody source operating at 290 K, radiances have been given as a fraction of the blackbody emission at this temperature, and plotted versus the tangent height. The altitude error is

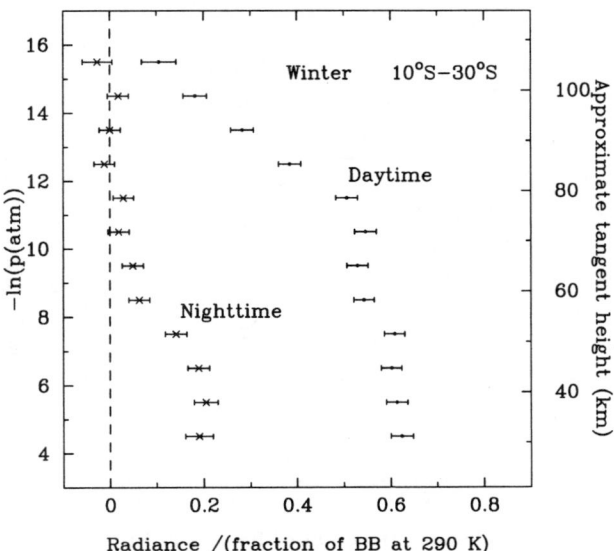

Fig. 8.14   Zonal and seasonal mean radiance profiles for night-time and daytime as measured by SAMS. The horizontal bars denote measurements and their 1-σ errors.

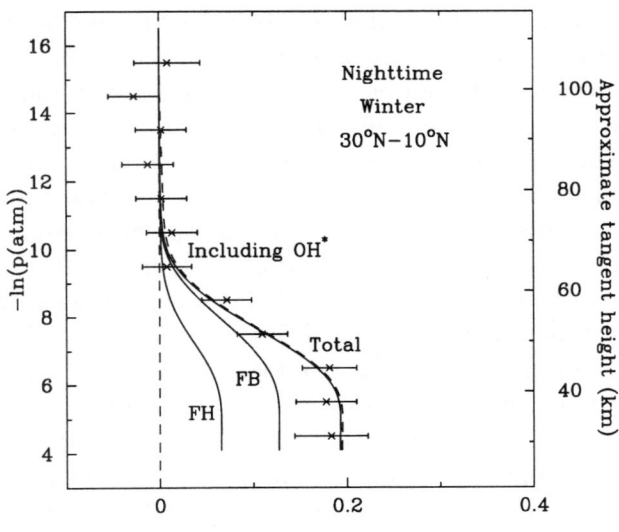

Fig. 8.15   Comparison between measured and calculated night-time radiances for the SAMS 4.3 μm channel, showing the contributions of the fundamental (FB) and first hot (FH) bands to the calculations (see Fig. 8.13).

only 0.37 km, but the vertical resolution is more like 1.2 scale heights (about 8.4 km), so the data are vertically averaged over that interval.

The night-time populations of the CO_2 levels excited in the ν_3 mode are in LTE up to approximately 65 km and slightly greater than LTE up to ~95 km, and then

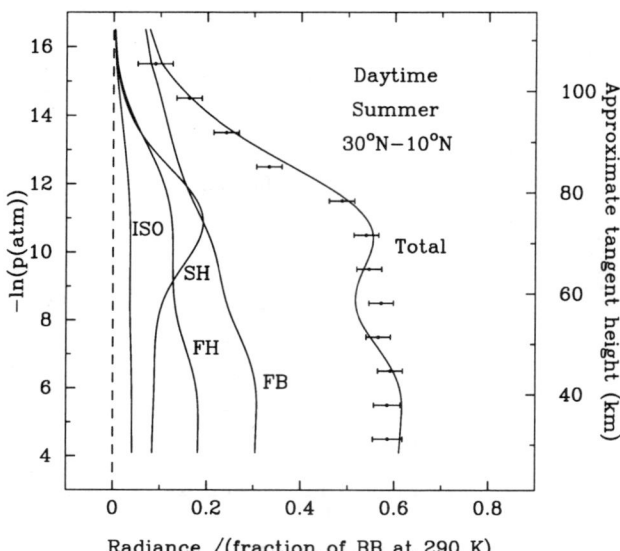

Fig. 8.16 CO_2 bands contributions to the daytime radiances for the SAMS $4.3\,\mu m$ channel.

smaller above (see Fig. 6.11). The daytime radiances, on the other hand, are very large above 65 km and show a variation with latitude, as would be expected because of the solar zenith angle dependence, as discussed in Sec. 6.3.6.3.

SAMS $4.3\,\mu m$ radiances are close to zero at night for altitudes above around 70 km. Fig. 8.15 shows the measured and calculated radiances for winter at several latitudes. Only two of the various bands near $4.3\,\mu m$ actually contribute significantly, the fundamental and the first hot bands. General agreement between observed and calculated radiances within the measurement errors is found in all cases. The effect of including the excitation of $N_2(1)$ from $OH(v \leq 9)$ (process 13 in Table 6.2) as predicted by Kumer *et al.* (1978) is below the sensitivity of SAMS.

A typical daytime profile is shown in Fig. 8.16, and the individual contributions of the six most important bands (see Fig. 8.13) are also displayed. The most prominent is the resonant transition of the most abundant CO_2 isotope, followed by fluorescence from $(0,1^1,1)$, $(0,2^0,1)$, $(0,2^2,1)$, and $(1,0^0,1)$, and the resonant emission from the $(0,0^0,1)$ isotopic levels. Fluorescence from the second hot band actually produces a larger net contribution than the fundamental over a narrow height range near 75 km.

Figure 8.17 shows the latitudinal variation of the measured and calculated radiance during Northern Hemisphere winter. The largest radiances at all heights are obtained for 10°S–30°S, where the Sun is overhead, and they decrease towards higher latitudes, where the Sun is closer to the horizon. Again, quite good agreement is found between model and data over most of the altitude range for all the different latitudes.

The variable parameters introduced in the calculation of the vibrational temperatures for the three latitude boxes are the kinetic temperature profile and the

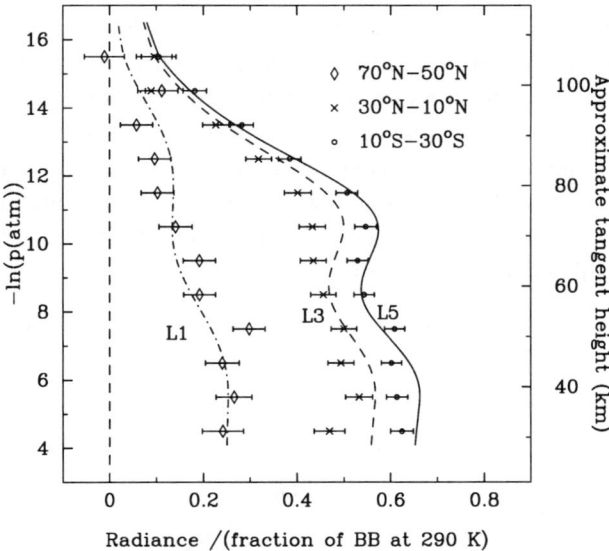

Fig. 8.17 Comparison between measured and calculated radiances for the SAMS 4.3 μm channel at different latitudes for the northern winter season. The radiances show the variation as a function of the solar zenith angle.

solar zenith angle χ, given by the expression

$$\cos(\chi) = \sin(\lambda)\,\sin(\delta) + \cos(\lambda)\,\cos(\delta)\cos\left[(t-12)\pi/12\right],\qquad(8.18)$$

where λ is the latitude, δ is the declination, and t is the local time of the tangent point location; averaging the cosine of the solar zenith angles over the latitudes in the box and over each day of observation, weighted by the number of atmospheric views taken by SAMS. The daytime vibrational temperatures of the upper levels of all bands are essentially determined only by solar illumination at altitudes above 65 km. Thus their vibrational temperatures at those heights are larger at lower latitudes than at near-polar latitudes. Below 65 km, however, the vibrational temperatures are dominated by thermal processes. They are also greater at lower latitudes in this season, but this is because the stratopause is warmer at equatorial latitudes in winter.

The solar zenith angle effect is also apparent when the radiances measured at the same altitude in different seasons are compared. For instance, at near-polar northern latitudes, the winter-time radiances are smaller than those measured during summer, and good agreement between observations and the predictions of the non-LTE model is again found.

8.3.5 CO_2 4.3 μm emission observed by UARS ISAMS

The ISAMS channel designed for determining the CO abundance in the upper stratosphere and lower mesosphere actually had many more CO_2 than CO lines in the spectral interval defined by its filter, which was located in the R branch of the $CO(1-0)$ band, centred at $2220\,cm^{-1}$ (near $4.6\,\mu m$) with half power points of 2170 and $2260\,cm^{-1}$. The pressure modulator preferentially selects the CO lines in order to determine the abundance of that species, but there is no such selection in the wideband signal. Thus, this signal is dominated by the emission from no less than 32 bands of CO_2 (see Table 8.2), with significant contributions from the $N_2O(\nu_3)$ and two $O_3(\nu_1+\nu_3)$ fundamental bands, and all of these contribute to the observed day-night differences (see Fig. 8.18).

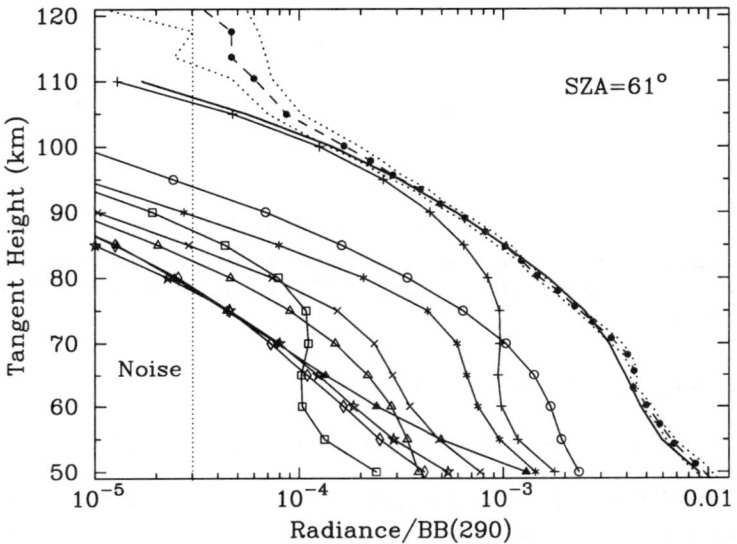

Fig. 8.18   Measured and calculated profiles of ISAMS wideband radiance near $4.6\,\mu m$ for September 29, 1991, $\lambda = 20°S$, $\chi = 61°$. Dashed line with solid circles represents measurements, and dotted lines are the measurement errors ($\pm 1\sigma$). Solid lines are the contributions of the bands in Table 8.2: 636 FB ($+$), 636 SH ($\circ$), 636 FH ($*$), 638 FB ($\times$), 626 FRH ($\triangle$), 626 SH ($\square$), 626 FB+FH+TH+628 FB ($\diamond$), 636 TH+FRH+638 FH (open stars), and O_3, N_2O, and CO bands (solid triangles). The vertical dotted line is the instrument noise [$\sim 3.4 \times 10^{-5}\bar{B}(290\,K)$].

The measurements show a high signal to noise ratio up to about 120 km in the daytime and up to 100 km at night-time. ISAMS measurements have advantages over the CIRRIS-1A space experiment because they extend over a longer period of time, and over rockets experiments, because ISAMS observed both the day and night-time atmosphere, as well as obtaining global and seasonal coverage. Another advantage over most previous experiments which observed CO_2 in non-LTE is that the kinetic temperature could be obtained from simultaneous $15\,\mu m$ radiance measurements.

Above about 50 km, the radiances are much brighter during the day, and vary strongly with the solar zenith angle (see Plate 2), indicating a non-LTE source. Lower down, the radiances are correlated with the kinetic temperature, which means they are of thermal origin. In addition to the variations due to the isotopic and hot

Table 8.2   Bands contributing to the ISAMS CO wideband channel.

CO_2 group[†]/ Molecule	No.	Isotope	Upper level	Lower level	$\tilde{\nu}_0$, cm^{-1}	Strength[*]
626 FB	1	626	00^01	000	2349.143	6
626 FH	2	626	01^11	01^10	2336.632	103
626 SH	3	626	02^01	02^00	2324.141	617
	4	626	02^21	02^20	2327.433	197
	5	626	10^01	10^00	2326.598	227
	6	626	02^01	10^00	2224.656	87
626 TH	7	626	03^11	03^10	2315.235	1924
	8	626	03^31	03^30	2311.667	2889
	9	626	11^11	11^10	2313.773	2296
626 FRH	10	626	04^01	04^00	2305.256	2757
	11	626	04^21	04^20	2302.963	6860
	12	626	12^01	12^00	2306.692	2367
	13	626	04^41	04^40	2299.214	9470
	14	626	12^21	12^20	2301.053	8110
	15	626	20^01	20^00	2302.525	3569
636 FB	16	636	00^01	000	2283.488	153
636 FH	17	636	01^11	01^10	2271.760	53
636 SH	18	636	02^01	02^00	2261.909	387
	19	636	02^21	02^20	2260.051	820
	20	636	10^01	10^00	2262.849	372
636 TH	21	636	03^11	03^10	2250.694	1053
	22	636	03^31	03^30	2248.356	1102
	23	636	11^11	11^10	2250.605	1040
636 FRH	24	636	04^01	04^00	2240.536	643
	25	636	04^21	04^20	2239.297	1301
	26	636	12^01	12^00	2242.323	4777
	27	636	04^41	04^40	2236.678	1686
	28	636	12^21	12^20	2283.570	1282
	29	636	20^01	20^00	2240.757	619
628 FB	30	628	00^01	000	2332.113	0.37
638 FB	31	638	00^01	000	2265.971	1.4
638 FH	32	638	01^11	01^10	2254.380	4.2
N_2O	33	446	0001	0000	2223.757	$\sim$50000
O_3	34	666	200	000	2201.157	$\sim$30
O_3	35	666	101	000	2110.785	$\sim$1130
CO	36	26	1	0	2143.271	2422

[*]After weighting by the filter response and dividing by the LTE population of the lower energy level, at 296 K in cm^{-1}/(mol cm^{-2})$\times 10^{21}$. [†] FB, FH, SH, TH, and FRH mean fundamental, first hot, second hot, third hot, and fourth hot bands, respectively.

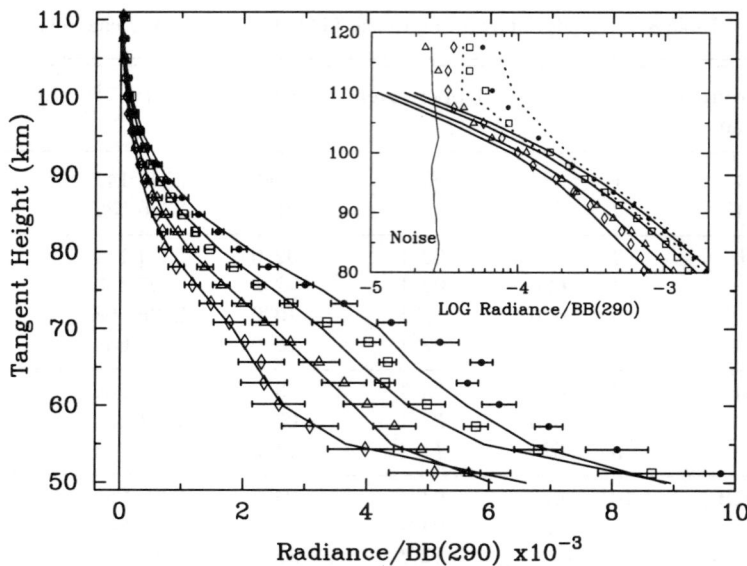

Fig. 8.19 Measured and calculated ISAMS wideband radiances near 4.6 μm for December 5, 1991, 10°S, $\chi = 25°$ (•); September 29, 1991, 20°S, $\chi = 61°$ (□); January 15, 1992, 25°N, $\chi = 82°$ (△); and July 20, 1992, 10°S, $\chi = 88°$ (◇). Error bars are the 1-σ standard deviation of measurements. Solid lines represent calculations. The inset represents the measurements at higher altitudes on a log scale. The two dotted lines denote the standard deviation (radiance $\pm 1\sigma$).

bands, the ISAMS observations allowed a measurement of the emission from the OH Meinel bands at night-time, since about 30% of the $v = 9$–8 band and about 5% of $v = 8$–7 fall within the ISAMS CO channel.

Figure 8.18 shows a comparison between model simulations and measurements, and the contributions of the different CO_2 bands to the ISAMS CO wideband channel, for a representative value of the solar zenith angle. Wideband measurements are relatively more sensitive to the CO_2 bands, and these take in the longer-wavelength tail of the 4.3 μm CO_2 complex, where many isotopic and hot bands that have seldom been measured are located. The fundamental of 636 is the most important at high altitudes, with the second hot band of 636 dominating below 70 km, with a large contribution from the weaker 636 hot bands and those of the other rare isotopes at stratospheric heights. The calculated total radiance agrees well with the measurements up to 105 km, except near 65 km, where a peak in the atmospheric temperature may not have been accurately registered in the ISAMS temperature retrieval. Fig. 8.19 shows the solar zenith angle dependence from nearly overhead conditions ($\chi = 25°$) to near twilight ($\chi = 88°$). The model calculations again follow the measurements fairly well, especially for the larger solar zenith angles.

In both plots, however, it can be seen that the measured radiances are significantly higher than the calculations above 105 km. The difference is too large to be caused by errors in the modelling of the major contributor, the CO_2 636 4.3 μm fundamental band, or by the uncertainty in the abundance of CO_2. More likely,

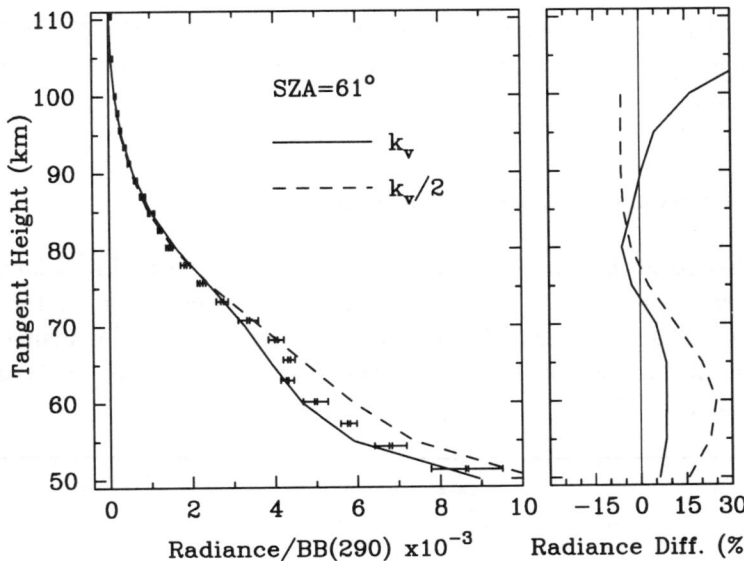

Fig. 8.20   Effect of the vibrational–vibrational exchange rate of the $v_3 = 1$ quanta between CO_2 $(v_1,v_2,1)$ and $N_2(1)$ states on the modelled ISAMS wideband radiance near $4.6\,\mu m$ for September 29, 1991, $\lambda = 20°S$, $\chi = 61°$. The right panel shows the differences between modelled $k_{vv}/2$ and k_{vv} radiances (dashed line), and between measured and modelled–k_{vv} radiances (solid line).

it is a contribution from the vibrationally excited NO^+ bands, which span from 2100 to 2400 cm^{-1}, and have also been detected in the measurements taken by the CIRRIS-1A experiment.

The effect of non-LTE on the vibrational temperatures of the various molecular levels was discussed at length in Chapter 6. The transfer of vibrational energy between $N_2(1)$ and the $CO_2(v_1,v_2,1)$ combinational levels (process 9 in Table 6.2), has a large effect on the CO_2 4.3 μm atmospheric limb radiances. Fig. 8.20 shows the effect on the radiances of reducing the nominal rate coefficient of process 9, k_{vv}, by a factor of 2. The slower rate predicts radiances up to 25% higher at tangent heights below about 70 km, as a consequence of the larger excitation of the high-energetic levels. In other words, solar energy is thermalized less efficiently before being emitted for smaller k_{vv} rates. The best overall fit to the ISAMS measurements results if the rate coefficient is reduced by about 20%, but this overestimates the radiance at larger solar zenith angles.

Below about 75 km the key process, apart from solar pumping and the V–V $N_2(1)$–CO_2 exchange, is excitation by $O(^1D)$ controlled by the efficiency of the $O(^1D)\rightarrow N_2(1)$ transfer and the highly variable atomic oxygen abundance. The ISAMS radiances are also sensitive to the CO_2 volume mixing ratio (VMR), which ceases to be constant with height above about 70 km. The simulation is good when the CO_2 VMR derived from previous SAMS 4.3 μm measurements in the upper mesosphere and lower thermosphere, updated to take account of the secular increase to 350 ppmv at altitudes below 75 km, is used. This profile is significantly lower than the mean of a number of rocket measurements in the 75–100 km region and only slightly smaller than the values derived from ATMOS data (see Sec. 8.10.3 and Fig. 8.32).

8.4 Observations of Ozone in Emission

Ozone emissions in the infrared near $10\,\mu m$ have been measured by a large number of experiments (see Table 8.1). Early measurements were taken by the ICECAP rocket experiment in the early 1970s, and were reasonably well explained by non-LTE models on the basis of the O_3 formation from recombination of O_2 and $O(^3P)$ (process 1 in Table 7.4).

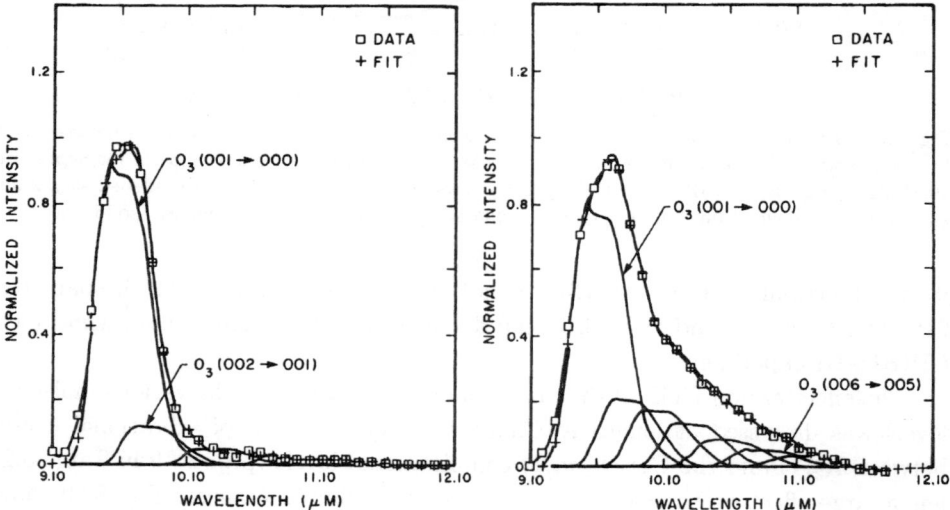

Fig. 8.21 SPIRE measurements of the O_3 $9.6\,\mu m$ radiance. Left: Spectral radiance at 98 km at night-time. The model fits show that the peak emission is mainly from the ν_3 fundamental, with a smaller contribution from the hot bands. Right: Spectral radiance at 90 km under sunlit conditions. The emission from the fundamental is little altered, but the hot band contributions are enhanced.

The SPIRE rocket observations of the atmospheric spectrum between wavelengths of 9 to $12\,\mu m$ included the ν_3 bands of ozone. By launching just before dawn, SPIRE was able to observe both the day and night-time radiances. Fig. 8.21 (left) shows the measured spectrum at 98 km altitude at night, and Fig. 8.21 (right) that at 90 km when the atmosphere was illuminated by the Sun, both fitted with the calculated radiance from a non-LTE model. The day-night difference is mostly due to the higher-order ν_3 bands, as might be expected, and the experimenters found that this difference vanishes below about 60 km, from which they deduced that LTE applies at these levels.

Several recent model studies of the non-LTE populations of the O_3 v_1, v_3, overtone, and combination levels have confirmed that non-LTE effects in the ν_1 and ν_3 bands of ozone are important for tangent heights above around 60 km. In particular, non-LTE corrections were needed above this height for the retrieval of ozone mixing ratio from the LIMS ozone channels at 9.6 μm, and the 10 μm O_3 daytime radiances measured by ISAMS are clearly enhanced relative to night-time in the same region (Plate 3). The enhancement is less than 5% for pressures greater than 1 mb but increases with decreasing pressure, yielding values as high as 50% of the night-time radiance at 0.1 mb.

The SISSI rocket experiment team achieved very good agreement between model calculations and measured ozone radiances in the upper mesosphere and thermosphere. These payloads, flown in 1990 and 1991, included a simultaneous measurement of the atomic oxygen abundance using a resonance fluorescence technique. The models assumed that the radiance is due to non-thermal emission from chemically pumped O_3 from the ozone formation reaction (process 1 in Table 7.4), and calculated the O_3 vibrational temperatures assuming that the production of vibrationally excited O_3 is restricted to the ν_3 mode, with the populations given by the 'zero surprisal' nascent distribution (see Sec. 7.3.2).

Calculations using the model described in Chapter 7 predict a daytime radiance enhancement in the ozone band which agrees with the ISAMS measured values within a few percent for pressures > 0.2 mb. In this case, the O_3 abundances used in the calculation were retrieved from the UARS Microwave Limb Sounder (MLS) measurements, assumed to be unaffected by non-LTE at the long wavelengths used by MLS to observe pure rotational transitions at around 1.5 mm. The principal uncertainties arise from the errors in the MLS retrieval, and from uncertainties in the nascent distribution and in the relaxation rate of $O_3(v_1,v_2,v_3)$. However, since ISAMS is most sensitive to the emission from the (0,0,2) and (0,0,1) levels and since the collisional relaxation of these is quite well understood experimentally, some conclusions about the nascent distribution can be drawn. The ISAMS results appear to show that the higher energy levels are populated in the ozone formation reaction.

Given a reasonable model for the source function, it becomes possible to measure ozone up to the mesopause with a modern limb-viewing remote sensing instrument. Fig. 8.22 shows calculated radiance profiles for the ozone channel in the High Resolution Dynamics Limb Sounder (HIRDLS) on the EOS-Aura mission, scheduled for launch in 2004. This channel, one of three used to measure ozone, extends from 984 to 1016 cm^{-1} and so is very similar to the ISAMS ozone channel. The figure shows the daytime non-LTE and LTE profiles separately, and also distinguishes the fundamental and hot band contributions of O_3 and the CO_2 10 μm bands to the non-LTE radiance. The predicted signal is above the instrumental noise level up to about 95 km when non-LTE pumping of the upper states is taken into account, but only up to about 75 km in LTE. This is an example of those cases, mentioned above, where non-LTE significantly improves the performance of a measurement system.

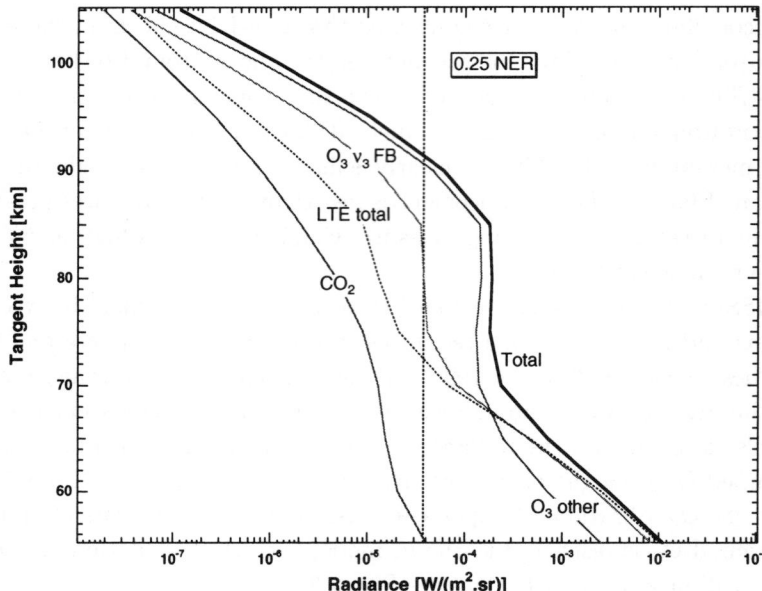

Fig. 8.22   Profiles of O_3 radiances in the 984–1016 cm^{-1} channel of the HIRDLS limb-viewing instrument, showing the contributions of the fundamental and hot bands, in and out of LTE. The contribution to the signal of the main contaminant, CO_2, is also shown. The vertical line represents a quarter of the instrumental noise, so useful single-profile measurements lie to the right of this line.

Other examples of the use of measurements of the O_3 non-LTE emission for retrieving its abundance in the mesosphere are the SABER and MIPAS instruments on the NASA TIMED and ESA ENVISAT missions, respectively.

8.5   Observations of Water Vapour in Emission

Early measurements of emission from H_2O at 6.3 μm were made by the SPIRE and MAP/WINE rocket experiments, and by LIMS on the Nimbus 7 satellite. The SAMS on the same spacecraft measured limb non-LTE daytime radiances from H_2O at 2.7 μm and derived the H_2O VMR from 30 to 100 km. More recently, the ISAMS data provides clear evidence of non-LTE in water vapour in two channels, one (like SAMS) using the pressure modulation technique, which is more sensitive to the fundamental (0,1,0–0,0,0) transition, and a wide band channel, which is more sensitive than the PMC channel to the first hot band (0,2,0–0,1,0).

Simultaneous measurements of the emission of the H_2O fundamental and of the O_3 abundance in the mesosphere can be used in principle to investigate the excitation of $H_2O(0,1,0)$ by V–V collisions with O_2, the latter previously excited by photodissociation of O_3. Any non-LTE enhancement of the mesospheric H_2O 6.3 μm emission is likely to be a sign of vibrationally excited O_2, since $O_2(1)$ and $H_2O(0,1,0)$ are strongly coupled by near-resonant V–V collisions (process 1 in Table 7.6). The

ISAMS measurements in the 6.3 μm band of water vapour indeed show systematic day-night differences above about 55 km, for both pressure modulated (PM) and wideband (WB) channels (see Fig. 8.23). The daytime signal is about twice that at night-time in the PM channel, and about 3 times in the in WB. These factors are too large to be accounted for by any reasonable change in the concentration of water vapour.

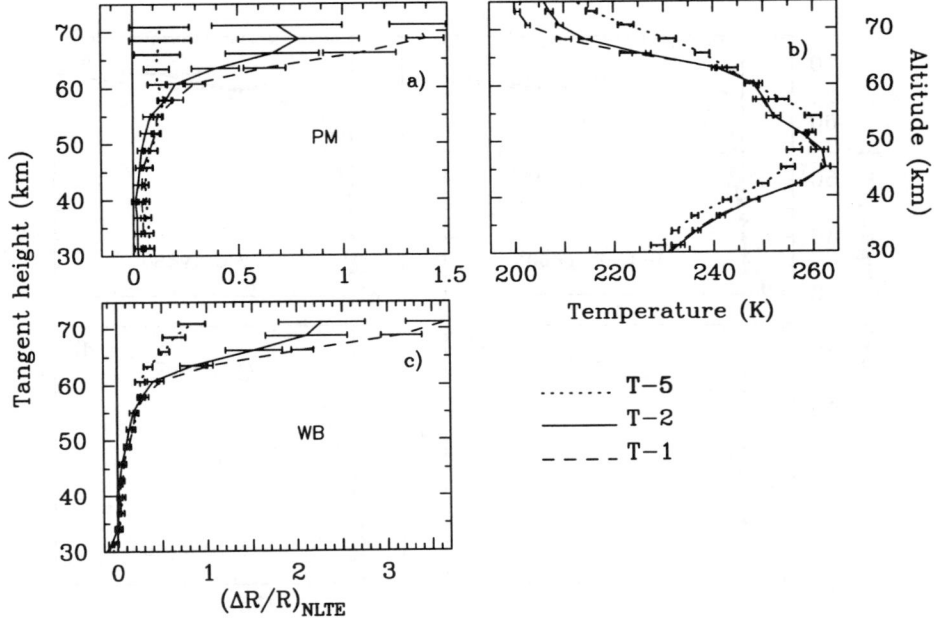

Fig. 8.23   The ISAMS H_2O 6.3 μm daytime radiance for several mesospheric kinetic temperature profiles [shown in frame b)]. Note the different scale in a) pressure modulated (PM) channel and c) wideband (WB) channel.

Figure 8.23 compares mean profiles of the daytime non-LTE radiance enhancement, $(\Delta R/R)_{\rm NLTE}$, defined as (Measured day non-LTE radiance$-$Estimated day LTE radiance)/Estimated day LTE radiance, for similar illumination conditions but different mesospheric temperatures. This illustrates the inverse dependence of the non-LTE daytime radiance enhancements on the kinetic temperature in the mesosphere, with larger enhancements for a colder mesosphere. On the other hand, the dependence of the non-LTE enhancements on solar zenith angle is of the same magnitude or smaller than the measurement errors. This lack of dependence on solar illumination is explained by the weak variation with solar zenith angle of the solar absorption in both the O_3 Hartley bands and the 2.7 μm H_2O bands.

   The model described in Chapter 7 includes the vibrational levels of atmospheric water vapour that give rise to emission at 6.3 μm and at 2.7 μm, so it can be used to analyse and interpret the ISAMS measurements. One of the most important processes is the near resonant V-V exchange between $H_2O(0,2,0)$ and $O_2(1)$. There is a variation of around an order of magnitude in the laboratory-determined rates for this process. During the day, $O_2(1)$ can be produced in the mesosphere after the photolysis of O_3 in the ultraviolet Hartley band. This excitation is counterbalanced by the thermal relaxation of $O_2(1)$ by collisions with $O(^3P)$, and affects only the $H_2O(0,1,0)$ population. Both the efficiency of the photochemical production of $O_2(1)$, ε, and the rate coefficient of the quenching with $O(^3P)$, k_{vt} (processes 23 and 21 in Table 7.6), are also very uncertain.

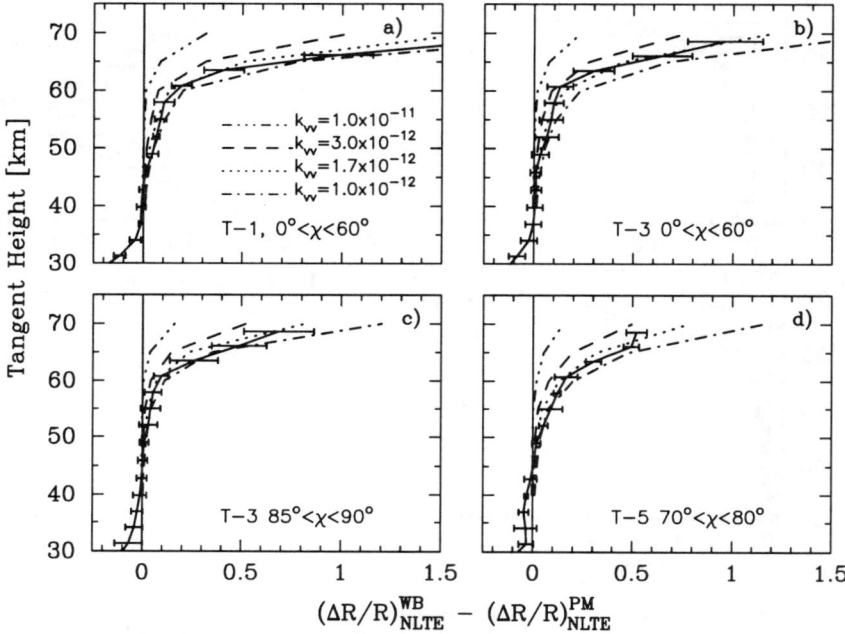

Fig. 8.24   Contribution of the H_2O first hot band to the non-LTE daytime radiance enhancements (see text) calculated for different rate coefficients of the vibrational coupling between $H_2O(v_2)$ and $O_2(1)$, k_{vv}. Solid line corresponds to the differences derived from ISAMS measurements. The four cases correspond to different solar zenith angles (χ) and mesospheric temperatures (see Fig. 8.23b).

As mentioned above, the ISAMS pressure modulator technique discriminates in favour of the fundamental ν_2 band while the wideband channel is sensitive to both the fundamental and the hot bands. The contribution of the fundamental band to the non-LTE daytime radiance enhancements is practically the same in both channels, so the hot band contribution is obtained by subtracting the two channels. Since the hot band, originating in $(0,2,0)$, is affected only by k_{vv} below about 70 km, the $(\Delta R/R)_{\mathrm{NLTE}}^{\mathrm{WB}} - (\Delta R/R)_{\mathrm{NLTE}}^{\mathrm{PM}}$ double difference gives the rate of the vibrational coupling between $H_2O(v_2)$ and $O_2(1)$ unambiguously. Once this is determined, the $(\Delta R/R)_{\mathrm{NLTE}}^{\mathrm{PM}}$ difference provides information about the factors affecting the $(0,1,0)$ level, in particular the quantum yield of $O_2(1)$ from O_3 photolysis and the rate coefficient of the thermal quenching of $O_2(1)$ with $O(^3P)$.

Figure 8.24 shows the contribution of the hot band to the non-LTE daytime radi-

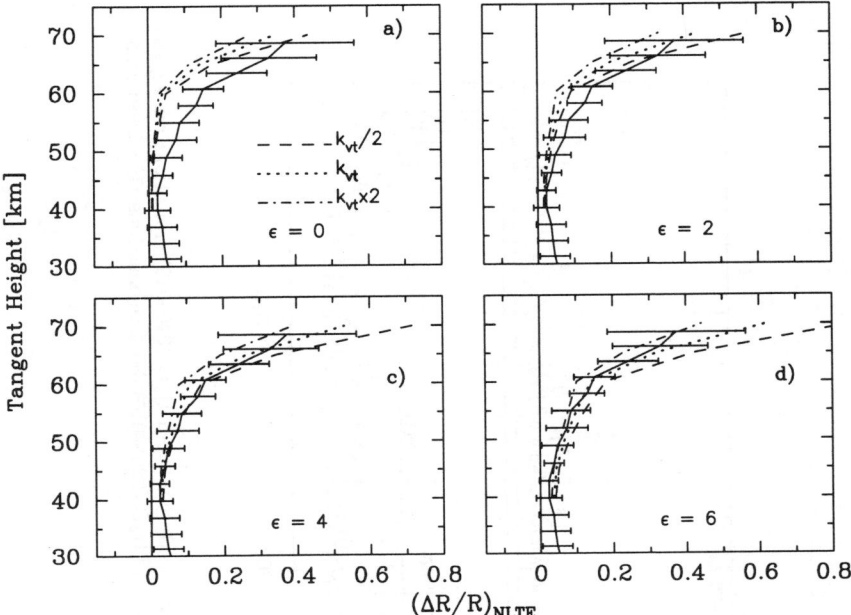

Fig. 8.25   Non-LTE daytime enhancements for the ISAMS H_2O pressure modulated (PM) channel calculated for the relaxation rate of $O_2(1)$ by $O(^3P)$, k_{vt}, in Table 7.6 (process 21) (dotted line), and k_{vt} multiplied and divided by 2 (dotted-dashed and dashed lines), for 4 values of the quantum yield for $O_2(1)$ from O_3 photolysis (ε). Solid line represents the differences derived from ISAMS measurements. This case corresponds to solar zenith angles between $0°$ and $60°$.

ance for four values of the coupling rate k_{vv}, covering the whole range of uncertainty. For larger k_{vv}, the population of (0,2,0) is smaller and hence the non-LTE enhancements are also smaller. The value of k_{vv} that best fits the measurements is close to the lowest value given in the literature, in the range $(1.0–3.0)\times10^{-12}$ cm^{3}s^{-1}, with a best value of 1.7×10^{-12} cm^{3}s^{-1}.

The quantum yield of $O_2(1)$ production after O_3 photolysis, ε, and the rate coefficient for the thermal quenching of $O_2(1)$ with $O(^3P)$, k_{vt}, have opposite effects on the population of $H_2O(0,1,0)$ in the 50–70 km region. However, the non-LTE daytime enhancement is more sensitive to the $O_2(1)$ quantum yield than to the quenching rate within its range of variability. The effect of these parameters on the populations of the vibrational levels of H_2O was discussed in Sec. 7.4.4. Fig. 8.25 shows how those changes affect the radiances measured by the ISAMS H_2O channel, for four values of the quantum yield, $\varepsilon = 0$, 2, 4, and 6, and for three values of the k_{vt} rate coefficient. The best overall fit occurs for the intermediate deactivation rate and a quantum yield of 4 (those in Table 7.6), although values as low as 2 and as high as 6 can also reproduce the observed non-LTE enhancements inside their error bars in most cases.

Non-LTE effects in mesospheric water vapour were also observed from the space shuttle by the CIRRIS-1A cryogenic Michelson interferometer (see Fig. 8.26). The

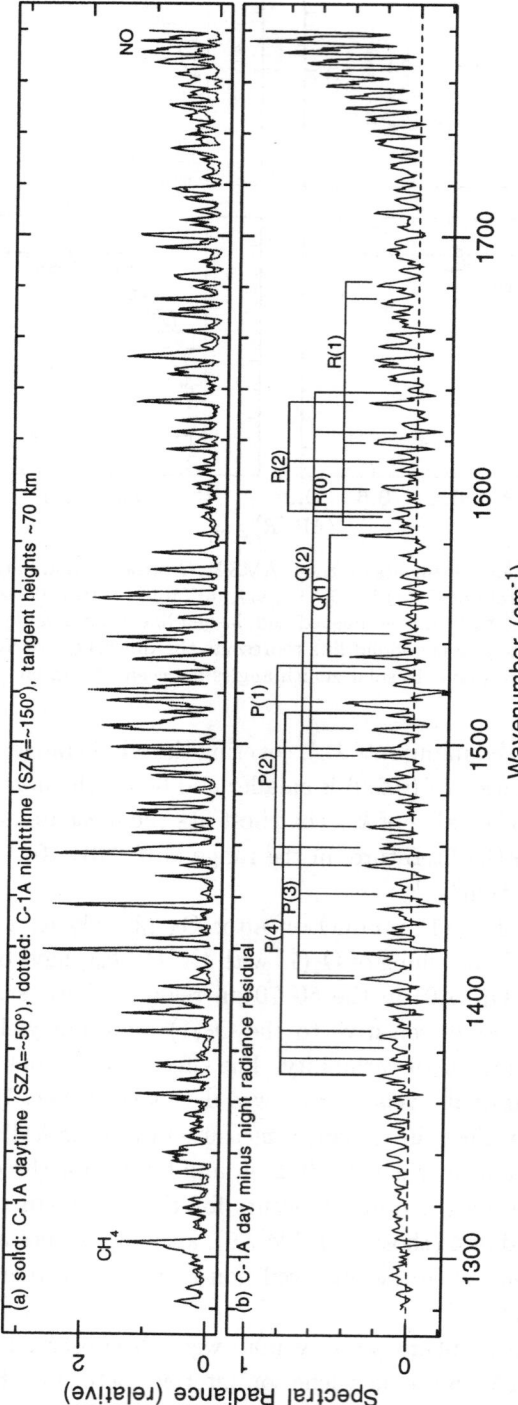

Fig. 8.26 (a) CIRRIS-1A observed day (solid) and night (dotted) spectra showing significantly enhanced H_2O (020–010) transitions during the daytime. (b) The residual of daytime minus night-time spectra, illustrating the contribution of H_2O (020→010) daytime emissions. The band origin gap and some of the P, Q and R transitions are identified.

spectra from 1280 to $1780 \, \mathrm{cm}^{-1}$ are dominated by four major emissions: (1) the $CH_4 \, \nu_4$ band centred at $1306 \, \mathrm{cm}^{-1}$; (2) the H_2O (0,1,0–0,0,0) band at $1595 \, \mathrm{cm}^{-1}$; (3) the H_2O (0,2,0–0,1,0) band at $1556 \, \mathrm{cm}^{-1}$ (daytime only); and (4) the NO(1–0) band at $1875 \, \mathrm{cm}^{-1}$. The day minus night radiance residual, as plotted in the bottom panel of Fig. 8.26, includes some of the low-J P, Q and R transitions in the (0,2,0–0,1,0) band. These measurements demonstrate the enhanced daytime emissions of H_2O (0,2,0–0,1,0) lines and are in general good agreement with the non-LTE model described in Sec. 7.4. The radiance ratios of the fundamental H_2O (0,1,0–0,0,0) and first hot (0,2,0–0,1,0) $Q(1)$ transitions during daytime correspond, when compared with non-LTE models, to an exchange vibrational rate k_{vv} of (0.9–1.5)$\times 10^{-12} \, \mathrm{cm}^3 \mathrm{s}^{-1}$ for the $H_2O(0,v_2,0)$–$O_2(1)$ V–V exchange in the mesosphere, which is comparable to the rate derived from ISAMS of $(1–3)\times 10^{-12} \, \mathrm{cm}^3 \mathrm{s}^{-1}$.

More recently, H_2O non-LTE $6.3 \, \mu\mathrm{m}$ emission has been observed by CRISTA with a moderate spectral resolution, which has also made possible to discriminate between the fundamental and hot band contributions of H_2O. The analysis of the measurements have shown that they favour a slightly larger k_{vv} vibrational rate coefficient for the $H_2O(0,v_2,0)$–$O_2(1)$ V–V exchange ($1.7–3.1\times 10^{-12} \, \mathrm{cm}^3 \mathrm{s}^{-1}$), a slightly higher quantum yield of $O_2(1)$ production from O_3 photolysis (4–6), and a slightly lower rate coefficient for the thermal quenching of $O_2(1)$ with $O(^3P)$ ($0.65–1.3\times 10^{-12} \, \mathrm{cm}^3 \mathrm{s}^{-1}$).

8.6 Observations of Carbon Monoxide in Emission

Rocket-borne instruments have observed CO under auroral conditions with high spectral resolution, while satellite sensors have covered the middle and upper atmosphere with good vertical resolution and on the global scale. As early as 1978, the SAMS instrument on Nimbus 7 made the first global observations of CO using measurements of the emission at $4.6 \, \mu\mathrm{m}$. The data had a poor signal to noise ratio, however, which meant that profiles in the 30–100 km region could be obtained only if complete seasons were integrated. Nevertheless, even daily mean radiance profiles showed the large difference between day and night due to non-LTE effects by day.

Later, the ISAMS instrument repeated the measurements with much higher sensitivity and vertical resolution. These relatively low-noise data could be used to retrieve global and seasonal CO abundances, using a non-LTE model for the source function similar to that described in Sec. 7.2. The results demonstrated for the first time the feasibility of using CO to trace the dynamics, not only of the mesosphere but also of the stratosphere, most notably in the polar vortex.

As we have seen above, the daytime atmospheric limb emission at mesospheric and stratospheric tangent heights in the region of the $4.6 \, \mu\mathrm{m}$ fundamental of CO is actually dominated by a number of strong $CO_2 \, \nu_3$ bands, and the $N_2O(\nu_3)$ and two $O_3(\nu_1+\nu_3)$ bands, while the emission from CO(1–0) is relatively unimportant (see also Plates 15 and 16). Even the ISAMS pressure modulator technique only makes the CO(1–0) emission dominate that of the other atmospheric constituents for tangent heights above about 50 km. The contribution from N_2O below 50 km is about 40% of the total signal, similar to that of CO itself, with the remaining ~20% due to O_3. The contribution of the CO_2 bands to the pressure modulated signal is small, particularly at night. The data analysis must take account of all these contributions.

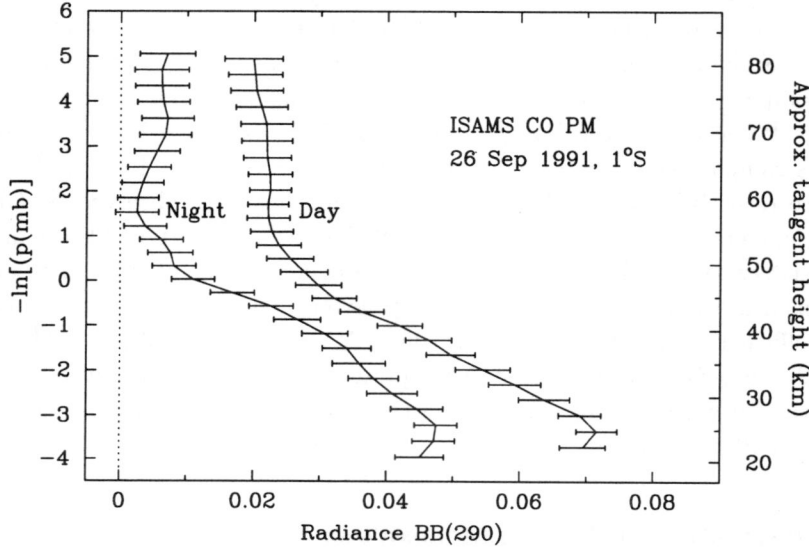

Fig. 8.27 Typical daytime ($\chi = 28°$) and night-time radiance profiles with the 1-σ uncertainty as measured by ISAMS in the pressure modulated CO channel near 4.6 μm for September 26, 1991 (1°S). Radiance units are mean filter-weighted Planck radiance at 290 K, $\bar{B}(290\,K)$.

Figure 8.27 shows that the daytime signal is greater than at night. Although CO is highly variable, it has a long lifetime and shows little diurnal variation. Thus, the enhancement must therefore be due to a large non-LTE excitation, presumably of CO(1), by the Sun. Using a model like that described in Sec. 7.2 to calculate the non-LTE populations, it is possible to retrieve CO VMR profiles in a similar way to the CO_2 retrieval described in Sec. 8.3.5. As for O_3, the non-LTE enhancement helps by extending the vertical range of the detectable CO radiance signal. The results (Plate 4) show the dramatic rate of increase in the abundance of CO with height, which occurs because of the dissociation of CO_2 by solar UV radiation, primarily in the thermosphere. A big increase in the CO mixing ratio is seen at all levels over the winter pole, due to the strong descending motion in the polar vortex which draws CO-rich air down from higher levels. CO is also produced, as a pollutant as well as from natural sources, in the troposphere and at the surface, and is a by-product of the oxidation of methane in the stratosphere. Its time-dependent overall distribution is therefore quite complicated, and still the subject of detailed study, for which these new global data, derived from observations of non-LTE emissions, are a powerful tool.

8.7   Observations of Nitric Oxide in Emission

Stratospheric nitric oxide has a key role in ozone photochemistry, and hence is an important target for remote sensing from space. However, because its abundance is very large in the lower thermosphere, the very bright non-LTE emissions from this overlying region tends to overpower the stratospheric and mesospheric limb radiances, making them difficult to separate out. The inversion of NO $5.3\,\mu$m radiances into middle-atmosphere NO profiles must take this into account, and employ a non-LTE source function to calculate the contaminating radiance from the higher levels. This is complicated, because of the large variability of the emission, due to the variability both of NO itself and also its thermospheric excitation sources, which include the kinetic temperature and the atomic oxygen abundance. The ISAMS measurements showed a very large variability in the $5.3\,\mu$m emission at stratospheric tangent heights correlated with variations in the solar activity, mainly as a result of the non-LTE contribution from the thermosphere. In addition, the NO(1) emission emanating from the stratosphere can itself be in non-LTE due to the excitation of NO(1) in its formation processes from NO_2 photodissociation and in the $NO_2 + O \rightarrow NO(1) + O_2$ reaction.

The fundamental vibration of NO at $1876\,\mathrm{cm}^{-1}$ is thermalized in collisions with O_2 and N_2 in the stratosphere and mesosphere. Vibrational-translational collisions with the former are the most important. Vibrational-translational exchange with N_2 has been measured to be 200 times less efficient. Since O_2 is abundant in the atmosphere, and $O_2(1)$ is in LTE below 65 km, NO(1) tend to remain close to LTE throughout the stratosphere, despite its non-thermal excitation mechanisms. At levels above 65 km, the absorption of solar radiation from above, the absorption of terrestrial radiation from below, and collisions with atomic oxygen all become important factors determining the population of NO(1). The flux of thermal radiation from the troposphere is a function of cloudiness and surface temperature, so that it varies by as much as an order of magnitude. Higher up, the main dependence is on atomic nitrogen [$N(^2D)$ and $N(^2S)$] and atomic oxygen abundances as well on the kinetic temperature, all of which strongly depend on solar activity, which of course is also very variable.

Nitric oxide is an interesting candidate for studies of emission at exceptionally low pressures because of its high abundance and large source function in the thermosphere, where the emission from most other species is too weak to be observed at all. The whole question of radiative cooling by nitric oxide and its role in the energetics of the upper atmosphere is also interesting and still largely unexplored.

Some early observations of NO in the fundamental ($5.3\,\mu$m) and in the first overtone ($2.8\,\mu$m) bands were obtained by the ICECAP rocket project, as zenith radiances under auroral conditions at 70–130 km altitude. The mechanism giving

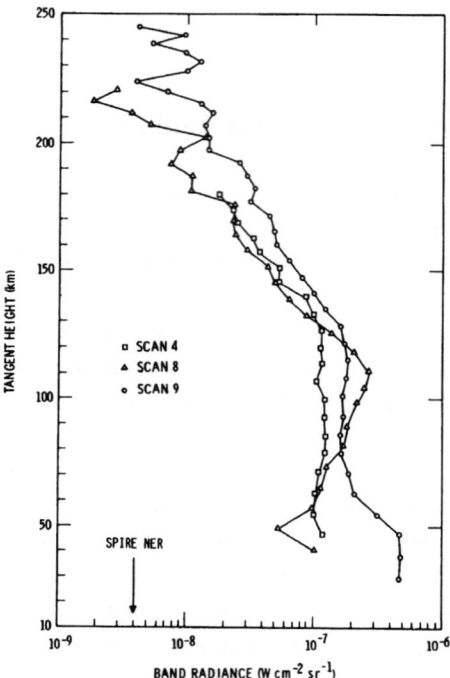

Fig. 8.28   Limb radiances profiles obtained by SPIRE for the NO fundamental band (5.08–5.64 μm) under non-auroral conditions. Scans 4 and 9 observed the limb in daylight. In scan 8, tangent heights below $\sim$130 km were in darkness; the maximum around 110 km might be due to a weak aurora.

rise to these emissions is thought to be $N(^2D) + O_2 \rightarrow NO(v \leq 12) + O$ (process 6 in Table 7.10) followed by spontaneous emission with $\Delta v = 1$ and $\Delta v = 2$, respectively. SPIRE obtained the day- and night-time radiance from 100–200 km (Fig. 8.28), observing strong thermospheric non-LTE NO emission, and allowing the retrieval of rotational temperature from 110–150 km, from which NO cooling rates and the $NO(1)$–$O(^3P)$ rate coefficient could be derived. This experiment confirmed the hypothesis proposed by Kockarts in 1980 that the NO fundamental band at 5.3 μm is the most important mechanism for cooling the atmosphere above about 130 km.

Measurements under auroral conditions of the 5.3 μm NO emission in the 70–125 km region were taken by the rocket-borne HIRIS interferometer with a resolution of 1.8 cm^{-1}. A strong NO signal with contributions from at least the first six vibrational levels was observed. From this and similar observations by the FWI, SPIRIT I and CIRRIS-1A instruments, it was concluded that the excitation mechanism was the chemiluminescent reaction $N(^2D) + O_2 \rightarrow NO(v) + O$, with additional excitation of $NO(1)$ by collisions with thermal atomic oxygen and aurorally enhanced $NO(v=0)$. The data also suggest that $NO(v)$ in the 100–125 km region preferentially relax in collisions with $O(^3P)$ rather than by radiative cascade. CIRRIS-1A found a large latitudinal enhancement, by a factor larger than 10, in the integrated radiances between high latitudes (under strong auroral conditions) and mid and low-latitudes. The daytime radiance (for similar magnetic latitudes and similar auroral activity) was larger (by a factor of 2–3) than the night-time radiance, due to the $N(^2D)$ produced by solar EUV radiation.

The NO $5.3\,\mu$m emission was also measured in the EBC, MAP/WINE and DYANA campaigns over the altitude range from 80 to 180 km, this time under non-auroral conditions. Simultaneously atomic oxygen abundance and kinetic temperature measurements allowed the derivation of NO $5.3\,\mu$m cooling rates, the three profiles showing good agreement between 120 and 170 km.

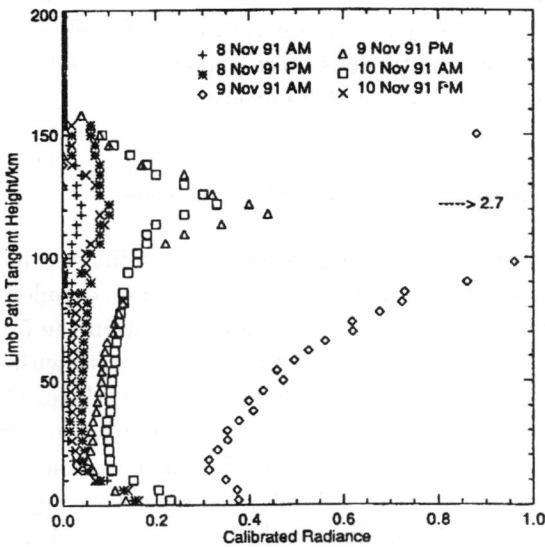

Fig. 8.29   Limb radiances at 35°N, 150°W from nitric oxide at $5.3\,\mu$m, as observed by ISAMS on three successive days in November 1991.

ISAMS observations of NO show remarkable variations in the limb emission in the (1–0) band with time (Fig. 8.29), related to variations in the influx of energetic particles from the Sun. The enhancement in emission is due in part to an increase in the production of NO, resulting initially from the dissociation of N_2 by collisions with the incoming particles, but also probably to enhancements in the population of NO(1) created during the chemical formation process and by thermal collisions.

The high spectral resolution of the CIRRIS-1A data made it possible to show that the rotational and spin degrees of freedom of thermospheric $NO(v>1)$ are not in equilibrium. The ratio of the spin orbit populations $NO(S=3/2)/NO(S=1/2)$ was found to be much lower than its LTE value at thermospheric altitudes. The rotational line strength distribution showed a subthermal part, a Maxwell–Boltzmann distribution with a rotational temperature of approximately 100 K below the kinetic temperature, and a superthermal part, with rotational temperatures of 3000–5000 K. Theoretical studies suggest that the 'hot' $NO(v>1)$ originates in the nascent rotational distribution of NO produced by chemical recombination of atomic nitrogen with molecular oxygen, while the subthermal part arises from collisions of NO with atomic oxygen. A further discussion on rotational non-LTE in NO is given in Sec. 8.9.

The CRISTA experiment also observed the global NO emission, in November 1994. Large enhancements (factors of 4–6) of the $5.3\,\mu m$ radiance were observed at high northern and high southern magnetic latitudes. According to the model described in Sec. 7.6, the radiative excitation of NO(1) is negligible above about 115 km, and the NO(1) population is controlled by the atomic oxygen abundance and the kinetic temperature. So, to derive NO abundance from measurements of its $5.3\,\mu m$ emission, the thermospheric temperature and $O(^3P)$ abundances have to be known. CRISTA made simultaneous measurements of the $O(^3P)$ $63\,\mu m$ emission, which is in LTE in this region, for this purpose.

8.8 Observations of Other Infrared Emissions

Several other species have been observed emitting radiances that are probably the result of non-LTE processes, but the data are sparse and detailed modelling, or even an appreciation of which processes are at work, is still in the future. This includes the NO_2 ν_3 bands near $6.2\,\mu m$. When nitric oxide reacts with ozone, it produces nitrogen dioxide which may be excited to high values of the v_3 vibrational quantum number: $NO+O_3 \rightarrow NO_2(v_3)+O_2$ (process 4 in Table 7.13). This mechanism was originally thought to be responsible for non-thermal emission in the $6.9\,\mu m$ channel of the LIMS instrument on Nimbus 7, intended for measuring water vapour. However, a reprocessing of the LIMS data in the late 90's showed that the enhancement was illusory. ISAMS had a similar channel for measuring water vapour, although much narrower and hence covering a small number of NO_2 ν_3 hot bands, and no evidence of a stratospheric non-LTE enhancement was found in this. However, a new visible airglow layer in the 35–55 km region has been reported, the origin of which may be the chemiluminescence reaction $NO+O_3 \rightarrow NO_2^*+O_2$, which is a potential source of the non-LTE excitation of the NO_2 $v_3 = 1$–7 levels.

ISAMS obtained the NO_2 abundance from the measurements of its infrared emission in the 1592–$1625\,cm^{-1}$ band in the 100–0.01 mb (16–80 km) height range. In this spectral region, there may be a non-LTE contribution from the first hot (0,0,2–0,0,1) band (see Sec. 7.7), but the pressure-modulation technique is insensitive to the hot bands and model calculations fitted the data within its measurement error without the need to invoke non-LTE effects.

Emission from the NO^+ ν_3 hot bands at $4.3\,\mu m$ in the upper mesosphere and lower thermosphere was first reported by the FWI experiment, observing during night-time auroral conditions, although this was not confirmed. It was suggested that NO^+ emission contributes to the ISAMS CO wideband signal above about 110 km. Recently, this emission was observed by CIRRIS-1A in the thermosphere at 100–215 km, confirming its non-LTE origin. The excitation source is not the process

$N_2^+ + O \rightarrow NO^+(v, J) + N(^2D)$, as previously thought, but $O^+ + N_2 \rightarrow NO^+(v, J) + N$. This is discussed further in the rotational non-LTE section below.

Fluorescence emission in the $2.7\,\mu m$ water vapour band was observed by the SAMS experiment in the daytime mesosphere and lower thermosphere. These measurements were used, assuming rates for the excitation (solar pumping) and relaxation processes of the emitting levels $H_2O(0,0,1)$ and $(1,0,0)$, to derive the water vapour abundance at high atmospheric levels where other measurements are sparse. The SPIRE experiment showed that CO_2 as well as H_2O makes a significant contribution in the 2.5–$2.9\,\mu m$ spectral range at altitudes of 50–$90\,km$, but that water vapour is larger. It also confirmed the large contribution of the CO_2 $2.7\,\mu m$ hot bands (see Table F.1) especially below $75\,km$.

The emission from $O(^3P)$ in the far infrared near $63\,\mu m$ was measured in the altitude range 80–$180\,km$ by the Polar High Atmosphere rocket campaign in March 1977. After several years of analysis it was finally concluded that the measured emission is in LTE in the thermosphere. Another experiment carried out four years later during solar maximum conditions, and thus at high thermospheric temperatures, found that the measured $63\,\mu m$ emission was compatible with both LTE and non-LTE assumptions. More recently the analysis of measurements by the SISSI rocket experiment, which include simultaneous atomic oxygen abundance data, clearly indicates that the $O(^3P)$ $63\,\mu m$ emission is in LTE from 80 to $160\,km$. The CRISTA experiment later used this information to derive the global latitude-altitude distribution of atomic oxygen in the thermosphere.

Other emissions in the near-IR as the OH emission in the 1–$4\,\mu m$ interval, and the O_2 near-IR at 1.27 and $1.58\,\mu m$ emissions have also been frequently measured but are not dealt in detail here. Some OH emissions measurements are described in the next section in connection with rotational non-LTE. We list in the 'References and Further Reading' section some good recent reviews of the OH and O_2 infrared emission measurements.

8.9 Rotational non-LTE

Most of this book deals with the role of non-LTE in determining the populations of vibrational energy levels, its consequences for the infrared emission from the atmosphere, and the quantitative effect on atmospheric remote sensing and energy balance. However, as we saw in Chapter 2, each vibrational level has a system of rotational energy levels associated with it. Because these are separated by energies which are typically two orders of magnitude less than the corresponding vibrational energy gaps, it is intuitively obvious that they reach thermal equilibrium among themselves much more easily. In the classical model of the molecule, we picture collisions as being more effective in influencing the rotation, rather than the vibration, of a molecule, and this simple picture holds up in reality.

However, there have been a number of reports in the literature of measurements which seems to indicate that rotational non-LTE is observable in the upper atmosphere, contrary to expectations based on simple theory. In general, the potential for rotational non-LTE depends on which J level we consider, since the Einstein coefficient for spontaneous emission of radiation from that level, $A_{J,J-1}$, is proportional to J^3. As briefly explained in Sec. 3.7, the height of non-LTE departure (in the absence of non-thermal collisions) is approximately that at which $k_t[M] \sim A$, where k_t is the thermal collisional rate and $[M]$ the collider (air) density. As an example, for the CO $J = 1 \rightarrow 0$ rotational transition at 2.53 mm, a density of only $[M] \sim 200 \, \text{mol cm}^{-3}$ is enough to keep the level in LTE. For the $J = 50 \rightarrow 49$ transition, the density required is much larger, $[M] \sim 3 \times 10^7 \, \text{mol cm}^{-3}$, but this still corresponds to the density at a very high level in the thermosphere. Hence we would not expect to find non-LTE in rotational levels of J smaller than ~ 50 in the *vibrational ground* state of CO except at altitudes above 200 km.

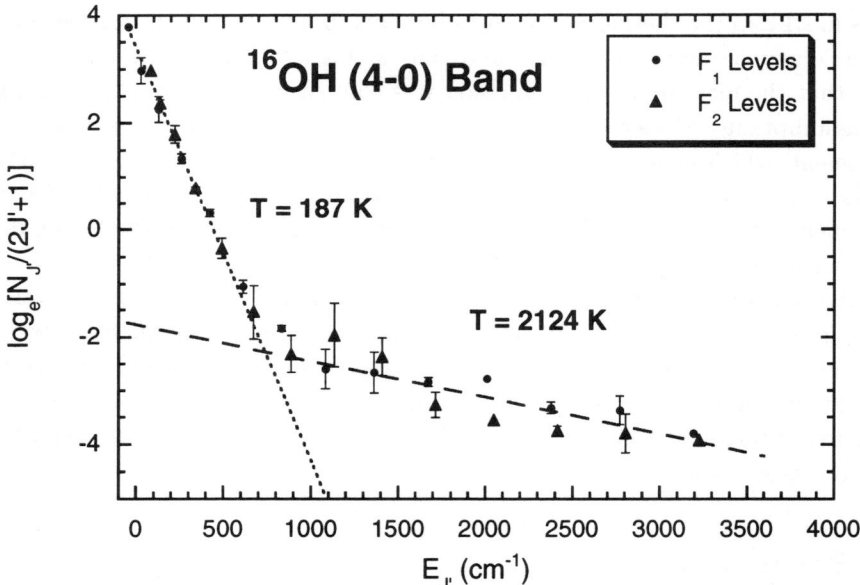

Fig. 8.30 Populations of OH rotational levels in the $v = 4$ vibrational state as function of their energy, observed in the nightglow from the ground at high spectral resolution. Two distributions for $J \leq 10$ and $J > 10$ at excitation temperatures of 184 K and 2124 K are clearly distinguished.

The situation may be somewhat different for the rotational levels of *vibrationally excited* states, however, since these have to adjust their rotational equilibrium after vibrational transitions and hence may be more prone to rotational non-LTE. There have been theoretical predictions and some experimental evidence for rotational non-LTE in $CO_2(0,0^0,1)$, CO(1) and NO(1) at $z \geq 120$ km; in the OH ($3 \leq v \leq 7$, $J \geq 7$) states, and in high-J (24–29) states in OH($v = 0$–2). Fig. 8.30 shows the populations of the OH rotational levels in the $v = 4$ vibrational state as a function of their energy. Two distributions for $J \leq 10$ and $J > 10$ at excitation temperatures of 184 K and 2124 K are clearly distinguishable. Since this emission originates from the region close to the mesopause where typical temperatures are close to 180 K, it is deduced that the levels with $J \leq 10$ have LTE populations, while the higher-J levels have populations greatly enhanced with respect to LTE. The photochemical reaction $O_3 + H \rightarrow O_2 + OH(v, J)$ is thought to be responsible for the enhancement. The CIRRIS-1A team recently reported pure rotation emission lines from highly rotationally excited OH for both night and daytime quiescent conditions.

CIRRIS-1A has also found evidence of rotational non-LTE in the NO vibrational states. Band head emissions from the highly excited rotational states $NO(v, J)$, with $v = 1$–7, 9, 10 and $J \approx 90$, have been observed in the fundamental band (1650–2050 cm^{-1}) between 115 and 205 km for quiescent (sunlit) and aurorally disturbed conditions. These exhibit a strong diurnal behaviour consistent with the rotational thermal component of lower vibrational bands, and are thought to be caused primarily by the reaction $N(^2D) + O_2 \rightarrow NO(v, J) + O$ (process 6 in Table 7.10), possibly supplemented by the reaction of 'hot' $N(^4S)$ atoms with O_2 (process 7b in Table 7.10).

Non-LTE from highly-excited vibrational levels of NO^+ were observed recently by CIRRIS-1A for the first time. Vibration-rotation band head emissions from $NO^+(v, J)$ with $J \geq 90$ for quiescent (sunlight) and aurorally disturbed conditions were seen at tangent height of 100–215 km, and were about one order of magnitude larger in the daytime than at night. The primary source of the daytime emission is thought to be the $O^+ + N_2$ ion-molecule reaction.

These examples of non-LTE rotational populations in OH, NO, and NO^+ somewhat change our classical view of the fate of the energy released in exothermic reactions in the upper mesosphere and thermosphere. It seems to be channelled preferentially into rotational excitation, which might be relaxed later into vibrational excitation, rather than directly into vibrational excitation.

Spectroscopic measurements with sufficiently high resolution can determine the rotational temperature, T_r, which describes the population distribution between rotational states within the same vibrational level, usually the ground state. If the gas is in LTE, T_r, T_v, and the kinetic temperature T_k, are, of course, all identical. From the preceding discussion, it follows that a determination of T_r is in effect a determination of the kinetic temperature under most conditions below 100 km altitude, and can be compared to a simultaneous measurement of T_v in order to study vibrational non-LTE. An example of this approach will be discussed below in Sec. 8.10.1.

8.10 Absorption Measurements

We have seen in the preceding sections how measurements of infrared *emission* from vibration-rotation bands in non-LTE can be compared to numerical models, and many features of the non-LTE processes at work elucidated and explained. Of course, emission measurements involve transitions from an excited level to a less excited one, usually to the ground state. Rather different information is acquired if we have access to transmission measurements, since in this case the spectral features arise as a result of the absorption of photons, and excitation from a lower to a higher state occurs. The amount of absorption measured is proportional to the population of the lower state of the transition, regardless (to a good approximation, i.e., neglecting vibrational partition function effects) of whether the population of the upper state is in LTE or not. Transmission measurements of the hot bands therefore give us in principle a rather direct way of probing the population of the vibrationally excited levels, and offer another way to test non-LTE models which can add to what we have learned from emission studies.

Two problems with transmission measurements are: firstly, the need for a source outside the atmosphere, and secondly the need to work at quite high spectral resolution. The Sun is obviously the source which gives the best signal-to-noise, although the Moon and more recently the stars, can also be used, at least in the visible, as for example in the case of the GOMOS experiment on ENVISAT. Solar measurements are, by definition, always at local sunrise or sunset, often an inconvenient constraint. The second problem arises because most of the molecules in the atmosphere, at any height level, are in the ground vibrational state. As we saw in Chapter 2, at room temperature and pressure, only about one in a million CO_2 molecules is in the first excited vibrational state, and of course this ratio is even smaller at lower temperatures and pressures, or for higher excited states. Thus, the transmission spectrum of the atmosphere is dominated by bands which correspond to transitions from the ground state. Measurements of these are useful for determining the atmospheric composition, but contain little information about non-LTE. For that, we need to observe hot bands or combination bands originating in an excited lower state, and determine its population. The fact that weak upper-state bands, which inevitably occur mingled with the spectra of much stronger ground-state bands, are the most interesting, dictates the need for high spectral resolution. When using the Sun as the source, the very high radiance means, in the near infrared at least, that high spectral resolution is not incompatible with good signal to noise ratios. Several instruments have been built and flown which operate on this principle. Our next case study address the most sophisticated of these to date, the Atmospheric Trace Molecular Spectroscopy (ATMOS) experiment.

8.10.1   *The ATMOS experiment*

ATMOS is a large and sophisticated infrared spectrometer which mounts on the pallet of the space shuttle. It obtains data on the atmosphere as an absorption path during the brief periods each orbit when the Sun, as seen from the orbiter, passes behind the limb of the Earth. The instrument itself is equal in performance to a very good laboratory spectrometer and so provides high-quality spectra across the range from 2 to 16 μm, which includes the vibration-rotation bands of many minor and trace species of interest in atmospheric physics and chemistry. These spectra have many applications, including searching for exotic molecules such as $ClONO_2$ or H_2O_2 which help to diagnose the chemical cycles going on in the stratosphere.

The measurements taken by ATMOS during the Spacelab 3 mission from April 30 to May 6, 1985, and on a number of subsequent flights of similar short duration, offered an unprecedented data set for testing non-LTE models and improving knowledge of the mechanisms responsible for the population of the $CO_2(0,1^1,0)$ level in the upper atmosphere. The high resolution spectra contain simultaneous, high vertical resolution (about 2 km) information on the kinetic temperature, the vibrational temperature of $CO_2(0,1^1,0)$ and the carbon dioxide mixing ratio. Moreover, ATMOS measurements were taken over two latitude bands with significantly different kinetic temperatures, and therefore gave additional information about the role that thermal processes play in populating $CO_2(0,1^1,0)$ in the lower thermosphere.

8.10.2 *CO$_2$ v_2 vibrational temperatures*

Calculated equivalent widths from ATMOS spectra, for selected strong, non-overlapping lines of CO_2 in the fundamental $(0,0^0,1 \leftarrow 0,0,0)$ and combination bands, e.g., $(0,1^1,1 \leftarrow 0,1^1,0)$, give populations for the ground state and the first excited level of the ν_2 vibration as a function of height over the range from 60 to 110 km. The differences between the simultaneously measured kinetic and $CO_2(0,1^1,0)$ vibrational temperatures, T_k–T_v, are shown in Fig. 8.31 for the Northern and Southern Hemispheres. The two temperatures are the same to within the experimental error, up to about 100 km.

Prior to this result, it was widely believed from model calculations that the ν_2 band would depart from LTE above an altitude of about 75 km. Although it was obvious that there were many uncertainties in the current knowledge of some of the model parameters, such a large divergence from the ATMOS results was remarkable, since the results of a wide range of radiative transfer models had produced the same general conclusion. It is important not just as a detail of the physics, but because the rate at which the upper atmosphere cools depends on the vibrational temperature of the $CO_2(0,1^1,0)$ level much more than any other level, so the question of the extent to which it departs from LTE is fundamental to any understanding of the upper mesosphere and lower thermosphere. As we will see in Chapter 10, this is important not only for the Earth's atmosphere but even more so for the energy balance and temperature structure of the upper atmospheres of Mars and Venus.

At this time, it was first realised that this puzzling discrepancy between measurement and theory can be explained by the role played by atomic oxygen. It was known already (Sec. 9.2.1.2) that oxygen atoms in the ground state are very much more efficient at deactivating the excited v_2 states of CO_2 than are collisions with the more abundant molecules like N_2, O_2, or CO_2 itself, and the process was incorporated in the models. However, the rate at which the deactivation occurs had been grossly underestimated. In the years previous to ATMOS measurements, Sharma and co-workers were continuously revising this rate to higher values based on their interpretation of SPIRE CO_2 15 μm rocket measurements, and Sharma and Wintersteiner finally came up in 1990 with a rate about 30 times the old value. At about the same time as the first ATMOS measurements, Shved *et al.* reported new laboratory measurements of the rate of this process at atmospheric temperatures, which were 7 times greater than the previous best value. Fig. 8.31 shows the fit with model calculations with these new values for the $CO_2(0,1^1,0)$-$O(^3P)$ deactivation coefficient, k_{CO_2-O}, with the latest values for the vertical profile of CO_2 and for the abundance of $O(^3P)$. The dominant effect is the very large change in the atomic oxygen coefficient, and it can be seen that the closest agreement with the ATMOS results is obtained with the value of Sharma and Wintersteiner.

Another result from Fig. 8.31 is that the T_k–T_v deviation is significantly different in the two hemispheres above around 105 km, being larger in the Northern (70 K) than in the Southern Hemisphere (40 K). A simple model for the population of $CO_2(0,1^1,0)$, based on the 'cooling-to-space' approximation with a diffusivity factor β, and considering that $CO_2(0,1^1,0)$ is mainly populated by collisions with $O(^3P)$ and depopulated by quenching with $O(^3P)$ and by radiative processes, gives

$$T_k - T_v = \frac{C T_k^2}{1 + CT_k},\tag{8.19}$$

where we have used the definition of T_v (Eq. 6.1), and C is given by

$$C = \frac{1}{E_v} \ln \left[1 + \frac{\beta A \mathcal{T}^\star}{4 \, k_{CO_2-O} \, [O(^3P)]} \right],\tag{8.20}$$

where E_v is the energy of the $(0,1^1,0)$ level, A is the spontaneous Einstein coefficient for the ν_2 transition, and $\mathcal{T}^\star$ is the escape probability function (see Sec. 5.2.1.2). From Eq. (8.19) it follows that

$$\Delta(T_k - T_v) = \frac{D}{(1 + D)} \Delta T_k,\tag{8.21}$$

where $D = C^2 T_k^2 + 2CT_k$, and the temperature dependence of k_{CO_2-O} has been neglected. Since C is positive, and consequently D, the deviation of the vibrational temperature of $CO_2(0,1^1,0)$ from the kinetic temperature generally (except when D is negligible) increases with increasing kinetic temperature. Thus, the larger T_k–T_v deviation observed by ATMOS SL3 in the Northern Hemisphere, where the atmosphere was warmer, can be qualitatively explained by this simple model. Eq. (8.21) also predicts that a given ΔT_k will induce a smaller $\Delta(T_k$–$T_v)$ change. The AT-MOS measurement of the kinetic temperature in the Northern Hemisphere at the 112 km level exceeds that in the same level of the Southern Hemisphere by 64 K. The change in $(T_k$–$T_v)$ between both hemispheres at the same height is smaller, only about 30 K, again consistent with the simple model prediction. It also follows from

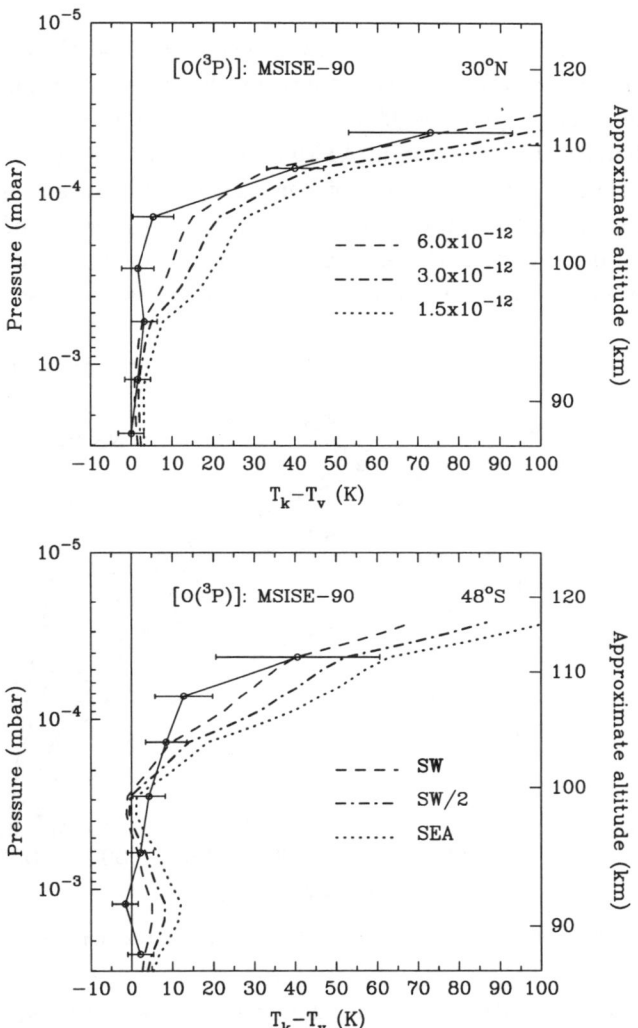

Fig. 8.31   Measured and calculated differences between the kinetic temperature and the vibrational temperature of $CO_2(0,1^1,0)$. Measurements are zonal mean values derived from ATMOS SL3 Northern (top) and Southern (bottom) Hemisphere occultations with their 1-σ uncertainties. The three calculations correspond to vibrational temperatures computed with the $k_{CO_2\text{-}O}$ rate coefficients of Sharma and Wintersteiner (SW), SW divided by 2, and that of Shved *et al.* (SEA).

Eq. (8.21) that when collisions dominates, $C \to 0$, and $T_k = T_v$; and when radiative processes dominates C is very large, and the change in $(T_k\text{-}T_v)$ is similar to that in the kinetic temperature.

8.10.3 *CO_2 abundance*

The high resolution ATMOS spectra can also be used to infer the CO_2 abundance in the mesosphere and lower thermosphere. Once derived the kinetic temperature from the ratio of transmittance in the rotational lines, the CO_2 abundance can be derived from the absolute transmittance of the lines. Normally the lines of the strong fundamental $(0,0^0,1-0,0^0,0)$ band are used for retrieving CO_2 in the upper atmosphere. CO_2 derived from ATMOS has the advantage over the emission measurements that it is virtually free of non-LTE effects, since it is measuring in absorption using a fundamental band (see Sec. 8.2.1). A mean of the two mid-latitude profiles derived from the ATMOS instrument on Spacelab 3 is shown in Fig. 8.32, together with other CO_2 measurements which we discuss below.

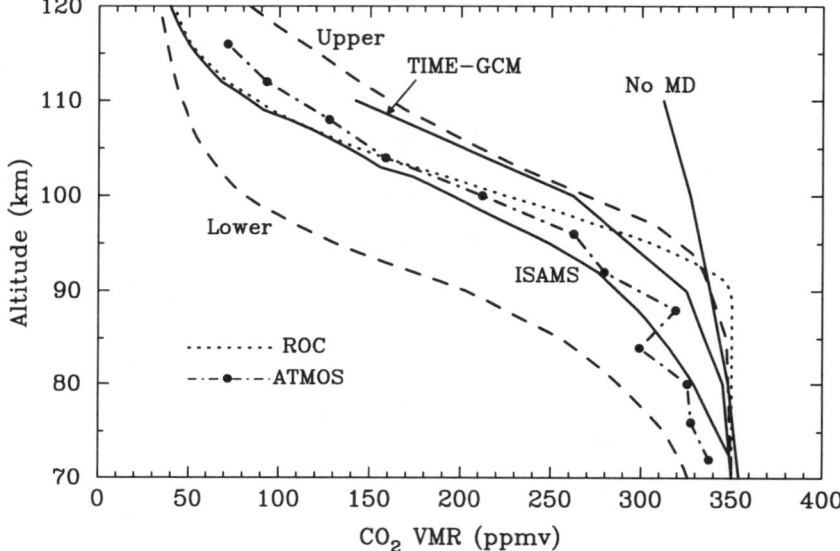

Fig. 8.32 Compilation of CO_2 volume mixing ratio profiles. See text for details of the profiles.

The volume mixing ratio of CO_2 in the upper mesosphere and lower thermosphere remains very controversial. The observations are mainly of three kinds: *in situ* rocket measurements; those inferred from limb emission radiances from rocket and satellites, such as the SPIRE rocket experiment, SAMS on Nimbus 7 and ISAMS on UARS (described in Sec. 8.3); and those taken by solar occultation spectrometers on Spacelab by two instruments, the Grille Spectrometer and ATMOS. Fig. 8.32 shows some of these measurements and illustrates the significant differences that exist between the results. 'ROC' is a mean compilation of rockets measurements, with the value in the mixed part of the atmosphere updated from typically 330 ppmv when they were measured in the 70's, to the more current value of 350 ppmv. 'Lower' is a mean CO_2 VMR derived from measurements by the Grille instrument, which had a low S/N ratio and hence large errors. This profile is considered a lower limit case in the sensitivity studies in Secs. 9.2.1.2 and 9.8.2. 'ISAMS' is the profile derived from the CO_2 4.3 μm emission measured by ISAMS up to about 100 km and extrapolated upwards with the rocket profile. ISAMS values are close to ATMOS and is considered to give the most representative profile of CO_2 for use in models. The figure also shows a profile which represents the upper limit allowed by the measurements, which is used in the sensitivity studies in Secs. 9.2.1.2 and 9.8.2.

Two theoretical profiles are also shown: one from the sophisticated TIME-GCM model, and the other from a simple model where molecular diffusion was neglected ('no MD'). The TIME-GCM result is for solstice (winter) conditions at mid-latitudes, where it matches the ATMOS and ISAMS measurements reasonably well. Slightly larger values are predicted by the same model at equinox, and smaller abundances are found in the polar winter, associated with the meridional circulation which produces a downwelling flux of thermospheric air, less rich in CO_2, into the polar winter mesosphere.

If diffusive separation is ignored ('No MD'), the model clearly overestimates the measured CO_2 VMR above about 95–100 km; this showing the important role of molecular diffusion at those heights. This suggests that TIME-GCM would best reproduce the ATMOS and ISAMS measurements if weaker eddy mixing (which is equivalent to enhance molecular diffusion) is assumed above about 80 km. This, however, is not in consonance with the expectation that breaking gravity waves should induce significant eddy mixing at this altitude; a conflict that remains to be resolved. In any case, what is clear from ATMOS and ISAMS CO_2, is that molecular diffusion is comparable to or larger than eddy diffusion as low as 80 km. This is in contrast with the conventional belief that the atmosphere is well mixed up to about 100 km, i.e., that the homopause (sometimes called the turbopause) is commonly located at about 100 km.

8.11 Simulated Limb Emission Spectra at High Resolution

ATMOS was able to observe the spectrum of the atmosphere at high spectral resolution because the Sun provides such a bright source. To do the same thing with the relatively feeble emission from the Earth's limb as a source was not possible until recently. Recent advances in detector technology and, in particular, in cryogenics, mean that experiments now under development will achieve this goal, in particular the MIPAS interferometer, designed to obtain high resolution infrared limb spectra from the European ENVISAT spacecraft. We present and describe in this section some synthetic high resolution $(0.025\,\mathrm{cm}^{-1})$ limb emission spectra of the atmosphere covering the mid-infrared range $(600$–$2400\,\mathrm{cm}^{-1})$ $(15$–$4.3\,\mu\mathrm{m})$ for some typical tangent heights of the stratosphere and mesosphere. The non-LTE–LTE difference spectra show the spectral regions and altitudes where non-LTE can be important and an estimate of its magnitude. These spectra include all the non-LTE populations described in Chapters 6 and 7, with the most up-to-date rate constants etc., and form a reference catalogue of the expected deviations from non-LTE across the atmospheric spectrum for the U.S. Standard Atmosphere (1976), representing mid-latitudes and daytime $(\chi = 0°)$ conditions. The atmosphere is considered to stop at

120 km altitude, except for the case of the NO emission in the range 1820–2000 cm^{-1} for which it was extended up to 200 km. As a yardstick for the importance of the non-LTE deviation, the noise equivalent spectral radiance (NESR) of the MIPAS instrument is plotted in these figures, representing the typical performance of a modern high spectral resolution instrument.

Plates 5 and 6 show the calculated LTE and non-LTE limb spectral radiances (top panel) and the non-LTE–LTE radiance difference (bottom panel) at low and high mesospheric tangent heights, 56 and 83 km respectively, for the 15 μm region. This region is mainly dominated by CO_2 ν_2 bands with some contribution from the O_3 ν_2 bands near 14.3 μm. As was shown in Chapter 6, the populations of the CO_2 v_2 levels start departing from LTE in the upper mesosphere, so the non-LTE radiance difference is very small in the lower mesosphere (see Plate 5). Non-LTE effects can be discerned in the fundamental CO_2 15 μm band, as this is optically thick at these tangent heights and a significant contribution comes from the lower thermosphere, where this band exhibits populations significantly lower than LTE. Weak non-LTE enhancements are also apparent in Plate 5: in the first hot bands around 618, 668, and 721 cm^{-1}, and in the third hot bands around 652, 668, and 683 cm^{-1}, due to populations larger than LTE in the upper mesosphere (see Sec. 6.3.6.1 and Table F.1). Note the different scales in the top and bottom panels, and the very small degree of non-LTE at these tangent heights.

The spectra in the same spectral region at higher altitudes (83 km) show similar features but of a larger relative magnitude. Small enhancements in the 628 and 627 fundamental isotopic bands around 662 and 665 cm^{-1}, respectively, can be appreciated. The non-LTE effects in the O_3 band close to 14.3 μm are too small to see, since the vibrational temperature remains close to T_k up to the lower thermosphere, as discussed in Sec. 7.3.6.

The kinetic temperature, T_k, is usually retrieved from the CO_2 15 μm bands; these results show that non-LTE effects in the lower mesosphere are generally smaller (in absolute magnitude) than the MIPAS noise level, with the exception of the fundamental band and very small contributions from the weak bands in their Q-branches. The non-LTE effect in the fundamental band is mainly caused by the large kinetic temperature in the lower thermosphere (100–120 km). We should note that this effect would be much smaller if we would consider the atmosphere extended up to 100 km only. In relative terms, the maximum non-LTE depletion is less than 3% at 56 km, increasing with altitude to around 30% at 83 km. Of course, this depends on the atmospheric conditions, since non-LTE effects are smaller for a warmer mesosphere (e.g., winter-like conditions), and larger for a cold mesosphere, similar to polar summer situations (Sec. 6.3.6.1). The retrieval of T_k at higher levels is also affected by the uncertainty in the CO_2 VMR, which ceases to be well mixed above about 80 km (Sec. 8.10.3).

Plates 7–9 show the atmospheric spectra in the 10 μm region. Stratospheric and mesospheric tangent altitudes are dominated by the O_3 ν_1 and ν_3 bands, and the

CO_2 laser bands, and show much larger non-LTE effects than at $15 \, \mu m$. These are very large in the CO_2 $10 \, \mu m$ laser bands even as low as the stratosphere, because of the departure from LTE of the emitting state $(0,0^0,1)$ in the daylight mesosphere, pumped by absorption of solar radiation at 4.3 and $2.7 \, \mu m$. The non-LTE enhancement is even larger at 65 km, and for this reason, these regions of the spectrum have usually been avoided for temperature retrieval. Conversely, since the emission from $(0,0^0,1)$ in the $4.3 \, \mu m$ fundamental band is optically very thick at mesospheric heights, the $10 \, \mu m$ emission can be used to obtain the population of $(0,0^0,1)$ and so learn about its non-LTE excitation processes in the mesosphere.

O_3 is usually retrieved from the $9.6 \, \mu m$ band, though wide band radiometers have to take care to avoid contamination by the nearby CO_2 bands, as well as by the non-LTE emission from the highly excited O_3 ν_3 hot bands. Plate 7 (bottom panel) shows that non-LTE effects in the ν_3 fundamental bands of O_3 are significant compared to the noise, although this is still less than 1% of the total signal at stratospheric tangent heights so non-LTE can usually be neglected for O_3 retrievals below about 50–60 km. The non-LTE radiance differences rise rapidly with increasing tangent height above the stratopause, and at shorter wavenumbers (see inset in Plate 9), in the latter case due mainly to the presence of the ν_3 hot bands.

The region from 1200 to 1500 wavenumbers is mostly dominated by CH_4 and H_2O bands, with some possible contribution of the NO_2 ν_3 hot bands at the short-wavelength end of this interval (Plates 11 and 12). The predicted non-LTE emission from the NO_2 $\Delta v_3 = 1$ hot bands has not yet been confirmed experimentally, and that from methane might be overestimated since CIRRIS-1A measurements suggest that non-LTE in CH_4 is very small up to 80 km. A significant non-LTE contribution is present due to water vapour.

The 1550 to 1750 wavenumber region in the lower atmosphere (Plate 13) is dominated by H_2O lines, and at the longer wavelengths by the fundamental and first hot NO_2 ν_3 bands. An instrument like MIPAS will clearly be able to detect such large non-LTE contributions; the H_2O lines have a good signal-to-noise ratio up to 71 km (see Plate 14) and non-LTE can be measured in both the fundamental and first hot bands of $H_2O(\nu_2)$.

Plates 15 and 16 show synthetic spectra for stratospheric and upper mesospheric tangent heights for mid-latitudes daytime conditions in the 4–$5 \, \mu m$ region. This is rich in non-LTE effects, since more energetic transitions are more likely to be in non-LTE because they are less efficiently thermalized and solar pumping is more important at shorter wavelengths. For CO and NO, which have only one strong vibrational transition in the mid-infrared, non-LTE effects are important over the whole altitude range, and in the case of CO the inclusion of non-LTE modelling into the retrieval scheme is difficult due to the dependence of the $CO(1)$ excitation temperature on the CO VMR itself (see Sec. 7.2).

In the stratosphere, the prominent band of $NO(1$–$0)$ show negative non-LTE effects which are primarily due to radiation from the upper thermosphere (above

120 km), where the vibrational excitation of NO(1) is much smaller than the very high thermospheric temperature. The non-LTE effects at stratospheric tangent heights from the layers above also applies to carbon monoxide, although in this case in the opposite sense. The large mesospheric gradient of the CO mixing ratio and the over-Boltzmann vibrational population of CO(1) (particularly in the daytime) makes the mesospheric contribution very important for the radiance observed at stratospheric tangent heights. In this spectral region we also see clearly the large contributions of the fundamental, isotopic, and hot bands of CO_2, again with non-LTE all the way down to the tropopause, and of the O_3 combination bands near 4.8 μm, also with a large non-LTE component (see inset in bottom panel of Plate 15).

Plate 16 shows a spectrum at 83 km. The large non-LTE effect from NO, CO and CO_2 bands is also clear at this altitude. Note the larger radiance from the CO_2 4.3 μm bands at 83 km than at 41 km, a consequence of the upper atmospheric excitation. At this spectral resolution, the contributions of the different CO_2 bands near 4.3 μm, i.e. the fundamental, isotopic, and first hot bands, can be distinguished (see inset in top panel).

Plate 10 shows the limb radiance spectra of the most important airglow emission for a tangent height of 60 km in the daytime. The limb path radiances have been obtained assuming optically thin conditions and using Eq. (8.16), i.e., no radiative transfer nor absorption along the line of sight was considered. This is an accurate approach for most of the bands and lines at this altitude with the possible exception of the O_2 atmospheric and near-IR bands. The strong emission of O_2 in the red and near-IR part of the spectrum and the very rich spectra of the OH Meinel bands cover a large part of the spectrum; all with much enhanced emission relative to that expected from LTE. The OH radiance is larger for night-time conditions by a factor of between 2 and 4 depending on the transition.

8.12   Simulated Nadir Emission Spectra at High Resolution

Now we consider the effects of non-LTE on the radiances measured in emission in the infrared using a nadir geometry (see Fig. 8.1). Normally these are of much smaller magnitude than those at the limb because the outgoing atmospheric radiation comes mostly from the lower atmosphere where, as we saw in Chapters 6 and 7, most of the vibrational levels of the atmospheric molecules are in LTE. We should recall that the treatment of non-LTE first appeared in atmospheric studies when the need for calculating accurate cooling rates in the upper atmosphere emerged. The subject achieved more importance with the advent of the limb sounding technique, particularly when more sensitive instruments became able to sound the very tenuous upper atmosphere, where non-LTE is most prevalent.

Nevertheless, there are some spectral regions in the infrared where the atmosphere is optically thick and the nadir radiance emanates from high regions where non-LTE applies. The CO_2 4.3 μm spectral region is one example of such a case.

The outgoing radiation in those bands from a sunlit atmosphere, calculated assuming both LTE and non-LTE conditions, is shown in Fig. 8.33. This utilises the equations developed in Sec. 8.2.1, except that the coordinate in (Eq. 8.1) for limb geometry is replaced by the vertical coordinate z (see Fig. 8.1), and we have to add the contribution of the lower boundary (Earth's surface or cloud top) (see, e.g., Eq. 3.20). The non-LTE populations are introduced in a similar way to the limb viewing case, through the r_v population factors and the vibrational temperatures.

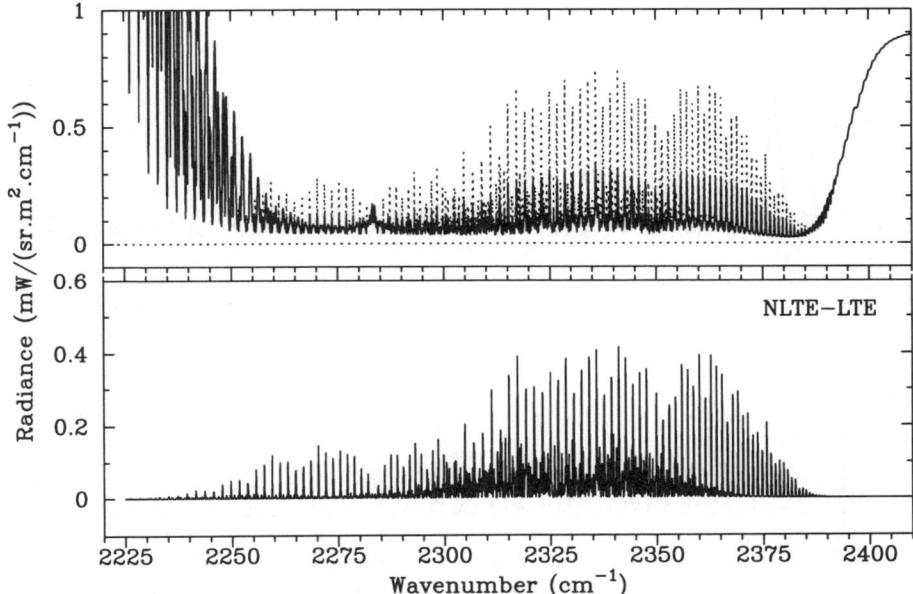

Fig. 8.33   Nadir spectral radiances in the CO_2 $4.3\,\mu m$ bands region at a resolution of $0.1\,cm^{-1}$ for daytime ($\chi = 0°$) conditions for the mid-latitude atmosphere. A surface temperature of 288 K was assumed and no cloud coverage. Top: Non-LTE (dotted) and LTE (solid) radiances. Bottom: Non-LTE–LTE radiance difference. The noise-equivalent spectral radiance of a current high resolution infrared instrument as MIPAS or TES is between 0.1 and 0.01 radiance unit.

The outgoing radiation at $4.3\,\mu m$ comes mostly from the region between 40 and 45 km in the strong lines of the 626 major CO_2 isotope. Many lines of the first hot and second hot bands are overlapped with the wings of the fundamental lines, so radiances in those weaker bands also comes from the upper stratosphere. Hence, the radiances in Fig. 8.33 correspond to the temperatures of this region, which is much colder than the surface. At smaller wavenumbers, between 2250 to $2300\,cm^{-1}$ where the 636 $4.3\,\mu m$ fundamental band lies, the radiance comes from lower altitudes, which are also colder than the surface, and even slightly colder than the upper stratosphere (see Fig. 1.1).

Figure 8.33 (bottom) shows the considerable importance of non-LTE emissions at these wavenumbers in the daytime. The most prominent non-LTE enhancement is in the centres of the lines of the strong 626 fundamental band, as would be expected since the emission comes from the highest levels (see, e.g., Fig. 6.12). The upper states of the first and second hot bands have even larger non-LTE deviations than the fundamental, at a given level, (see Fig. 6.12) but since these bands are weaker the emission comes from deeper layers and the non-LTE contribution to the spectrum is relatively less important. It would be weaker still were it not for the overlap with the line wings in the stronger fundamental band since, if these were not present, the emissions from the hot bands would come from the troposphere.

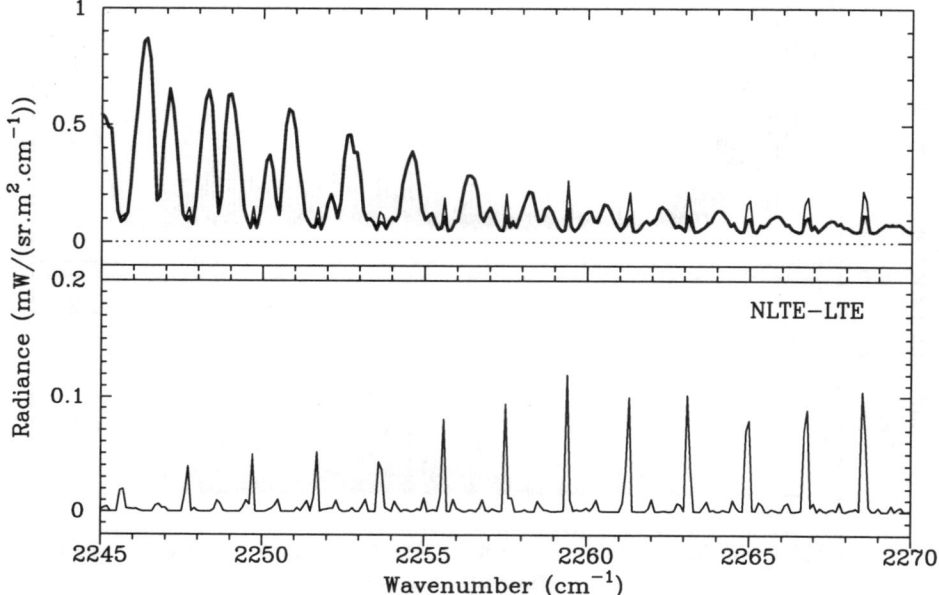

Fig. 8.34 The nadir spectral radiances shown in Fig. 8.33 in a larger scale. Thick line: LTE radiance; thin line: non-LTE radiance.

The dependence of the non-LTE enhancement on the strength of the line is illustrated in Fig. 8.34, which presents at a larger scale some lines of the 4.3 μm band of the 636 isotope. In this region, as wavenumber increases, line strength increases and the non-LTE–LTE radiance difference in the line centre increases. The emission in the line centre of the stronger lines comes from the upper layers where the non-LTE deviation is larger.

Another case where non-LTE can be important in nadir radiances is when the emitting species has a large abundance in the upper regions and a significant non-thermal source of excitation, as is the case for CO or NO. Fig. 8.35 shows a nadir spectrum in the region of the NO 5.3 μm band at a resolution of 0.001 cm^{-1}, calculated assuming the atmosphere extends up to 200 km. The strongest features are due to H_2O lines in the troposphere, but, superimposed on them, we can also see the contribution of NO lines emitting under non-LTE in the thermosphere (120–200 km) (see Sec. 7.6). These are apparent only because of the high spectral resolution of

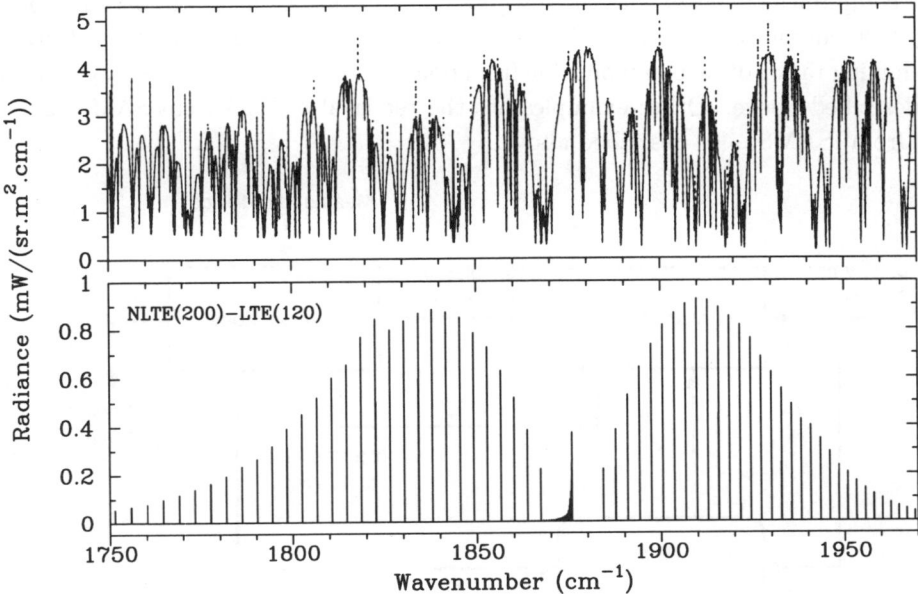

Fig. 8.35 Nadir spectral radiances in the NO 5.3 μm region at a resolution of 0.001 cm^{-1} for mid-latitude daytime conditions. A surface temperature of 288 K was assumed and a cloud-free atmosphere. Top: Non-LTE with atmosphere's top at 200 km (dotted); and LTE with atmosphere's top at 120 km (solid). No significant differences exist between non-LTE and LTE with atmosphere's top at 120 km. Bottom: Non-LTE(200 km)–LTE(120 km) radiance difference.

the simulations; if it is degraded to that of a current instrument like MIPAS, i.e. 0.025 cm^{-1}, the maximum difference in the strongest lines decreases to about 4 times the noise-equivalent spectral radiance of the instrument.

Nadir observations showing non-LTE are quite rare and rather marginal. The SPIRIT III radiometer on the MSX (Midcourse Space Experiment) satellite showed enhancements in the CO_2 4.3 μm radiance similar to those in the simulations shown above, and NIMS (Near-Infrared Mapping Spectrometer) observations during the flyby of the Galileo spacecraft over the Earth in December 1990 took nadir spectra covering the entire CO_2 4.3 μm region which showed a daytime enhancement of around 40%, in line with model predictions. Mostly, however, unlike limb radiances, the weak signal from the upper atmosphere, the poor vertical resolution, and the large background contribution from the Earth's surface, mean that the nadir radiance is much less useful for studying non-LTE processes.

8.13 Non-LTE Retrieval Schemes

Non-LTE models have already been used in retrieval codes for retrieving the species abundances from measurements of their atmospheric emissions. The cases developed so far include mostly those bands which are simpler to model, with manageable computational costs, and those whose non-LTE mechanisms are better known. Fig. 8.36 summarises the procedure for such a retrieval scheme, used for deriving the mixing ratios of carbon monoxide from non-LTE radiances measured by ISAMS, as described above. Other examples are the retrieval of H_2O from SAMS and of NO from ISAMS, also described above.

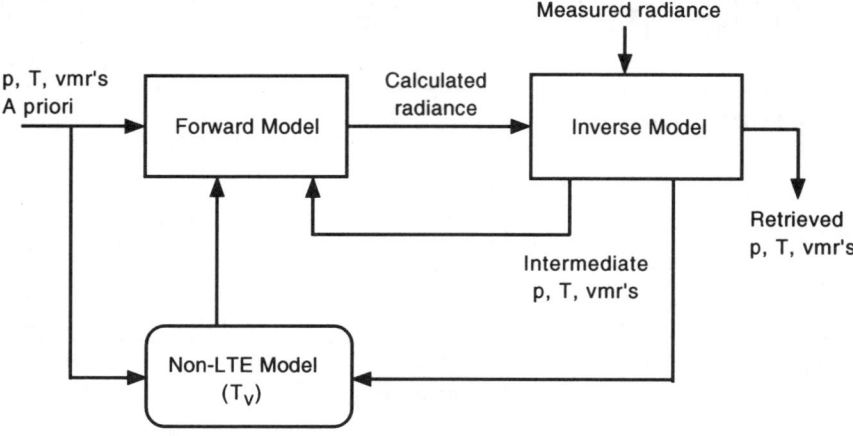

Fig. 8.36   Diagram of the retrieval scheme developed for the retrieval of carbon monoxide abundances from ISAMS observations of non-LTE emission.

The fast computers now available allow us to include non-LTE models within the forward models of the retrieval scheme. A big effort is currently underway to extend this approach, making it possible to sound more parameters in the middle and upper atmosphere with infrared measurements. For example, most of the geophysical parameters derived from SABER wideband radiances, including pressure-temperature, CO_2, O_3, H_2O and NO, use operational non-LTE retrieval codes. Even temperature can be retrieved under non-LTE, although, obviously, only in those regions where the populations of the emitting states are still partially in LTE, otherwise the information on kinetic temperature would be small.

In addition to fast computers, the high information content available in a high resolution spectrum (including non-LTE emissions from different levels of a given molecule, some of which may be in LTE and others not) have given rise to the development of highly sophisticated schemes in which the non-LTE part of the forward model and some of its parameters are treated as unknown in the retrieval process and derived along with the usual temperature and species information. An example of this kind is the retrieval of NO from MIPAS spectra (c.f. Sec. 7.6). This includes the dependence of the non-LTE population on the NO VMR itself, due to chemical production of $NO(v > 1)$ by NO_2 photolysis in the stratosphere and the reaction $N(^4S) + O_2$ in the thermosphere. It also includes a radiative transfer code which allows for vibrational, rotational and spin non-LTE in the calculation of limb radiance over the entire atmospheric range (see Fig. 8.37). This allows

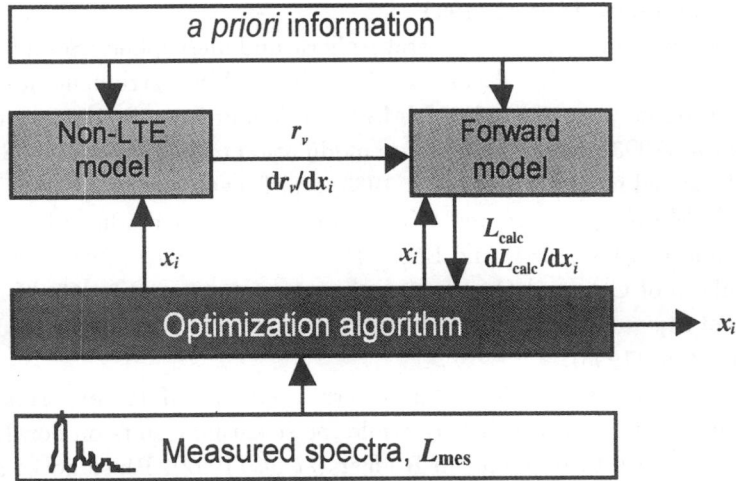

Fig. 8.37 The non-LTE retrieval scheme for simultaneous retrieval of abundances species and non-LTE model parameters. L is the measured or modelled limb spectral radiance, x_i are the retrieval vector elements, and r_v the non-LTE populations.

the simultaneous determination of the NO abundance and the nascent vibrational distribution of NO_2 after photolysis, an important non-LTE parameter, from limb emission spectra of the NO fundamental band. The approach can be generally applied to the inversion of any non-LTE affected trace species where there is a limited amount of spectral information on the non-LTE state distribution. For example, to the O_3 VMR, and the rate constants of several V–V and V–T processes affecting the O_3 vibrational state distribution.

8.14 References and Further Reading

The rocket-based experiments which observed non-LTE in CO_2 and several other species are discussed by Stair *et al.* (1975, 1983, 1985); Kumer *et al.* (1978); Nadile *et al.* (1978); Caledonia and Kennealy (1982); Philbrick *et al.* (1985); Sharma and Wintersteiner (1985, 1990); Ulwick *et al.* (1985); Brückelmann *et al.* (1987); Picard *et al.* (1987); Winick *et al.* (1987); Wintersteiner *et al.* (1992); Nebel *et al.* (1994); and Ratkowski et al. (1994). For CO_2 15 μm observations by CIRRIS-1A see Wise *et al.* (1995).

The population of $CO_2(v_3)$ by excited OH, through $N_2(1)$ as an intermediary is explained by Kumer *et al.* (1978) and Kumer and James (1983).

A review of the use of CO_2 15 and 4.3 μm emissions for deriving the CO_2 VMR in the upper mesosphere and lower thermosphere can be found in López-Puertas *et al.* (2000). A detailed analysis of the information content and non-LTE effects on CO_2 15 μm limb emissions can be found in Bullitt *et al.* (1985) and Edwards *et al.* (1993). The effects of the vibrational partition function on the limb radiance have been studied by Edwards *et al.* (1998).

The objectives and results of the Stratospheric and Mesospheric Sounder (SAMS) experiment are reviewed by Taylor (1987). The SAMS instrument has been described by Drummond *et al.* (1980), LIMS by Gille and Russell (1984), and ISAMS by Taylor *et al.* (1993), while the pressure modulation technique used by SAMS and ISAMS is discussed in detail by Taylor (1983) and Edwards *et al.* (1999). The analysis of the SAMS temperature and CO_2 $4.3\,\mu$m measurements have been reported by Barnett and Corney (1984) and López-Puertas and Taylor (1989), respectively.

The analysis of CO_2 and O_3 $10\,\mu$m emissions as measured by CLAES and the importance of the excitation of the $CO_2(0,0^0,1)$ state by $O(^1D)$ is described in detail by Edwards *et al.* (1996).

The Atmospheric Trace Molecular Spectroscopy (ATMOS) experiment is described by Farmer and Norton (1989), while the scientific results on non-LTE studies with ATMOS are to be found in Rodgers *et al.* (1992), Rinsland *et al.* (1992) and López-Puertas *et al.* (1992b).

For a discussion of the inferred abundance of CO_2 in the mesosphere and lower thermosphere, see Taylor (1988) and the recent review by López-Puertas *et al.* (2000). Observational evidence for the fast rate coefficient of $CO_2(0,1^1,0)$ in collisions with $O(^3P)$ is discussed by Sharma and Wintersteiner (1990), Shved *et al.* (1991), López-Puertas *et al.* (1992b), Pollock *et al.* (1993), and Mertens *et al.* (2001).

ICECAP measurements of O_3 emissions were reported by Stair *et al.* (1974) and analysed by Yamamoto (1977) and Gordiets *et al.* (1978). For SPIRE rocket observations of ozone see Stair *et al.* (1985) and Green *et al.* (1986), for SISSI see Grossmann *et al.* (1994), and for both see Manuilova and Shved (1992). Non-LTE corrections to the model used for the retrieval of ozone mixing ratio from the LIMS ozone channel at $9.6\,\mu$m are described by Solomon *et al.* (1986) and Pemberton (1993). A revised version of Mlynczak's model has been used to analyse the CIRRIS-1A data by Zhou *et al.* (1998). Non-LTE models for the populations of the $O_3(v_1,v_2,v_3)$ overtone and combination levels have been developed by the following: Rawlins *et al.* (1985, 1993); Mlynczak *et al.* (1990a,b), Fichet *et al.* (1992), Manuilova and Shved (1992), Pemberton (1993), Manuilova *et al.* (1998), Koutoulaki (1998), and Martín-Torres and López-Puertas (2002).

Non-thermal emission in measurements made using the $6.9\,\mu$m channel of the LIMS instrument on Nimbus 7 are discussed by Kerridge and Remsberg (1989). SAMS limb daytime $2.7\,\mu$m radiances were used to deduce H_2O concentrations in the 35–$100\,$km region by Drummond and Mutlow (1981). Non-LTE effects on ISAMS measurements are discussed by López-Puertas *et al.* (1995), Goss-Custard *et al.* (1996) and Zaragoza *et al.* (1998). Non-LTE models of the population of the $H_2O(0,1,0)$ and $(0,2,0)$ levels have been reported by Manuilova and Shved (1985) and López-Puertas *et al.* (1995).

Vibrational and rotational non-LTE in the limb radiance from the $4.6\,\mu$m band has been studied by Oelhaf and Fischer (1989), Winick *et al.* (1991) and Kutepov *et al.* (1993a). Measurements of CO abundances derived from the ISAMS CO emission at $4.6\,\mu$m have been reported by López-Valverde *et al.* (1993, 1996), and Allen *et al.* (2000). Emission by CO isotopes in the $4.7\,\mu$m band in the upper mesosphere as observed by the CIRRIS-1A experiment have been reported by Dodd *et al.* (1993).

Degges (1971) predicted that NO($v = 1,2$) will be in non-LTE in the mesosphere and in the lower thermosphere; for more recent treatments see Caledonia and Kennealy (1982), Kaye and Kumer (1987), and Sharma *et al.* (1993). The latter also discusses the evidence for rotational non-LTE in thermospheric NO. ICECAP and HIRIS measurements of NO are described by Stair *et al.* (1975, 1983, 1985). Analyses were carried out by Rawlins *et al.* (1981) and Zachor *et al.* (1985). Observations of NO at $5.3\,\mu$m by the FWI instrument were reported by Picard *et al.* (1987) and those by SPIRIT I by Adler-Golden *et al.* (1991).

The measurements of NO $5.3\,\mu$m emission during the EBC campaign are discussed by Ulwick *et al.* (1985); and a comparative study of the NO $5.3\,\mu$m data measured in the EBC, MAP/WINE, and DYANA (SISSI) campaigns can be found in Grossmann *et al.* (1994). The NO $5.3\,\mu$m emission measured by the CIRRIS-1A experiment has been analysed by Sharma *et al.* (1993), Smith and Ahmadjian (1993), Armstrong *et al.* (1994), Lipson *et al.* (1994), and Wise *et al.* (1995). For CRISTA NO $5.3\,\mu$m emission measurements see Grossmann *et al.* (1997a). Early ISAMS NO results are presented by Ballard *et al.* (1993). For the effect of NO on the energy balance, see Kockarts (1980), Caledonia and Kennealy (1982), Zachor *et al.* (1985), and Grossmann *et al.* (1994).

Non-LTE versus LTE in atomic oxygen $63\,\mu$m emission is discussed by Kockarts (1970), Grossmann and Offermann (1978), Iwagami and Ogawa (1982), Sharma *et al.* (1994) and Grossmann and Vollmann (1997). More recently this emission has been measured globally from space by the CRISTA experiment (Grossmann *et al.*, 1997). Feasibilities for retrieving $O(^{3}P)$ abundances and temperature in the thermosphere by using the 63 and $147\,\mu$m emissions are discussed by Zachor and Sharma (1989) and Sharma *et al.* (1990).

The evidence for chemically-pumped hot band emission from high v_3 vibrational levels of nitrogen dioxide in LIMS $6.9\,\mu$m measurements has been discussed by Kerridge and Remsberg (1989). Daytime non-LTE in hydrogen chloride (HCl) and hydrogen fluoride (HF) is discussed by Kumer and James (1982) and Kumer *et al.* (1989). Recent reviews of the OH emission in the 1–$4\,\mu$m interval and of the O_2 near-IR at 1.27 and $1.58\,\mu$m emissions can be found in Winick *et al.* (1985), Solomon (1991), López-González *et al.* (1992b), Sivjee (1992), and McDade (1998); see also Chamberlain (1995).

Rotational non-LTE in OH is discussed by Smith *et al.* (1992), Pendleton *et al.* (1993), and Cosby *et al.* (1999); and in NO(v, J) by Espy *et al.* (1988), Rawlins *et al.* (1989), Smith and Ahmadjian (1993), Sharma *et al.* (1993), Sharma *et al.* (1996a-c, 1998), and Sharma and Duff (1997). Evidence of rotational non-LTE in

NO^+ is described by Smith *et al.* (2000) and discussed by Duff and Smith (2000). Armstrong *et al.* (1994) presented evidence for rotational non-LTE in the high-J levels of $CO(v = 1)$. Theoretical predictions and experimental evidence for non-LTE in rotational levels of vibrationally excited states of $CO_2(0,0^0,1)$, $CO(1)$ and $NO(1)$ are reported by Kutepov *et al.* (1991); Sharma *et al.* (1993); Dodd *et al.* (1994), and Funke and López-Puertas (2000).

The non-LTE retrieval code for retrieving CO abundance from ISAMS is described in López-Valverde *et al.* (1991). A non-LTE retrieval code that simultaneously retrieve VMR's and vibrational temperatures profiles has been developed and applied to CRISTA measurements by Timofeyev *et al.* (1995, 1999). Non-LTE retrieval codes for deriving species abundances from SABER measurements can be found in Zhou *et al.* (1998, 1999) and Mertens *et al.* (2001). The non-LTE retrieval code for retrieving NO abundance and non-LTE parameters is described in Funke *et al.* (2001).

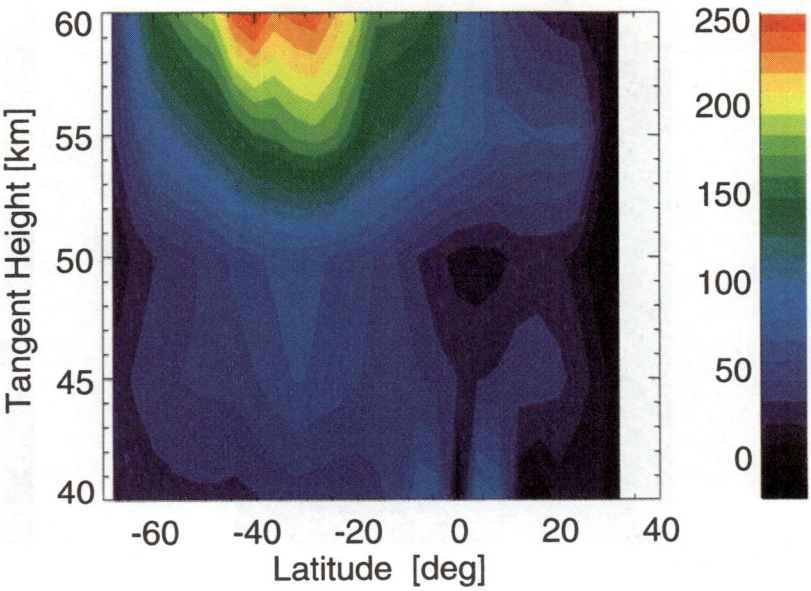

Plate 1   Diurnal difference in CLAES $10\,\mu$m radiance as a percentage of the night-time value [(day-night)/night%]. Day 92-02-12.

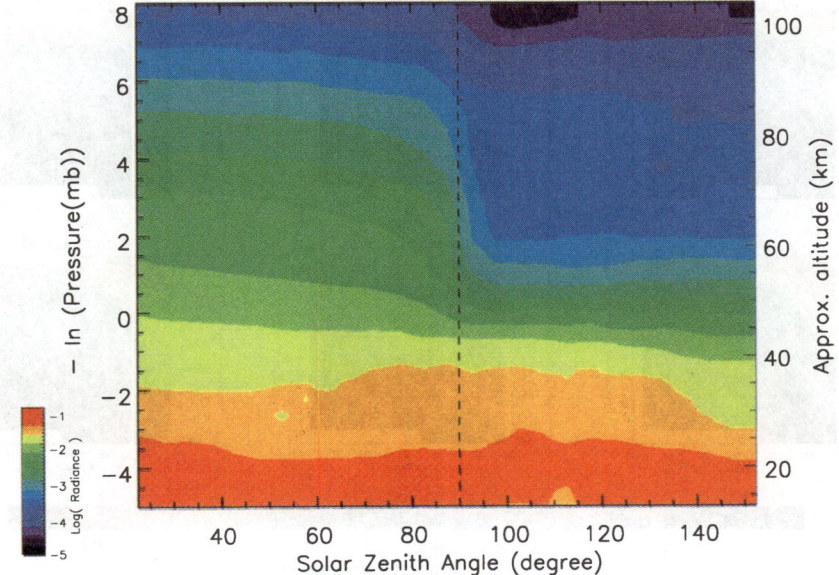

Plate 2   Variation of the ISAMS wideband radiances near $4.7\,\mu$m with solar zenith angle and with altitude for January 13, 1992. Radiances have been averaged in boxes of $\sim$2.5 km in altitude and $5°$ in zenith angle and plotted on a log scale in units of $\bar{B}(290\,$K$)$.

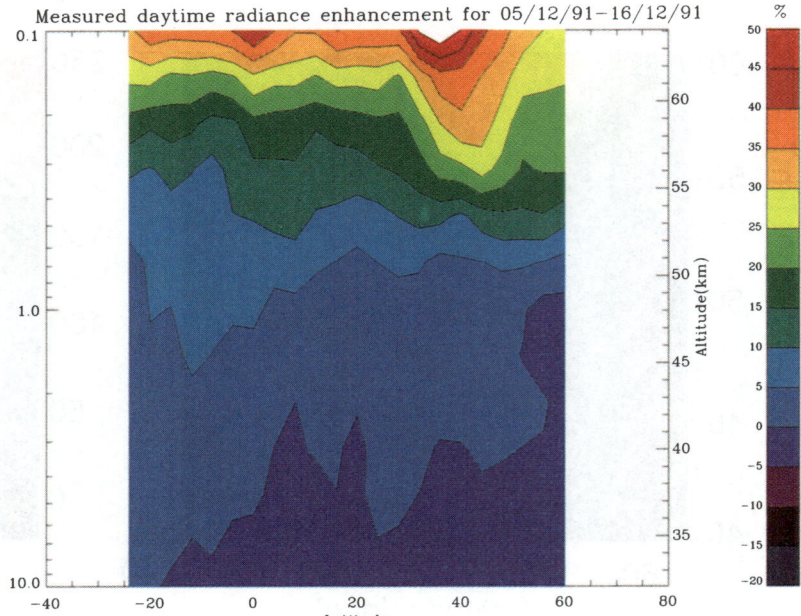

Plate 3   Daytime non-LTE radiance enhancement [(Measured day non-LTE–Estimated day LTE)/Estimated day LTE] from ozone at 9.6 μm as observed by ISAMS. Estimated daytime LTE was calculated with daytime temperature from other ISAMS channel and daytime O_3 from MLS.

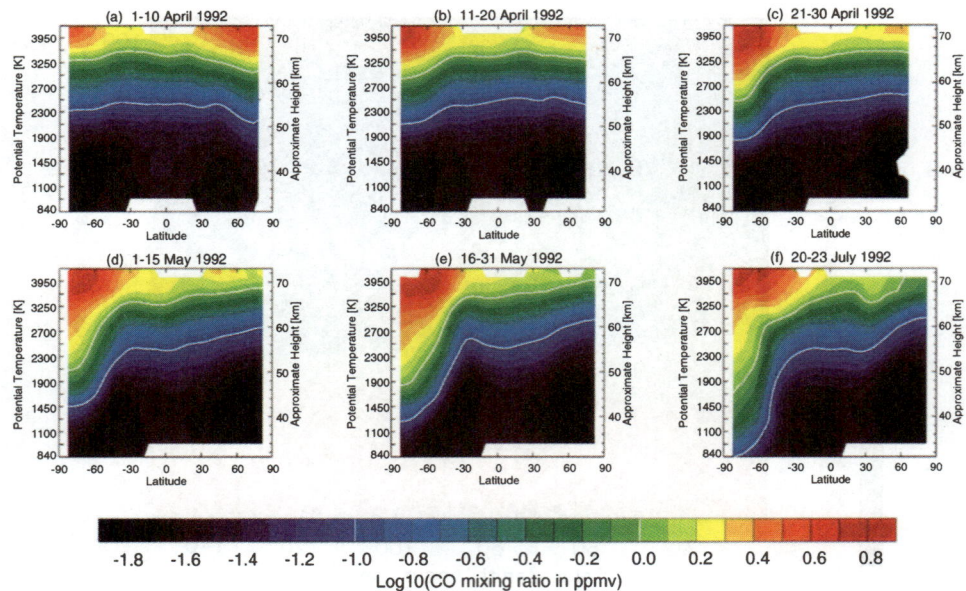

Plate 4   Zonal mean CO VMR averaged over 6 time periods from April to July 1992 derived from ISAMS 4.7 μm observations. The contours are $\log_{10}$ of the CO mixing ratio in ppmv. The 0.1 and 1.0 ppmv contours are highlighted in white for illustrative purposes. Note the seasonal changes, especially at high latitudes.

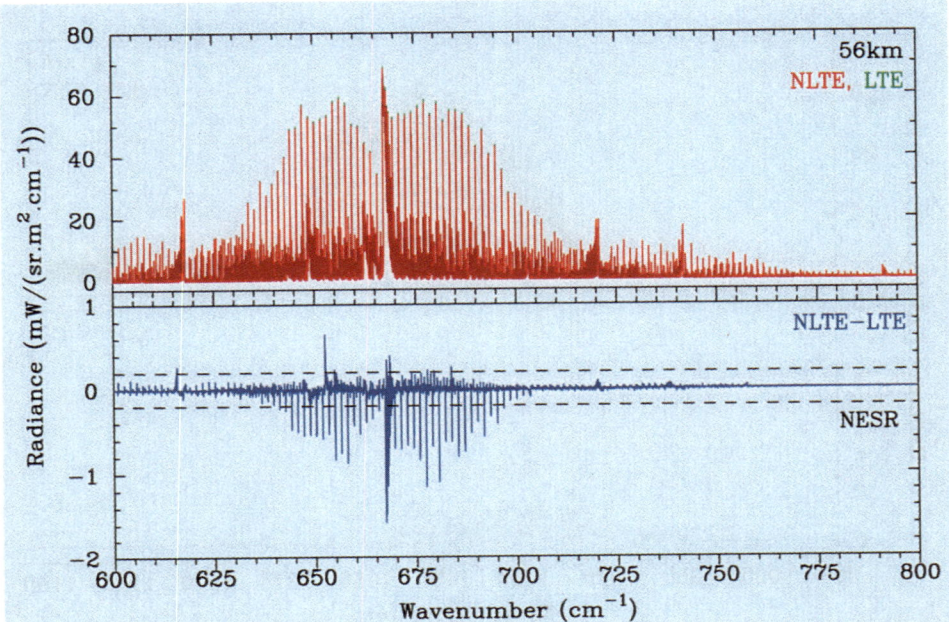

Plate 5   Synthetic limb radiances for the 15 μm region at a tangent height of 56 km in a daytime mid-latitude atmosphere. Spectral resolution of 0.025 cm^{-1}. Top: Non-LTE (red) and LTE (green) radiances, hardly distinguishable. Bottom: Non-LTE–LTE radiance difference. NESR is the noise equivalent spectra radiance of MIPAS.

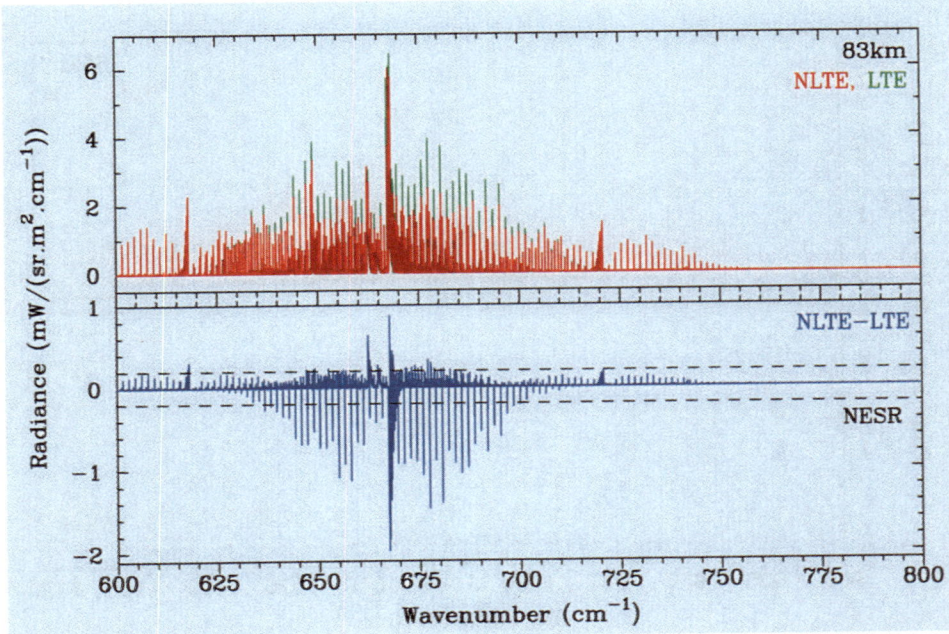

Plate 6   As plate above but for a tangent height of 83 km. Top panel: the LTE radiance at the edges is masked by the larger non-LTE emission.

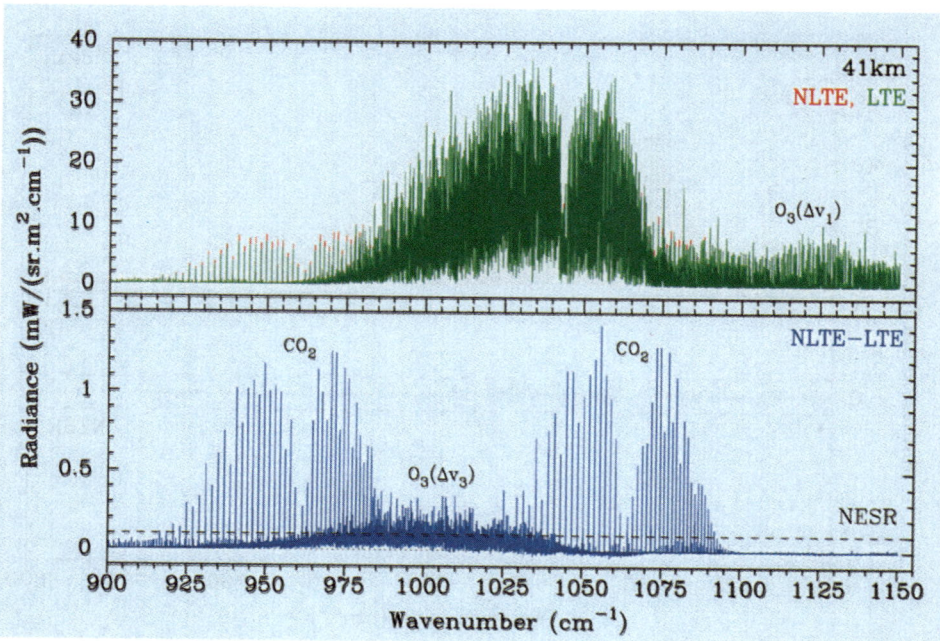

Plate 7   Synthetic limb radiances in the $10\,\mu$m region at a tangent height of 41 km for a daytime mid-latitude atmosphere. Spectral resolution of $0.025\,\mathrm{cm}^{-1}$. Top panel: Non-LTE (red) and LTE (green) radiances, hardly distinguishable. Lower panel: Non-LTE–LTE radiance difference.

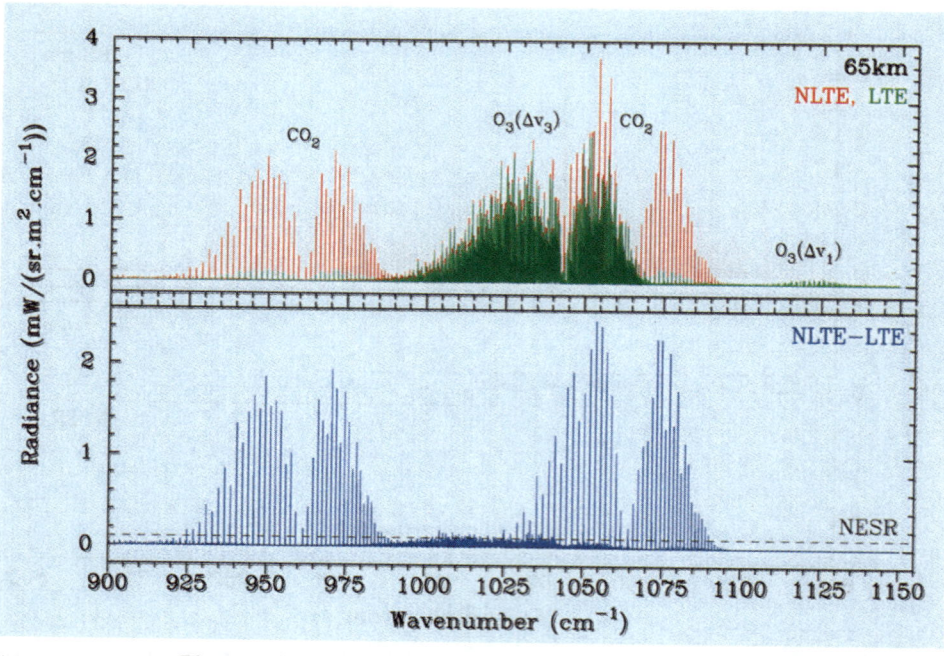

Plate 8   As plate above, but for a tangent height of 65 km.

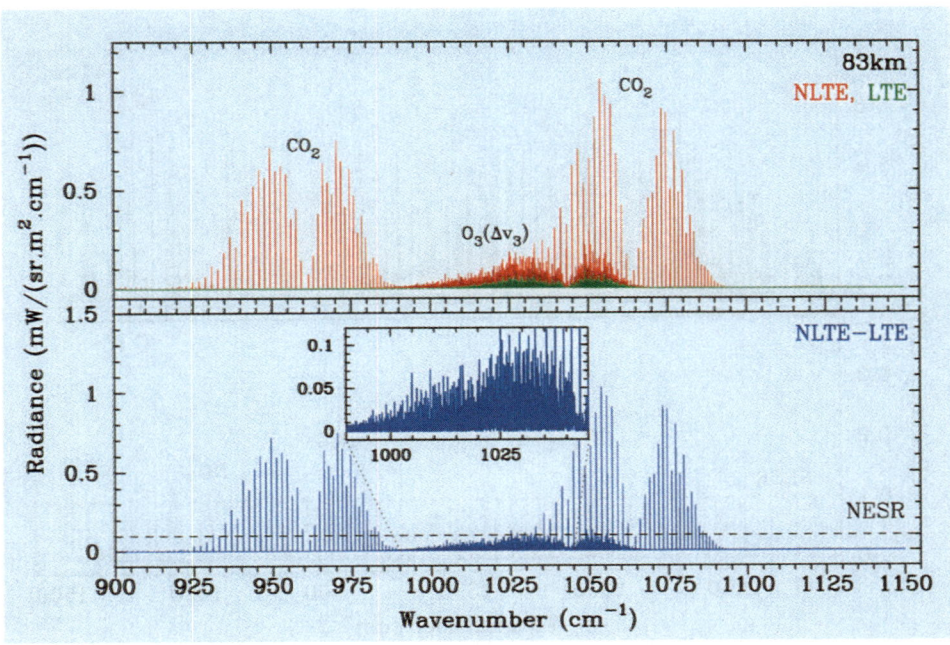

Plate 9   As Plate 7, but for a tangent height of 83 km.

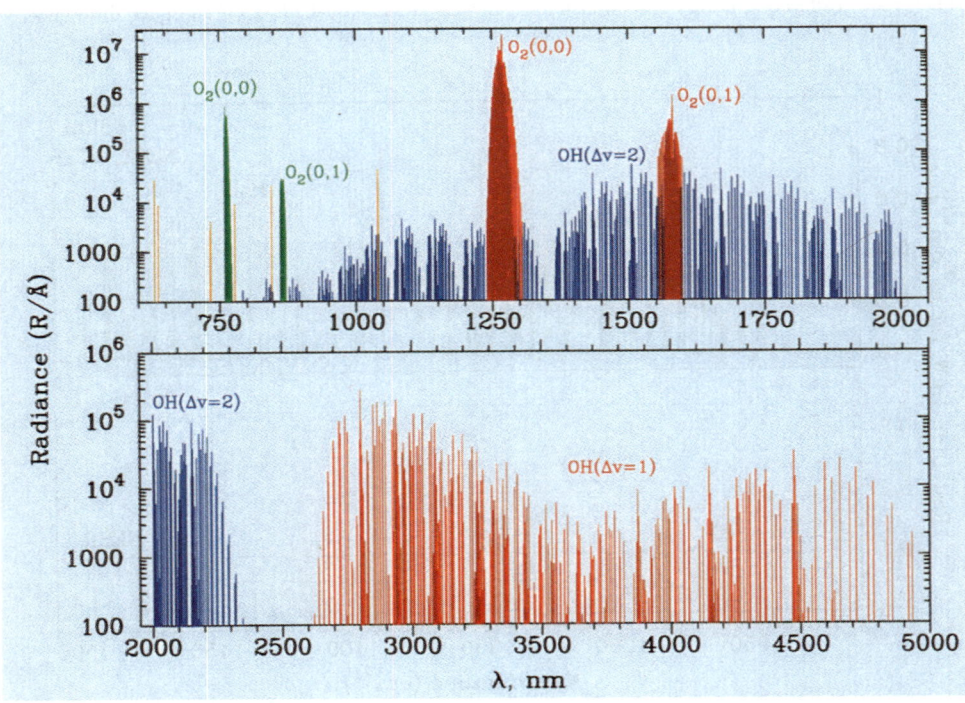

Plate 10   Limb spectral radiance of the major airglow emissions in the visible and near-infrared at a tangent height of 60 km for mid-latitudes daytime conditions. Spectral resolution of 0.1 nm.

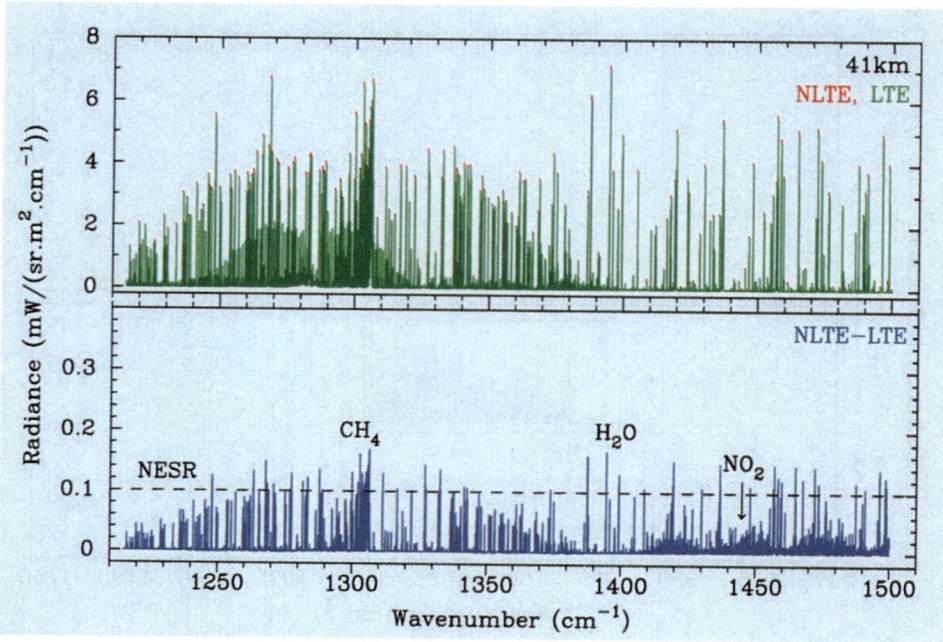

Plate 11   Synthetic limb radiances in the 6.5–8 μm region at a tangent height of 41 km for a daytime mid-latitude atmosphere. Spectral resolution of 0.025 cm⁻¹. Top panel: Non-LTE (red) and LTE (green) radiances, hardly distinguishable. Lower panel: Non-LTE–LTE radiance difference.

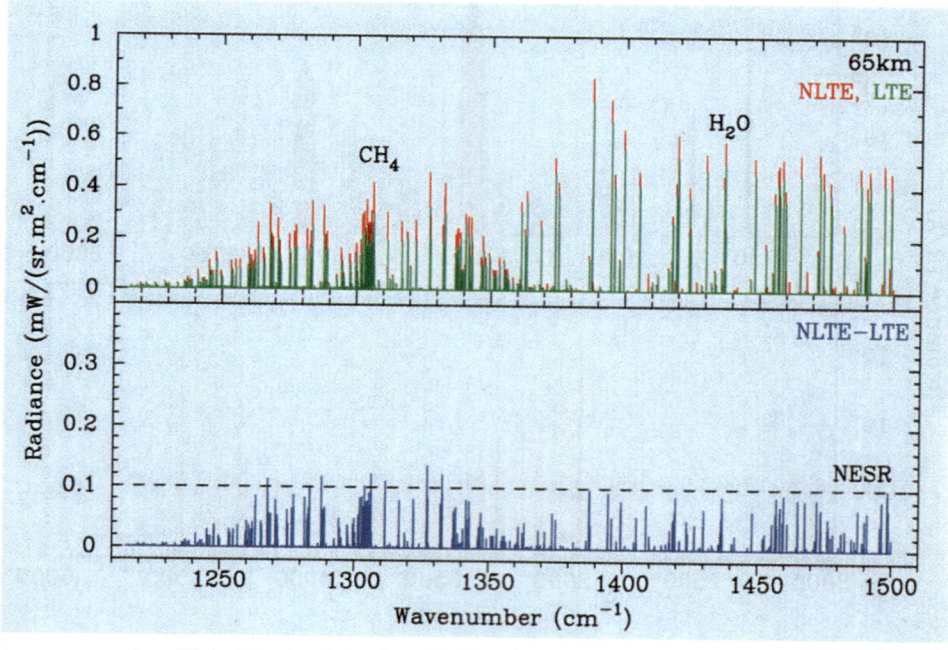

Plate 12   As plate above but for a tangent height of 65 km.

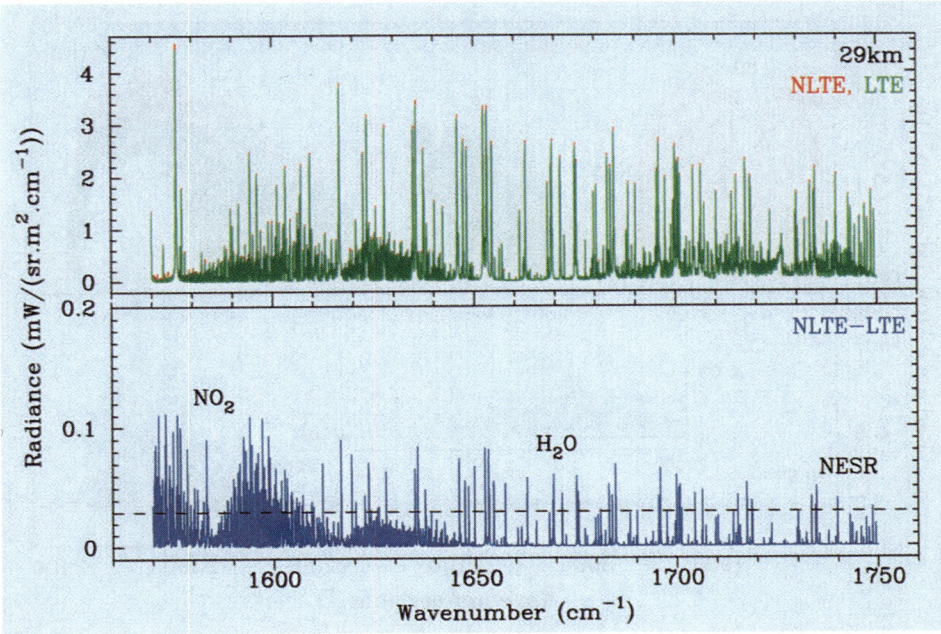

Plate 13   Synthetic limb radiances in the 5.5–6.5 μm region a tangent height of 29 km for a daytime mid-latitude atmosphere. Spectral resolution of 0.025 cm^{-1}. Top panel: Non-LTE (red) and LTE (green) radiances, hardly distinguishable. Bottom panel: Non-LTE–LTE radiance difference.

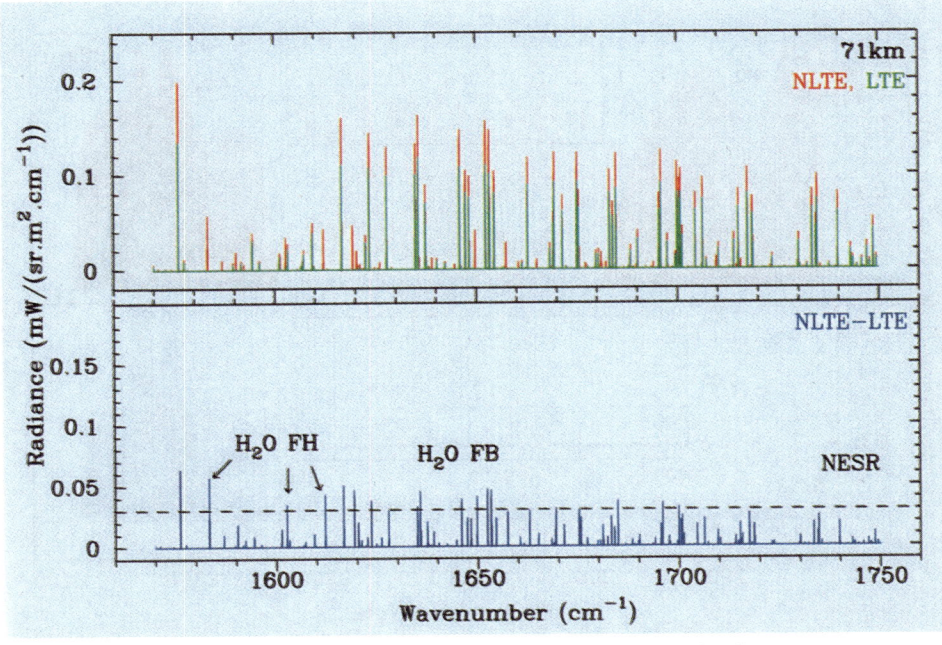

Plate 14   As plate above but for a tangent height of 71 km.

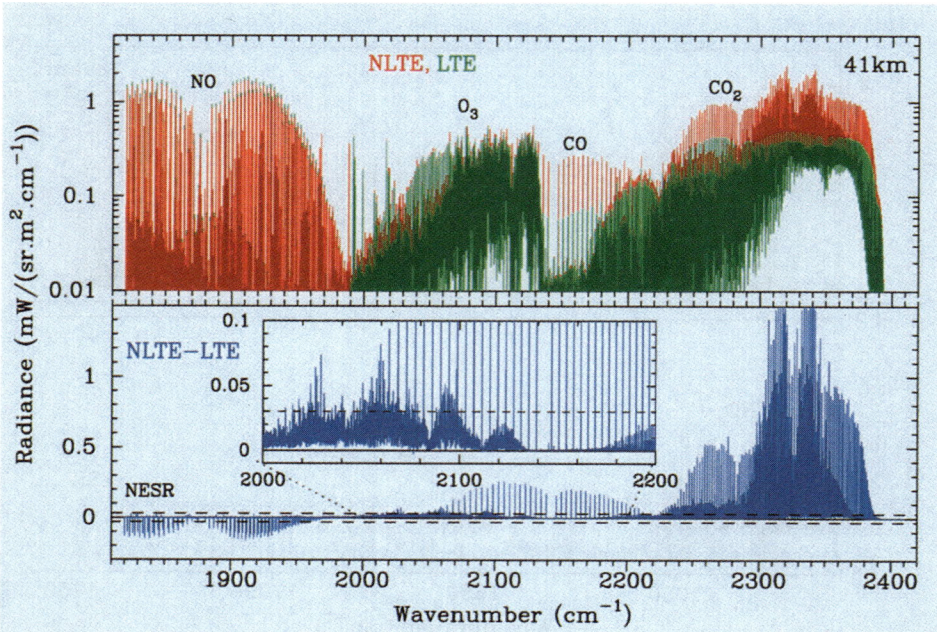

Plate 15 Synthetic spectra in the 4–5 μm region at a tangent height of 41 km. Note the log scale. Spectral resolution of 0.025 cm^{-1}. Top: Non-LTE (red) and LTE (green) radiances. Note that LTE radiances are larger at $\tilde{\nu} < 1990$ cm^{-1}. Bottom: Non-LTE–LTE radiance difference. The inset shows the non-LTE–LTE radiance differences in O$_3$ bands on an expanded scale.

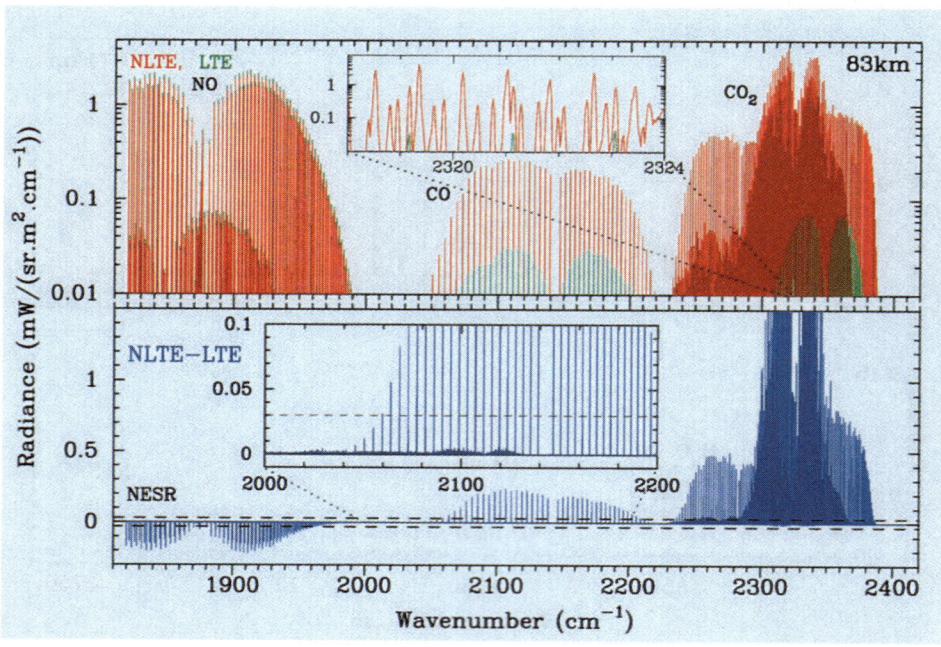

Plate 16 As plate above but for a tangent height of 83 km.

Chapter 9

Cooling and Heating Rates

9.1 Introduction

In this chapter, we use the theory and the models developed in the previous chapters to investigate the energy balance of the middle and upper atmosphere affected by non-LTE and the scale of the effect on the infrared radiative cooling and heating processes. We should note that these are only a part of the total energy balance of these regions: we will not be dealing with other sources such as transport of sensible heat, breaking of waves, chemical heating, etc., nor even all radiative sources and sinks. An example of the latter which is beyond the scope of this book is the important heating effect of the absorption of solar radiation by O_2 and O_3 in the UV and visible parts of the spectrum.

The reduction in collisional excitation which occurs in non-LTE situations leads to less efficient transfer between thermal kinetic energy and radiation, and hence affects the rate at which the atmosphere cools to space. Less obviously, it affects the way in which the energy from the Sun is absorbed as well: the lack of frequent collisions prevents the energy absorbed from the solar radiation from being efficiently thermalized, and so can have an influence on the net heating rates. A quantitative understanding of non-LTE is therefore of value not only to improve our knowledge of processes in the upper atmosphere and to derive accurate temperature and constituent abundances from remote sensing measurements, but also to understand the energy budget of this region. It follows that non-LTE processes will have a considerable effect on the structure and dynamics of the upper atmosphere. In particular, computer models used to study the global circulation of relative low-pressure regions like the mesosphere need to incorporate a non-LTE radiation code, or large errors can result.

We will discuss radiative cooling from the vibrational states first, followed by the near-infrared heating processes. The most important bands contributing to cooling in the infrared are the CO_2 15 μm emissions, and next the O_3 bands near 9.6 μm, followed by the far-infrared pure rotational (beyond 20 μm, 0–500 cm^{-1}) and the vibration-rotation 6.3 μm H_2O bands. For H_2O, the cooling produced by the pure rotational bands is significantly larger than that due to the 6.3 μm bands. The NO 5.3 μm emission is dominant in the thermosphere; and the $O(^3P)$ 63 μm emission is also important in this region.

The effect of non-LTE on heating is mainly through the resonant scattering of incoming solar radiation in the CO_2 and H_2O bands near 2.7 μm, and CO_2 at 4.3 μm. The heating rate due to absorption of solar radiation in the near-infrared would be much larger if the photons were immediately and locally thermalized, as they would be if the excited levels were kept in LTE. Before it is thermalized, part of the absorbed energy is re-radiated through fluorescence processes, usually at a different wavelength. Also, the absorbed energy can be redistributed among other internal energy states of the molecule, before it is finally converted to kinetic energy. In addition we also consider the case where energy from excited states of other constituents can be transferred into the system under consideration. For example, from $O(^1D)$, which is not itself active in the infrared spectrum but which can provide energy during collisions to those which are, like CO_2 and H_2O.

In these calculations we will consider the global distribution of the parameters, that is, the whole range of seasonally-varying conditions of temperature, composition (especially important for such highly variable species as atomic oxygen), and solar illumination. We will also consider the effect of the uncertainties and variability in some atmospheric parameters as the CO_2 and $O(^3P)$ abundances on typical profiles of the cooling and heating rates.

As in previous chapters, the calculation and understanding of heating and cooling rates requires the knowledge of some key parameters of the non-LTE models. In particular, the rate coefficients for many important reactions need to be known with precision, and their dependence on temperature understood. As we have seen, many rates are poorly known and sometimes little studied, so the results we will be describing for cooling and heating are subject to considerable uncertainties. For some cases we quantify these uncertainties by introducing in the non-LTE models the uncertainties in the rates. This is a considerable advance, even so, since until recently few studies of atmospheric energy budget were in a position to take non-LTE into account at all. The situation is gradually improving due in part to the availability of satellite measurements to constrain the results and reduce the uncertainties in the models. Examples relevant to heating and cooling rate calculations, some of which were discussed in Chapter 8, are the measurements of the 4.3 μm atmospheric emission and the CO_2 VMR by the SAMS and ISAMS instruments; ATMOS information on $k_{CO_2\text{-}O}\,[O(^3P)]$; CLAES data on the $O(^1D)$ excitation mechanisms; and CIRRIS-1A and CRISTA on NO and $O(^3P)$ coolings.

9.2 CO_2 15 μm Cooling

We use again in this chapter the non-LTE radiative transfer model described in Chapter 6, which accounts for coupling between energy levels by collisional processes (vibrational to translational and vibrational to vibrational) and by radiative processes, including absorption by solar radiation and the exchange of photons between atmospheric layers. The Curtis matrix method is used to solve the radiative transfer and statistical equilibrium equations simultaneously for all of the relevant energy states and transitions between them. For CO_2, the model includes the levels of four different isotopes, coupling with the vibrational levels of N_2 and O_2, and pumping of CO_2 (via N_2) during collisions with vibrationally excited hydroxyl radicals and electronically excited oxygen atoms.

The heating rate is normally calculated by the expression (c.f. Eq. 3.60)

$$h(z) = 4\pi S n_a(z) \left[\bar{L}_{\Delta\nu}(z) - J_{\nu_0}(z) \right],$$

where $J_{\nu_0}(z)$ is the non-LTE source function described in Chapter 5. When V–V processes are involved, the net radiative energy absorbed or emitted by a given band does not always go directly into kinetic energy. Some of it may end up in excited states where it can be re-radiated. Thus, it must be kept in mind that the term usually called the 'heating' (or cooling) rate in the radiative transfer equation may not always represent the amount of absorbed (or emitted) energy that is converted into (or taken from) kinetic energy.

We showed already in Chapter 5 that the heating rate can be simplified under certain conditions. Taking for example the case of the CO_2 15 μm fundamental band in the thermosphere, the excitation of its upper level $(0,1^1,0)$ is dominated by thermal collisions with $O(^3P)$ (process 2 in Table 6.2) while the major loss is spontaneous emission. Radiative transfer plays a minor role in this region and hence can be neglected. Using, for example, the 'total-escape' approximation (see Eq. 5.6), the cooling rate, $q = -h$, is given by

$$q = \frac{[CO_2]}{Q_{\text{vib}}} [O(^3P)] \, g_2 \, k_{CO_2\text{-O}} \, \exp\left(-\frac{h\nu_0}{kT} \right) h\nu_0, \qquad (9.1)$$

where Q_{vib} in the vibrational partition function, $k_{CO_2\text{-O}}$ is the rate coefficient for collisions between $CO_2(0,1^1,0)$ and $O(^3P)$, $g_2 = 2$ is the degeneracy factor of $(0,1^1,0)$, and the brackets mean number densities. Alternatively, taking the 'cooling-to-space' approximation (see Sec. 5.2.1.2), the cooling rate is given by $q_{\text{CTS}} = q \times (T^\star/2)$, where $T^\star$ is the probability of photons to escape to space. q is given in units of energy (erg or J) per unit of volume (cm^{-3}) and time (s^{-1}), while it is more usual to express heating or cooling in terms of the rate at which an atmospheric parcel will change its temperature keeping its pressure constant, with units such as K/day. To do that, q is divided by the atmospheric density and heat capacity at constant pressure, e.g, by $[N] \, M \, c_p$, where $[N]$ is the total number density, M is the mean molecular weight, and c_p is the specific heat at constant pressure. Hence, the cooling rate in K/day depends on the VMR of CO_2, rather than its number density, and, to a lesser extent, on the relative abundance of $O(^3P)$ through its contribution to the atmospheric heat capacity and the mean molecular weight.

At this point it is useful to consider how the cooling rate is related to the radiance measurements taken by emission instruments on satellites, particularly in the optically thin case. In Sec. 8.2.2 we derived a simple expression for the latter, (Eq. 8.16). Integrating that equation over the band' spectral interval we have

$$L(x_{\rm obs}) = \frac{h\nu_0}{4\pi} A_{21} \int_{x_{\rm s}}^{x_{\rm obs}} n_2(x)\,{\rm d}x, \qquad (9.2)$$

from which, by performing a simple Abel inversion, we get $A_{21}n_2(z)$, the number of $h\nu_0$ photons emitted by the atmosphere per unit volume and time. This quantity, usually called the *volume emission rate*, coincides with the expression given above for the cooling rate (Eq. 9.1) since, for the case of $CO_2(0,1^1,0)$ under these conditions, the population of the upper emitting state n_2, is given by

$$[CO_2(0,1^1,0)] = \frac{[CO_2]}{Q_{\rm vib}} \frac{[O(^3P)]\, g_2\, k_{CO_2\text{-}O}}{A_{21}} \exp\left(-\frac{h\nu_0}{kT}\right). \qquad (9.3)$$

Hence, from a measurement of the limb radiance under optically thin conditions, we can derive to a first approach the rate of loss of heat directly in units of energy per volume and time. However, with that measurement alone we cannot get the cooling rate in K/day since for this we also need the atmospheric number density, mean molecular weight and heat capacity, which requires a knowledge of the atmospheric pressure and temperature, and the $O(^3P)$ concentration. This analysis applies equally to the NO(1) 5.3 μm cooling which is discussed in Sec. 9.5.

From Eq. (9.1) we see that the CO_2 15 μm cooling rate in the thermosphere depend primarily on the following four parameters:

(i) the kinetic temperature;
(ii) the CO_2 abundance (number density or volume mixing ratio);
(iii) $k_{CO_2\text{-}O}$, the rate coefficient for the collisional deactivation of $CO_2(0,1^1,0)$ by atomic oxygen; and
(iv) the $O(^3P)$ number density.

The first and last of these are very variable and so we must consider a range of global and seasonal conditions. Furthermore, the models and measurements of CO_2 abundance show significant discrepancies even at well-studied latitudes and times of year (see Sec. 8.10.3).

We have seen in previous chapters that atomic oxygen is very effective in deactivating the bending mode of CO_2 and must be taken into account. The fast rate coefficient for collisions between $CO_2(0,1^1,0)$ and $O(^3P)$ makes the cooling due to the CO_2 15 μm emission a major term in the energy budget of the lower thermosphere, comparable with the role played by dynamical processes. However, the value of the coefficient at atmospheric temperatures is uncertain and the published values do not agree with each other. The rate employed in Chapter 6 was determined from ATMOS spectra, and is intermediate between the value determined in

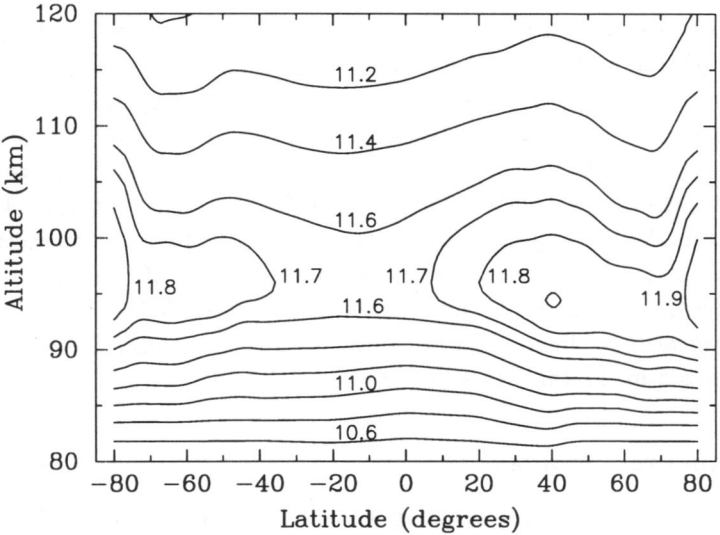

Fig. 9.1   The global distribution of atomic oxygen number density used in the model for December (solstice). The values are given in $\log_{10}$ of number density in $mol\,cm^{-3}$.

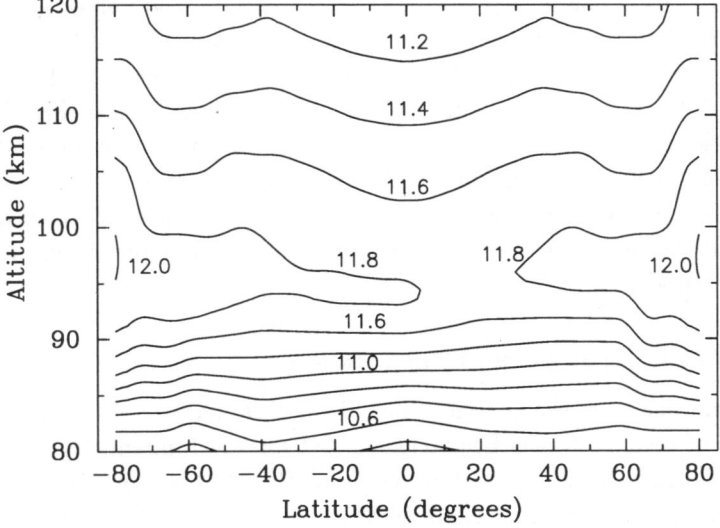

Fig. 9.2   The global distribution of atomic oxygen used in the model for March (equinox). The values are given in $\log_{10}$ of number density in $mol\,cm^{-3}$.

the laboratory (which is consistent with the SISSI radiances) and that inferred from an analysis of the CO_2 $15\,\mu m$ emission measured by SPIRE.

Uncertainties in the concentration of $O(^3P)$ have the same effect on the cooling rates as the $k_{CO_2\text{-}O}$ rate coefficient, especially above 90 km. They also have a significant impact on the solar near-IR heating of CO_2, since the vibrational energy transferred from $CO_2(v_1,v_2,1)$ to $N_2(1)$ is thermalized by collisions of $N_2(1)$ with

$O(^3P)$ (process 15 in Table 6.2) (see Sec. 9.8.2). To incorporate this properly, we need a complete latitude- and altitude-dependent model of the atomic oxygen abundance in the higher mesosphere and lower thermosphere, such as that of Rees and Fuller-Rowell. This is fitted to experimental data from the Solar Mesosphere Explorer (SME) satellite in the 80–86 km region, where the model systematically underestimates the measurements. The resulting $O(^3P)$ concentration for December (solstice) and for March (equinox) are shown in Figs. 9.1 and 9.2, respectively.

Finally, the rate coefficient for the vibrational exchange of CO_2 v_3 quanta with the first excited state of molecular nitrogen, process 9 in Table 6.2, is also important. The current uncertainty in the knowledge of this rate is within factors of 0.2 and 2 of the value listed in the table (see Sec. 9.8.2).

9.2.1 *Cooling rate profiles*

At night, the energy radiated to space by CO_2 in the 15 μm bands is all thermal in origin and so cools the atmosphere, although, as will be shown later, net heating can still result at cool high levels due to the absorption of part of the flux from warmer regions below. During the day, there is an additional heating from the solar energy absorbed by the CO_2 bands near 2.7 and 4.3 μm, as discussed below in Sec. 9.8.

The general behaviour of the CO_2 15 μm cooling rate is not the same in the stratosphere and mesosphere as it is in the lower thermosphere. In the latter, the cooling is mainly due to the ν_2 fundamental band, which is optically thin, and, as shown in the previous section and in more detail below, largely depends on the CO_2 and $O(^3P)$ concentrations and on the rate coefficient for the de-excitation of the bending mode of CO_2 by $O(^3P)$. In the mesosphere, the fundamental band is optically thick and hence the radiative cooling in the weaker isotopic and hot bands become more important.

9.2.1.1 *Cooling in the stratosphere and mesosphere*

Figures 9.3 and 9.4 show the contributions of the CO_2 bands to the cooling. They show profiles of cooling versus height calculated for the fundamental (FB), the 3 first hot bands and the 7 second hot bands of the 626 isotope (HOT), and the fundamental bands of the minor isotopes 636, 628, and 627 (ISO) (see Table F.1). All of these profiles are calculated for the same model parameters as in Chapter 6, using the CIRA 86 temperature profiles for polar summer (December at 80°S) and polar winter (December at 80°N) conditions (see Fig. 1.1).

The main result is that the weaker bands contribute as much as the strong fundamental band to the cooling rate in this region, because the weaker bands are optically thinner and the photons escape to space more easily, while the fundamental band becomes optically thin only at much higher altitudes in the lower thermosphere. Note how the hot (mainly the first hot) bands induce a cooling rate larger than that of the fundamental band in the region between the stratopause

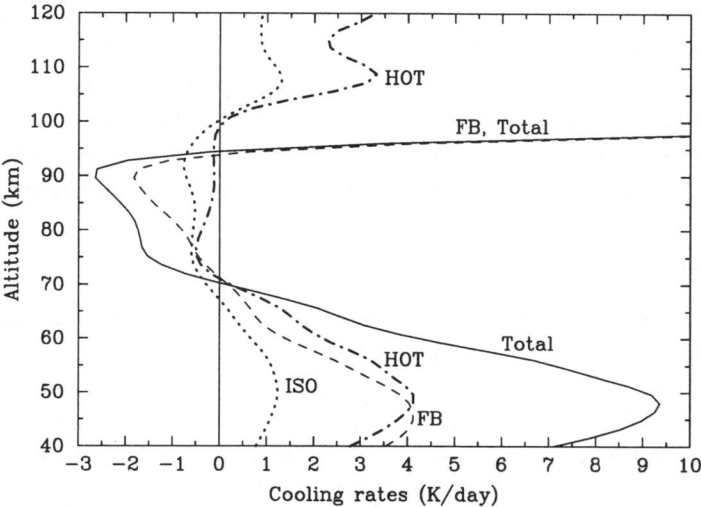

Fig. 9.3   Cooling rates of the CO_2 $15\,\mu m$ bands for summer conditions.

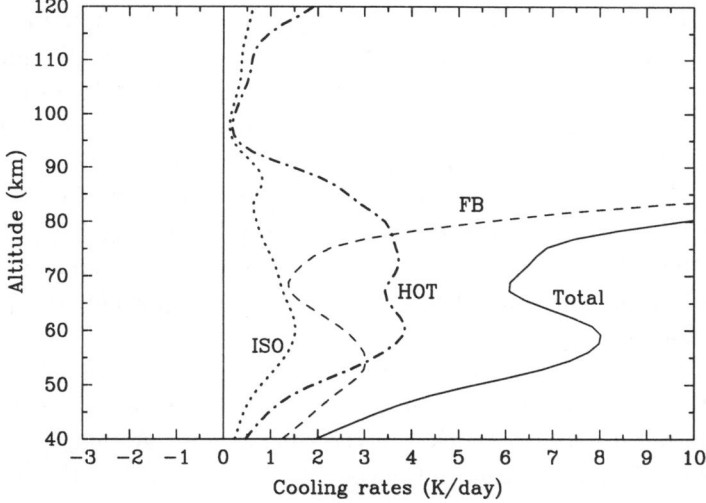

Fig. 9.4   Cooling rates of the CO_2 $15\,\mu m$ bands for winter conditions.

and the 75 km level. The contributions of these weak bands, relative to the fundamental, depend quite strongly on the temperature structure. For example, the relative contribution of the hot bands in the polar winter mesosphere, at around 65 km (Fig. 9.4), is more than twice the fundamental component and much larger than in the same region in the polar summer (Fig. 9.3).

The three minor isotopic bands also radiate away a significant amount of energy throughout the mesosphere. The large cooling near the winter pole is due to emission from the Doppler wings, driven by the warmer mesospheric temperatures which occur at that time. The net heating of the summer polar mesopause is due to photons from the lower mesosphere, which are trapped there by the large negative temperature lapse rate in the mesosphere and thermalized. The heating of the weak bands is of the same order as that produced by the fundamental.

The CO_2 4.3 μm bands produce a maximum cooling of about 0.2–0.35 K/day around the stratopause, and that of all other CO_2 bands is negligible.

9.2.1.2 *Cooling in the thermosphere*

The cooling in the thermosphere is dominated by the 15 μm fundamental band, with the hot (mainly the first hot) bands also contributing significantly, especially in the lower part of the region (Figs. 9.3 and 9.4). One of the most outstanding feature is the much greater cooling in the lower thermosphere for summer conditions, compared to winter (Figs. 9.5 and 9.6). The large $k_{CO_2\text{-}O}$ rate coefficient and the high temperatures force the fundamental band close to LTE in the thermosphere, and the big seasonal difference in cooling rates is mostly due to the much larger temperature over the summer pole (see Fig. 1.1). The rapid increase of cooling with height above 95–100 km is also due to the rapid increase of temperature. At altitudes above 120–130 km (Figs. 9.7 and 9.8), the CO_2 VMR and the $O(^3P)$ concentration both fall off, and the cooling gradually diminishes despite the high temperatures.

The detailed dependence of the cooling rate on the CO_2 VMR can also be seen in Figs. 9.5 and 9.6, from calculations of the cooling rates for the extreme CO_2 profiles of Fig. 8.32, for polar summer and polar winter conditions. It is interesting to note that the cooling rates in the 85–95 km region are very similar despite the fact that the CO_2 profiles in this region are rather different (Fig. 8.32). For a larger CO_2 VMR, the number of photons emitted at the lower thermosphere ($z \geq 100$ km) and absorbed by the layers just below is larger, and this partially compensates for the local cooling.

As we would expect from the effect of $O(^3P)$ on the vibrational temperatures (Sec. 6.3.6.4), the abundance of atomic oxygen and the value of the $k_{CO_2\text{-}O}$ coefficient also have an important effect on the cooling rate. A change of a factor of 4 in the rate coefficient can change the cooling rates by 100% in the thermosphere (Figs. 9.7 and 9.8). This, plus a further 50% due to an estimated factor of two variability in the concentration of $O(^3P)$, represents the range of plausible variability of these factors and the expected range of changes in the cooling rate produced by CO_2 15 μm in the thermosphere. The change in the cooling (or heating) of the 85–95 km region due to the variation in $k_{CO_2\text{-}O}$ is much larger than that due to the uncertainty in the CO_2 concentration.

9.2.2 *Global distribution*

The global variability of the cooling rate produced by the CO_2 15 μm bands for night-time is shown as a function of latitude, altitude and season in Figs. 9.9 and

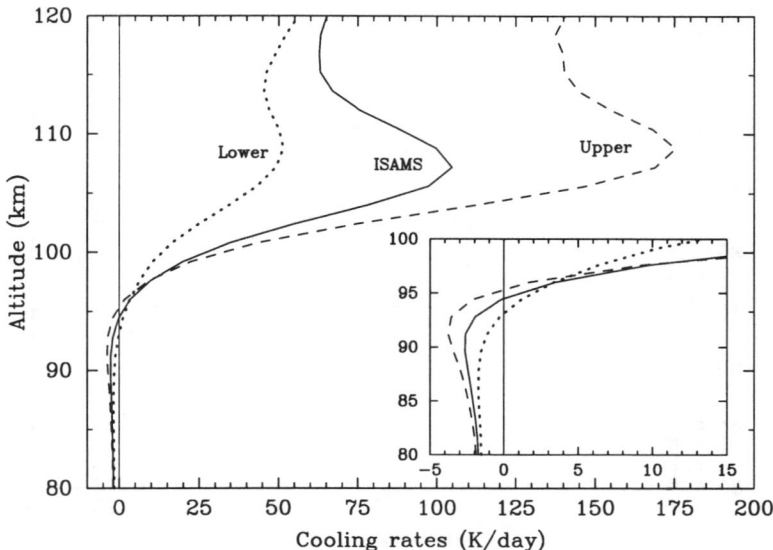

Fig. 9.5   Effect of the CO_2 VMR (see Fig. 8.32) on the cooling rates of the CO_2 $15\,\mu m$ bands for summer conditions. The inset shows the cooling around the mesopause at a larger scale.

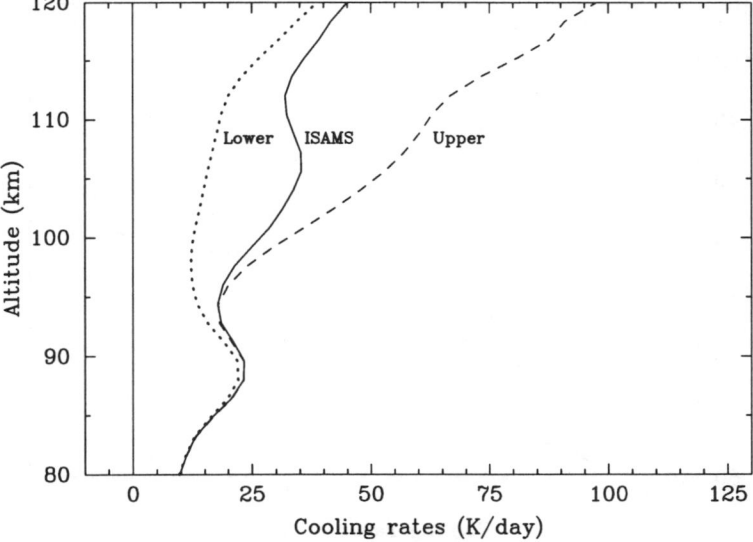

Fig. 9.6   Effect of the CO_2 VMR (see Fig. 8.32) on the cooling rates of the CO_2 $15\,\mu m$ bands for winter conditions.

9.10. They principally reflect the temperature structure differences (see Figs. 1.2 and 1.3) between December (solstice) and March (equinox) conditions. At solstice, near the stratopause, we find a cooling rate of 9 K/day over the summer pole, decreasing to a minimum of 6 K/day over the equator and increasing again near

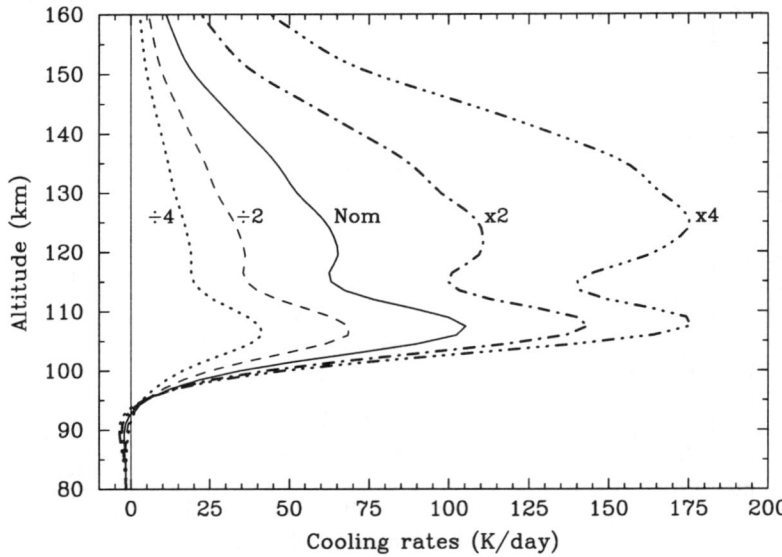

Fig. 9.7 Effect of the $k_{CO_2-O}[O(^3P)]$ product on the cooling rates of the CO_2 15 μm bands for summer conditions. Note the extended altitude scale in this and next figure.

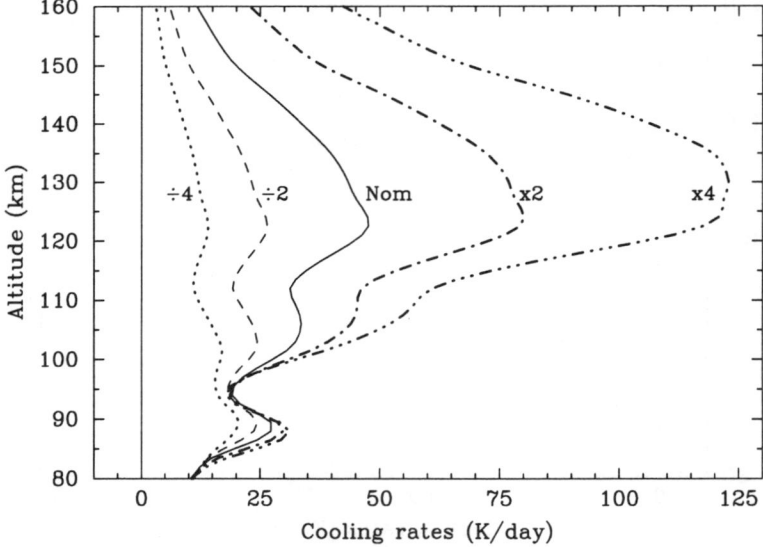

Fig. 9.8 Effect of the $k_{CO_2-O}[O(^3P)]$ product on the cooling rates of the CO_2 15 μm bands for winter conditions.

the winter pole to 8 K/day. In the mesosphere, the cooling is smaller and mostly by the weak bands; it has a minimum value of 1 K/day over the equator, slightly decreases towards the summer pole, and increases toward the winter pole. Around the mesopause, the total cooling in winter is about 20 K/day but the atmosphere is

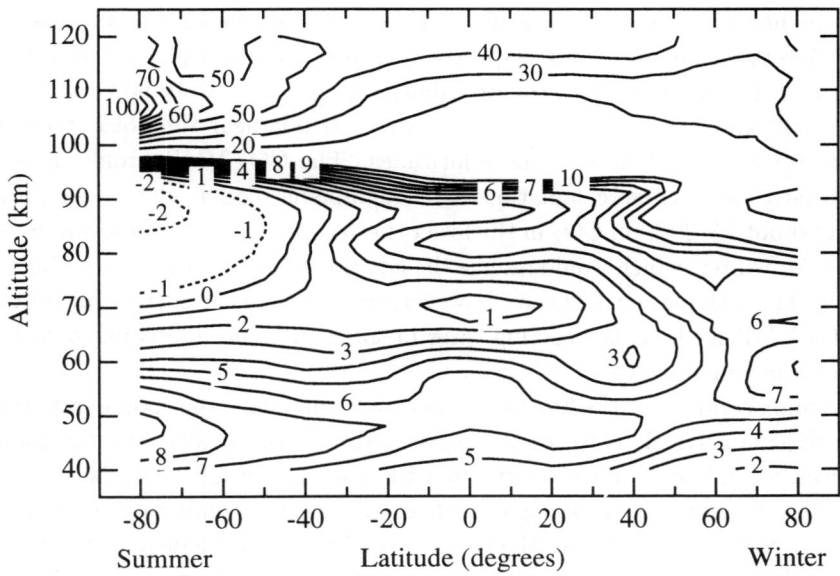

Fig. 9.9 Global altitude-latitude distribution of the cooling rate (K/day) induced by the CO_2 $15\,\mu m$ bands for night-time and solstice conditions (December) for the CIRA 1986 temperature structure (Fig. 1.2). The dashed lines represent heating.

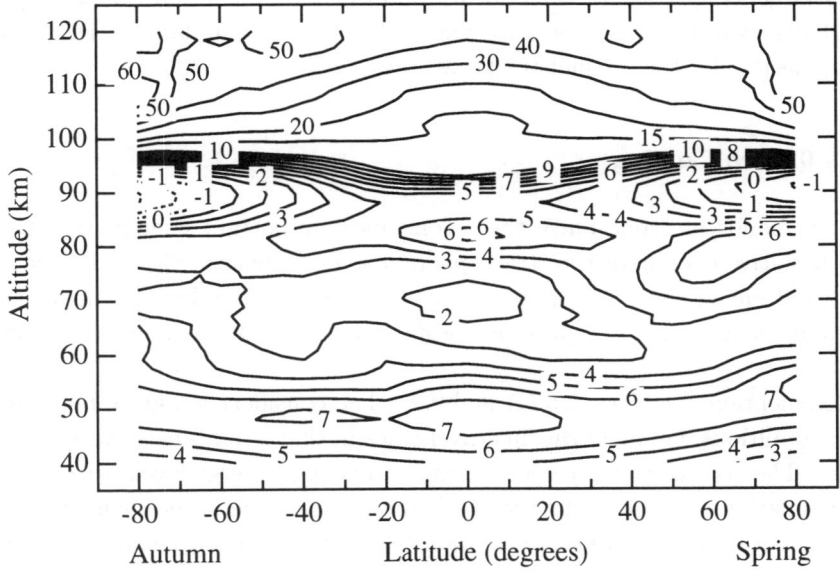

Fig. 9.10 Global altitude-latitude distribution of the cooling rate induced by the CO_2 $15\,\mu m$ bands for night-time and equinox conditions (March) for the CIRA 1986 temperature structure (Fig. 1.3). The dashed lines represent heating.

heated at the summer pole by as much as $3\,K/day$.

As we have already seen, the lower thermosphere exhibits very large differences between the hemispheres (30 K/day and 70 K/day around the 110 km altitude level near the winter and summer poles, respectively) due to the atmospheric temperature structure and, to a lesser extent, the distribution of $[O(^3P)]$. Note how the cooling in the 110–120 km region tends to increase with latitude near the poles, reflecting the higher $O(^3P)$ abundances at these latitudes. The CO_2 VMR profile does not vary with latitude in these calculations, although some models show a significant depletion (about 15–20%) in CO_2 in the lower thermosphere/upper mesosphere (80–100 km) polar winter, which would reduce the latitudinal gradient in the cooling rate of Fig. 9.9. These changes in CO_2 are, however, smaller than the current uncertainty in the CO_2 VMR itself, so the cooling rates in this region are uncertain by at least a factor of 2 in consequence.

For equinox conditions, the kinetic temperature and hence the cooling rate tends to be nearly symmetrical, as we would expect. Around the stratopause the cooling rate is around 7 K/day at most latitudes, with a maximum of 7.5 K/day over the tropics. The mesospheric rates again reflect the kinetic temperature structure, with greater cooling over the relatively warm region in the vicinity of 60°N. The atmosphere is heated in a small region around the mesopause over the poles, the rate being higher in the Southern Hemisphere, but still smaller than at solstice. In the lower thermosphere (∼120 km), the cooling rate is slightly larger near the South pole (∼60 K/day) where the atmosphere is warmer. Note that this cooling rate is as large as that at the same location during summer when the atmosphere is around 40 K warmer; this is because the smaller $O(^3P)$ abundance (Figs. 9.1 and 9.2) compensates for the warmer temperatures.

9.3 O_3 9.6 μm Cooling

The 9.6 μm bands of ozone make the second most important contribution to the infrared radiative cooling of the stratosphere and mesosphere. Figs. 9.12 and 9.13 show calculations of the O_3 cooling rate for the mid-latitude day and night, polar summer and polar winter temperature profiles with the O_3 abundances shown in Fig. 9.11.

As was mentioned in Secs. 3.6.5.1 and 9.2, when considering a given band under non-LTE conditions, not all of the flux divergence in that band is necessarily heating or cooling. The O_3 molecule is a good example. The high energy levels are excited in the recombination reaction leading to O_3 formation, as some of the chemical energy is partially transferred to internal vibrational energy. The energy emitted by these levels is not actual 'cooling', but rather part of the chemical energy which is emitted away instead of being thermalized. Then, a clear distinction needs to be maintained between the 'classic' non-LTE problem, involving radiative and thermal collisional processes, and the case when non-LTE chemiluminescence is also involved.

Figure 9.12 shows the cooling rate in LTE for the fundamental ν_3 band of ozone in the 9.6 μm region, and for this and all of the other O_3 bands in that spectral region

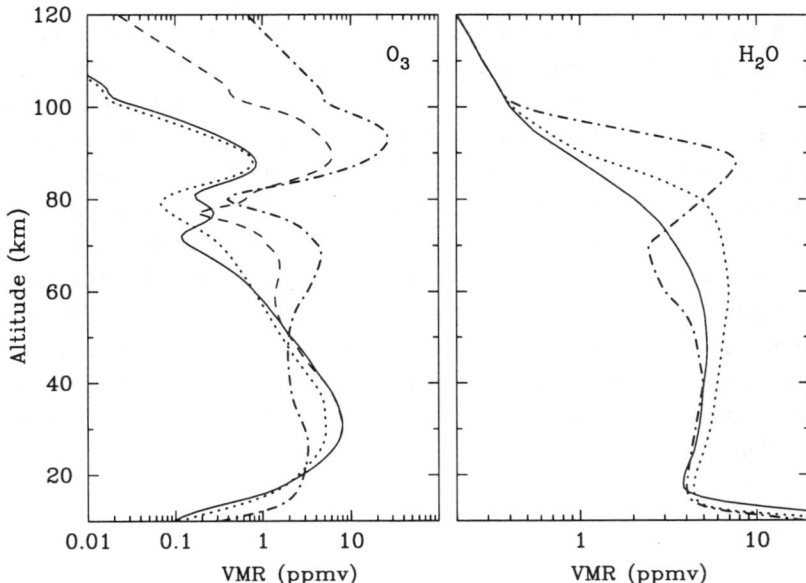

Fig. 9.11   Volume mixing ratio of O_3 and H_2O for mid-latitudes day (solid) and night (dashed), polar winter (dot–dash) and polar summer (dotted). The O_3 VMR's are taken after the Garcia and Solomon model.

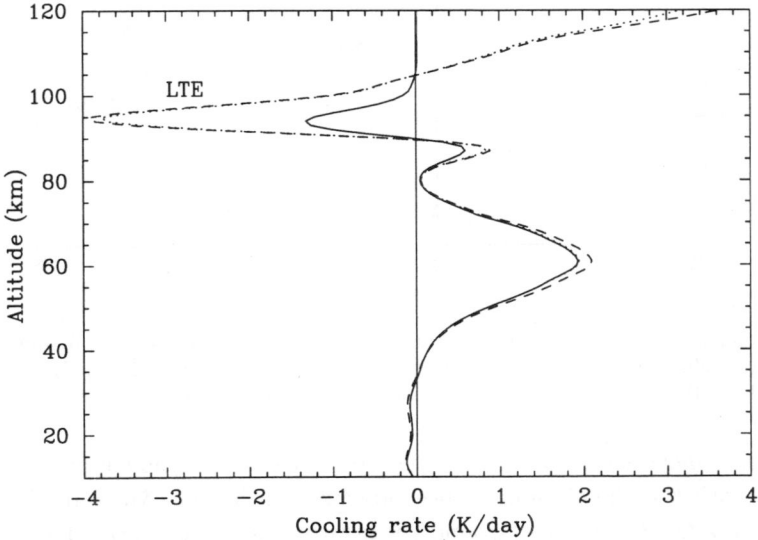

Fig. 9.12   Cooling rates of the O_3 9.6 μm bands for polar winter conditions. The solid curve is the non-LTE cooling rate of the $(0,0,1\rightarrow0,0,0)$ band; the dotted line (overlapped with the solid line below 80 km and nearly overlapped with dashed line above) is the LTE $(0,0,1\rightarrow0,0,0)$ cooling; and the dashed curve is the contribution of all bands near 9.6 μm (ν_1 and ν_3 fundamental and hot bands) in LTE.

combined, all for polar winter conditions. Except for the stratosphere and upper mesosphere (particularly around the stratopause and in the lower stratosphere), the cooling is dominated by $(0,0,1-0,0,0)$ (ν_3), with very small contributions from $(1,0,0-0,0,0)$ (ν_1) and the various hot bands. The figure also shows the cooling rate by the $(0,0,1-0,0,0)$ band under non-LTE conditions caused by radiative processes only, i.e., excluding the small contribution from the O_3 formation reaction. We see that LTE is a good approach below about 80 km, but substantially overestimates the mesopause heating. The LTE heating is larger for the following reason. The absorption of photons from the upper stratosphere and mesosphere is the same in LTE as in non-LTE, but the local upper mesospheric cooling is underestimated in LTE because it does not account for the emission to space produced by the extra O_3 molecules excited by absorption of the upwelling radiation around the mesopause (T_v is larger than T_k).

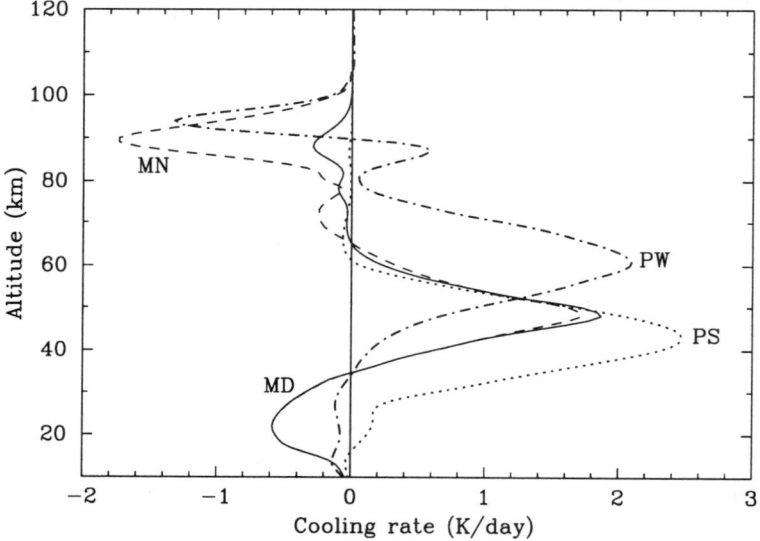

Fig. 9.13 Cooling rates of the O_3 9.6 μm bands for non-LTE conditions caused by radiative processes only (chemiluminescence processes neglected). Mid-latitude day (MD, solid), mid-latitude night (MN, dashed), polar winter (PW, dot–dashed) and polar summer (PS, dotted).

Figure 9.13 shows the radiative cooling rates of the O_3 9.6 μm bands for non-LTE conditions, with the chemiluminescence processes neglected. Much of the cooling occurs within a 20–30 km layer around the stratopause, just above which both the temperature and O_3 volume mixing ratio decrease rapidly. The maximum cooling rate is approximately 2 K/day near the stratopause. The cooling in the mesosphere is very small and, as for the CO_2 bands, they produce a net heating rate around the cold mesopause. This heating is larger for night-time and polar winter conditions, when O_3 is more abundant, and can be near 2 K/day at night (corresponding to 1 K/day globally) which is considerable compared to the other sources and sinks of energy in this region.

As in the case of the CO_2 15 μm bands, the radiative cooling rates for the O_3 bands in Fig. 9.13 generally follow the temperature profile, although the effect of the greater variability of O_3 VMR is also apparent. The maximum cooling of around 2.5 K/day occurs for summer in the polar stratosphere. The second peak in the cooling at around 85 km in polar winter is due to the local maximum in O_3 VMR (see Fig. 9.11). The day and night-time mid-latitude (solid and dashed) curves have the same temperature profile but rather different O_3 VMR's (see Fig. 9.11). The cooling rate at the stratopause is decreased slightly at night as a consequence of an increase in the O_3 abundance at the 70 km level and above, which inhibits the local cooling near the temperature maximum at the stratopause.

Figure 9.13 also shows weak heating in the lower stratosphere, which is more pronounced at mid-latitudes where the troposphere is warmer. This region, just below the O_3 maximum, is heated by radiation emitted from the warmer troposphere and the surface below. In consequence, the heating in this region is sensitive to the tropospheric O_3 abundance: when more ozone is present in the troposphere, this partially blocks the upward flux in the ozone bands, leading to less absorption in the lower stratosphere and hence less heating.

The heating rate near the mesopause depends on the temperature and O_3 VMR, both locally and in the layers below. A warmer temperature in the upper stratosphere and lower mesosphere produces increased heating at the mesopause. When the stratospheric peak in the O_3 VMR is smaller, the atmosphere is more transparent, exposing the mesopause to the lower, colder, stratospheric regions. On the other hand, more O_3 at the mesopause produces more absorption and hence more heating. Higher mesopause temperatures tend to increase the heating, since the faster collisions which result are more effective in converting the absorbed radiation into kinetic energy of the molecules. The opposite is true for the vibrational populations, which are larger for a colder mesopause temperature.

The heating rates shown in Fig. 9.13 represent the combined effects of the temperature and O_3 profiles. It appears that the larger O_3 abundances near the mesopause dominate, giving larger heating rates for night-time and polar winter conditions. Comparing polar winter to mid-latitude night, the former has larger mesopause O_3 VMR's and warmer mesopause temperatures, but the temperature is colder at the stratopause and the O_3 VMR stratospheric peak is reduced. The net effect is a slightly smaller heating for polar winter conditions. At the illuminated mesopause, the heating is smaller, becoming almost negligible at the summer pole, as shown in the figure. In the polar summer mesopause, the upwelling flux is large, mostly because of the warmer stratospheric temperature, but the mesopause temperature is so low that collisions are not efficient enough to thermalize the energy absorbed.

9.4 H_2O 6.3 μm Cooling

In contrast with its role in the troposphere, water vapour contributes little to the radiative energy budget of the stratosphere and mesosphere, because of its relatively low concentration in these regions. Model results for the radiative cooling by the 6.3 μm fundamental band are shown in Fig. 9.14 for four atmospheric conditions (mid-latitude day and night, polar summer and polar winter conditions), using the volume mixing ratios shown in Fig. 9.11. The contribution of the 6.3 μm band is smaller, typically around 30%, than that of the pure rotational H_2O band, whose emission extends beyond 20 μm (0–500 cm^{-1}).

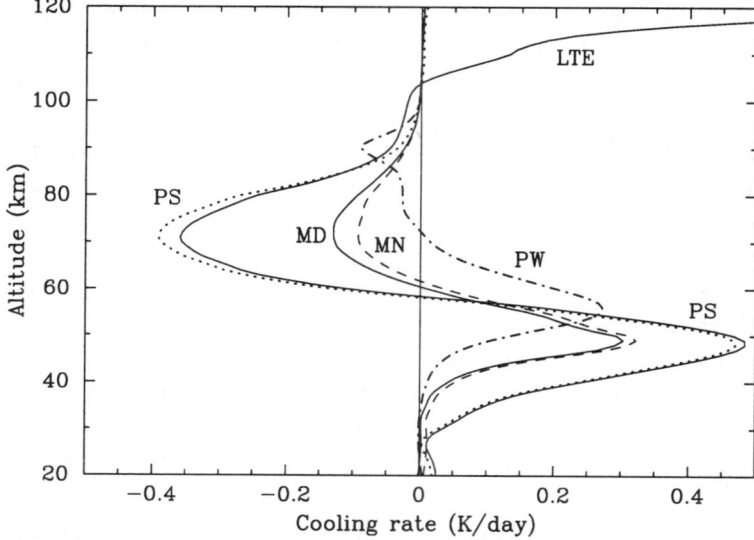

Fig. 9.14   Cooling rates of the H_2O 6.3 μm bands for mid-latitudes day (MD, solid) and night (MN, dashed), polar winter (PW, dot–dash) and polar summer (PS, dotted: non-LTE, and solid: LTE).

The cooling rate due to the 6.3 μm band reaches a maximum of about 0.3 K/day at mid-latitudes, while the region near 70 km exhibits net heating induced by the absorption of radiation from the lower regions around the stratopause. Like CO_2, the H_2O cooling pattern reflects the temperature structure and is most important for polar summer conditions, when cooling at the stratopause reaches around 0.5 K/day, and heating at the 70 km level around 0.4 K/day. As can be seen in Fig. 9.14 by comparing the non-LTE (dotted line) and LTE (solid) polar summer cases, the effect of non-LTE is very small up to 100 km but large above. The heating at around 70 km, which would be overestimated by an LTE calculation for O_3, is underestimated in the water vapour case because of the heating produced by solar absorption at 6.3 μm and energy transferred from $O_2(1)$ (see Sec. 7.4). Hence, at mid-latitudes in the daytime the heating in the upper mesosphere is slightly larger than at night. For the same reasons, the cooling in the stratosphere is slightly reduced.

9.5   NO 5.3 µm Cooling

Emission from the fundamental band of nitric oxide at $5.3\,\mu$m is the dominant radiative mechanism for cooling the thermosphere between 120 and 200 km. Because the band is optically thin throughout the atmosphere, and NO(1) is mainly excited by thermal collisions with $O(^3P)$ in the thermosphere, the cooling q produced by NO(1) under non-LTE conditions can be expressed by an equation similar to Eq. (9.1) for CO_2, i.e.,

$$q = [NO]\,[O(^3P)]\,k_{NO\text{-}O}\,\exp\left(-\frac{h\nu_0}{kT}\right)h\nu_0, \tag{9.4}$$

where the vibrational partition function is negligible. As before, q has to be divided by the atmospheric number density and heat capacity to obtain units of K/day.

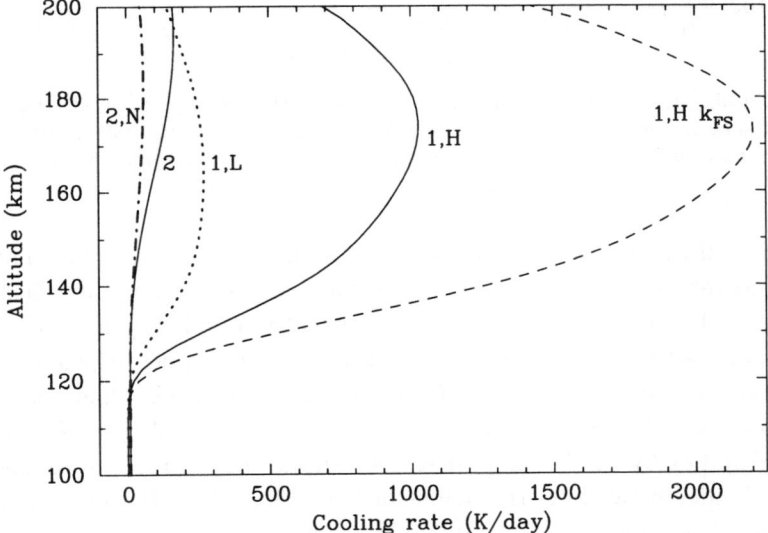

Fig. 9.15   Cooling rates of the NO(1–0) and (2–1) $5.3\,\mu$m bands for mid-latitudes, daytime conditions. The calculations are for high solar activity and the nominal $k_{NO\text{-}O}$ rate of Dodd *et al.* (1999) (see Sec. 7.6), except as noted. Symbols '1' and '2' mean the (1–0) and (2–1) bands. Curve '2-N' neglects the production of NO(2) from $N+O_2$. Curve '1-L' represents the cooling of (1–0) for low solar activity; and '1,H k_{FS}' for high solar activity and the higher $k_{NO\text{-}O}$ rate of Fernando and Smith (1979).

The cooling rate calculations require a knowledge of the $NO(1)+O(^3P)$ relaxation rate, $k_{NO\text{-}O}$, which, as discussed in Sec. 7.6, may be uncertain by a factor larger than 2. Fig. 9.15 compares cooling rates calculated with rate coefficients which span this uncertainty. According to Eq. (9.4), the cooling rate changes by about the same factor.

In addition to this dependence on the relaxation rate, the cooling rate depends directly on the NO mixing ratio, on the $O(^3P)$ number density and on the kinetic

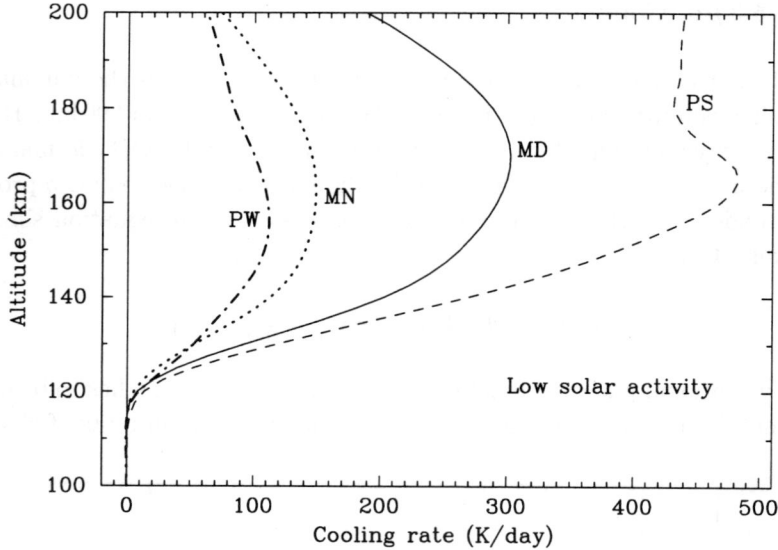

Fig. 9.16   Diurnal, seasonal, and latitudinal variabilities of the cooling rate of NO in the 5.3 μm bands for low solar activity and the k_{NO-O} rate of Dodd *et al.* (1999). 'MD', 'MN', 'PW', and 'PS' stand for mid-latitudes daytime, mid-latitudes night, polar summer, and polar winter conditions, respectively.

temperature, all three of which are very variable in the thermosphere. This variability is present in several facets, i.e., diurnal changes, seasonal and latitudinal changes, as well as changes associated with the 11-year solar cycle. The diurnal change for mid-latitudes is about a factor of 3, while from pole to pole under solstice conditions is can be as much as a factor of 5 (Fig. 9.16). The change associated with the solar cycle is also of considerable magnitude (factor of 5, see Fig. 9.15).

The energy lost by the NO(2–1) first hot band at 5.3 μm is much smaller than that of the fundamental, but still significant (see Fig. 9.15). However, most of it does not come from the kinetic energy but from the chemical energy released in the $N + O_2$ reaction (Fig. 9.15). As mentioned before, this fraction of the energy is not true cooling, since it does not come directly from the kinetic energy of the molecules or atoms. In any case, this is an energy loss of the atmosphere and as such should be taken into account in the atmospheric energy budget. This amounts to a loss rate as large as 100 K/day in the middle thermosphere for mid-latitude, daytime, high solar activity conditions. Analogously, as we will see in Sec. 9.8, we should also discount from the NO energy loss, the energy emitted by NO after being excited by absorption of solar radiation. This is, however, not significant. Being very purist, one may even say that the fraction of the energy lost by NO in the thermospheric layers due to the absorption of the tropospheric flux is not actually 'cooling', since it does not come from the *local* kinetic energy of atmospheric molecules. That distinction, however, is seldom considered when studying the atmospheric energy balance.

9.6 $O(^3P_1)$ $63\,\mu$m Cooling

Atomic oxygen emits in the far-infrared at 63 and 147 μm due to magnetic dipole transitions within the fine structure in the ground electronic state (see Fig. 7.30). The 63 μm emission makes an important contribution to the cooling of the thermosphere, while that at 147 μm is more than an order of magnitude weaker. The $O(^3P_1)$ level, in which the 63 μm line originates, is thought to be in LTE up to at least 200 km. The line is optically thick, and it is necessary to include radiative transfer to compute the $O(^3P_1)$ 63 μm cooling accurately below 150 km. In particular, $O(^3P_1)$ slightly heats the mesopause by absorption of the 63 μm radiation from the troposphere. There is very little $O(^3P)$ in the regions below, hence the atmosphere is optically thin at that wavelength. A calculation of this effect is shown in the inset of Fig. 9.17, for a surface temperature of 300 K for mid-latitudes and 250 K for polar conditions. The heating is found to be about 0.25 K/day at 100 km.

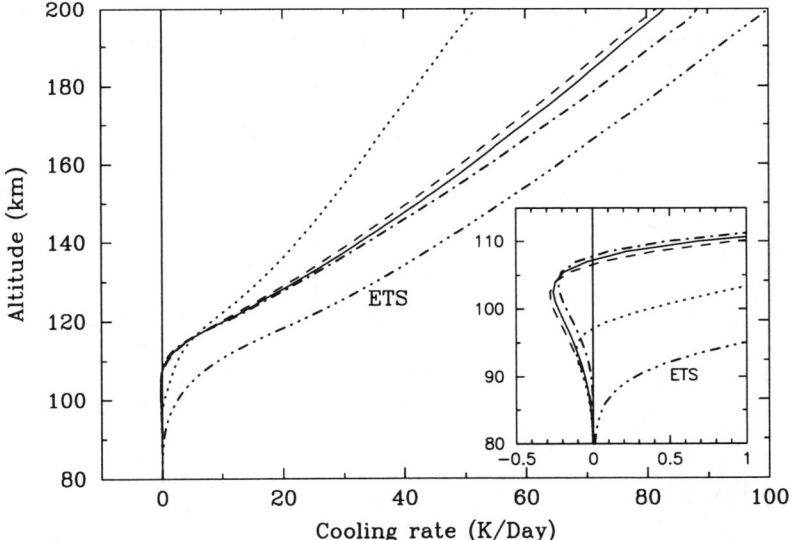

Fig. 9.17   Cooling rates of the $O(^3P_1)$ 63 μm emission for mid-latitudes daytime (solid), mid-latitude night-time (dashed), polar winter (dot–dashed), and polar summer (dotted). The 'ETS' (dot–dot–dash) line shows the total 'escape-to-space' (or optically thin) approximation for mid-latitude daytime. The inset shows the heating in the lower thermosphere in a larger scale.

Figure 9.17 shows the cooling rate of the $O(^3P_1)$ 63 μm emission for four different conditions of the thermosphere. The cooling is quite large, particularly in the upper thermosphere where it competes in magnitude with NO 5.3 μm (see previous section and Fig. 9.18). The rate increases with altitude because, when it is expressed in K/day, it is proportional to the $O(^3P)$ volume mixing ratio which increases with height in the thermosphere. This is also the reason for the seasonal variations. The kinetic temperature has only a small direct effect on the cooling rate because the small separation of the energy levels means that the fraction $[O(^3P_1)]/[O(^3P)]$ in LTE changes very little across the range of temperature variations.

Figure 9.17 also shows the importance of including radiative transfer in the computation of the cooling rate of the 63 μm line. The 'ETS' curve shows the calculation of the cooling rate assuming total 'escape-to-space' (also called the optically thin approach) (see Eq. 5.3) which, assuming LTE, e.g., $l_t \gg A_{12}$, is given, in units of $\mathrm{erg\, cm^{-3}\, s^{-1}}$, by

$$q = 3\,\frac{[O(^3P)]}{Q}\,\exp(-c_2\tilde{\nu}_0/T)\,A_{12}\,h\nu_0, \tag{9.5}$$

where $\tilde{\nu}_0 = 158.5\,\mathrm{cm^{-1}}$, Q is the partition function given by $Q = 5+3\exp(-c_2\tilde{\nu}_0/T)+\exp(-c_2E_0/T)$, c_2 is the second radiation constant, and E_0 is the energy of the $O(^3P_0)$ state ($226.5\,\mathrm{cm^{-1}}$, see Fig. 7.30). It can be seen that the 'escape-to-space' approximation seriously overestimates the cooling in the thermosphere, and does not reproduce the small heating feature in the lower thermosphere.

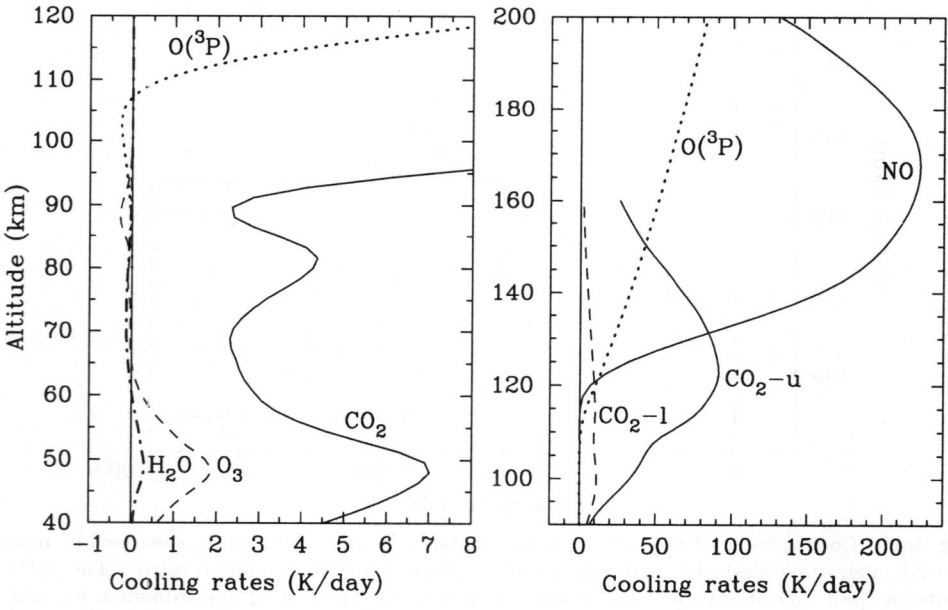

Fig. 9.18 Cooling rates of the major infrared emitters for mid-latitude conditions. The range of values for CO_2 shows the current uncertainty in the $k_{CO_2\text{-}O}$ collisional rate. NO cooling is for the day/night mean, low solar activity conditions, and the rate of Dodd *et al.* (1999).

9.7   Summary of Cooling Rates

Figure 9.18 summarises the chapter so far by comparing the cooling rates of all the major infrared emitters discussed, for typical mean mid-latitudes conditions in the U.S. Standard 1976 atmosphere. In the stratosphere, mesosphere, and lower thermosphere the CO_2 15 μm emissions dominate the cooling, while O_3 9.6 μm and H_2O 6.3 μm emissions are the second and third in importance in the stratosphere. Although not included in the figure, the far-infrared emission by H_2O is also significant. NO 5.3 μm emission dominates the cooling in the thermosphere, with important contributions from the CO_2 15 μm fundamental band and from the $O(^3P_1)$ 63 μm line. Note that the cooling by CO_2, NO and $O(^3P)$ in the thermosphere are very variable, as discussed in previous sections, and can be significantly different from those in the right panel of Fig. 9.18.

9.8   CO₂ Solar Heating

Carbon dioxide is the most important constituent of the middle atmosphere for infrared heating, as well as cooling. The bands near 2.0 μm, 2.7 μm and 4.3 μm all contribute significantly and, as we have seen in Chapter 6, are in non-LTE from the middle atmosphere upwards. They also interact strongly with the energy levels of other species, especially N_2, O_2, and atomic oxygen.

Figure 9.19 shows schematically the pathways followed by the solar energy initially absorbed by the relevant CO_2 bands, which are listed in Table F.1.

The $n v_2 + v_3$-reservoir is formed by the energy levels which are excited by the absorption of sunlight at 2.0 and 2.7 μm by ground-state CO_2 and at 4.3 μm by the excited $CO_2(0,1^1,0)$ and $(0,2,0)$ states, e.g. the $CO_2(lv_1,mv_2,v_3)$ states, with $2l + m = n = 1$, 2, 3, and 4. The similar levels without the asymmetric v_3 vibration, the $CO_2(lv_1,mv_2,0)$ states, constitutes the $n v_2$-reservoir. The v_3-reservoir comprises the $(0,0^0,1)$ level of the four most abundant isotopes of CO_2, and the first vibrational level of the nitrogen molecule.

The combination level reservoir, $n v_2 + v_3$, exchanges energy with the v_3-reservoir through V–V collisions, mostly with N_2 but also, to a lesser extent, through the less common CO_2 isotopes. A small fraction of the vibrational energy, the excess of the v_3 quanta over that required to excite $N_2(1)$, goes into kinetic energy during these V–V collisional interactions. A considerable fraction of the energy going into these reservoirs is ultimately re-emitted to space, mainly near 4.3 μm. After the combination levels lose their v_3 vibration by near-resonant V–V collisions or by radiation, the remaining $n v_2$ vibrational energy goes to enhance the $n v_2$-reservoir.

Vibrational energy is exchanged between the $n v_2$- and v_3-reservoirs through several processes. Any CO_2 isotope excited in the asymmetric stretching mode $(0,0^0,1)$ can redistribute its vibrational energy in collisions with air molecules, resulting in relaxation to the $(0,3,0)$, $(0,2,0)$ or $(0,1^1,0)$ levels. Also, $N_2(1)$ may excite the first vibrational level of O_2 and in a subsequent collision with CO_2 excite the $(0,2,0)$ state or transfer the energy to $H_2O(0,1,0)$. The excess of vibrational energy before and after the collisions is converted into thermal energy; for example, this is the principal path converting v_3 vibrational energy into kinetic energy below 80 km. At

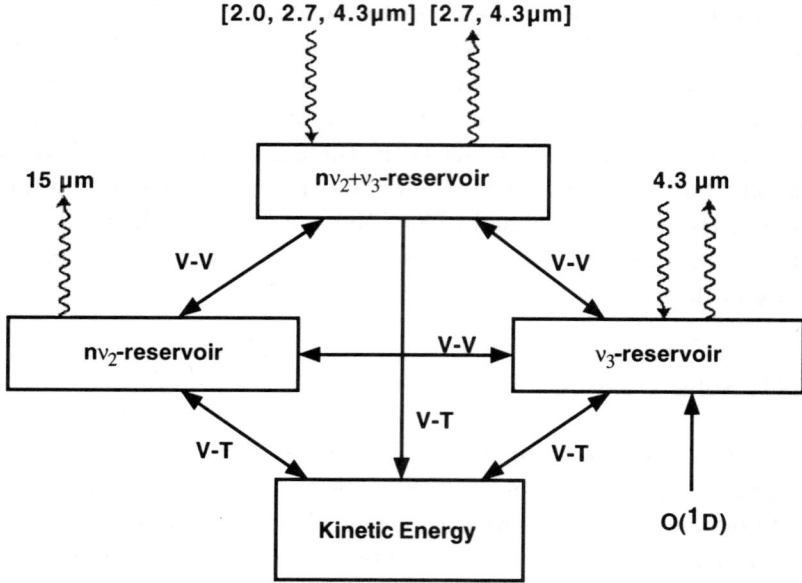

Fig. 9.19   Diagram of the pathways followed by the solar energy absorbed by CO_2 near-infrared bands. For descriptions of the 'reservoirs', see text.

higher altitudes, the exchange between the v_3-reservoir and thermal energy occurs mainly by deactivation of $N_2(1)$ in collisions with atomic oxygen. The energy going into the nv_2-reservoir is partially lost to space by the CO_2 15 μm hot bands, mainly by the first hot band. The rest is thermalized in collisions with N_2 and O_2 below 80 km, and with atomic oxygen above this altitude.

   In the following sections we calculate the amount of energy exchanged by each of the important CO_2 infrared bands, again using the model described in Chapter 6. The energy rates are given for winter conditions (December) at mid-latitudes (30°N), integrated over a 24-hour period, using the CIRA zonal mean kinetic temperatures.

The 2.7 μm bands

Carbon dioxide is excited to its $(1,0,^0,1)$ and $(0,2,^0,1)$ levels by the absorption of sunlight near 2.7 μm. For the mean solar illumination conditions of the winter hemisphere at mid-latitudes, radiation at these wavelengths penetrates down to around 90 km.

   The curve labelled 'I' in Fig. 9.20 shows the rate of energy locally absorbed from the solar radiation by those two bands (given by Eq. 4.22 integrated over v). The decrease above the mesopause reflects the falling abundance of CO_2, following the volume mixing ratio profile used in the calculations. Curve 'RT' is the divergence of the radiative flux (with sign reversed) in these bands, normally called 'heating' rate, e.g., h_{12} in Eq. (3.60). In other words, it is the absorbed solar energy after

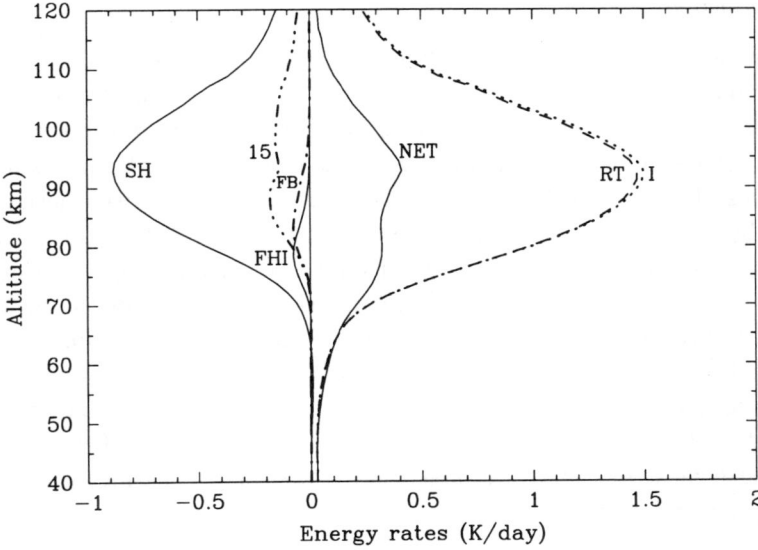

Fig. 9.20 Heating rate ('NET') and energy rate profiles integrated over one day in the winter Northern Hemisphere at 30°N, for the solar radiation absorbed in the CO_2 2.7 μm bands. See text for details. The energy rates are expressed in K/day for easy comparison with the heating rate.

including radiative transfer and all the collisional processes in these bands. This represents the energy potentially available for heating. Curves 'I' and 'RT' are very similar showing that little energy is lost from the system due to emission at 2.7 μm. The $(1,0^0,1)$, $(0,2^2,1)$ and (02^01) states are more likely to radiate the v_3 part of their energy than to re-radiate at 2.7 μm, since the spontaneous emission rates for the 4.3 μm second hot bands are around a factor of 100 larger than those for the 2.7 μm bands (see Table F.1). The photons emitted by the 4.3 μm hot bands are not re-absorbed because the populations of the absorbing levels $(1,0^0,0)$, $(0,2^2,0)$ and $(0,2^0,0)$, are relatively small, and also, the individual spectral lines of the 4.3 μm first hot, fundamental and isotopic bands do not overlap significantly in the middle atmosphere. Hence the importance of the energy lost (re-emitted to space) in the 4.3 μm second hot bands, as shown by the curve labelled 'SH' in Fig. 9.20. A considerable fraction of the energy absorbed goes into the v_3-reservoir through collisions with N_2. Some of this is also re-emitted back to space near 4.3 μm, mainly by the fundamental band of the principal isotope of CO_2, curve 'FB', and by the first hot and isotopic bands, curve 'FHI'.

When the $(1,0^0,1)$, $(0,2^2,1)$ and $(0,2^0,1)$ levels lose their v_3 quantum, whether by emitting to space directly or by transference to the v_3-reservoir, the remaining lv_1 and mv_2 quanta enhance the nv_2-reservoir. A significant fraction of this energy is radiated to space, primarily by emission at 15 μm, in the first hot bands (curve '15'). Overall, the insolation near 2.7 μm induces an average net heating rate (curve 'NET') of around half a degree per day in the mesosphere and lower thermosphere, making this band the most important contributor to the solar heating rate by CO_2.

The 2.0 μm bands

The absorption of solar energy near $2.0\,\mu m$ by CO_2 gives rise to the excitation of the vibrational levels $(2,0^0,1)$, $(1,2^0,1)$ and $(0,4^0,1)$, enhancing the energy of the $n v_2 + v_3$-reservoir. The rate of absorption of solar energy by these three bands is small but not negligible in comparison with the contribution of other bands. Most of the energy locally absorbed at altitudes above $\sim 70\,km$ is lost (mostly emitted to space) by the emission of v_3 quanta by the $(2,0^0,1)$, $(1,2^0,1)$ and $(0,4^0,1)$ levels. Some is also re-emitted near $15\,\mu m$ by hot bands. The sum of the energies emitted by the $4.3\,\mu m$ and $15\,\mu m$ hot bands represents the major loss for the region above $80\,km$. Part of the energy absorbed is also emitted again by the $4.3\,\mu m$ fundamental, isotopic and first hot bands, and by the $15\,\mu m$ bands from the $n v_2$-reservoir. As a result, a large part of the insolation at $2.0\,\mu m$ above $70\,km$ is re-radiated. However, at lower altitudes, where the pressure is higher, collisions are sufficiently frequent to thermalize most of the vibrational energy, producing heating of up to about $0.05\,K/day$ around $70\,km$.

The 4.3 μm bands

Solar radiation near $4.3\,\mu m$ is absorbed by several CO_2 bands. We first consider the absorption by the $4.3\,\mu m$ second hot bands, i.e. the processes $(1,0^0,1\leftarrow 1,0^0,0)$, $(0,2^2,1\leftarrow 0,2^2,0)$ and $(0,2^0,1\leftarrow 0,2^0,0)$. The amount of energy initially absorbed from solar radiation by these bands is modest, reaching a maximum of $0.05\,K/day$ at around $60\,km$, because of the low populations of the lower states, located at around $1350\,cm^{-1}$. The solar energy initially absorbed at quite low altitudes, in the 50–70 km region, is mostly thermalized in collisions and little energy is re-emitted, either in the $4.3\,\mu m$ absorbing bands or in any other band. As for the $2.0\,\mu m$ bands, their contribution alone is not very important, but becomes significant when the contributions of all the weak CO_2 near-infrared bands are added together.

Absorption also occurs near $4.3\,\mu m$ due to the $(0,1^1,1\leftarrow 0,1^1,0)$ first hot transition. The rate of energy absorption from solar radiation by this band is significant in the upper mesosphere and lower thermosphere, peaking around $85\,km$. A major part of the energy absorbed is lost by emission by the same transition that absorbs. The remainder is redistributed through V–V collisions in the v_3-reservoir and, as a result, a significant loss takes place through the $4.3\,\mu m$ fundamental and isotopic bands. Re-emission in the $15\,\mu m$ bands is not significant since the absorbed radiation does not excite directly any of the bending or symmetric stretching modes. After subtracting all the losses from the initially absorbed energy, for winter at mid-latitude, the absorption by this band contributes a maximum averaged heating rate of $0.15\,K/day$ around $70\,km$.

The fundamental band of the 626 isotope at $4.3\,\mu m$ is the strongest infrared band of CO_2. The energy absorbed locally from solar radiation by this band averaged over a full day for the winter at mid-latitudes corresponds to a maximum heating rate of

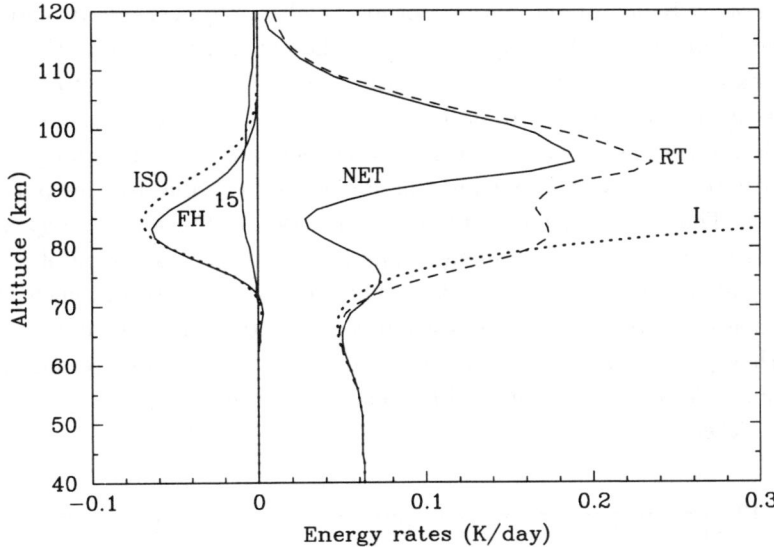

Fig. 9.21 Heating rate ('NET') and energy rate profiles integrated over one day in the winter Northern Hemisphere at 30°N, for the solar energy absorbed in the CO_2 4.3 μm fundamental band. See text for details. The energy rates are expressed in K/day for easy comparison with the heating rate.

10 K/day around 105 km. However, most of this is lost above 80 km by re-emission in the same band, as can be observed in Fig. 9.21, where the curve labelled 'RT', which includes not only the initial solar absorption by this band (curve 'I'), but also the subsequent transfer of energy by V–V and V–T collisional processes and by radiative re-emission and re-absorption processes, is much smaller than curve 'I'. In addition, a large part of the absorbed energy in the 70–100 km region escapes in the weaker first hot and isotopic ν_3 bands, which have been excited by V–V collisions in the ν_3-reservoir through $N_2(1)$ as an intermediate agent. Note the importance of these cooling rates, curves 'FH' and 'ISO' in Fig. 9.21, which re-emit almost all of the energy absorbed around 85 km.

Between approximately 65 and 80 km, the divergence of the radiative flux (with sign reversed) of this band, curve 'RT', exceeds by a small amount the energy initially absorbed from solar radiation, the excess being due to energy transfer processes from higher in the atmosphere by multiple emission and absorption. Below 60 km, collisions are frequent enough to thermalize all of the absorbed energy. On balance, the absorption of solar radiation by the 4.3 μm fundamental band provides a net, diurnally-averaged heating rate which is most important in the lower thermosphere, reaching a maximum value close to 0.2 K/day just below 100 km for the northern mid-latitude winter.

Most of the energy absorbed above 70 km in the 4.3 μm fundamental bands of the 636, 628 and 627 minor isotopes of CO_2 is re-emitted in the 4.3 μm isotopic bands themselves. The remainder is mostly lost by emission in the fundamental and first hot ν_3 bands of the major isotope at altitudes between 70 and 100 km. The isotopic bands contribute appreciably to the net heating rate in the 60–85 km region, peaking at 0.15 K/day around 75 km for the northern mid-latitude winter.

The weak first hot and isotopic bands of CO_2 near 4.3 μm make a significant contribution to the energy balance of the mesosphere and lower thermosphere, since they contribute a relatively important fraction of the CO_2 solar heating in the upper mesosphere. They also re-emit an important fraction of the solar energy absorbed by the 4.3 μm and 2.7 μm fundamental bands, due to the exchange of v_3 quanta with $N_2(1)$ at slightly higher altitudes. Their net effect is to diminish the total averaged heating rate in the 75 to 100 km region and to increase it below 75 km. That is, they produce net lost of energy at altitudes where they are optically thin but collisions are still important, and heating at lower altitudes where they become optically thicker.

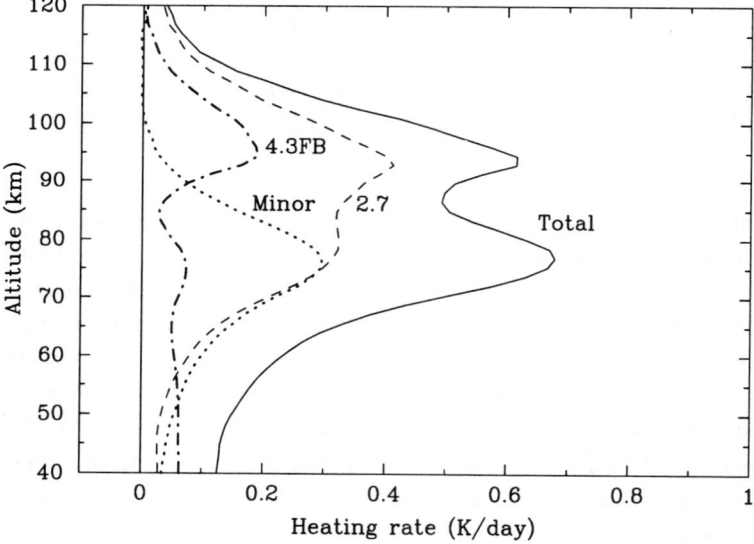

Fig. 9.22 Heating rate profiles integrated over one day in the winter Northern Hemisphere at 30°N, for the energy initially absorbed in the CO_2 2.7 μm (2.7), 4.3 μm fundamental band of the main isotope (4.3FB), and the 2.0 μm and 4.3 μm isotopic and hot bands ('Minor').

In summary, despite the considerable amount of energy re-radiated, the absorption of sunlight by all of the near-infrared CO_2 bands, averaged over a full day, contributes a significant amount of heating (see Fig. 9.22) to the energy budget of the mesosphere and lower thermosphere which is significant in comparison with sources such as ozone and molecular oxygen ultraviolet heating. The heating rate profile has a double-peaked structure, the upper peak being due to the 4.3 μm fundamental and 2.7 μm bands, and the lower one to the 2.7 μm and the weaker 2.0 μm and hot and isotopic 4.3 μm bands.

9.8.1 *Thermalization of the* $O(^1D)$ *energy*

The electronic internal energy of $O(^1D)$ is a potentially significant source of thermal energy in the mesosphere and lower thermosphere. The question is, how is this energy converted into heat? One possible pathway is: $O(^1D) + N_2 \rightleftharpoons O(^3P) + N_2(1)$ (process 12 in Table 6.2), as illustrated in Fig. 9.19.

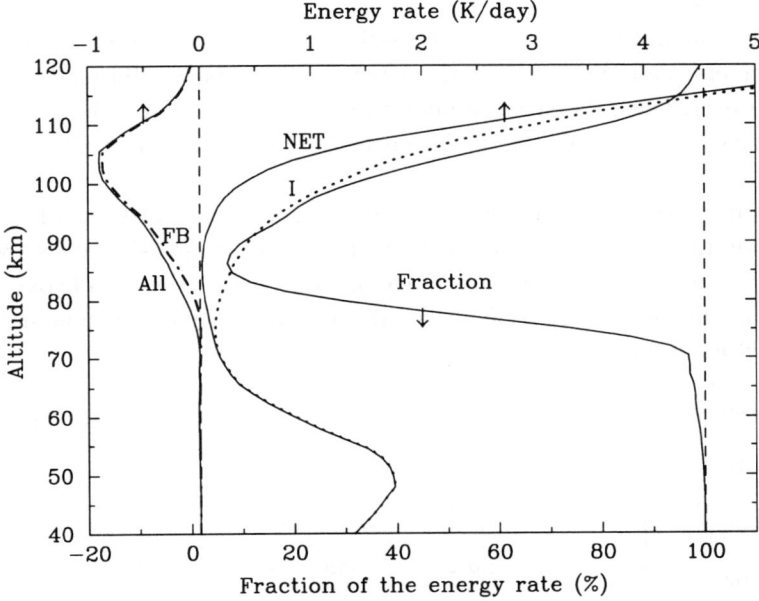

Fig. 9.23 Rate of transfer of electronic energy from $O(^1D)$ to the v_3-reservoir. For curves 'All', 'FB', 'NET' and 'I' refers to the upper scale. The curve 'Fraction' (lower scale) is the fraction of the $O(^1D)$ energy entering the v_3-reservoir that is thermalized. More details in the text.

The profile of the amount of $O(^1D)$ energy going into the v_3-reservoir in this way is shown in Fig. 9.23, curve 'I'. This is calculated for the rate coefficient of process 12 given in Table 6.2, with an efficiency of 25% and the $O(^1D)$ number density profile of Harris and Adams (1993). The part of that energy that is thermalized is shown as curve 'NET'. Below 75 km, the V–T processes are very efficient and all of the energy transferred results in heating. Above this altitude, the principal loss process for the injected energy is the 4.3 μm fundamental band of the major isotope of CO_2 (curve 'FB') and a small contribution by the other weak bands. The maximum fraction of energy originally from $O(^1D)$, which is emitted to space by this band, occurs around 85 km. At that altitude the pressure is high enough to make V–V collisions efficient in transferring the electronic energy from $O(^1D)$ to $N_2(1)$ and then to $CO_2(0,0^0,1)$. At the same time, it is sufficiently low to make the $(0,0^0,1-0,0,0)$ transition optically thin, allowing v_3 photons to escape to space. At altitudes above 100 km, the pressure is low, V–V collisions infrequent, and little energy is transferred from $N_2(1)$ to $CO_2(v_3)$, but most of it is thermalized by collisions between $N_2(1)$

with $O(^3P)$. The 'Fraction' curve shows that most of the $O(^1D)$ energy entering the $N_2(1)$–$CO_2(v_3)$ system is thermalized below about 70 km and above 110 km but in the intermediate region a large fraction is emitted to space. This fraction is fairly independent of the $O(^1D)$ concentration profile.

9.8.2 *Uncertainties in the heating rates*

The kinetic temperature influences the heating rates through its effect on the spectral parameters, and the temperature dependence of some collisional rate coefficients. However, when its effect through these parameters upon the solar heating rates is calculated, it turns out to be negligible, except for a small effect below 60 km.

The atomic oxygen abundance is more important. Calculations of the daily-mean heating rates for the winter Northern Hemisphere at mid-latitudes are shown in Fig. 9.24. These cover a plausible range of variations of the atomic oxygen concentration in the lower thermosphere, obtained by multiplying the nominal mixing ratio by factors of 0.2 and 2.0.

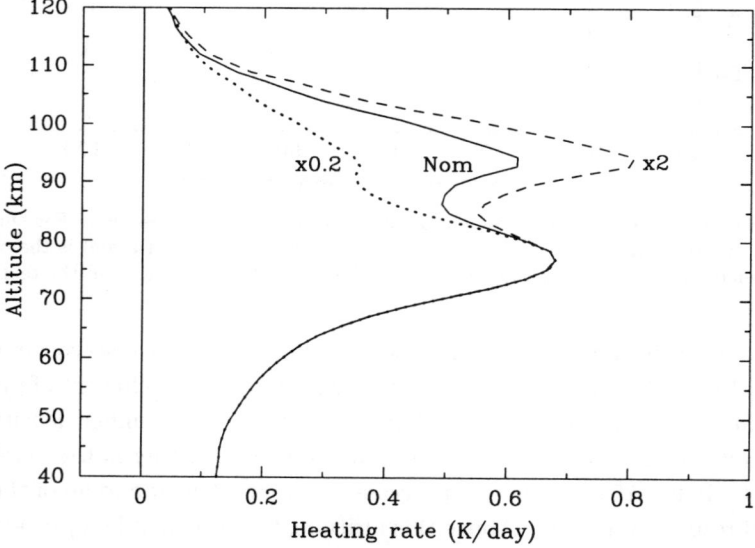

Fig. 9.24   The effect of the atomic oxygen abundance on the solar heating rate by CO_2 bands.

The fact that the deactivation of $N_2(1)$ by $O(^3P)$ is the principal mechanism for conversion of the vibrational energy of the v_3-reservoir into thermal energy above 80 km, explains the considerable enhancement of the heating rates above that level that results when a larger atomic oxygen abundance is assumed. This increase in $O(^3P)$ has the opposite sign to the enhancement of the 15 μm cooling rates produced by a similar augmentation of $[O(^3P)]$, and tend to moderate that important cooling in the energy budget of the lower thermosphere.

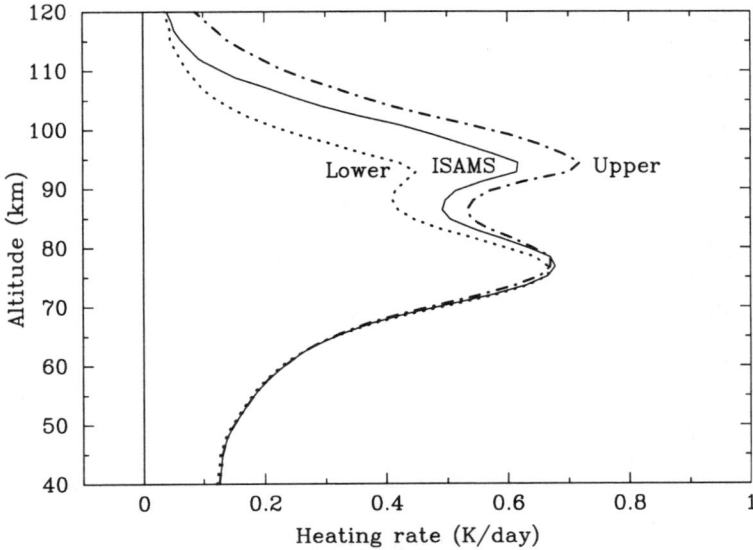

Fig. 9.25   Effect of the CO₂ mixing ratio on the heating rate by the near-IR CO₂ bands integrated over one day in the winter Northern Hemisphere at 30°N. The labels correspond to the CO₂ VMR profiles in Fig. 8.32.

The atmospheric heating produced by the absorption of solar radiation by the CO₂ infrared bands is proportional to the CO₂ volume mixing ratio, which is uncertain by about 100% in the lower thermosphere. Fig. 9.25 illustrates this with calculations of the net heating rates, integrated over one day in the winter Northern Hemisphere at 30°N, for three of the CO₂ profiles from Fig. 8.32.

The vibrational relaxation rate of v_3 quanta in collisions with N₂ (process 9 in Table 6.2), k_{vv}, has an important effect upon the CO₂ 4.3 μm limb radiances (see the analysis of SAMS and ISAMS data in Chapter 8) and a similar effect on the net heating rate would therefore be expected. Fig. 9.26 shows the net heating rates (again integrated over one day in the winter Northern Hemisphere at 30°N) calculated with the value of the rate coefficient in Table 6.2 reduced and increased by factors of 0.2 and 2.0 respectively, to reflect the current uncertainty.

The net heating rate is less at every level when a slower energy transfer rate is used. The reduction between 60 and 80 km is due principally to less heating by the 2.7 μm bands (see Fig. 9.20). When a slower rate is introduced, the flow of v_3 quanta from the nv_2+v_3- to the v_3-reservoir is smaller. A larger fraction of the solar energy absorbed is re-emitted by the CO₂ 4.3 μm second hot bands, and less is thermalized. The reduction above 90 km is mainly due to the 4.3 μm band of the CO₂ major isotope (see Fig. 9.21). The lower rate of exchange of v_3 quanta between $CO_2(0,0^0,1)$ and $N_2(1)$ means fewer v_3 quanta are thermalized through the process $N_2(1) + O(^3P) \rightleftharpoons N_2 + O(^3P)$.

There is an intermediate region near 90 km where the decrease in the heating rate is small. This is mainly due to heating by the 4.3 μm fundamental of the

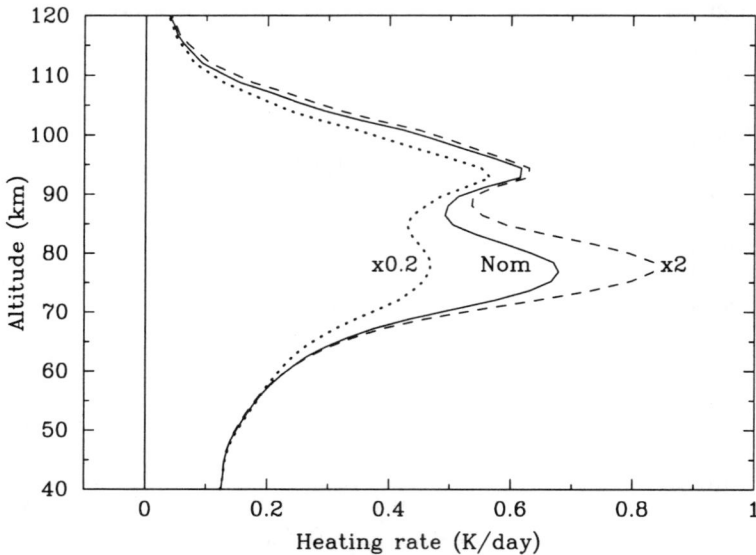

Fig. 9.26   Effect of the vibrational energy exchange rate of $CO_2(v_1,v_2,1)$ and $N_2(1)$ on the heating rate by the near-IR CO_2 bands integrated over one day in the winter Northern Hemisphere at 30°N.

main CO_2 isotope, which is normally kept small in this region by re-emission in the 4.3 μm first hot and isotopic bands (see Fig. 9.21). When k_{vv} is diminished, fewer v_3 quanta are transferred from $(0,0^0,1)$ of the main isotope to the same level of 636, 628 and 627 or to the $(0,1^1,1)$ state of 626, and so less energy is re-emitted by the weak bands originating in these levels. This effect tends to cancel out the other two.

9.8.3   *Global distribution*

Having calculated the vertical profile of solar heating rate by the CO_2 infrared bands, and considered the principal uncertainties, the next step is to look at the seasonal and latitudinal variations. Figs. 9.27 and 9.28 show the latitude and altitude distribution of heating, for solstice and equinox conditions respectively. The vertical profiles at each latitude are 24-hour averages of profiles calculated every hour in local time. The solar zenith angle as a function of the local time, the latitude, and the season is given by Eq. (8.18). The zonal mean kinetic temperatures are those compiled in the CIRA 1986 atmosphere (Figs. 1.2 and 1.3). The CO_2 mixing ratio profile and the rate of vibrational energy exchange of v_3 quanta between $CO_2(lv_1,mv_2)$ and $N_2(1)$, are the same as those used above and in the CO_2 model described in Chapter 6.

Unlike O_3 and O_2, little of the solar energy absorbed by CO_2 is stored as chemical or electronic energy, so the energy absorbed is not significantly modified by advective transport. Thus, any differences between the rate of solar absorption and the heating rate are due to radiative transfer, and the latitude variations of solar heating by CO_2 infrared bands are controlled by solar illumination. This is

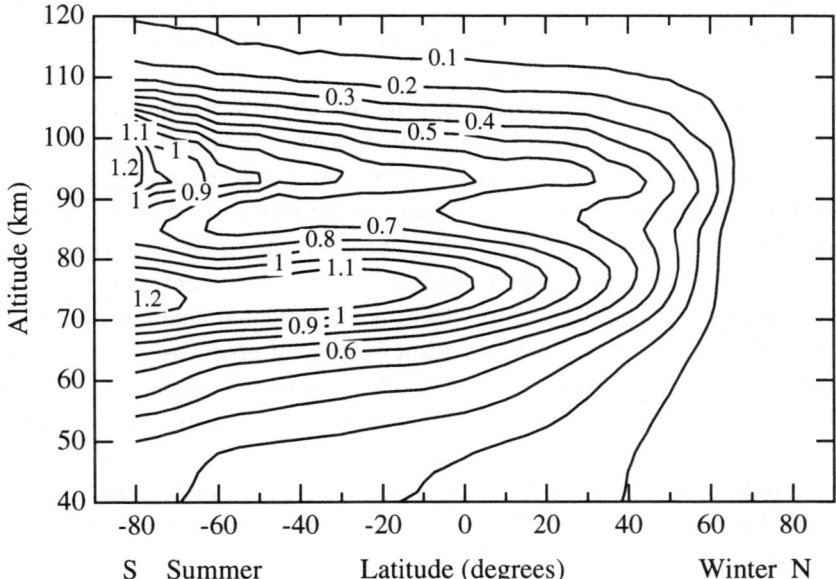

Fig. 9.27 The latitude and altitude distribution of the daily-averaged net solar heating by the CO_2 near-infrared bands for the northern winter solstice.

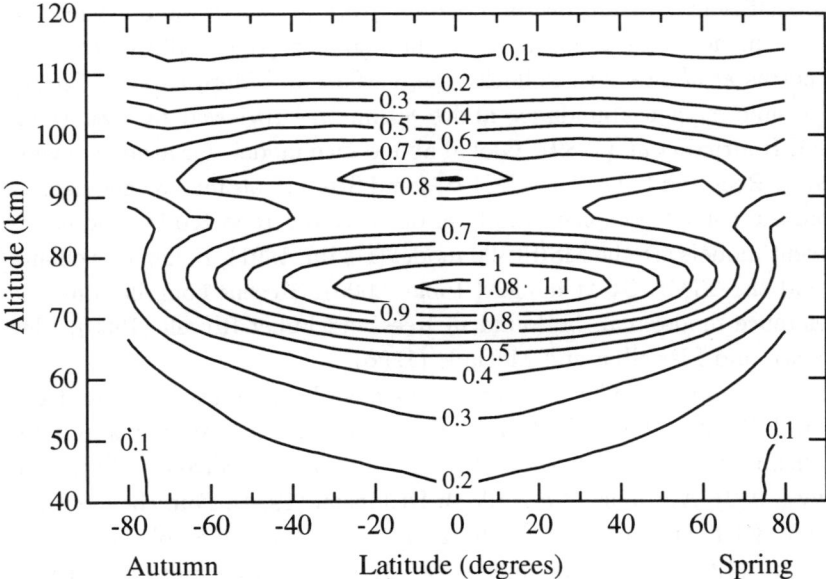

Fig. 9.28 The latitude and altitude distribution of the daily-averaged solar heating by the CO_2 near-infrared bands for equinox conditions.

clearly seen in the symmetric distribution of averaged heating rates under equinox conditions (Fig. 9.28), where the insolation is also symmetric. The distribution of heating rates for solstice conditions (Fig. 9.27) on the other hand, is not symmetric around the summer tropic, but is shifted towards the summer pole, where the mean insolation is larger than at the tropic.

The heating rates exhibit another feature which is also due to variations in the solar zenith angle. In Fig. 9.27, at altitudes near 75 km, the gradient of the heating rate with latitude is larger near the winter tropic than near the summer pole. However, at higher altitudes ($\sim$95 km), the situation is reversed. This shows the effect of solar elevation, sunlight being absorbed at deeper atmospheric levels when the Sun is overhead and higher when it is closer to the horizon.

The changes in the heating rate induced by the seasonal changes in the kinetic temperature through the collisional, principally those of processes 3, 4, 5, 9, and 15 in Table 6.2, are not very important, especially around the mesopause level where the heating rates are largest. Consequently, this effect is not apparent in Figs. 9.27 and 9.28. The global distributions do show, as expected, the double-peak structure seen in a single profile (Fig. 9.22), with maxima near 75 and 95 km peaking at around 1 K/day for equinox conditions at the tropics and around 1.2 K/day for solstice at the summer pole.

9.9   References and Further Reading

The rate coefficient for collisions between $CO_2(0,1^1,0)$ and atomic oxygen, $k_{CO_2\text{-}O}$, is discussed in detail by Sharma and Wintersteiner (1990), Shved $et\ al.$ (1991), López-Puertas $et\ al.$ (1992b), Pollock $et\ al.$ (1993), and Mertens $et\ al.$ (2001).

The atomic oxygen abundance above 80 km was reviewed by Van Hemelrijck (1981) and Rodrigo $et\ al.$ (1989). Global abundance models for atomic oxygen were produced by Rees and Fuller-Rowell (1988) and fitted to atomic oxygen densities derived from the Solar Mesosphere Explorer observations reported by Thomas (1990). Other recent models are the MSISE-90 model (Hedin, 1991), Fuller-Rowell and Rees (1996), and the TIME-GCM model (Roble, 2000). Recent compilations and proposed reference models can be found in Rees and Fuller-Rowell (1993), Llewellyn $et\ al.$ (1993), and Llewellyn and McDade (1996).

Radiative cooling by CO_2, O_3 and H_2O have been treated in detail by Murgatroyd and Goody (1958), Rodgers and Walshaw (1966), Williams and Rodgers (1972), Dickinson (1973), Kutepov (1978), Kutepov and Shved (1978), Wehrbein and Leovy (1982), Apruzese $et\ al.$ (1984), Dickinson (1984), Fomichev $et\ al.$ (1986), Haus (1986), Kiehl and Solomon (1986), Zhu (1990), Zhu and Strobel (1990), López-Puertas $et\ al.$ (1992a), Wintersteiner $et\ al.$ (1992), Akmaev and Fomichev (1992), Zhu $et\ al.$ (1992), and Fomichev $et\ al.$ (1993, 1998); and they have been reviewed by Dickinson (1975), London (1980), and Dickinson $et\ al.$ (1987).

Radiative cooling by NO 5.3 μm has been treated by Kockarts (1980), Zachor $et\ al.$ (1985), Gordiets and Markov (1986), Sharma $et\ al.$ (1996b), and more recently by Funke and López-Puertas (2000).

The effects of the enhanced greenhouse gases on the cooling of the upper atmosphere can be found in Roble and Dickinson (1989), Rishbeth and Roble (1992), and Rishbeth and Clilverd (1999).

The proposal that the 63 μm atomic oxygen emission is an important mechanism for cooling the atmosphere above 200 km originated with Bates (1951). Kockarts and Peetermans (1970) predicted that this emission produces heating in the lower thermosphere as a consequence of the absorption of radiation from the troposphere and surface. Gordiets *et al.* (1982) derived $O(^3P)$ 63 μm cooling rates for day and night and compare them with those by CO_2 at 15 μm and NO at 5.3 μm.

Near-infrared heating, by carbon dioxide in particular, has been studied by Houghton (1969), Williams (1971), Shved *et al.* (1978), Fomichev and Shved (1988), López-Puertas *et al.* (1990), and Mlynczak and Solomon (1993). For a discussion of ozone cooling rates, see Kuhn and London (1969), London (1980), Kiehl and Solomon (1986), Gille and Lyjak (1986), Mlynczak and Drayson (1990a,b), Mertens *et al.* (1999), and Mlynczak *et al.* (1999). The question of whether the middle atmosphere is in radiative equilibrium has been treated by Fomichev and Shved (1994).

Chapter 10

Non-LTE in Planetary Atmospheres

10.1 Introduction

The study of non-LTE molecular emissions and radiative transfer is not, of course, restricted to the Earth's atmosphere. The different compositions, pressure, temperature structure, and solar illumination conditions found among the terrestrial and giant planets offer different scenarios where the relative importance of the physical processes at work in non-LTE situations can be different from those in our terrestrial atmosphere. Studies of these varied situations, as in other fields of atmospheric science, often help to improve our appreciation of the physics involved and hence our understanding of radiative processes in all atmospheres, including the Earth's. To cite just one example, Dickinson's early work on non-LTE radiative transfer in the atmosphere of Venus found that the weak isotopic and hot bands of CO_2 play a major role in the infrared radiative balance in the middle atmosphere of that planet. Later, he extended the work to Earth's atmosphere and found that these weak bands are responsible for a significant fraction of the cooling of the stratosphere and lower mesosphere.

Because non-LTE situations usually occur in the more tenuous higher regions of an atmosphere, observations are rare even for the relatively well-studied planetary atmospheres of Venus and Mars. There are a few ground based measurements of CO_2 at $10\,\mu m$ and CO at $4.7\,\mu m$ on Mars and Venus, and of CH_4 at $3\,\mu m$ on Titan, while some spaceborne instruments, such as the Near Infrared Mapping Spectrometer (NIMS) on the Galileo spacecraft and the Short Wave Spectrometer on the Infrared Space Observatory (ISO), have made measurements in which non-LTE effects on Venus and Mars can be seen. The Composite Infrared Spectrometer (CIRS) on Cassini/Huygens, due at the Saturn system in 2004, and the Mars Atmospheric Climate Sounder, expected to launch in 2005, have the potential to make limb measurements on the atmospheres of Titan and Mars respectively. The data to be obtained will be among the richest in information about non-LTE processes in non-Earth planetary atmospheres so far obtained, and will require sophisticated models for their interpretation.

There have been a number of theoretical studies of non-LTE in planetary atmospheres, mainly devoted to specific problems such as the energy balance of the upper atmospheres. The models generally use a number of simplifications and often do not have the detail required to obtain non-LTE populations of the vibrational levels involved in the determination of temperature and minor constituents abundances. Mars has been treated in more detail than any other planet except the Earth, followed by Venus. To date, very little work, either experimental or theoretical, has been done on the outer planet atmospheres, where methane and the other hydrocarbons, rather than CO_2 and H_2O, tend to be the gases of interest.

In this chapter we will deal with the best-studied atmospheres, Mars and Venus, first. These are not only the best studied cases, they obviously have most in common with non-LTE in the Earth's atmosphere, and can draw on material already discussed in the earlier chapters. Both are CO_2 rich, and have many features in common, but with major differences in the pressure and temperature regimes, and some small but important variations in the minor constituent abundances as well. For instance, the higher pressures on Venus make the CO_2 bands optically much thicker than on Mars, so that different radiative processes can dominate, and the $O(^3P)$ volume mixing ratio is higher on Venus, which alters the detailed energy balance of the upper atmosphere. After Mars and Venus we deal more briefly with the gas giant planets, as typified by Jupiter, and consider the unique and interesting case of Titan, the giant satellite of Saturn with its thick nitrogen-methane atmosphere. Finally, we conclude with some remarks about non-LTE processes in cometary atmospheres.

10.2   The Terrestrial Planets: Mars and Venus

The Martian and Venusian atmospheres are nearly pure CO_2, resulting in a more complex radiative transfer behaviour for this gas than for the Earth, where the abundance is only 0.036%. A larger range of band strengths needs to be considered for realistic simulations of the radiative energy balance and non-LTE populations. Hence, many more transitions have to be included, in particular those higher-order lines and bands which are too weak to be important on Earth but which do contribute when the gaseous abundance is higher. For Mars and Venus, it is frequently necessary to cover more than 5 orders of magnitude in line strength. Also, the frequency integration has to pay particular attention to the overlapping of lines and bands in the pure CO_2 case.

Another important difference from the Earth occurs because the near-resonant V–V energy exchange between vibrationally excited and ground state CO_2 molecules is much faster than between CO_2 and N_2 or O_2. This results in a faster redistribution of the solar energy absorbed in the near-infrared $(1-4\,\mu m)$ to the isotopic and weak bands near 15 and $4.3\,\mu m$, with the result that an important fraction of the solar energy absorbed is re-emitted at $15\,\mu m$. Thus, on one hand, because of the large CO_2 density, the absorption of solar energy takes place at a faster rate, but on the other hand, the faster relaxation of this energy to CO_2 levels connected by optically thin bands causes a larger fraction of the absorbed energy to be re-emitted, so the efficiency of thermalization is smaller. Overall the former process dominates, and solar heating by near-infrared CO_2 bands is even more important on Mars and Venus than it is on the Earth. Also, because the absorbed solar radiation undergoes rapid V–V collisional relaxation, and because the energy is trapped in the strong and moderate bands, many CO_2 vibrational levels become excited, and the models have to include a large number of pathways between a large number of states in order to calculate the populations accurately. If great care is not taken to include all relevant processes, it is easy to reach incorrect conclusions when interpreting remotely sensed measurements, or when calculating solar heating rates.

The Martian atmosphere contains significant amounts of molecular nitrogen, water vapour and carbon monoxide, all species that have an important role in radiative transfer on the Earth, either directly or through coupling with carbon dioxide. In a nearly pure CO_2 atmosphere, however, the vibrational exchange of CO_2 v_3 quanta with N_2, which is so important in the terrestrial atmosphere, becomes negligible, due to the faster exchange of v_3 quanta among CO_2 molecules and the long radiative lifetime of $N_2(1)$.

Both Mars and Earth experience larger day/night and solar cycle changes in the global mean thermospheric temperature than does Venus. The small upper atmospheric temperature variation on Venus occurs because the peak in the strong radiative cooling by CO_2 at $15\,\mu m$ occurs at the same altitude as the maximum in the solar EUV heating, and efficiently radiates the heating to space. On Mars, the maximum in solar heating occurs at a higher altitude than the peak cooling and hence much of the heat must be conducted downwards before radiating to space. The result is a larger change in the amplitude of the thermospheric temperature, both diurnal and with the solar cycle. The difference has its origin in the fundamental planetary parameters, acceleration due to gravity and heliocentric distance. Because Venus is closer to the Sun, CO_2 is more strongly photolized and hence produces larger $O(^3P)$ abundances and more efficient cooling. Also, the higher gravity on Venus means the scale height is smaller, with the result that the EUV heating is absorbed in a more concentrated region than on Mars. The two factors together makes the strong $15\,\mu m$ cooling and the solar EUV heating peak very closely in altitude. Part of the large variation in the Earth's thermospheric temperature has a completely different cause, being due to the influence of its considerable magnetic field on heating by charged particle precipitation, a feature which both Venus and Mars lack.

One thing all three planets have in common is the ignorance of many of the key rate coefficients for vibrational energy exchange. Once again, this will continue to limit how well we can determine the vibrational temperatures of many of the levels until new laboratory experiments are devised to fill the gaps in our knowledge of these important and fundamental molecular properties. The rates which are needed only for Venus and Mars, and which are less relevant for the Earth, tend to be particularly poorly known, especially those for the less common isotopes of CO_2.

10.3   A Non-LTE Model for the Martian and Venusian Atmospheres

In this section we describe a non-LTE model for Mars and Venus which is based on that developed in Chapter 6 for the Earth's atmosphere. We will not describe every process in detail again, but emphasise those aspects peculiar to these atmospheres. As before, the model must treat the exchange of photons between different layers and the local collisional processes by solving the radiative transfer and the statistical equilibrium equations for a large set of vibrational levels and transitions. As noted above, we must include a larger number of levels and bands than we needed for the terrestrial case: these are listed in Table F.1. In addition to those considered for the Earth, shown in Fig. 6.1, the second and third overtone of the bending and symmetric stretching modes of the minor isotopes 636, 628, and 627 and the bending and the asymmetric-stretching levels of the much less abundant 637, 638, 728, and 828 isotopes all have to be included. For the reasons explained in the previous section, the model also has to incorporate the combination levels which are excited after the absorption of solar radiation in the near-infrared (1.2–2.7 μm). CO is the second infrared emitter, after CO_2, and is also included in the model with its fundamental band (1–0) and, as for the same reason as for CO_2, with its first overtone band (2–0) near 2.35 μm. The number of vibrational levels and the transitions between them is so large that the whole system of coupled equations has to be arranged into subsystems or groups, and solved one group at a time.

As for the Earth, the radiative transfer equation is set up to include spontaneous emission, absorption of solar radiation, and the exchange of photons among the atmospheric layers and between these and the planet's surface. For some bands, however, like those of CO_2 at wavelengths shorter than 2.7 μm, the exchange of photons between layers is much smaller than the excitation due to the direct absorption of solar radiation. Then, the radiative transfer equation can be simplified, since the properties of a given layer depend only on local conditions. For bands at intermediate wavelengths, where exchange between layers is not negligible, an approximation can be used which assumes an effective radiative loss of $AT^*/2$, where A is the Einstein spontaneous emission coefficient, and T^* is the probability of a photon escaping to space defined by Eq. (5.7) (see Sec. 5.2.1.2 and ff.). This is equivalent to assuming that photons are emitted isotropically in all directions, with the half that travels upwards escaping from the atmosphere with probability T^*, while the half emitted downwards is exactly balanced by the absorption of photons emitted from the layers below and travelling in the opposite direction. The accuracy of this approach depends on the band and on the atmospheric temperature structure and composition. It is not, for example, a good approximation for the Earth's daytime

CO_2 4.3 μm and CO 4.7 μm emissions.

The combinational levels excited by absorption of solar radiation in the near-IR relax also through other bands, usually quicker than in those responsible for their excitation. Hence it is necessary to include in their statistical equilibrium equations radiative relaxation (usually just by spontaneous emission) mainly in the $\Delta v_3 = 1$ and $\Delta v_2 = 1$ hot bands.

10.3.1 *Radiative processes*

Radiative transfer plays a major role in determining the populations and heating rates in a CO_2 atmosphere. This can be treated using the Curtis matrix method, as described in Chapter 5. At the lower boundary the source function is equal to the Planck function at the temperature of the bottom atmospheric layer. At the upper boundary, the gradient of the source function with height can be assumed to be the same as that in the adjacent lower layer. The flux transmittances that comprise the Curtis matrix can be calculated using the quasi-line-by-line histogramming algorithm described in Sec. 6.3.3.1 or using the line-by-line integration schemes described in Sec. 4.3.1, employing the Curtis-Godson approximation and a Voigt line shape. As in previous models in Chapters 6 and 7, an spectral line database is needed. HITRAN 92 was used in the calculations presented here (see Sec. 4.5 for other spectroscopic databases). Overlapping between bands is not included, but that between lines of the same band is introduced approximately by using the random band model. This approximation overestimates the cooling rates at low altitudes, but has no effect on the populations of the levels since they are in LTE.

In the Curtis matrix formulation, the absorption of solar radiation appears in the radiative transfer equation as a separate term which represents the flux divergence induced by the radiant energy coming from sources outside the atmosphere. The flux divergence induced by the absorption of solar radiation by a given band is given by Eq. (4.22) integrated over frequency. The bands included in the calculation of solar heating in the Martian and Venusian atmospheres are listed in Table F.1.

10.3.2 *Collisional processes*

Not only is the effect of collisional interactions on the energy level populations much larger in predominantly CO_2 atmospheres than on the Earth, the vibrational-vibrational (V–V) exchanges that take place between CO_2 molecules are also very efficient, compared to vibrational-translational (thermal or V–T) collisions, especially for the transfer of v_3 quanta. The main processes affecting the CO_2 states involve collisions between vibrationally excited CO_2 and ground state CO_2 and $O(^3P)$, while collisions with the minor constituents N_2 and CO are negligible, since the importance of a collisional process directly depends on the number density of the colliding molecules. It also follows that processes which require both molecules to be in an excited vibrational state can be neglected, since each population will be relatively small compared to the ground state. For example, the number of CO_2 molecules excited in the ν_2 mode, with a characteristic energy of $667\,cm^{-1}$, under LTE conditions at $150\,K$ is 0.003 times smaller than in the ground level.

We again distinguish between V–T and V–V processes. A useful simplification is to assume that the most likely V–T processes are those in which only a single quantum is exchanged. This is a good approximation when the energy transferred, ΔE, is much greater than kT, as is generally the case in the Martian and Venusian atmospheres where temperatures are typically around $150\,K$, i.e. $kT \approx 100\,cm^{-1}$.

Other rules for determining the most important collisional processes are:

(i) V–V interactions are much faster than V–T for polyatomic molecules;

(ii) V–V processes are especially efficient for near resonant cases, i.e. very small ΔE; and

(iii) The V–V collisional processes in which minimum changes in quantum numbers occur are the most likely.

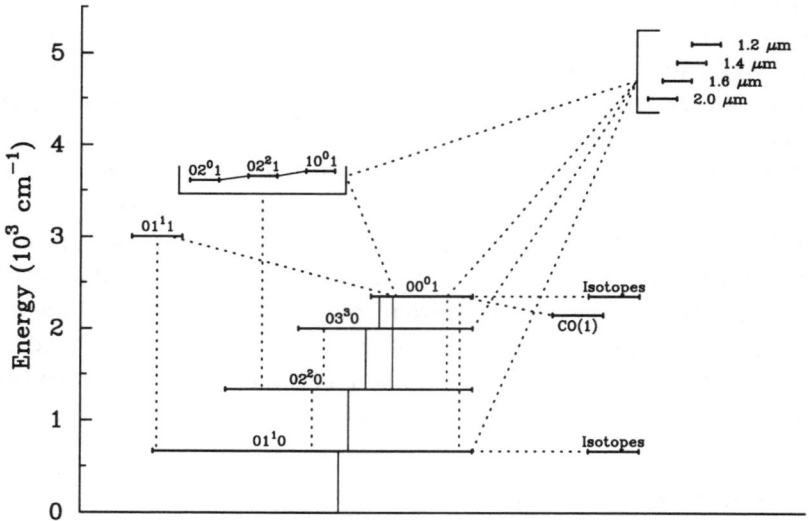

Fig. 10.1 Collisional energy exchange paths allowed in the Martian and Venusian models. Solid lines are V–T and dotted lines V–V processes.

The resulting set of collisional processes for the most important vibrational levels of the 626 isotope of CO_2 is shown in the schematic diagram of Fig. 10.1. The V–T and V–V exchanges are the vertical-solid and dotted lines respectively. For the other important isotopes 636, 628, and 627, the main difference is that a smaller number of combination levels at 1.2, 1.4, 1.6 and $2.0\,\mu m$ needs to be included (see Table F.1). A summary of the important V–T and V–V collisional processes is presented in the following sections.

Table 10.1 Principal collisional processes in the atmospheres of Mars and Venus.

Process	M	Rate coefficient[†]	Reference
VT1: $CO_2^i(0,0^0,1) + M \rightleftharpoons CO_2^i(0,v_2,0) + M$ $k_a: v_2=1; k_b: v_2=2; k_c: v_2=3$	CO_2	$k_{1a}=7.3\times10^{-14}\exp(-8.5A + 8.65B)$ $k_{1b} = 0.18k_{1a}, k_{1c} = 0.82k_{1a}$	Bauer et al. (1987) Lepoutre et al. (1977)
	N_2	$k_{2a}=2.2\times10^{-15}+1.14\times10^{-10}\exp(-76.7/\sqrt[3]{T})$	Bauer et al. (1987)
	CO	$k_{3a}=1.7\times10^{-14}\exp(-4.48A + 5.36B)$	Starr and Hancock (1975)
	$O(^3P)$	$k_{4a}=2\times10^{-13}(T/300)^{1/2}$	Buchwald and Wolga (1975)
VT2: $CO_2^i(0,v_2,0)+M \rightleftharpoons CO_2^i(0,v_2-1,0)+M$ $k_a: v_2=1; k_b: v_2=2; k_c: v_2=3$	CO_2	$k_{5a}=4.2\times10^{-12}\exp(-29.9A + 30.4B)$ $k_{5a}=3.3\times10^{-15}$ for $T < 175$ $k_{5b} = 2.5k_{5a}; k_{5c}=1.5k_{5b}$	Lunt et al. (1985) LVLP94* LVLP94
	N_2, CO	$k_{6a}=2.1\times10^{-12}\exp(-26.6A + 22.3B)$ $k_{7a}=3\times10^{-12}, k_{7b}=2k_{7a}, k_{7c}=1.5k_{7b}$	Lunt et al. (1985) López-Puertas et al. (1992b)
	$O(^3P)$		
VT3: $CO_2^i(1,0^0,1) + M \rightleftharpoons CO_2^i(0,2^0,1) + M$ $CO_2^i(1,0^0,1) + M \rightleftharpoons CO_2^i(0,2^2,1) + M$ $CO_2^i(0,2^2,1) + M \rightleftharpoons CO_2^i(0,2^0,1) + M$	CO_2, N_2 CO_2, N_2 CO_2, N_2	$k_{8a}=1.6\times10^{-12}$ $k_{8b}=5.0\times10^{-12}$ $k_{8c}=5.0\times10^{-12}$	Orr and Smith (1987) Orr and Smith (1987) Orr and Smith (1987)
VT4: $CO(1) +O(^3P) \rightleftharpoons CO + O$		$k_9=1.4\times10^{-5}\exp(-109.6A + 148.6B)$ $k_9=2.3\times10^{-14}$ for $T < 265$	Lewittes et al. (1978) LVLP94
VV1: $CO_2^i(v_1,v_2,v_3)+CO_2^j \rightleftharpoons CO_2^i(v_1,v_2,v_3-1)+CO_2^j(0,0,1)$ $(i,j = 1, 2, 3, 4; i \neq j$ if $v_1=v_2=0$ and $v_3=1)$		$k_{10}=v_3 3.6\times10^{-11}\sqrt{T}\exp(-\Delta E/26.3)$ but $=v_3 6.8\times10^{-12}\sqrt{T}$ for $\Delta E <42\,\mathrm{cm}^{-1}$	Shved et al. (1978) Shved et al. (1978)
VV2 a: $CO_2^i(0,0^0,1)+CO_2 \rightleftharpoons CO_2^i(0,2,0)+CO_2(0,1^1,0)$		$k_{11}=3.6\times10^{-13}\exp(-16.6A + 17.7B)$ $k_{11}=8.8\times10^{-15}$ for $T \leq 175$	Lepoutre et al. (1977) LVLP94
b: $CO_2(0,0^0,1)+CO_2^{j=2-4} \rightleftharpoons CO_2(0,2,0)+CO_2^j(0,1^1,0)$		$k_{12}=k_{11}$	LVLP94
VV3 a: $CO_2^i(0,v_2,0)+CO_2 \rightleftharpoons CO_2^i(0,v_2-1,0)+CO_2(0,1^1,0)$ $(i = 1\text{–}4; i = 2\text{–}4$ if $v_2=1)$		$k_{13a}=k_{13b}/2$ $(v_2=1)$ $k_{13b}=2.5\times10^{-11}$ $(v_2=2)$ $k_{13c}=(3/2)k_{13b}$ $(v_2=3)$	Orr and Smith (1987) LVLP94 LVLP94
b: $CO_2(0,v_2,0)+CO_2 \rightleftharpoons CO_2(0,v_2-1,0)+CO_2(0,1^1,0)$ $(i = 2\text{–}4,$ with $v_2=2, 3)$		$k_{14b}=k_{13b}$ $(v_2=2)$ $k_{14c}=(3/2)k_{14b}$ $(v_2=3)$	LVLP94 LVLP94
c: $CO_2^i(v_1,v_2,0)+CO_2 \rightleftharpoons CO_2^i(v_1',v_2',0)+CO_2(0,1^1,0)$ $(i = 1\text{–}4; m =2v_1+v_2;$ with $m=m'+1>3)$		$k_{15}=(m/m')(k_{13b}/2)$	LVLP94
VV4: $CO_2(0,0^0,1) + CO \rightleftharpoons CO_2 + CO(1)$		$k_{16}=1.6\times10^{-12}\exp(-11.7A + 7.76B)$	Starr and Hancock (1975)

[†]Rate coefficient for the forward sense of the process in $\mathrm{cm^3 s^{-1}}$. $A = 10^2/T$; $B = 10^4/T^2$. T is temperature in K. $i,j = 1, 2, 3, 4$ for isotopes 626, 636, 628, and 627, respectively. *López-Valverde and López-Puertas (1994a).

10.3.2.1 *Vibrational-translational processes*

Table 10.1 lists four groups of vibrational-translational or thermal processes affecting the CO_2 and CO levels in the Martian and Venusian atmospheres. The VT1 group contains processes which redistribute the vibrational energy of the asymmetric ν_3 mode into the ν_1 and ν_2 modes. The most important of these is the thermal relaxation of $CO_2(0,0^0,1)$ by CO_2 itself, instead of by N_2 as dominates for Earth. Due to the lack of measurements, we have to use the same mechanisms and rate coefficients for all isotopes, and it seems reasonable to assume that they behave similarly during collisions.

The VT2 processes in the table correspond to thermal excitation and de-excitation of the bending-stretching or ν_2 levels, including only single quantum transitions. The case of $\nu_2 = 1$ is the most important because, as discussed below, the upper levels (0,2,0) and (0,3,0) are more influenced by V–V than by V–T processes. Fortunately, the thermal relaxation rate of $CO_2(0,1^1,0)$ by CO_2 itself (k_{5a} in Table 10.1) has been the subject of a number of experimental determinations and is quite well known. Collisions between $CO_2(0,1^1,0)$ and $O(^3P)$ are just as important in the upper atmospheres of Mars and Venus as on Earth. Three V–T processes (VT3) affect the daytime populations of the $(0,2^0,1)$, $(0,2^2,1)$, and $(1,0^0,1)$ levels, the first and the last states being directly excited by solar absorption at $2.7\,\mu$m. The thermal deactivation of CO(1) by atomic oxygen (VT4) can be important in the thermosphere.

10.3.2.2 *Vibrational-vibrational processes*

Table 10.1 also lists the four most important groups of vibrational-vibrational collisional processes operating in the Martian and Venusian atmospheres. The first, VV1, represents the exchange of one v_3 quanta with the same or a different isotope of CO_2. This process is similar to the Earth, where the role played there by N_2 is taken on Mars and Venus by CO_2 itself. These processes are also responsible for the V–V relaxation of the v_3 quanta of the highly energetic CO_2 combination levels excited after solar absorption in the near-infrared. Hence, they determine not only the CO_2 vibrational populations, but also the thermalization of the absorbed solar energy.

The processes VV2 represent the re-arrangement of the ν_3 vibrational mode into the ν_1 and ν_2 modes of the four major CO_2 isotopes. We need only consider the main isotope as the deactivating molecule, because of its larger abundance. The excited states resulting after the collisions are highly uncertain. It is normally accepted, however, that those processes involving the smaller number of quanta (and the smaller amount of energy) are the most probable.

Process VV2b is the excitation of the $(0,1^1,0)$ level of the minor CO_2 isotopes by V–V relaxation of the highly populated $(0,0^0,1)$ state of the major 626 isotope. The backward direction of either reaction can be neglected, according to the rule stated above for reactions which require both collision partners to be excited above the ground state. Its effects, as a production mechanism for excitation of the main isotope and for loss in the less abundant isotopes, are very small in comparison with other collisional processes.

Process VV3 corresponds to the very efficient vibrational exchange of v_2 quanta between the different CO_2 isotopes. The deactivation of the (0,2,0) and (0,3,0) levels of the major isotope by 636, 628, and 627, process VV3b, is also important for exciting the $(0,1^1,0)$ level of these minor isotopes. Again, the backward direction can be neglected.

The vibrational relaxation of levels with $2v_1 + v_2 > 3$ and $v_3 = 0$, process VV3c, occurs after the absorption of solar radiation in the near-infrared bands and the subsequent loss of their v_3 quanta by collisional and radiative processes.

Collisions between CO_2 and CO are important for the excitation of the CO levels, but not for the deactivation of CO_2. The vibrational relaxation of $CO_2(0,0^0,1)$ in collisions with CO is the most important process for populating CO(1) (process VV4). Collisions of CO with CO_2 in its higher excited levels, and with the asymmetric mode of the minor CO_2 isotopes, are less important because of their lower abundances.

We have described so far the major features of the non-LTE model. In the next sections we describe the most salient aspects of the non-LTE populations and cooling and heating rates resulting after applying this model to the Martian and Venusian atmospheres.

10.4   Mars

In presenting model results for non-LTE in the Martian atmosphere, we begin with the simpler night-time case and consider daytime later. Following the routine which is now familiar from dealing with the Earth, the populations are again expressed in terms of the vibrational temperature T_v, defined by Eq. (6.1), relative to the ground state of the molecule. With this definition, $T_v = T_k$ when the level is in LTE, while $T_v \neq T_k$ when the population of the excited level is different (either larger or smaller) from that corresponding to Boltzmann statistics at T_k.

As for Earth, one of the most important applications of a non-LTE radiative transfer model is to study the radiative cooling and heating rates induced by the infrared emissions in the atmosphere. The strong CO_2 bands make very large contributions to these rates, and in fact dominate the energy balance in the middle atmosphere. In Secs. 10.4.5 and 10.4.7, we calculate the cooling rates and flux divergences of the bands at $15\,\mu m$, and the heating produced by the absorption of solar radiation by the near-infrared ($1.2–4.3\,\mu m$) bands. By combining those calculations a global mean radiative equilibrium temperature profile of the middle and upper atmosphere will be obtained.

10.4.1   *Reference atmosphere*

Figure 10.2 shows the daily-mean temperature profile for summer and mid-latitude conditions on Mars specified in the COSPAR reference atmosphere. The most noticeable aspects are the generally colder temperatures, with respect to the Earth, the absence of the stratosphere and the stratopause, and the much colder thermospheric temperatures. Mars has a thin atmosphere, with a mean surface pressure which varies with season in the range of 5 to 10 mb as the main constituent, carbon dioxide, condenses onto the winter pole and sublimes again in the spring. The detailed composition of the Martian atmosphere is shown in Table 10.2 and Fig. 10.3.

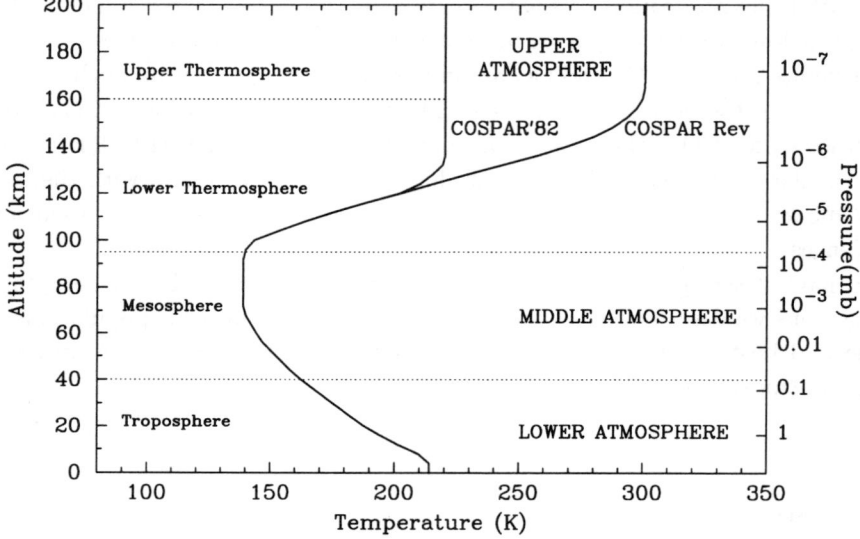

Fig. 10.2   The COSPAR'82 and the daily mean revised COSPAR (Kaplan, 1988) Martian kinetic temperature profiles. The range of pressures is from 5.2 mb at the surface to 2.27×10^{-8} mb at a height of 200 km.

10.4.2   *Night-time populations of CO_2 and CO*

Figure 10.4 shows the night-time vibrational temperatures of the first three v_2 levels and the first v_3 level of CO_2–626, and that of the first excited level of carbon monoxide, CO(1). The breakdown of LTE for the CO_2 $(0,1^1,0)$, $(0,2,0)$, and $(0,3,0)$ states of the 626 isotope takes place around 85 km. The Martian atmosphere is optically thick in the 15 μm fundamental band up to about 110 km, so the mean free path of photons is very short below this altitude. Consequently, the net radiative losses are very small and, in principle, a few thermal collisions would be enough to maintain $(0,1^1,0)$ in LTE up to these high altitudes. As mentioned before, however, the predominant collisions on Mars are V–V exchanges, in particular, the fast interaction of $(0,1^1,0)$ with $(0,2,0)$, which dominate up to 100 km, and with the minor isotopes

Table 10.2   Atmospheric composition at the surface of Mars.

Species	Mixing ratio, %
Carbon Dioxide (CO_2)	95.32
Nitrogen (N_2)	2.7
Argon (Ar)	1.6
Oxygen (O_2)	0.13
Carbon Monoxide (CO)	0.09
Water (H_2O)	0.03
Neon (Ne)	0.00025
Krypton (Kr)	0.00003
Xenon (Xe)	0.000008
Ozone (O_3)	0.000003

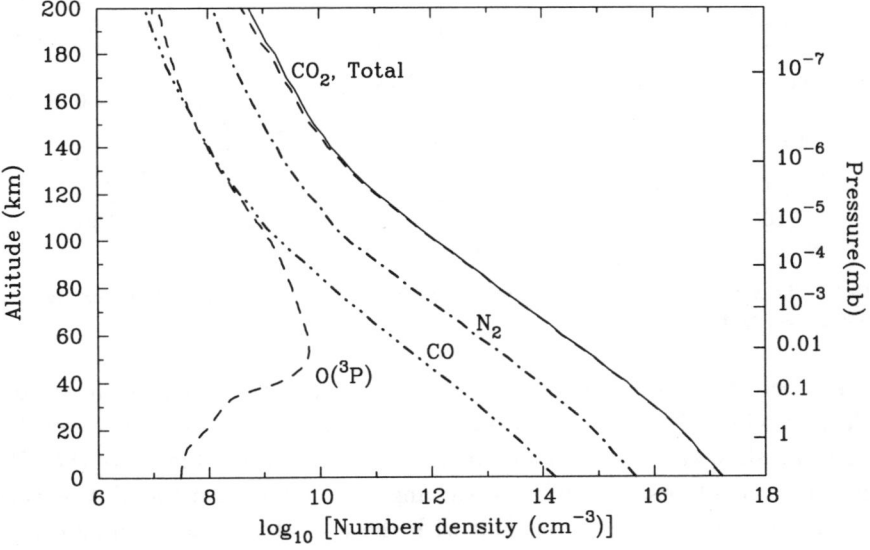

Fig. 10.3   The neutral composition of the Martian atmosphere from the non-stationary one-dimensional model for equinox conditions, mid-latitude, moderate solar activity, and clear atmosphere (small amounts of dust) developed by Rodrigo *et al.* (1990).

$(0,1^1,0)$ levels, relevant up to 115 km. Since the bands originating from these levels are much weaker than the 15 μm 626 fundamental band, photons emitted by them escape to space more easily. So the combined effects of the fast V–V collisions plus the easier escape to space in the weak 626 hot and minor isotopic fundamental bands make the $(0,1^1,0)$ state to depart from LTE at lower altitudes, around 85 km.

It is interesting to compare the heights of departure from LTE of the $(0,1^1,0)$ level in Mars and in the Earth. The pressure at the height of LTE departure (85 km) on Mars is $\sim 1.5 \times 10^{-4}$ mb, significantly lower than that on the Earth (90 km),

$\sim 1.5 \times 10^{-3}$ mb, (see Fig. 6.2 in Sec. 6.3.6.1). The optically thicker Martian atmosphere prevents photons to escape to space and hence allow LTE up to lower pressures. Note that if V–V exchanges with the overtone and minor isotopic v_2 levels would be smaller, LTE would occur in Mars even up to lower pressures. On Venus, this band departs from LTE at around the same atmospheric pressure as on Mars ($\sim 1.5 \times 10^{-4}$ mb), but the height is around 120 km (see Fig. 10.21).

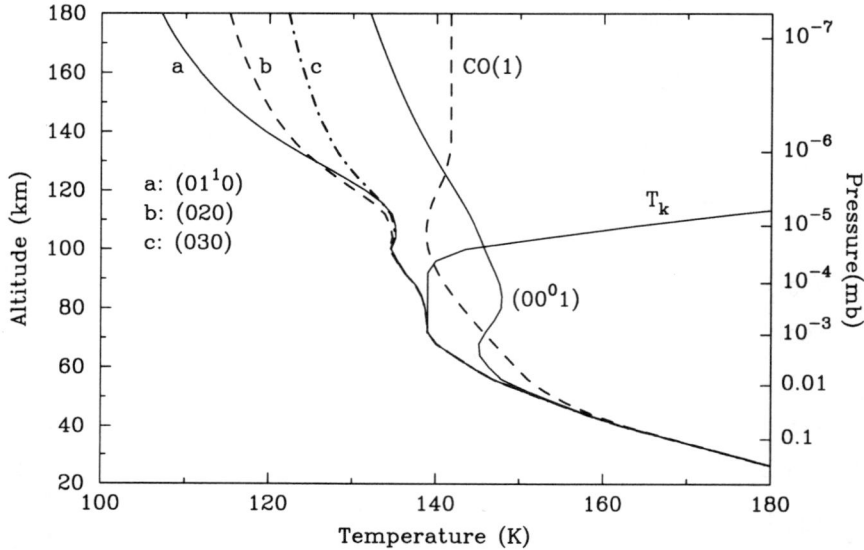

Fig. 10.4   The night-time vibrational temperatures of three v_2 levels and the first excited state of v_3 of CO_2–626, and of CO(1).

The effect of radiative transfer on the level populations can be illustrated by calculations which deliberately leave out the exchange of photons between layers. Figs. 10.5a,b show the vibrational temperatures of the $(0,1^1,0)$ levels of the 626 and 628 isotopes calculated without this exchange, excluding also the V–V exchanges among the $(0,1^1,0)$ isotopic levels and with the v_2 states. The nominal case from Fig. 10.4 is plotted for comparison. The large differences between curves A and B show the importance of radiative transfer between atmospheric layers for the vibrational temperatures of these levels and the altitude where they depart from LTE. When the exchange between layers is neglected, as it is, for example, when the 'escape-to-space' approach discussed in Sec. 5.2.1.2 is used (curve B), the vibrational temperature starts departing from LTE around 50 km. This corresponds to the height where radiative and V–T collisional deactivations are similar, as we would expect. When radiative transfer is included, but not V–V collisions (curve A), the $(0,1^1,0)$ vibrational temperature starts departing from LTE at the height where net emission in the 15 μm fundamental band becomes larger than the net V–T production, or about 105 km.

The weaker 628 15 μm fundamental band becomes optically thin at lower alti-

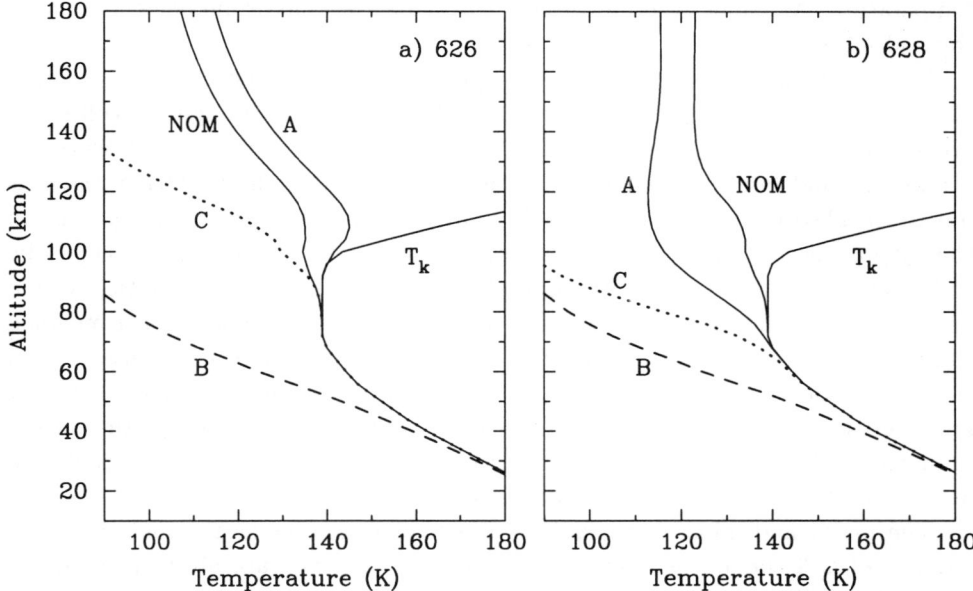

Fig. 10.5 Vibrational temperatures of the $(0,1^1,0)$ levels of the 626 (left) and 628 (right) isotopes calculated with (A) and without (B) radiative exchange between layers. C is a calculation with the 'cooling-to-space' approach. Cases A, B, and C do not include V–V exchanges by v_2 quanta.

tudes than the 626 isotope, and so, as shown in Fig. 10.5b, the $(0,1^1,0)$–628 level starts departing from LTE at a correspondingly lower level. A comparison of curves A and NOM in Figs. 10.5a,b shows that the inclusion of V–V coupling leads the $(0,1^1,0)$–626 state to break from LTE at lower altitudes, but has the opposite effect for the 628 isotope.

We mentioned in Sec. 5.2.1.2 that the altitude at which a given band departs from LTE is rather well approximated by that at which the specific collisional losses, l_t, are equal to the approximate net radiative losses, given by $(A/2)\mathcal{T}^\star$, where A is the Einstein coefficient and $\mathcal{T}^\star$ is the probability escape function. Curves C in Figs. 10.5a,b correspond to the vibrational temperatures obtained using this approach and neglecting V–V collisions. As can be seen, these vibrational temperatures depart from LTE at an altitude very close to those obtained when radiative transfer is included. This demonstrates the validity of this approximation for estimating the altitude of non-LTE breakdown without the need to carry out complex calculations.

Another effect of the strong coupling between the v_2 levels is that they all have similar night-time vibrational temperatures up to altitudes as high as ~ 120 km (see Fig. 10.4). Higher up, in the upper thermosphere, radiative processes take over and the absorption of radiation from lower layers is the most important excitation mechanism, with spontaneous emission the main loss. There, the vibrational temperatures tend to be independent of height and be much lower than the kinetic temperature. All bands eventually follow this radiative equilibrium behaviour above some altitude.

Figure 10.4 also shows the night-time population of the first excited v_3 level of 626. This state remains in LTE up to about 60 km, mainly through the collisional processes linking the v_2 and v_3 levels (VT1 and VV2a). Detailed analysis of the production and loss terms show that the enhancement in the 60–90 km region is due to the absorption of upwelling radiation mainly in the 10 μm bands, but also in the 4.3 μm isotopic and hot bands, whose energy is then transferred to $(0,0^0,1)$ by V–V collisions. Above 90 km, the isotopic bands at 4.3 μm become optically thin and depopulate the $(0,0^0,1)$ level of the major and minor isotopes by net emission to space. Between 90 and 130 km, the overall V–V transfer of v_3 quanta is from the major isotope to the minor isotopes, i.e., in the opposite direction to that at lower altitudes. In the upper thermosphere, spontaneous emission in the 4.3 μm fundamental band, with minor excitation by absorption of the upwelling radiation at 4.3 and 10 μm, controls the population of the v_3 level.

The major and minor isotopes have nearly the same v_2 level populations and altitude of LTE breakdown (around 85 km), because they are all coupled by V–V collisions. Between 100 and 140 km, the rarer isotopes tend to have lower v_2 and v_3 populations than 626, because they are optically thinner and can emit more effectively. Above about 140 km the situation reverses, as the absorption from below starts to dominate. This behaviour is similar to that exhibited by the $(0,2,0)$ level of the main isotope (see Fig. 10.4). For the v_2 overtone levels of the minor isotopes the absorption of upwelling radiation becomes relatively more important. Thus, for example, the vibrational temperatures of the $(0,2,0)$ and $(0,3,0)$ levels of 628 are larger than those of 626 throughout the thermosphere.

The first excited level of carbon monoxide is excited and de-excited in collisions with atomic oxygen (process VT4) and is collisionally coupled to the $(0,0^0,1)$–626 state through process VV4. Collisions are of little importance above about 110 km where radiative processes take over. The VV4 process dominates over radiative transfer in the 4.7 μm band below 55 km, keeping the CO(1) level close to LTE up to this altitude (see Fig. 10.4).

Above 55 km, the large absorption of radiation from the warmer lower layers gives rise to a super-Boltzmann population in CO(1). Around 65 km, where the 4.7 μm band begins to be optically thin, the escape of photons to space increases, leading to a substantial decrease in its population (see Fig. 10.4). Above 130 km, the increase in the kinetic temperature increases Doppler broadening which makes the absorption of photons emitted from the lower atmosphere more effective.

10.4.3 *Daytime populations*

As on the Earth, daytime conditions on Mars are more complex than at night because of the need to consider the excitation by solar pumping of a very large number of energetic molecular states, and the subsequent 'cascading' through lower energy levels, until the energy is eventually thermalized.

10.4.3.1 *Direct absorption from the solar flux*

When solar radiation is available it is the main excitation mechanism for both the CO_2 and the CO vibrational levels in the Martian atmosphere. The first step towards the calculation of the level populations, and the solar heating rates, is the evaluation of the energy directly absorbed from the solar flux at every altitude. Scattering in the atmosphere and reflection by the planet's surface are much less important, and can usually be neglected.

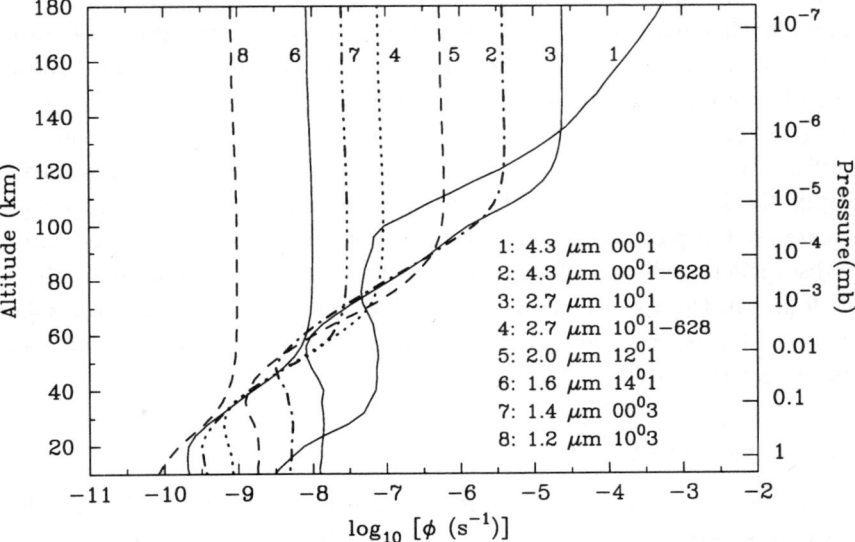

Fig. 10.6 Photoabsorption coefficients, expressed in number of photons per CO_2 molecule in the ground state and per second, for several fundamental bands of the main and 628 isotopes, for Sun overhead conditions and average Mars heliocentric distance. The upper level of the transition is indicated.

The rate at which energy is initially gained from the solar flux depends on the photoabsorption coefficient, which is given by Eq. (4.22) integrated over frequency. Some details of these calculations for the Earth are given in Sec. 6.3.3.2. We recall that for large solar zenith angles (χ), the atmospheric sphericity has to be included (see Sec. 4.2.3), that these rates are usually expressed per molecule in the ground state of the most abundant isotope, and that caution is required when dealing with the photoexcitation rates due to absorption of solar radiation by the hot and isotopic bands. Typical results for these rates are shown in Fig. 10.6 for several bands (including strong, weak, fundamental, and minor isotopic) of CO_2.

At the top of the atmosphere, most of the profiles have almost constant values which depend on the solar flux at the frequency of the band centre, and the strength of the transition. The only exception is the strongest CO_2-626 4.3 μm fundamental band, which is not optically thin even at the highest thermospheric layers shown. The non-fundamental bands show additional dependencies on the non-LTE populations of their lower vibrational levels (see below). As we would expect, the stronger bands are responsible for most of the absorption at the highest altitudes. The photoabsorption coefficients start declining at heights where the optical thickness of the atmospheric column above is such that the optically thin approximation is no longer valid; the largest attenuation of the solar flux takes place at these altitudes. This is important for understanding the heating effect of each band on the atmosphere, since the weaker the band, the lower this region occurs. It follows that at some heights the direct absorption of solar energy by very weak bands can be more important than the absorption in the stronger transitions. Fig. 10.6 shows, for example, that many 4.3 μm bands with very different strengths (see Table F.1) give very similar direct solar absorptions in the region between 40 and 100 km.

The photoabsorption coefficient of the 626 4.3 μm fundamental shows some interesting features. As a consequence of its large optical thickness, the solar flux is significantly attenuated by 100 km, and below 80 km the absorption takes place mainly in the Lorentz wings of the lines, where some flux remains, instead of in the saturated Doppler core. This produces a small but noticeable increase in the photoabsorption coefficient of this band in the lower mesosphere. A similar transition is found in the weaker bands, but at lower altitudes. In the troposphere the absorption takes place mainly in the far wings, and the photoabsorption coefficient decreases near the surface.

10.4.3.2 *Populations of CO_2*

Figure 10.7 shows the vibrational temperatures of the v_2, v_3, and v_2+v_3 levels of CO_2–626 for sun-overhead conditions and the COSPAR kinetic temperature profile. By day, the $(0,1^1,0)$ and $(0,2,0)$ levels stay close to LTE some 15 km higher than during the night. This result is however fortuitous since the daytime enhancement in their populations in that region just compensate for the night-time declination. So, above 85 km, during daytime, non-LTE processes dominate and this could actually manifests itself if we have a different mesospheric temperature profile. The $(0,3,0)$ levels start to depart from LTE around about 5 km lower than at night, and have higher populations. The day-night differences in the $v_2 = 1$–3 levels are larger than in the terrestrial atmosphere. For example, they are between 15 and 35 K at an altitude of 120 km while the maximum day/night changes in the Earth, occurring at the summer mesopause, are between 12 and 20 K for the $(0,2,0)$ and $(0,3,0)$ levels, and $(0,1^1,0)$ is hardly affected.

The reason for these daytime enhancements is the rapid V–V relaxation from the high v_2+v_3 combination levels, excited by absorption of solar radiation in the near-infrared. In particular, the $(0,3,0)$ levels are mainly populated by absorption in the 1.6 and 2.0 μm bands; the $(0,2,0)$ levels by the 2.7 μm bands; and the $(0,1^1,0)$

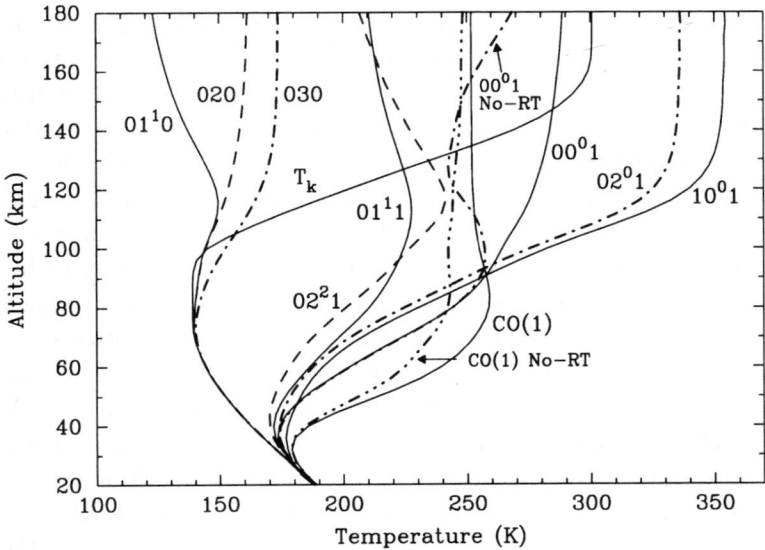

Fig. 10.7 The daytime vibrational temperatures of the v_2, v_3 and v_2+v_3 combinational levels of CO_2-626 and of CO(1) for Sun-overhead conditions. T_v's for $CO_2(0,0^0,1)$ and CO(1) calculated without radiative transfer are also shown.

level by its V–V coupling with the (0,2,0) states plus a small contribution from the deactivation of overtone levels pumped in the 1.6 and 2.0 μm bands.

The $(0,0^0,1)$ level departs from LTE by day some 10 km below the night-time level. It has a much larger non-LTE population than it does at night, since it benefits from energy cascading down to the lower levels from a number of highly energetic combination and overtone levels of CO_2. Absorption in the 2.7 μm bands and in the 4.3 μm isotopic bands (with subsequent V–V transfer to the main isotope) dominates in the upper mesosphere and lower thermosphere (85–125 km), with the fundamental at 4.3 μm taking over above about 120 km. Solar pumping in the weaker near-IR bands (1.2–2.0 μm), is the strongest excitation of $(0,0^0,1)$ between 50 and 85 km, where the solar flux in the strong 4.3 and 2.7 μm bands is greatly reduced.

The Martian atmosphere is optically thick at 4.3 μm up to very high altitudes, about 180 km. Hence, radiative transfer plays an important role in the redistribution of v_3 quanta between the atmospheric layers. This effect is illustrated in Fig. 10.7 where it is shown that the population of this level is much smaller above about 90 km if radiative transfer is neglected (curve 'No-RT'). Emission in the CO_2 laser bands is an important loss process of $(0,0^0,1)$ in the mesosphere and lower thermosphere of Mars.

The $CO_2(v_1,v_2,v_3)$ combinational levels with $2v_1+v_2 \leq 2$ and $v_3 = 1$ have several common features. They all interact strongly with the $(0,0^0,1)$–626 state by exchanging v_3 quanta during collisions, and their vibrational populations departs from LTE at around 30 km, increasing with altitude up to the lower thermosphere

(Fig. 10.7). The $(0,1^1,1)$ state is mainly excited by solar absorption in the $4.3\,\mu m$ first hot band (see Table F.1). Below $100\,km$, it partially transfers its energy to $(0,0^0,1)$ in V–V collisions, while above that altitude, the net V–V transfer takes place in the reverse direction. The $(0,2^0,1)$ and $(1,0^0,1)$ levels also exhibit very high vibrational temperatures in the upper and middle atmosphere where they are strongly excited by absorption of solar radiation in the $2.7\,\mu m$ fundamental bands. In the lower mesosphere and troposphere, the vibrational temperatures are similar to the fundamental band at $4.3\,\mu m$ because of efficient V–V coupling with $(0,0^0,1)$. The fact that of solar absorption at $2.7\,\mu m$ is forbidden in the transition between the ground state and $(0,2^2,1)$ makes the vibrational excitation of this level much smaller than for $(0,2^0,1)$ and $(1,0^0,1)$. However, all three levels are strongly inter-coupled by collisional processes (VT3 in Table 10.1) below about $60\,km$. Higher up, the $(0,2^2,1)$ state is pumped by solar energy in the $4.3\,\mu m$ second hot band.

The vibrational temperatures of the higher overtone and combination CO_2 levels $[CO_2(v_1,v_2,v_3)$ with $2v_1+v_2 > 2$ or $v_3>1]$ reach very high values in the upper atmosphere in order to achieve a balance between spontaneous emission and the strong solar excitation. Collisional deactivation of these levels is an important source of excitation for lower CO_2 levels, and the eventual thermalization of a significant fraction of this energy contributes significantly to the solar heating of the atmosphere.

The $(0,1^1,0)$ level of the minor isotopes shows almost identical vibrational temperatures to that of the main isotope up to about $100\,km$. V–T collisions dominate all production below about $30\,km$, while from there up to the thermosphere, V–V exchanges with the $(0,1^1,0)$ state of 626 are the principal excitation mechanism. The $(0,2,0)$ and $(0,3,0)$ levels of the CO_2 minor isotopes also have similar vibrational temperatures to the corresponding levels of the main isotope. The most important excitation mechanism of the $(0,2,0)$ minor isotopic levels is the absorption of solar radiation in the $2.7\,\mu m$ minor isotopic bands. The $(0,3,0)$ isotopic levels are mainly populated by solar absorption in the isotopic near-IR bands at $2.0\,\mu m$. There is also an important net excitation through V–T transfer from their respective $(0,0^0,1)$ isotopic levels between 70 and $90\,km$.

The vibrational temperatures of the $(0,0^0,1)$ minor isotopic levels are all equal up to about $80\,km$, because of the fast near-resonant V–V exchange of v_3 quanta between them. Their major production mechanism throughout the whole atmosphere, but particularly between 80 and $120\,km$, is the absorption in their respective $4.3\,\mu m$ fundamental bands. This excitation is so large that it leads to a net V–V energy transfer from the minor isotope $(0,0^0,1)$ levels to that of 626 up to about $120\,km$. Contrary to what happens with the $(0,0^0,1)$ level of 626, the near-IR minor isotopic bands do not produce a significant enhancement of the $(0,0^0,1)$ isotopic levels.

10.4.3.3 *Population inversions*

Several of the CO_2 vibrational levels that experience large daytime excitations in the mesosphere and thermosphere exhibit population inversions (that is, situations where the upper state of an allowed transition is more populated than the lower). In particular, the population of $(0,0^0,1)$–626 is larger than those of the less energetic $(0,2,0)$ and $(0,3,0)$ levels above about 80 and 50 km, respectively (see Fig. 10.23). The first of these inversions, which reach a maximum relative population enhancement $[(0,0^0,1)]/[(1,0^0,0)]$ of about a factor of 2, is the most interesting because of the existence of permitted transitions between $(0,0^0,1)$ and the $(0,2^0,0)$ and $(1,0^0,0)$ levels (see Fig. 6.1). The resulting laser emission at $10\,\mu$m has been detected in the atmosphere of Mars by telescopes on the Earth equipped with heterodyne spectrometers. The temperature has been derived from the width of the measured lines, and their radiances were used to retrieve the population of the upper $(0,0^0,1)$ level. The amplification factor for laser emission was found to be significantly large, and it even has been suggested that natural amplification could be used in programmes for communicating with extraterrestrial intelligence.

10.4.3.4 *Population of CO(1)*

Direct absorption of solar radiation in the $4.7\,\mu$m band is responsible for the high vibrational temperature of $CO(1)$ during the day (Fig. 10.7). This large excitation makes radiative transfer very important and much larger by day than at night. For example, the absorption at the top of the atmosphere of photons emitted lower down, mainly between 80 and 100 km, is similar in magnitude to the direct solar absorption. Fig. 10.7 also shows the vibrational temperature of $CO(1)$ calculated by neglecting the exchange between layers, to illustrate how the daytime vibrational temperature is seriously underestimated without it.

Unlike night-time, the absorption of emission from the surface is negligible during the day. Similarly, solar absorption in the overtone at $2.3\,\mu$m is negligible for the $CO(1)$ population. Below about 90 km the $CO(1$–$0)$ band becomes optically thick, and the energy from the solar flux and the downwards emission from thermospheric layers is preferentially deposited near or just below this height. This is the reason for the maximum in the $CO(1)$ vibrational temperature in the upper mesosphere.

Vibrational-vibrational exchange leads to a net collisional transfer from $CO(1)$ to $CO_2(0,0^0,1)$ below about 80 km, although the low concentration of CO means that this has little effect on the population of the CO_2 level. In comparison with this fast V–V exchange, V–T collisional production and loss of $CO(1)$ are also negligible.

10.4.4 *Variability and uncertainties in the populations of CO_2*

The results in the previous section, in particular the altitude of the departure from LTE, which is very important for temperature sounding from limb infrared measurements, have so far been discussed for only a single mean kinetic temperature profile. Since usual space experiments aim to measure globally, it is useful to know how expected variations in the kinetic temperature profile can affect the altitude of LTE-breakdown. Numerical experiments in which the model calculations are repeated with the atmosphere below 60 km around 20 K colder show little change in the departure from LTE of the v_2 bending levels, basically because the 15 μm fundamental is optically thick up to about 100 km and the v_2 overtone levels are strongly coupled with $(0,1^1,0)$–626 up to that altitude. So plausible variations in the Martian tropospheric temperature do not significantly affect the height at which the bands commonly used for temperature sounding start departing from LTE.

However, the $(0,0^0,1)$ level, emitting at 4.3 μm, shows large differences in its non-LTE behaviour with kinetic temperature. The local maximum in its vibrational temperature at about 80 km is greatly reduced when the troposphere is colder, and its departure from LTE is significantly smaller up to about 100 km.

The surface temperature of Mars experiences large daily and seasonal changes and it is interesting to know whether this has any influence on the departure from LTE of the CO_2 levels. Varying the temperature at the lower boundary by –75 K and +25 K, which covers the 150–250 K range measured by the Viking lander, shows a negligible change on the vibrational temperatures of most of the v_2 levels, and only a small variation in those of $CO_2(0,0^0,1)$, the CO_2 v_2 isotopic levels, and $CO(1)$.

Some of the rate coefficients of the collisional processes in Table 10.1 are poorly known, in particular the V–V exchange rates between the CO_2 excited levels. Therefore we need again to look at the sensitivity of the results to these uncertainties, both to bound the likely errors, and to understand the relative importance of the different physical processes at work. New rate coefficients regularly become available and so the best place to find up-to-date sensitivity studies is in the current literature. Here we will provide only a brief summary of some of the main points.

The rate coefficients for the most important V–T processes (VT2 in Table 10.1) are known better that the V–V rates, but the reported values still differ by a factor of 4. The main effect of varying the coefficient within this range is to change the night-time vibrational temperatures of the $CO_2(0,1^1,0)$ levels of all isotopes by about 10 K above 100 km. The T_v's of the $v_2 = 2$ and $v_3 = 3$ levels are also significantly affected ($\sim$10 K) in the lower thermosphere. All the vibrational temperatures increase when the V–T rate is faster. The LTE breakdown altitude changes very little.

In view of the important role played by V–V collisional processes in determining the populations of the bending modes, it is unfortunate that only the rate of exchange of v_2 between the $(0,1^1,0)$ and $(0,2,0)$ levels (VV3 in Table 10.1) has been directly measured. The rate coefficient of processes involving the overtone or isotopic levels is probably uncertain by an order of magnitude. The night-time vibrational temperatures of the $v_2 = 1$–3 levels of the main isotope diverge some 30 km higher when a rate faster by this amount is used. In the lower thermosphere, the vibrational temperature of $(0,1^1,0)$–626 increases for a slower V–V rate, those of the $(0,2,0)$ and $(0,3,0)$ states remain essentially unchanged, while those of the $(0,1^1,0)$ minor isotopic levels decrease.

The effect on the daytime vibrational temperatures is different because now the energy flows from the high-energy solar pumped states to the lower energy levels.

Thus, for the slower V–V rate, the T_v of the (0,3,0) states is largely increased (by a maximum of 20 K) in the 70–170 km range and that of (0,2,0) (by up to 10 K) in the 100–170 km interval. The T_v's of minor isotopic $(0,1^1,0)$ states are, on the other hand, decreased by a similar amount in that region, while the population of $(0,1^1,0)$–626 shows only small changes.

The rate coefficients for the V–V collisional processes (VV1 in Table 10.1) which provide coupling among the v_3 levels of CO_2, are also uncertain by perhaps a factor of ten in either direction. At night, however, the vibrational temperatures of all v_3 fundamental and combined levels of all isotopes are only slightly sensitive to such a change. In the daytime, an increase of one order of magnitude in the $k_{vv}(v_3)$ rate between the $(v_1,v_2,1)$ (with $2v_1+v_2=2$) levels and $(0,0^0,1)$ reduces the vibrational temperatures of the former by up to $\sim$35 K) in the mesosphere and 25 K in the lower thermosphere. The other important V–V transfer of v_3 quanta occurs between the $(0,0^0,1)$ levels of the various isotopes. When the rate is ten times faster, the populations of the minor isotopic levels are reduced by approximately 5 K between about 80 km and 120 km, and that of the main isotope is increased by 10 K in approximately the same region.

We turn now to the redistribution of vibrational energy between the ν_3 and ν_2 modes of CO_2 through V–T and V–V processes (VT1 and VV2 in Table 10.1). The rate coefficient for the V–V exchanges are known to within a factor of 2, and a change of this amount affects the vibrational temperatures only slightly. For V–T interactions between these vibrational modes, on the other hand, the rate coefficients can be uncertain by a factor of 4, especially for the case of collisions with atomic oxygen. The effect of an increase in the V–T rate is to decrease the vibrational temperature of the $(0,0^0,1)$ level by a maximum of 5 K at 80 km. The (0,2,0) and (0,3,0) levels are less affected and stay close to LTE at altitudes where the change in the $(0,0^0,1)$ level population is largest.

10.4.5 *Cooling rates*

We turn now to study the radiative energy balance of the CO_2 emissions. As would be expected, the strong CO_2 bands make very large contributions to the cooling and heating rates, and dominate the energy balance in the middle atmosphere of Mars. We discuss first the cooling rates of the 15 μm bands, and then the heating produced by the absorption of solar radiation in the near-infrared (1–5 μm).

A calculation of the total cooling rate obviously requires summing the radiative flux divergence of all of the bands active in the atmosphere. However, we must again be careful to note that some of the energy emitted by an excited level in a given band often comes from other excited states (e.g., through V–V exchanges or radiative relaxation from more energetic levels), so the flux divergence of the considered band does not entirely represent a net loss of kinetic energy of atmospheric molecules. Having said that, for a more fluid discussion, we will call the flux divergence of the

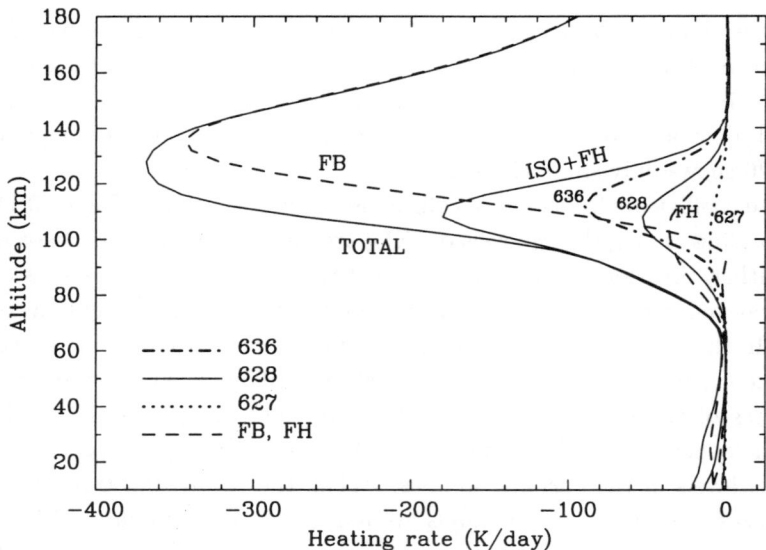

Fig. 10.8   Cooling rates of the CO_2 15 μm bands (in K per terrestrial day) for the revised COSPAR temperature profile (Fig. 10.2). 'FB' and 'FH' denote the fundamental, and first hot bands of the major 626 isotope, respectively. 'ISO+FH' represents the sum of the fundamental bands of 636, 628, and 627, and the first hot of 626. The curve 'TOTAL' represents the cooling rate at 15 μm, including the weak 626 second hot bands.

vibrational band as 'cooling rate'.

Figure 10.8 shows the contributions and the total cooling rate produced by the most important CO_2 15 μm bands (see Table F.1) for the revised COSPAR kinetic temperature profile (see Fig. 10.2), assuming no solar illumination. The profile is an intermediate 'daily mean' between the day- and night-time cases. The cooling rate is very large in the upper mesosphere and particularly in the lower thermosphere, with values larger than 300 K/day between 115 and 145 km. These rates are very dependent upon the atomic oxygen abundance and the kinetic temperature, being larger for a warmer or more oxygen-rich atmosphere (see next section).

The cooling rate in the 15 μm fundamental band of the major CO_2 isotope (FB) is very small below about 100 km because the band is optically thick in this region. There is a small heating around 95 km for this particular temperature profile, due to the absorption of radiation emitted from lower, warmer layers. Above that altitude, the band becomes optically thin and the cooling increases very rapidly until at very high layers, where the low pressure reduces the collisional excitation of $(0,1^1,0)$–626 and the cooling rate falls off.

The minor isotopic and first hot bands, being optically thinner than the fundamental, account for most of the cooling rate at lower levels, between 60 and 120 km. Between 80 and 120 km, the V–V exchange of energy between $(0,1^1,0)$–626 and the $(0,1^1,0)$–isotopic and $(0,2,0)$–626 levels is very efficient and the energy gained by $(0,1^1,0)$–626 in V–T collisions, mainly with CO_2 in the mesosphere and lower thermosphere and with $O(^3P)$ above, is transferred to these levels and subsequently emitted.

The cooling rates of the minor isotopic bands show all similar shape profiles, but they become gradually smaller, and peak at lower altitudes, as their respective bands' strengths decrease. In the upper thermosphere, all these isotopic and hot bands show a small heating of about $2\,K/day$; again due to absorption of radiation from below, mostly from the upper mesosphere.

Other $15\,\mu m$ bands, and the CO_2 transitions which occur in other spectral regions, give negligible contributions to the cooling rate of the Martian middle and upper atmosphere.

10.4.6 *Variability and uncertainties in the cooling rates*

The kinetic temperature structure affects the net emission of an optically thick transition both locally, by determining the emission of radiation at that altitude, and non-locally, by the absorption of radiation emitted at other atmospheric layers with different temperatures. Three height regions can be considered separately: the lower atmosphere where LTE prevails; the intermediate region in non-LTE where absorption of radiation from the layers below is important, and the optically thin non-LTE upper atmosphere.

In the LTE region, below about $60\,km$, the two most important factors affecting the cooling rates are the local kinetic temperature and its vertical gradient. For the first factor, the cooling rates for two different temperature profiles with similar gradients scale approximately by the ratio of the Planck function at $15\,\mu m$. In general, radiative transfer produces a cooling (heating) where the temperature vertical gradient is smaller (larger) than the radiative equilibrium gradient. For example, in the Martian troposphere (below about $20\,km$), temperature profiles with gradients in the range of $-(2.5\text{--}3)\,K/km$ can produce local heating. In the near isothermal $65\text{--}80\,km$ region of the COSPAR profile, only the temperature dependence through the Planck function is important.

Above $85\,km$, the dependence of the cooling on temperature is also positive (e.g., a larger cooling for a larger temperature) but, as the atmosphere becomes optically thinner, the probability of photon escape to space is larger, so the kinetic energy is more efficiently removed and the dependence of the cooling on temperature is stronger. In this region, where the departure from LTE is not very large, absorption of the upwelling radiation emanated in the lower regions, mainly by the hot and minor isotopic bands, is a significant source of excitation.

Above about $120\,km$ essentially all of the cooling is due to the net emission in the fundamental band of the 626 isotope (see Fig. 10.8). This band is optically thin in this region and the upper level population is dominated by radiative loss and collisional excitation. Under these conditions, the cooling rate depends exponentially on the kinetic temperature, according to the expression (c.f. Secs. 5.2.1.2 and 9.2):

$$q = \frac{[CO_2]}{Q_{vib}} \left\{ k_{CO_2\text{-}O}[O(^3P)] + k_{CO_2\text{-}CO_2}[CO_2] \right\} g_2 \exp\left(-\frac{h\nu_0}{kT} \right) \frac{T^\star}{2} h\nu_0. \quad (10.1)$$

The temperature is also important through its effect on the variation in the density or scale height. This is most important in the thermosphere, where the cooling rate is proportional to the $O(^3P)$ and CO_2 number densities (see Eq. 10.1). For a cooler lower thermosphere the atmosphere contracts, giving smaller cooling rates. The changes are significant above about 100 km, reaching typical values of $\sim 50\,K/day$ for a temperature change in the lower thermosphere of about 40 K.

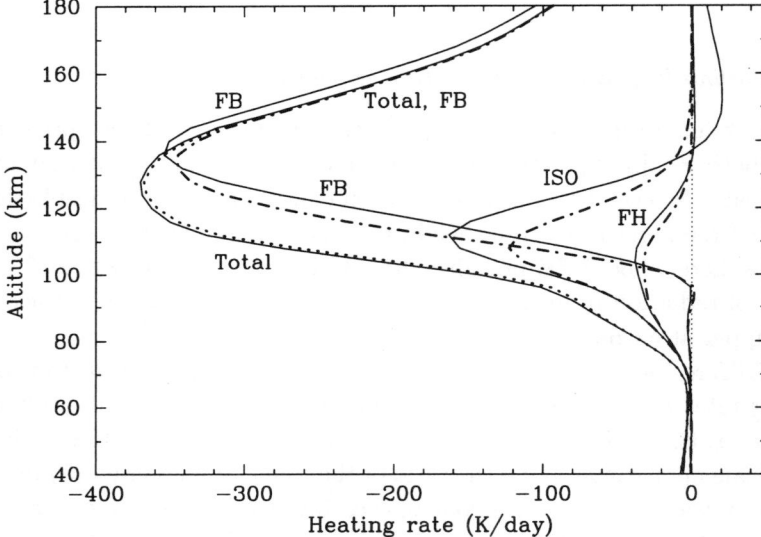

Fig. 10.9   Effect of the V–V(v_2) rate coefficient on the 15 μm cooling rate for the COSPAR revised temperature profile. Solid and broken curves denote cooling rates for the increased and decreased rate coefficients by a factor of 2, respectively.

Since we have seen that the rate of V–V transfer of v_2 quanta plays a major role in determining the Martian $CO_2(v_2)$ populations, it would be expected also to affect the 15 μm cooling rates. Fig. 10.9 shows the effects of perturbing this rate by a factor of 2. As discussed in Sec. 10.4.2, the net energy transfer goes from the 626-fundamental band towards the weak isotopic and hot bands between 60 and 130 km and in the opposite direction above. An increase in the exchange rate therefore results in an increase in the emission in the weaker 15 μm bands up to about 130 km, and a decrease above, where the cooling actually becomes heating for the faster V–V rate due to the absorption of the increased emission from below. This cannot be considered as actual heating, however, since most of the energy is emitted in the 15 μm fundamental band. Taken altogether, the cooling rates of the individual bands show important variations above 100 km, but the total cooling shows only a small increase ($\sim$10 K/day) in the upper mesosphere and lower thermosphere for the faster V–V rate.

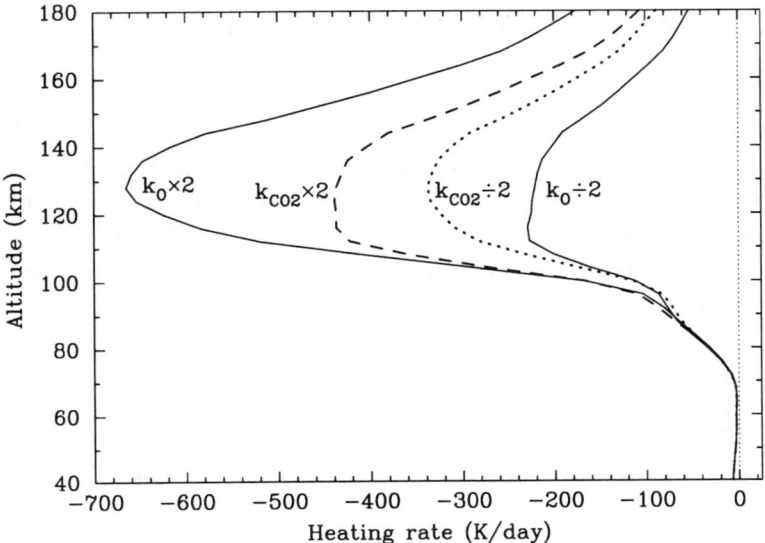

Fig. 10.10   Effect of the rate coefficient of V–T collisions of $CO_2(v_1,v_2,0)$ levels with CO_2 (k_{CO_2}) and with $O(^3P)$ (k_0) on the cooling rate at 15 μm for the COSPAR revised temperature profile. Cooling rates are shown for the increased and decreased rate coefficients by a factor of 2.

Figure 10.10 shows the effect on the cooling rates of modifying the rate coefficients of the V–T collisions of the CO_2 bending-symmetric (v_2) levels with CO_2 and with $O(^3P)$ (processes VT2 in Table 10.1) by a factor of 2. Both effects become noticeable above about 85 km, the altitude where most of the 15 μm bands become optically thin, and are particularly important above 110 km, where the 626 fundamental band becomes less optically thick. The effect of the $O(^3P)$ rate is larger than that for CO_2, except between 85 and 100 km, because atomic oxygen is very efficient in transferring kinetic energy. The biggest factor in quantifying the uncertainty due to $O(^3P)$ is probably not the reaction rate, but its volume mixing ratio. This varies, in response to solar cycle and thermospheric temperature variations, by about 100% between 100 and 120 km and more than a factor of 5 above about 140 km.

10.4.7   *Heating rates*

The photoabsorption coefficient used to calculate the energy absorbed from the solar flux were discussed in Sec. 10.4.3.1. A large amount of the energy initially absorbed may be re-emitted by fluorescence, and only a fraction converted into thermal energy. Several of the CO_2 bands, especially those at 15 μm, emit significant amounts of radiation at night, and this increases during the day. Thus, the calculation of the solar heating rate at a given altitude involves not only the incoming solar energy locally absorbed, but also the emission to space and the net balance resulting from the exchange of photons between layers at all wavelengths. As before, the fraction of the absorbed energy re-emitted can be obtained from the differences in the flux divergences calculated with and without solar illumination.

We already discussed for the Earth some examples of the paths followed by the energy initially absorbed in the CO_2 near-IR bands (see Fig. 9.19). For Mars, the thermalization processes are qualitatively the same but differ quantitatively, and some processes that are negligible on Earth are important here. As an example we will discuss the thermalization of solar energy absorbed in the strong $4.3\,\mu m$ fundamental band, and will describe later the other important bands contributing to Martian solar heating.

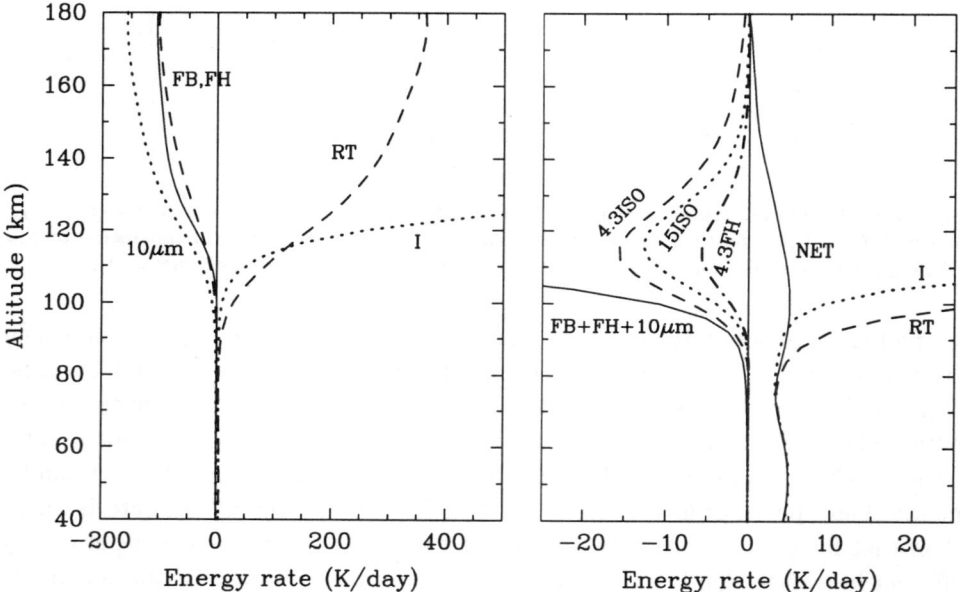

Fig. 10.11 Heating rate and contributions of the absorption of solar radiation in the CO_2–626 fundamental band at $4.3\,\mu m$ for overhead Sun conditions and average Mars heliocentric distance. 'I' is the local absorption rate of solar energy; 'RT', is the rate after considering spontaneous emission and radiative transfer in this band; 'FB' and 'FH' are the rates of energy re-emitted in the $15\,\mu m$ 626 fundamental and first hot bands; '10 μm', that re-emitted at $10\,\mu m$; '4.3ISO', '15ISO' and '4.3FH', those re-emitted in the isotopic bands at 4.3 and $15\,\mu m$ and in the 626 first hot band at $4.3\,\mu m$, respectively. 'NET' is the solar heating rate. Units are K per terrestrial day.

Figure 10.11 shows the absorption of solar radiation at $4.3\,\mu m$, the energy rates of the different pathways its follows, and the net heating rate that it produces. The rate at which energy is locally absorbed from the solar flux in the thermosphere (curve 'I') is very large, reaching nearly $5000\,K/day$ near the top of the atmosphere. However, when spontaneous emission and radiative transfer between layers is taken into account, the flux divergence (with sign reversed), curve 'RT', is much smaller than the initial absorption above 120 km. For this strong band, there is an unusual region located between 85 and 110 km, where the available energy for thermalization is larger than the local absorbed solar energy, which is due to the additional absorption of downwelling photons emitted from higher atmospheric layers. Below 75 km the curves 'I' and 'RT' come together as the net exchange of radiation between layers becomes negligible (see Sec. 10.4.3.2).

In the highest atmospheric layers, some of the absorbed energy is emitted in the two bands near 10 μm. These are, in fact, the most important relaxation mechanism of $(0,0^0,1)$. As a consequence, a significant amount of the locally absorbed energy goes into the $(0,2,0)$ levels. These states partially relax through emission in the 15 μm FH bands, and transfer the rest of the energy to $(0,1^1,0)$. At the same time the $(0,1^1,0)$ state re-emits a fraction of that energy in the 15 μm fundamental band ('15FB'). The remaining energy goes partially to thermal energy and partially enters the $(0,1^1,0)$ minor isotopic system through V–V collisions. At very high altitudes, the 15 μm FH and FB bands, together with the emission at 10 μm, are able to re-emit almost all of the energy available after considering the radiative processes in the 4.3 μm FB. In the lower thermosphere, the re-emission in these bands is much smaller because we have less available energy and because v_2 and v_3 V–V collisions become more frequent. These collisional processes lead to increased emissions by the 4.3 μm isotopic and first hot bands, and by the isotopic bands at 15 μm, which produce a significant reduction of the absorbed solar radiation in the 100–125 km region. Below about 75 km, the absorbed solar energy is entirely thermalized, mainly through collisions relaxing the v_2 levels, but relatively few solar photons reach this region and the heating rate is small. From the above it is found that the solar heating in the 4.3 μm fundamental band is about 5 K/day or less at most altitudes.

After considering similar processes for the solar absorption in the 4.3 μm first hot and isotopic bands (see Table F.1), it is found that the solar heating produced by the first hot band is similar to that of the fundamental, while that of the isotopic is significantly larger in the upper mesosphere and lower thermosphere (70–100 km), with a maximum of 25 K/day around 90 km. The solar flux in these weaker bands penetrates into deeper atmospheric layers, where collisions are more frequent and the absorbed solar energy is more efficiently thermalized. The heating rates induced by the second hot bands at 4.3 μm (see Table F.1) are negligible, since the number of absorbing molecules, those excited in the $(0,2,0)$ state, is very small.

The heating rate induced by absorption in all the 4.3 μm bands is shown in Fig. 10.12. Most of the heating occurs between 60 and 110 km, with a maximum of 40 K/day at around 85 km.

Solar radiation is also absorbed in several bands near 2.7 μm (see Table F.1). The heating rate produced by the fundamental has a broad peak of $\sim$50 K/day extending from 85 to 115 km, while the 2.7 μm minor isotopic and hot bands show

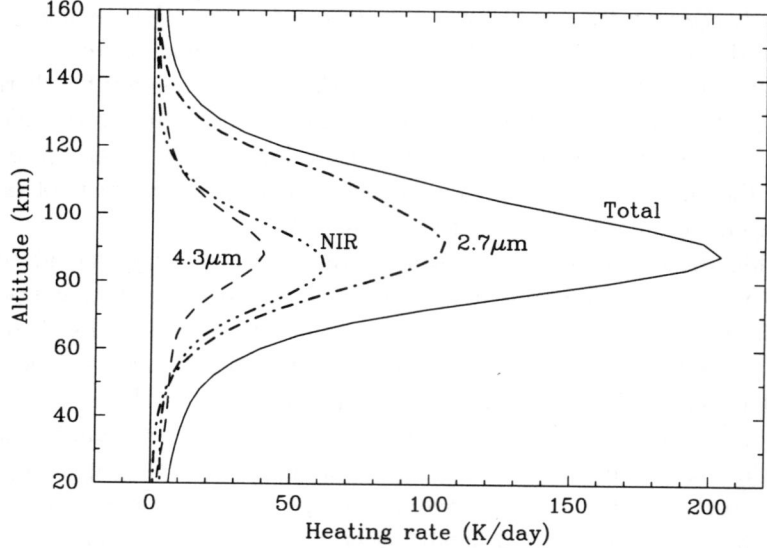

Fig. 10.12   Solar heating rates of the Martian atmosphere, Sun's overhead conditions, and average heliocentric distance of Mars. 'NIR' is the contribution of the near-IR bands.

smaller maximum values of $\sim$40 K/day and 5 K/day, located at lower altitudes and so dominating the heating of the mesosphere. The heating by all the 2.7 μm bands reaches 100 K/day at 90 km, more than twice that produced by the 4.3 μm bands at that altitude (Fig. 10.12). This is due to the fact that the two v_2 quanta in the upper levels of the 2.7 μm bands are more efficiently thermalized than the asymmetric v_3 quanta in the 4.3 μm bands.

A significant amount of solar heating also takes place in the many near-infrared bands in the 1.2–2.0 μm spectral range (curve 'NIR'), covering a height range spanning from about 50 to 110 km, with a maximum of about 60 K/day at 85 km.

10.4.8   *Variability and uncertainties in the heating rates*

Temperature has only a small effect on the solar heating, its influence being mainly through its effect on the atmospheric scale height and, consequently, on the absorber amount. In this way, tropospheric temperature variations influence the solar heating rate in the mesosphere and lower thermosphere, producing larger changes than temperature variations in the thermosphere. For example, a thermospheric temperature 150 K colder than the nominal profile produces a 10 K/day ($\sim$5%) change in the solar heating peak and shifts its altitude by just 1 km. However, a tropospheric profile 30 K colder than nominal lowers the altitude of the solar heating peak by about 8 km and decreases it by about 15 K/day ($\sim$7.5%). It follows that the seasonal pressure variation at the Martian surface significantly affects the net heating of the upper atmosphere, particularly the height of the heating peak.

Of the various uncertainties in rate coefficients, the largest effect on the heating

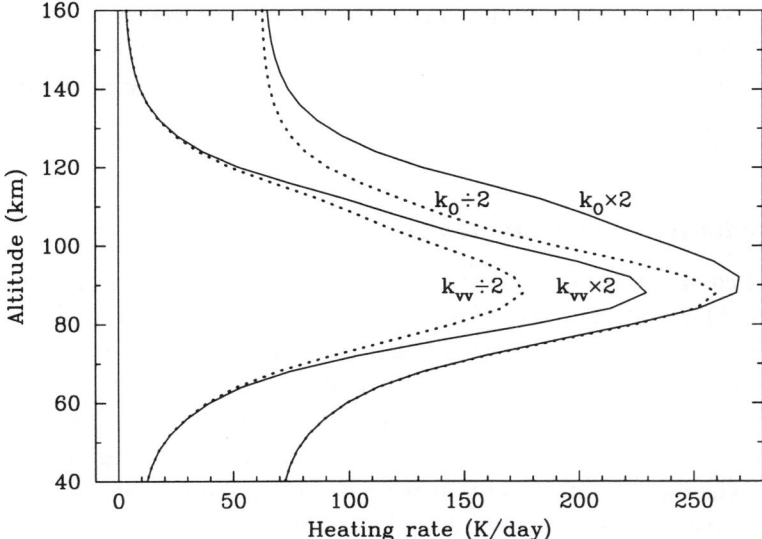

Fig. 10.13   Effect of the rate coefficient for the V–V relaxation of v_3 quanta in the high-energetic CO_2 levels (k_{vv}), and of thermal collisions of $CO_2(v_2)$ with $O(^3P)$, $k_{CO_2\text{-}O}$ (k_0), on the solar heating rates. The heating rates for the sensitivity of $k_{CO_2\text{-}O}$ are shifted by 60 K/day.

rate are found for changes in the V–V deactivation rate of the v_3 quanta of the high-energy levels, and in the V–T collisions of v_2 levels with atomic oxygen. Those for V–T collisions of CO_2 v_2 with CO_2, the rearrangement between the v_3 and v_2 quanta, and the V–V v_2 and v_3 couplings (excluding the v_3 relaxation of the high-energy combination levels), all have much smaller effects. The reason is basically that the former are fast enough to compete with radiative processes in the relaxation of the highly excited levels in the thermosphere. The others affect the level populations at lower altitudes where most radiative transitions are optically thick and they have little effect on the solar heating.

The effect on the solar heating rate of increasing and decreasing the $CO_2(0,1^1,0)$–$O(^3P)$ rate coefficient by a factor of 2 is shown in Fig. 10.13. This is important above about 90 km, where the atomic oxygen abundance is significant, producing larger heating for the increased rate, with changes of as much as 100% above about 100 km. Again, the effect of the variable and uncertain $O(^3P)$ abundance in the Martian thermosphere is probably even larger.

The figure also shows the changes produced by increasing and decreasing the V–V(v_3) deactivation rate of the combination levels by a factor of 2, which approximately covers the current dispersion among the measurements. For those changes in the rate, the heating changes in the same direction, by 30 K/day between 80 and 100 km. An increased rate means that the excitation energy of the higher CO_2 levels is transferred more rapidly to the lower states, where it is thermalized and the heating increases.

These changes in the heating rates are nevertheless smaller than those induced by the pressure and by the solar illumination conditions. This allows to parameterize the non-LTE heating rates rather easily and accurately only as a function of pressure and solar zenith angle, which is very useful for its inclusion in general circulation models (GCMs).

10.4.9 *Radiative equilibrium temperature*

Heating and cooling rates can be used to compute the daily averaged temperature profile of the Martian atmosphere for which radiative equilibrium applies at every level. This is done by iteration until a balance is found between the 15 μm cooling and the near-IR solar heating rates, using half of the heating rate for a solar zenith angle of 60° as an estimate of the daily average. The temperature so obtained can be expected to be realistic only for the global mean, and only in the region between 40 and 125 km, where we might expect that the thermal structure is mainly governed by infrared radiative processes. Above 125 km, EUV heating and molecular conduction start becoming important; and below 40 km, we would expect CO_2 infrared cooling and heating rates tend to give way to the effects of dust and dynamics.

The solar heating and thermal cooling rates corresponding to various situations are presented in Fig. 10.14a, and the resulting radiative equilibrium temperature profiles in Fig. 10.14b. The local maximum in the temperature profile around 72 km (the mesopeak), is produced by the large solar heating that takes place at these altitudes. It can be seen that the calculated profiles are all substantially different from the COSPAR reference temperature profile, which is representative of mean measured conditions. Even when the profile is forced to be the same as the COSPAR model below 40 km, where radiative equilibrium is not expected to apply, the profile diverges rapidly above this altitude. Also, although there are significant differences in the cooling and heating rates between different non-LTE models (compare dot-dashed and dotted curves in Fig. 10.14a), the two models show radiative equilibrium temperatures much larger than that of COSPAR, which suggests that the atmosphere of Mars is not in radiative equilibrium at any level. This is not too surprising, since the role of waves and other dynamical processes needs to be taken into account. Also, there is considerable variability in the profile, with factors such as airborne dust loading which make comparisons between global mean models and individual measurements complicated. There is also considerable scope for improvement in our understanding of the infrared energy balance of the Martian atmosphere in several areas, an urgent need for more detailed measurements of the infrared emissions, the variability of the atmospheric temperature, and studies of the influence of water vapour, dust and dynamical processes. Finally, there is the usual need for laboratory measurements of V–T and V–V collisional rate coefficients, in this case at the low Martian atmospheric temperatures. The most important are the rates for thermal deactivation of $CO_2(0,1^1,0)$ by $O(^3P)$ and by CO_2, and of the V–V(v_3) relaxation of high-energetic combination levels.

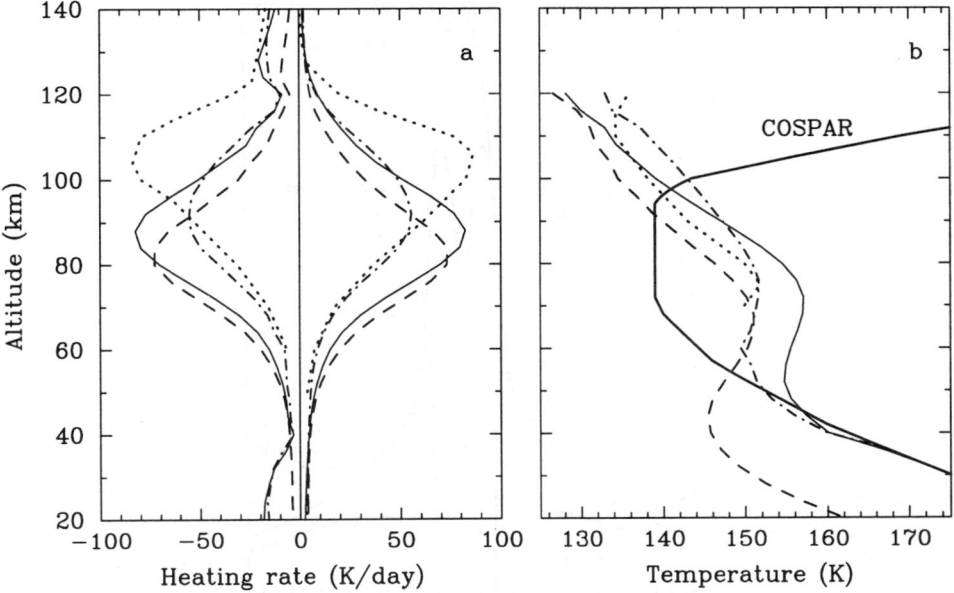

Fig. 10.14   (a) Infrared cooling and near-IR solar heating rates for radiative equilibrium conditions
in the mesosphere. Solid lines: troposphere fixed to the COSPAR profile; dashed, troposphere in
radiative equilibrium; dot–dashed, simulation with model's parameters from Bougher and Dickinson (1988) for solar minimum conditions; dotted, results from these authors for the same conditions.
(b) Radiative equilibrium temperatures for the cases in (a). The COSPAR temperature profile is
also plotted for reference.

10.5   Venus

Much of what has been said above about Mars also applies to Venus, because
it also has a nearly pure CO_2 atmosphere, although the prevailing pressures and
temperatures are generally much higher. Less work has been done on non-LTE
models for Venus in recent years, partly because the studies are generally induced
by plans for space experiments intended to make relevant measurements, and Mars
is now the centre of attention in this respect. A decade or two ago, the situation
was reversed, and a great deal of what is now known about non-LTE conditions
on Venus still depends on pioneering work done then, especially by Dickinson. His
approach was first to calculate a global mean radiative equilibrium temperature
profile, including all processes thought to be of importance for the global heat
budget. He pointed out the particular importance of including the combination and,
especially, the isotopic bands of CO_2. As we have seen for the case of Mars, these
weaker bands have a disproportionate effect because radiative transfer of energy
can take place over much longer distances than in the strong fundamentals of the
commonest isotope. Only bands with a strength at least five orders of magnitude
less than the strongest bands can be neglected. This effect is illustrated in Fig. 10.15,
which shows the contribution of a single spectral line to the cooling rate from the

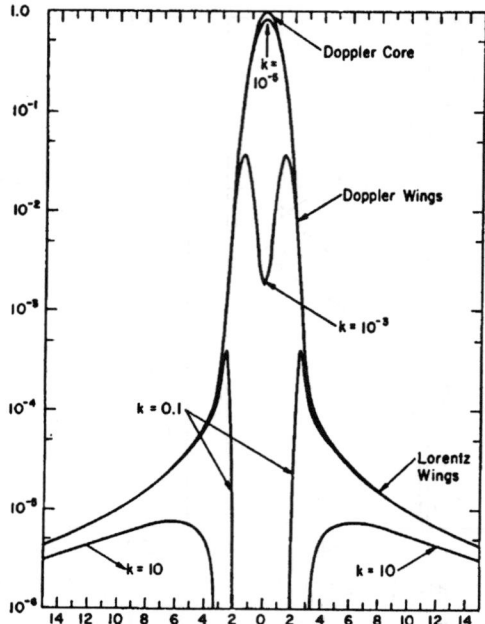

Fig. 10.15 The outer line is the emission profile of a 15 μm band line at the 10 μbar level on Venus. The other lines are the angle-integrated contributions to the net cooling from that level of a line with the strength k indicated. The abscissa is in units of the Doppler width.

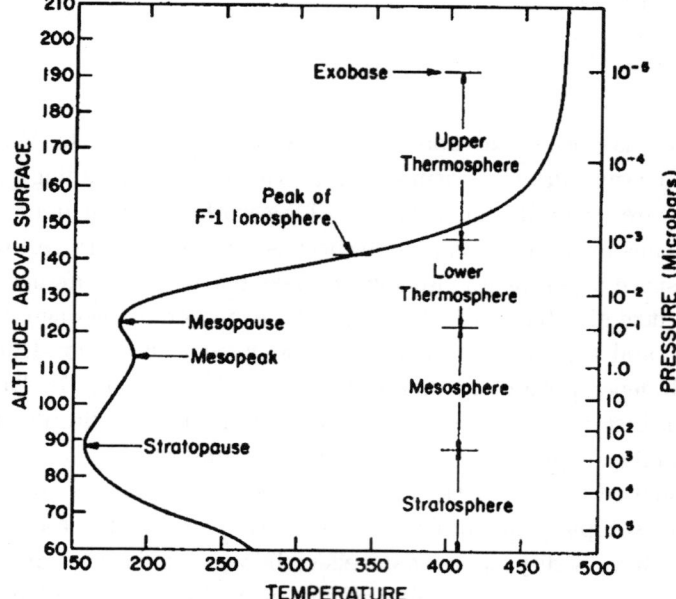

Fig. 10.16 A radiative equilibrium vertical temperature profile for Venus, calculated including non-LTE heating and cooling budgets for the CO_2 near infrared and 15 μm bands.

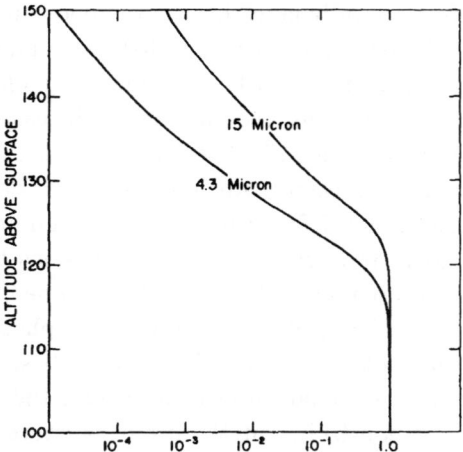

Fig. 10.17 First excited vibrational level populations, as a fraction of the Boltzmann value, for the v_2 and v_3 fundamentals of the main isotope of CO_2 on Venus.

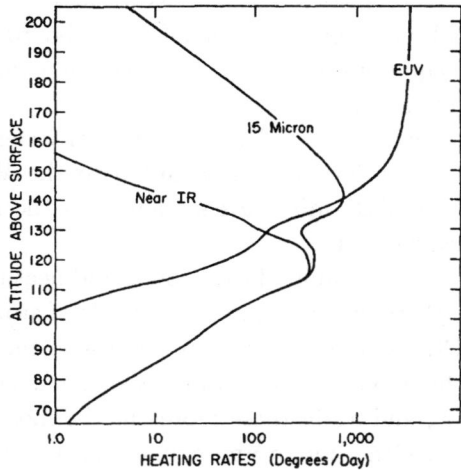

Fig. 10.18 Global mean net heating and cooling rates on Venus, showing the contributions of extreme ultraviolet (EUV) and near-infrared heating separately.

$10\,\mu$b level. Even at this low pressure, the core of a strong line is completely absorbing and cooling to space is negligible, except in the line wings. A weak line, on the other hand, emits strongly to space from its Doppler core.

Figure 10.16 shows the radiative equilibrium temperature profile derived by Dickinson, with the names which he proposed for the different regions, seeking to maintain an analogy with the Earth.

Figure 10.17 shows the departure from LTE of the $v_2 = 1$ and the $v_3 = 1$ levels as a function of height. Finally, Fig. 10.18 shows the net heating by UV and IR radiation separately, and the net cooling at $15\,\mu$m. When molecular conduction is also included, these terms are in balance for the temperature profile in Fig. 10.16. It can be seen that the lower regions are in radiative balance when only infrared heating and cooling are considered, and UV heating and molecular conduction can be disregarded. At the highest levels, the opposite is the case.

In a later model, Dickinson added V–V exchange terms, tested the 'cooling-to-space' approximation against a scheme which included transfer between layers, and studied the susceptibility to non-LTE of temperature sounding experiments which were at that time (1975) in preparation for space missions to Venus. He showed that the 15 μm fundamental band could be considered to be in LTE up to pressures as low as 1 μbar, but that at lower altitudes most of the photons escaping to space originated in the hot and isotopic bands. This meant basically that 'wide-band' measurements centred in the fundamental could sound the Venusian mesosphere without taking account of non-LTE, but the use of pressure modulators to sound the lower thermosphere would require a more complex interpretation.

Battaner and colleagues later had the additional advantage of being able to revise the model atmosphere used in the light of the Pioneer Venus entry probe measurements of 1979–1980, and to compare the results of their calculation with emission measurements by the infrared radiometer on board the Pioneer Orbiter. Their dayside radiative equilibrium temperatures were in good agreement with the observations in the lower thermosphere.

In the following sections, the non-LTE radiative transfer model described in Sec. 10.3, and applied above for Mars, is adapted to the Venusian atmosphere, and a similar study of the CO_2 infrared emissions and the radiative equilibrium temperature profile is carried out.

All previous studies which computed the cooling and heating rates for the Venusian and Martian atmospheres stress the important role of atomic oxygen: the thermal collision between CO_2 and $O(^3P)$ is a key process in the cooling of the upper layers of a CO_2 atmosphere. The collisional rate coefficients used here are the same as in the Martian model, and again these represent the most important source of uncertainty in the model. Other minor approximations are: 1) induced emission is neglected; 2) the rotational levels associated with given vibrational levels are considered to be in LTE among themselves; 3) the intermolecular collisions between CO_2 and other atmospheric compounds, N_2, O_2, CO and $O(^3P)$, are incorporated assuming that only single quantum transitions are important; 4) the CO_2 $(0,2^0,0)$, $(0,2^2,0)$ and $(1,0^0,0)$ levels, which are very close in energy, are considered to be in mutual equilibrium, and described by an equivalent level $(0,2,0)$; and the same has been assumed for the $(0,3^3,0)$, $(0,3^1,0)$, $(1,1^1,0)$ levels, grouped as the $(0,3,0)$ state. An iterative process is required to solve for the inherent non-linearities, like those arising from radiative transfer in hot bands and those from the V–V exchanges.

10.5.1 *Reference atmosphere*

Most of the observational information we have about the Venus atmosphere comes from the Pioneer Venus (PV) mission of the late 1970s. More recently, high resolution infrared spectra were obtained by the Near-IR Mapping Spectrometer (NIMS) during the Galileo fly-by in 1991, but temperatures were derived only between 60 and 100 km. Due to the significant day-night variation, we need to consider separate day and night-time reference temperature profiles, both taken from the PV measurements and shown in Fig. 10.19. The abundances of CO_2, CO, and $O(^3P)$ appear in Fig. 10.20. Note that CO_2 is the major constituent only below $\sim$140 km.

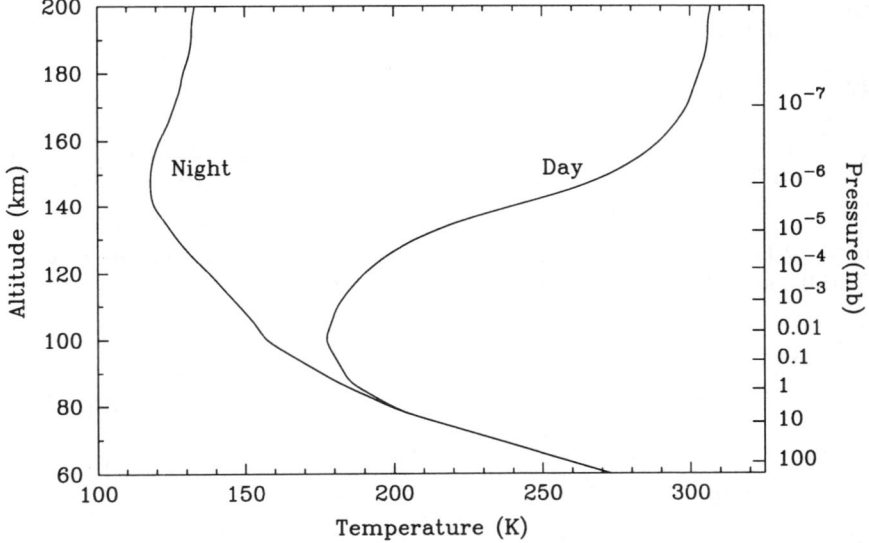

Fig. 10.19 Venus reference model temperatures. Night-side and day-side profiles from the empirical model of Hedin *et al.* (1983).

10.5.2 *Night-time populations of CO_2*

The set of CO_2 vibrational levels used is the same as that studied in the Martian atmosphere and listed in Table F.1. The definition of vibrational temperature is also the same, e.g., referred to the ground level of the molecule, to show the populations of the energy levels (see Eq. 6.1). Fig. 10.21 shows the vibrational temperatures for night-time conditions of the $(0,1^1,0)$ and $(0,0^0,1)$ levels of the 626 and 628 isotopes of CO_2, and the $(0,2,0)$ and $(0,3,0)$ levels of the principal isotope.

The departure from LTE of all the v_2 levels occurs at an altitude of around 120 km, at about the same pressure level as the LTE departure in the atmosphere of Mars ($\approx 0.15\,\mu$b), but some 35 km higher because of the higher atmospheric densities on Venus. The behaviour of the vibrational temperatures in the layers immediately above the LTE separation is also similar to Mars, with smaller vibrational than

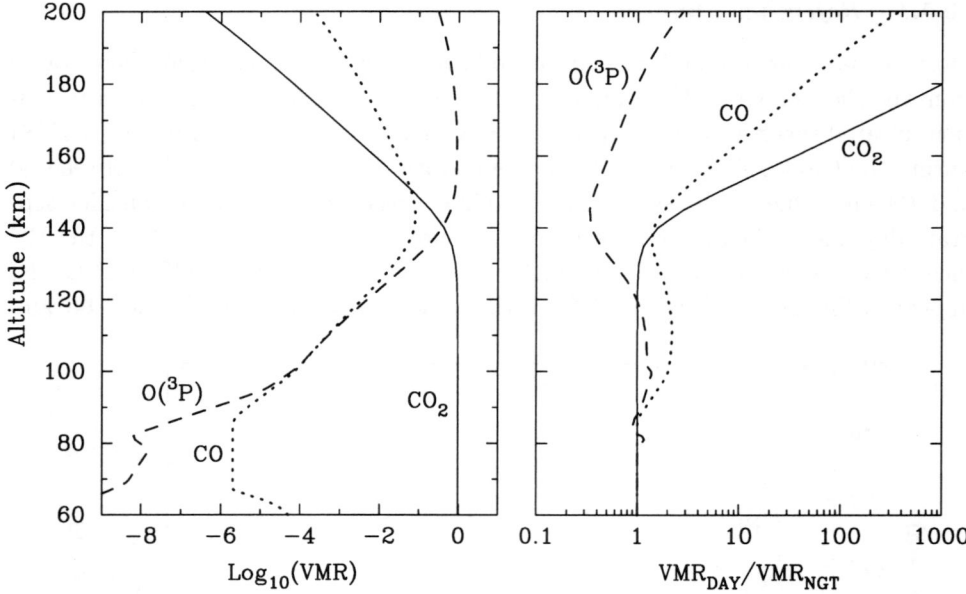

Fig. 10.20   Volume mixing ratios of the main compounds of the Venusian atmosphere (after Hedin *et al.*, 1983). (a) Night-time conditions; (b) day/night ratio.

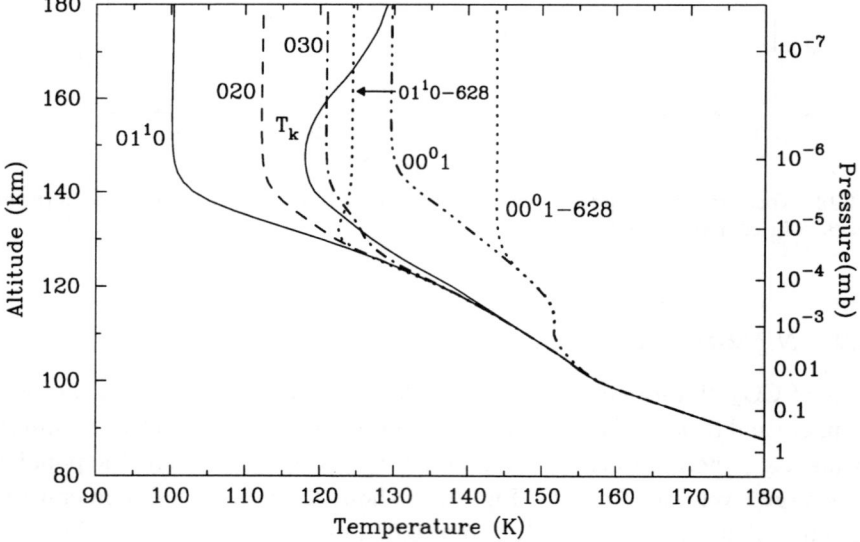

Fig. 10.21   Night-time vibrational temperatures of the $(0,1^1,0)$, $(0,2,0)$, $(0,3,0)$ and $(0,0^0,1)$ levels of the main isotope of CO_2 and the $(0,1^1,0)$ and $(0,0^0,1)$ of the 628 isotope. T_k represents the kinetic temperature.

kinetic temperatures. This LTE departure is caused by emission to space in the $15\,\mu$m bands, as they become optically thin and the radiative losses by spontaneous emission overcome collisional processes.

The population of the $(0,1^1,0)$ level is controlled by V–V energy transfer, again like Mars. Due to this strong collisional coupling, the altitude of LTE departure is similar for the v_2 levels of all the isotopes. Higher in the thermosphere, their vibrational temperatures follow a decreasing tendency towards a constant value at the top of the atmosphere, typical of level populations governed exclusively by radiative processes. The vibrational temperatures for the upper states of the weaker bands are larger because photons from lower and warmer layers reach these high altitudes. A difference with respect to the Martian thermosphere is that the constant vibrational temperature is reached on Venus only four scale heights above the altitude of LTE departure, due to the depletion of CO_2 above 140 km by diffusive separation.

Figure 10.21 also shows the vibrational temperatures of the $(0,0^0,1)$ levels of the 626 and 628 isotopes. The v_3 levels of all of the isotopes behave in this way, departing from LTE around 110 km, and maintaining the same vibrational temperature up to 125 km due to the strong V–V collisional processes. Above this, each level strikes its own balance between the absorption of photons from lower layers, mainly in the fundamental transition at $4.3\,\mu$m, and emission. The larger optical thickness on Venus means that, unlike Mars, the $10\,\mu$m laser bands are not involved significantly in maintaining the $(0,0^0,1)$ populations in the non-LTE region.

10.5.3 *Daytime populations of CO$_2$*

Figure 10.22 shows the vibrational temperature of several levels of the CO_2 626 and 628 isotopes for daytime conditions.

The altitude of LTE departure of the v_2 levels is about 120 km, similar to nighttime. The populations are higher during the day, however, partly because of the larger daytime kinetic temperature of the lower thermosphere and partly because of energy cascading from higher energy levels excited by the Sun. The excitation of $(0,3,0)$ comes mainly from the bands at 1.2 and $2.0\,\mu$m, after collisional and radiative relaxation of their v_3 quanta, while the $(0,2,0)$ levels are populated by relaxation of the $2.7\,\mu$m levels via emission in the $4.3\,\mu$m second hot bands. The peak in the $(0,1^1,0)$ vibrational temperatures around 135 km is due to V–V coupling with the $(0,2,0)$ and $(0,3,0)$ levels. This transfer is not strong enough to keep all the v_2 levels of the 626 isotope coupled above 135 km, nor those of the minor isotopes above 120 km.

The $(0,0^0,1)$ level of the main isotope departs from LTE at about 90 km, 10 km lower than at night, and has very large populations above that altitude. Both effects are due to strong absorption in the $4.3\,\mu$m fundamental band. The corresponding levels of the minor isotopes have similar vibrational temperatures up to about 120 km, where the strong V–V coupling between them ceases to dominate and they tend to individual values, independent of height, which are determined by

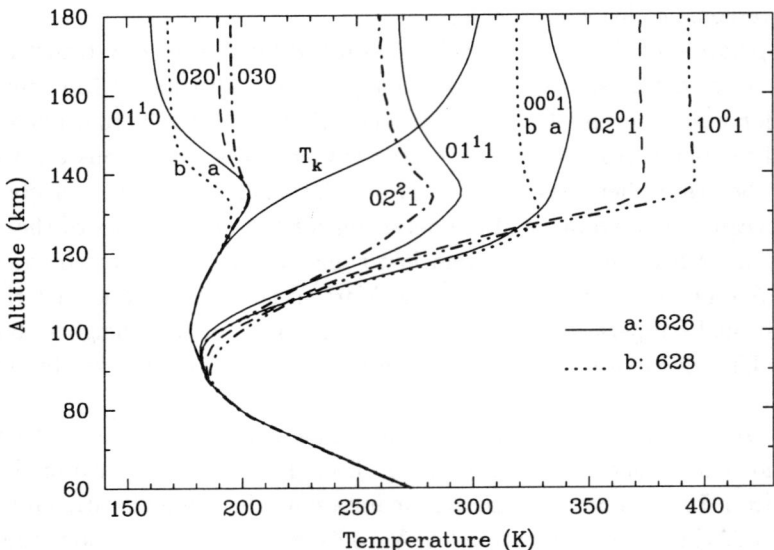

Fig. 10.22   Vibrational temperatures of CO_2–626 levels and of the $(0,1^1,0)$ and $(0,0^0,1)$ levels of CO_2–628 (dotted lines) during daytime conditions ($\chi = 0°$).

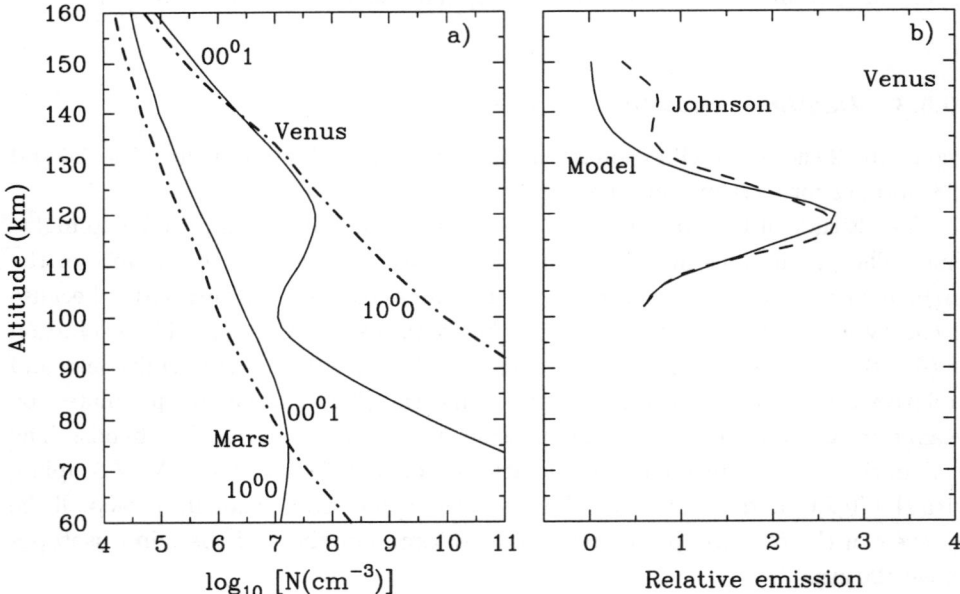

Fig. 10.23   a) Populations inversion between the $(0,0^0,1)$ and $(1,0^0,0)$ levels of CO_2 in the Martian and Venusian atmospheres. b) The number of photons at $10\,\mu m$ emitted by the Venusian atmosphere measured by Johnson *et al.* (1976) and predicted by the model described here.

their photoabsorption coefficients. All these processes are qualitatively similar to the Martian case.

The larger daytime temperature in the middle atmosphere of Venus than in Mars, causes a higher population in the $CO_2(1,0^0,0)$ level thus moving the inversion population $[(0,0^0,1)]/[(1,0^0,0)]$ to higher altitudes, above about 140 km (see Fig. 10.23a). The 10 μm emission in the laser bands has been observed on Venus as well as on Mars: Fig. 10.23b compares the measurements to the non-LTE model for Venus. The agreement is good in the main peak. Since this emission comes from the $(0,0^0,1)$ state, this means that we understand the population of this level, at least in that region. However, there is a secondary peak in the data which is not reproduced by the model and which remains unexplained.

10.5.4 *Cooling rates*

The cooling rate produced by the night-time CO_2 15 μm emission on Venus for the particular thermal structure in Fig. 10.19 is shown in Fig. 10.24. The maximum cooling of about 130 K/day occurs between 120 and 130 km and shows a double peak, caused by the fact that the contributions of the main and minor isotopes peak at different altitudes. Again, as in the case of Mars, the emission from the minor isotopes would be much less were it not for the large energy transfer by V–V coupling between their $(0,1^1,0)$ levels and that of 626.

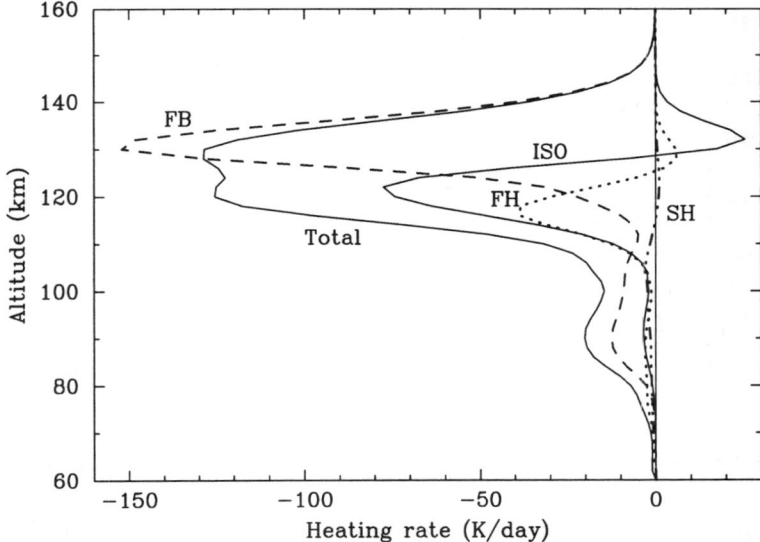

Fig. 10.24   Contributions and total cooling rate (in K per terrestrial day) produced by the CO_2 15 μm bands for the night-time kinetic temperature profile in Fig. 10.19.

On a matter of technique, note that the quasi-line-by-line histogramming algorithm for the calculation of the atmospheric transmittances, discussed in Sec. 6.3.3.1 and Sec. 10.3, cannot be used for Venus below about 90 km ($\approx$0.3 mb), because it treats the overlapping between lines too crudely. The absorptions are systemati-

cally overestimated, and very large cooling rates result, particularly for the weak
hot and isotopic bands. This was not a problem in the Martian case, because the
overlapping between lines in the upper troposphere is smaller due to the lower density and temperature. The cooling rates in Fig. 10.24 were calculated using a proper
line-by-line scheme to eliminate this source of error.

10.5.5 *Heating rates*

The solar heating rates show one major difference from Mars, in that the absorption
coefficients decrease slower with height in the upper atmosphere because the Venus
thermosphere is depleted in CO_2 above 140 km. At lower altitudes, the photoabsorption coefficients also decline when the bands become optically thick. Between 90
and 110 km, bands of very different strengths give similar absorption rates, although
peaked at different heights. The penetration of the solar flux into the atmosphere is
larger at shorter wavelengths, where most of the weak bands are located, allowing
them to make important contributions to the heating rate, as they do on Mars.

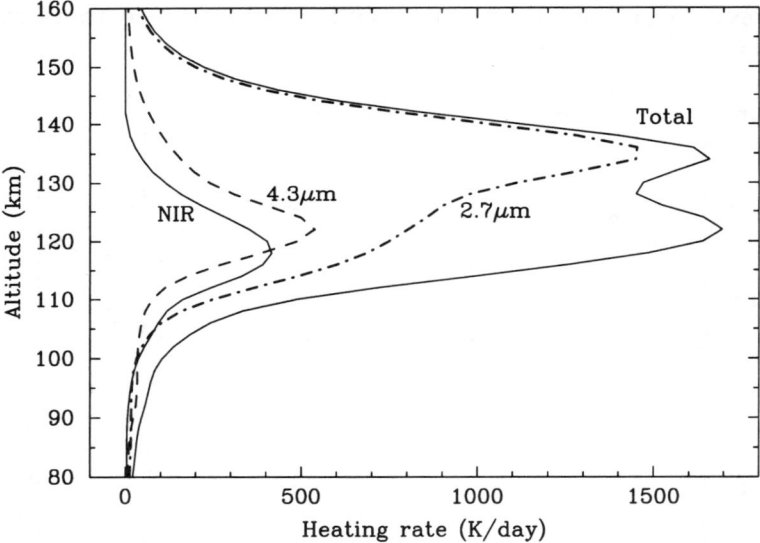

Fig. 10.25 Solar heating rates of the Venusian atmosphere for a solar zenith angle of 60°.

The total solar heating rate for a solar zenith angle of 60° appears in Fig. 10.25,
with the contributions from the different near-IR bands shown separately. As for
Mars, much of the energy initially absorbed from solar radiation is re-emitted, and
radiative transfer and V–V collisions play major roles.

The energy initially absorbed in the weaker CO_2 bands at 1.2–2.0 μm (curve
'NIR') is partially re-emitted mainly by relaxation in the 4.3 μm and 15 μm hot
bands. The remainder provides heating with a peak of about 400 K/day over a broad
region between 110 and 130 km, making it one of the most important contributor

to the heating rate in the lower thermosphere. The overall heating by the $4.3\,\mu$m bands reaches a maximum of around $500\,$K/day at $120\,$km, while the $2.7\,\mu$m heating peaks at $1400\,$K/day near $130\,$km, with a broad tail below $115\,$km due to the weaker more penetrating isotopic and hot bands. The weak bands contribute more to the total heating rate, compared to the fundamentals, than they do on Mars, because of the larger atomic oxygen abundances on Venus.

The maximum total solar heating amounts to more than $1500\,$K/day between 120 and $135\,$km with a double peak due to the different heights of the individual maxima. Significant heating is found on Venus only above about $100\,$km and below about $150\,$km (2×10^{-2} and 7×10^{-7} mb, respectively), the corresponding heights on Mars being 50 and $130\,$km (10^{-2} and 10^{-6} mb) (Fig. 10.12), i.e. at similar pressure levels. Above these heights, all of the solar energy absorbed is re-emitted before it can be quenched by thermal collisions. In the lower atmosphere, the solar flux is too attenuated to have much effect.

10.5.6 *Radiative equilibrium temperature*

To compute a radiative equilibrium temperature profile, we change the temperature at every altitude to achieve energy balance between the CO_2 solar heating and thermal cooling. As explained in Sec. 10.4.9, the iterative process converges since the cooling is very dependent on the local temperature, while the solar heating is largely insensitive. Since the energy balance of the Venusian thermosphere is not controlled by the CO_2 IR balance above about $130\,$km, and below $80\,$km convection begins to dominate, the radiative equilibrium temperature profile is likely to be realistic only in the intermediate region at best.

The global mean radiative equilibrium temperature in Fig. 10.26 ('GM') was computed using a diurnally averaged solar heating rate, calculated for a solar zenith angle of $60°$ and divided by 2 (see Sec. 10.4.9). This shows a maximum around $110\,$km produced by the solar absorption in the near-IR and in the weak bands at 4.3 and $2.7\,\mu$m. The change of slope above $\sim125\,$km is due to the solar heating in the $2.7\,\mu$m bands.

The differences between day and night-time conditions in Venus are particularly extreme, given its slow rotation. A purely radiative model predicts extremely cold temperatures in the night-time thermosphere (not shown in the figure), even colder than those measured by Pioneer Venus (curve 'Night'). This discrepancy between the measurements and the pure-radiative prediction is probably due primarily to adiabatic heating by dynamics, rather than errors in the radiative transfer calculations and their rate coefficients, although these are also quite large (see below). As an interesting exercise, we can simulate the dynamical redistribution of energy between the hemispheres, by incorporating a fraction of the solar heating in the night-time run and reducing the solar heating during the daytime by the same amount. Reasonable results (curves 'N' and 'D' in Fig. 10.26) are obtained when

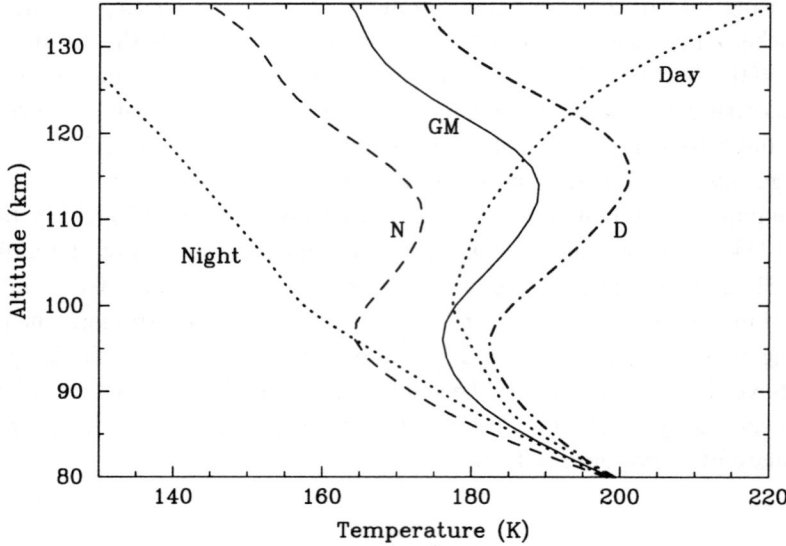

Fig. 10.26   Radiative equilibrium temperature profiles for dayside (D), global mean (GM), and nightside (N). The day and nightside temperatures were calculated assuming redistribution by dynamics of 30% of the daytime heating (see text). The reference temperature profiles of Fig. 10.19 are also plotted (dotted lines).

30% of the CO_2 solar heating rate is assumed to be transferred from day to night by dynamical processes. The day-night difference is nearly constant between 100 and 130 km and amounts to about 20 K.

10.5.7   *Variability and uncertainties*

Again, we need to consider how the results from model calculations are sensitive to errors introduced by uncertainties in the various rate coefficients. One of the most important uncertainties on Earth and Mars is the contribution of atomic oxygen, because of its very efficient interactions with carbon dioxide, especially in (de-)activating some of the most populous levels, and its high variability. We would expect the same to be true on Venus, where, as we have seen, the atomic oxygen abundance is particularly high because of the intense solar UV flux. Figs. 10.27 and 10.28 show some calculations of the effect of changing the rate for vibrational-translational collisions between CO_2 and atomic oxygen (or the abundance of the latter, which has the same effect) by a factor 4.

At night, the effect of a larger rate is principally to shift the altitude of LTE departure about 2 km upwards, with hotter vibrational temperatures above the breakdown height. Uncertainties in other V–T collisions do not have any comparable effect on the v_2 level populations, although a change in the V–V v_2 rate coefficient by an uncertainty factor of 100 (possibly overestimated) has clear effects on the coupling between the v_2 overtone and isotopic levels in the same sense that was

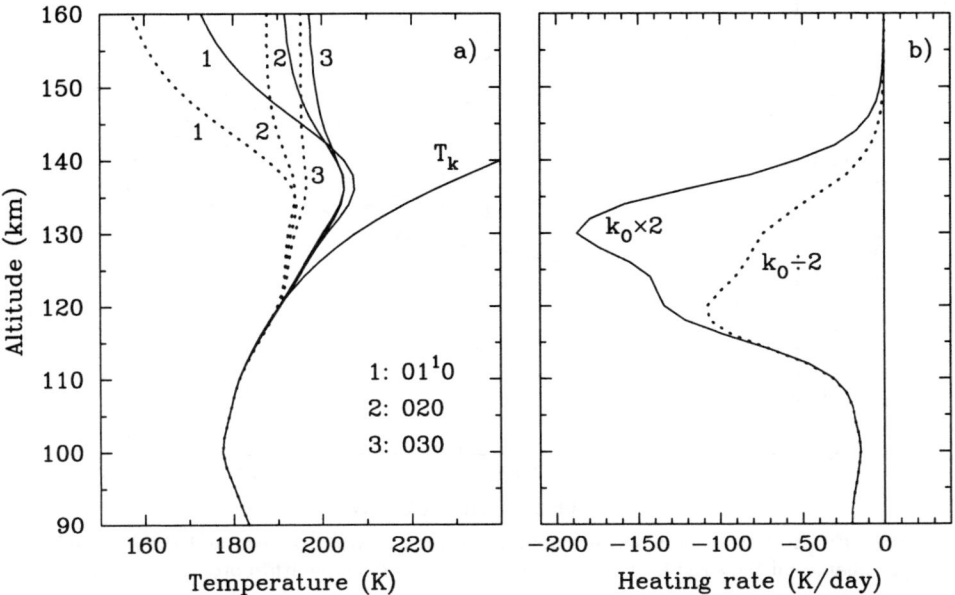

Fig. 10.27 Effect of the $k_{CO_2\text{-}O}$ rate (or the atomic oxygen abundance) on: (a) the daytime vibrational temperature of the CO_2 626 v_2 levels; and (b) on the CO_2 15 μm cooling rate. Solid lines: rate coefficient increased by a factor of 2; dotted lines: reduced by the same factor.

described for Mars (Sec. 10.4.4).

By day, an increase in the $k_{CO_2\text{-}O}$ rate (or in the atomic oxygen VMR) produces an increase in the vibrational temperatures of the v_2 levels which is larger than at night because of the larger kinetic temperature (Fig. 10.27a). The cooling rate is also strongly affected by the $k_{CO_2\text{-}O}$ rate and the $O(^3P)$ VMR, as shown in Fig. 10.27b. The effect is particularly large between 120 and 140 km where the strongest 15 μm band, the 626 fundamental, becomes optically thin. A difference with Mars is that the cooling rate decreases very quickly above about 140 km due to the larger depletion of CO_2 on Venus. The changes in the daytime populations and cooling rates induced by other rate coefficients, such as the V–V rates of exchange of v_2 and v_3 quanta, are in the same direction and of similar magnitude as for Mars.

Figure 10.28 shows the effect on the heating rate of changing the VMR of atomic oxygen (or $k_{CO_2\text{-}O}$) by a factor of 4. The heating is larger when the rate is increased, and the change is greater than on Mars because thermal collisions between $O(^3P)$ and the depleted CO_2 are relatively more important. Bellow 120 km (80 km on Mars), the effect becomes small due to the fall off in atomic oxygen.

The effect of changing the rate coefficient for V–V collisional deactivation of v_3 quanta by a factor 4 in Venus is also large and also produces more heating for a faster rate (Fig. 10.28). The largest effect takes place in the lower part of the solar heating region, between 110 and 130 km. It falls off above 135 km as pressure decreases and V–V collisions become infrequent.

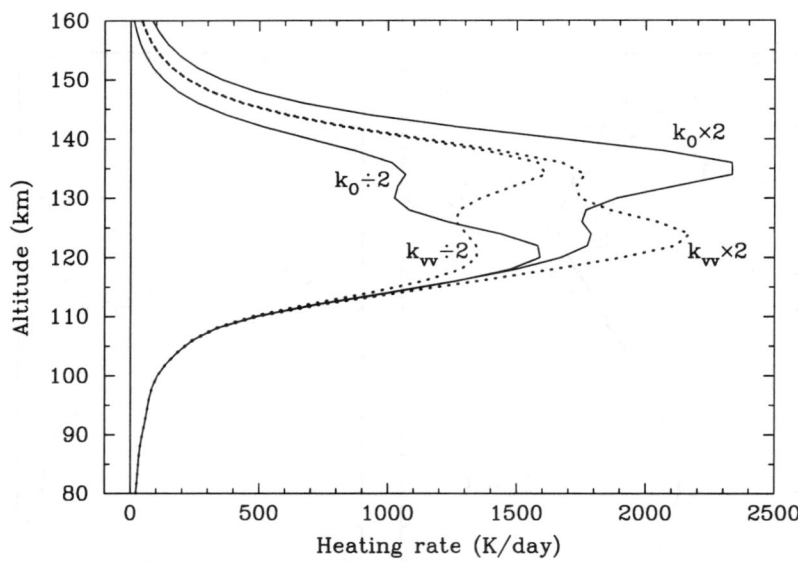

Fig. 10.28 Effect of the atomic oxygen abundance (or the $k_{CO_2\text{-}O}$ rate) and the k_{vv} exchange rate of v_3 quanta on the solar heating rates for the daytime temperature profile.

The solar heating obviously depends on the solar zenith angle. Apart from the expected reduction in the heating for larger solar zenith angles, the most remarkable feature is that the highest peak found for $\chi = 0°$, mainly due to the 2.7 μm fundamental bands, disappears at large solar zenith angles. As the Sun moves close to the horizon the solar absorption in this band moves upwards by about 20 km, to a level where the 15 μm bands are optically thin and the energy is all re-emitted away.

10.6 Outer Planets

The giant gas planets of the outer Solar System are composed mainly of hydrogen and helium, and so present a quite different sort of radiative transfer problem. Nevertheless, we are interested in their energy balance and in carrying out remote sensing of their composition and temperature structure, just as for the Earth-like planets.

In general terms, the problems are similar, with the major constituents being radiatively-inactive (except at high pressures), so that the non-LTE processes of interest involve the quantum levels of the more abundant minor constituents. In the case of Jupiter, Saturn, Uranus, Neptune and Titan the most important gas in the latter category is methane, which has a role rather analogous to that of carbon dioxide in the Earth's atmosphere. Ammonia is the primary condensable species in the observable part of Jupiter's atmosphere, and in many ways act as water does on Earth, with its rich infrared spectrum, high latent heat, and tendency towards cloud formation. Water itself is undoubtedly important on Jupiter, but only at deep (and mostly unobserved) levels, because of the vertical structure which placed the condensation temperature of H_2O down around the several bar pressure level. Ammonia tends to condense much higher, at around 0.25 bars, while methane, on Jupiter at least, is found only in the gaseous phase.

As well as being the key radiatively active molecule in the thermal balance of the upper atmospheres of the giant planets and Titan, methane is the most useful species for vertical temperature sounding, using the ν_4 band near 7.7 μm. The application of non-LTE methods to these problems is, however, scarcely developed. The following account is based mainly on the work of Appleby, who studied the radiative equilibrium temperatures of CH_4 in the upper atmospheres of Jupiter, Saturn, Uranus, and Neptune; and Yelle, who performed a similar study for the atmosphere of Titan.

Table 10.3   Methane, acetylene, and ethane vibrational bands thought to be of importance for the energetics of the atmospheres of the outer planets. After Appleby (1990) and Yelle (1991).

Band	Band centre (cm^{-1})	Degeneracy	A, s^{-1}	Relative LTE population at 100 K
CH_4 1.7 μm group				
$2\nu_3$	6005	6	47.4	1
$\nu_2+\nu_3+\nu_4$	5861	18	25.9	24
$\nu_1+\nu_2+\nu_4$	5775	6	2.2	27
$\nu_3+2\nu_4$	5585	18	27.94	1264
$4\nu_4$	5218	15	8.48	206878
CH_4 2.3 μm group				
$\nu_2+\nu_3$	4546	6	23.78	1
$\nu_3+\nu_4$	4313	9	25.82	43
$\nu_1+\nu_4$	4224	3	2.12	52
$\nu_2+2\nu_4$	4123	12	4.32	880
$3\nu_4$	3914	10	6.36	14827
CH_4 ν_3 group				
$2\nu_2$	3072	3	0.16	1
ν_3	3019	3	23.7	2
$\nu_2+\nu_4$	2828	6	2.2	66
$2\nu_4$	2610	6	4.24	1530
CH_4 ν_4 group				
ν_2	1533	2	0.08	1
ν_4	1306	3	2.12	40
C_2H_2 ν_5	729		4.75	
C_2H_6 ν_9	821		0.37	

Table 10.3 shows the bands of methane included by Appleby, and Fig. 10.29 shows the solar excitation and V–V and V–T relaxation paths followed in his model. The vibrational sublevels within each group are assumed to be in LTE with each other, and relaxation to take place by transfer to the nearest overtone of v_4 followed

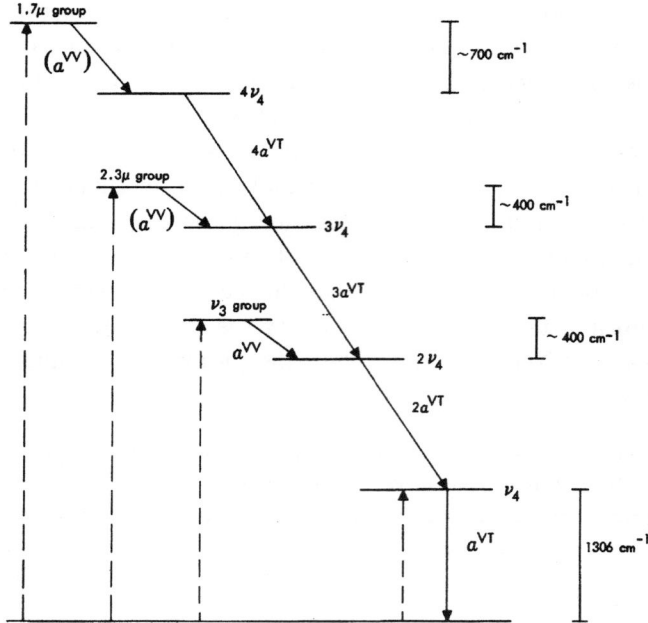

Fig. 10.29   Solar pumping and V–V and V–T relaxation of the methane levels in Appleby's model of the outer planet atmospheres.

by emission of one or more photons at wavelengths near the fundamental frequency of that vibration ($1306\,\mathrm{cm}^{-1}$), as shown in the diagram. A key assumption was employed: the model does not take into account the full coupling between CH_4 v_4 and the higher-energy vibrational states. This, in the light of the importance of similar effects in CO_2 in the Martian and Venusian atmosphere, could change quantitatively the populations obtained by Appleby, although it may not change the main qualitative conclusions about the departures from LTE.

Figure 10.30 shows the results for a 'nominal' set of assumed values. As usual, considerable uncertainty is introduced by a lack of good information on the rate coefficients for the various pathways, including their temperature dependence. The low temperatures of the outer planet atmospheres exacerbates the problem. Nevertheless, a general conclusion emerges: that non-LTE becomes important only at pressure levels around and below $10\,\mu$bar. In this very high atmospheric region, photochemical dissociation of methane occurs and other hydrocarbons are produced in significant amounts, thus further complicating the problem. Differences in the 'extreme' non-LTE model, produced by including the estimated uncertainties in the collisional rates in 'nominal' model, reach a maximum difference of about $\pm 20\,\mathrm{K}$ in the CH_4 v_4 vibrational temperature at the $0.1\,\mu$b pressure level, near the highest level represented in the models.

Other important conclusions are that the transition to non-LTE occurs rapidly: for all models $J/B < 0.5$ at pressures below $1\,\mu$b. Also, at a given pressure level,

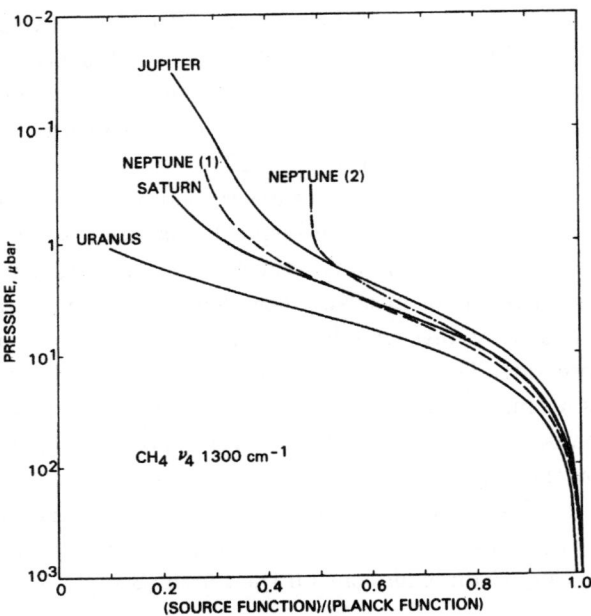

Fig. 10.30 The ratio of the source function to the Planck function (J/B) for the CH$_4$ ν_4 band for standard models of the atmospheres of the four giant planets. The main difference between Neptune 1 and Neptune 2 is that the latter includes aerosol heating.

there is a progression away from LTE moving from Jupiter to Saturn to Uranus which arises primarily from the decreasing stratospheric temperatures and, for Uranus relative to Jupiter and Saturn, from its much smaller CH$_4$ stratospheric mixing ratio. Neptune does not follow this trend because of the higher CH$_4$ VMR assumed (0.02, vs 7×10^{-4} for Jupiter and Saturn, and 2×10^{-5} for Uranus). A similar effect occurs in the non-LTE departure of the CO$_2(0,1^1,0)$ level in the Earth's lower thermosphere, which varies with the stratospheric temperature (see Sec. 8.10.2).

Recently, observations of Jupiter and Saturn in the 3.2–3.4 μm region with the Infrared Space Telescope have provided the first detection of the infrared fluorescence of methane in the giant planets. The measured 3.3 μm emission apparently comes from the CH$_4$ fundamental ν_3 band as well as from the combination levels excited by solar absorption at 2.3 and 1.7 μm. By using a similar non-LTE model as that of Appleby but allowing V–V cascading, Drossart *et al.* were able to derive a relaxation rate for the CH$_4$ v_3 levels, which is close to that for the v_4 levels, hence confirming the altitude of LTE departure of the ν_3 bands in Jupiter and Saturn as predicted by Appleby.

10.7 Titan

In his study about the thermal structure of the middle and upper atmosphere of Titan, Yelle developed a non-LTE model for the principal vibration-rotation bands of CH_4, C_2H_2, and C_2H_6. The rotational band of HCN at $713\,cm^{-1}$ is thought to play a minor role in Titan's upper atmosphere energy balance, although its pure rotational transitions (this asymmetric molecule has a permanent dipole moment) may be important in governing Titan's exospheric temperature. As pure rotational lines, however, these are considered to be in LTE.

The non-LTE model for CH_4 includes absorption of solar radiation by the CH_4 near-IR bands (see Table 10.3) and subsequent relaxation to the lower v_4 levels. Non-LTE plays a major role in Titan's atmosphere. In particular, non-LTE emissions in the infrared bands of methane, acetylene and ethane are responsible for locating the mesopause at pressures higher than where it would be if these emissions were not present. Titan's temperature structure (see Fig. 10.31), although much colder, has a similar shape to the Earth's, including a stratosphere extending from the tropopause at $\sim$130 mb to 1 mb, with a positive temperature gradient, caused mainly by the absorption of solar radiation by aerosols, the role played by ozone in the Earth. The v_9 band of C_2H_6 actually heats the mesopause by absorbing the upwelling radiation from the warmer stratospheric levels, a process similar to that involving CO_2 15 μm emissions on the Earth (Sec. 9.2.1.1).

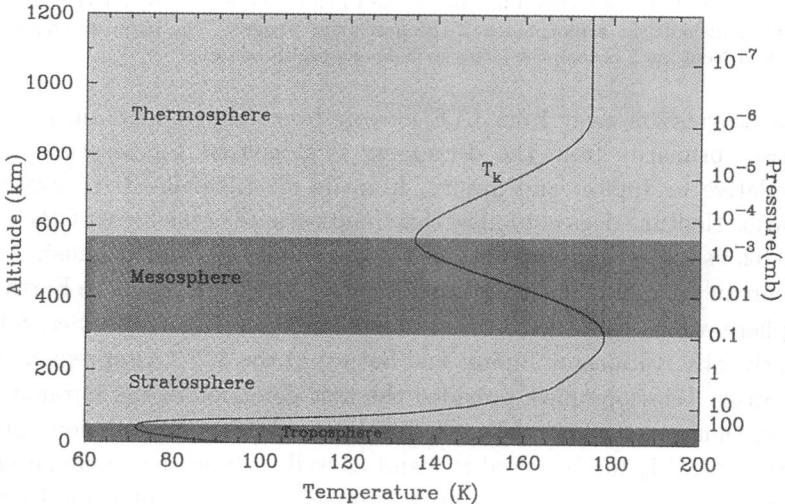

Fig. 10.31   Titan temperature structure. Temperatures from 0–200 km were obtained from Voyager 1 occultation measurements (Lindal *et al.*, 1983) and those from 200–1260 km were taken from the engineering model of Yelle *et al.* (1997). After Wilson and Atreya (2000).

The CH_4 v_4 band in Titan's atmosphere is remarkable in that its photons escape from the atmosphere over a broad spectral range ($400\,cm^{-1}$), so the extent of the band is large compared to the frequency dependence of the Planck function. This can be treated with the Curtis matrix method by dividing the band into smaller spectral regions and calculating Curtis matrices for each of those intervals. Yelle treated this problem using the Rybicki method, discussed in Secs. 5.4.3 and

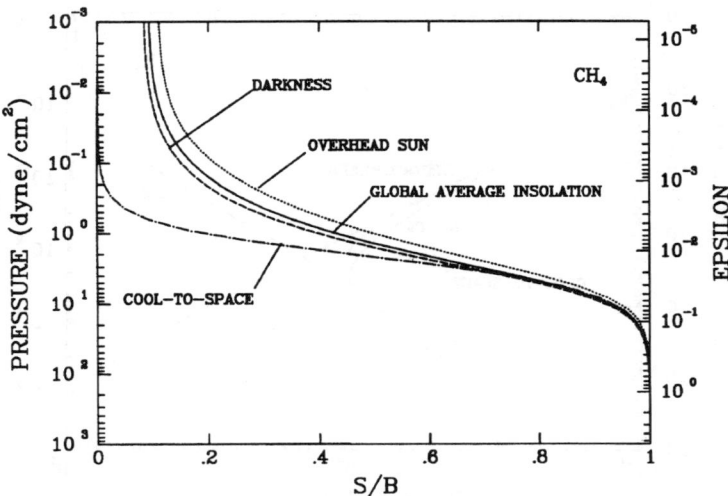

Fig. 10.32 The ratio of the source function to the Planck function for the CH$_4$ ν_4 band for an isothermal Titan's atmosphere at 175 K with a CH$_4$ mole fraction of 2% for three different illumination conditions. 1 dyne cm^{-2} = 1 μbar. Epsilon = l_t/A_{21} is the ratio of the collisional to the radiative de-excitation rates, as defined in Sec. 3.6.4.

5.8. He showed that the CH$_4$ ν_4 band starts departing from LTE at $\sim$10 μbar (10 dyne cm^{-2}) (Fig. 10.32). This level of LTE departure occurs at pressures where collisional deactivation is already a factor 10 smaller than the spontaneous emission rate. This is reminiscent of the LTE departure of the CO$_2$ 15 μm fundamental band in Mars and Venus, where the trapping of radiation is so large than just a few collisions are enough to keep the levels in LTE. As discussed in Sec. 5.2.1.2, the 'cooling-to-space' approximation gives a good estimate of the height of departure from LTE, even for optically thick bands.

Figure 10.33 shows the cooling rates obtained by the non-LTE model (with solar excitation excluded) and with the assumption of (i) LTE; (ii) the 'thermospheric' approximation; and (iii) the 'cooling-to-space' approximation for the CH$_4$ ν_4 band. The 'thermospheric' approach is the name usually given in the outer planets literature to the 'escape-to-space' (or 'total-escape') approximation discussed in Sec. 5.2.1.2. We recall that in this approach the two major assumptions are: (i) that the band is sufficiently optically thin for photons to escape freely to space; and (ii) radiative excitation is negligible compared to collisional excitation. This approach gives accurate cooling rates for the CO$_2$ 15 μm band in the Earth's thermosphere, where the low CO$_2$ VMR makes it optically thin and the thermosphere is so hot that thermal excitation is much larger than radiative excitation. This situation does not hold, however, for methane in Titan's lower thermosphere, where the mixing ratio is several percent and the temperature rise in the thermosphere is small. Fig. 10.33 also shows (c.f. Sec. 5.2.1.2 for the Earth) that the LTE approximation is grossly inadequate at low pressures, while the 'thermospheric' (escape-to-space) approximation is not appropriate at high pressures. As on Mars (Fig. 5.5), there

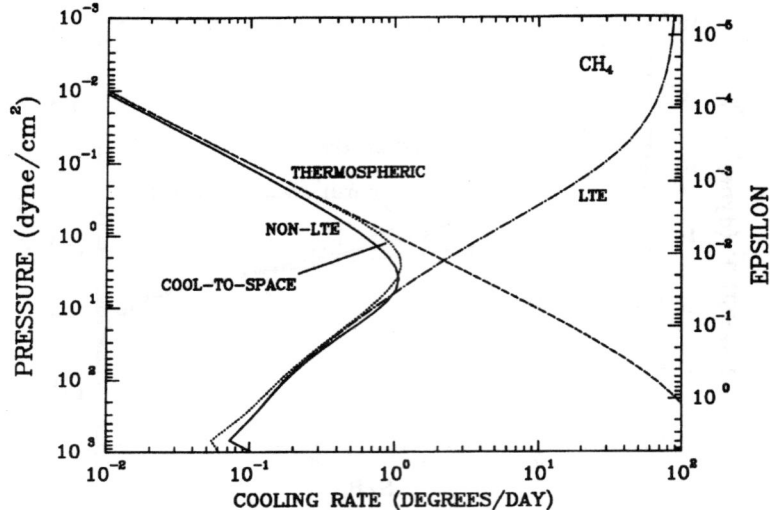

Fig. 10.33   Cooling rate by the CH_4 ν_4 band for an isothermal Titan's atmosphere at 175 K with a CH_4 mole fraction of 2% calculated by Yelle's non-LTE model and two common approximations: 'cooling-to-space' and 'total-escape' (thermospheric). The LTE calculation is also shown.

is a region on Titan (10–100 μbar) where the LTE cooling is smaller than that in non-LTE. This occurs because the LTE cooling is offset by the heating from the (unrealistically) large LTE source in the region above. The 'cooling-to-space' approximation matches the non-LTE calculations quite well at high and low pressures but is inaccurate by as much as 50% in the changeover region around 1 μbar.

For the C_2H_2 ν_5 band, the source function begins to fall below its LTE value also at ~10 μbar. The C_2H_2 concentration is smaller than that of methane, while its ν_5 band strength is similar to that of the CH_4 ν_4 band, and its lines are concentrated in a narrower spectral region. Overall, the probability of photons escaping to space is similar and so these bands depart from LTE at similar pressures. The C_2H_6 ν_9 band is optically thinner and hence departs from LTE at higher pressures, ~30 μbar.

Figure 10.34 shows the contributions to the energy budget of Titan's mesosphere and lower thermosphere of the CH_4 ν_4, C_2H_2 ν_5, and C_2H_6 ν_9 bands. Near the lower boundary at 300 km, C_2H_6 is the dominant coolant despite its relatively weak band strength (in comparison with C_2H_2) and its low abundance (in comparison with CH_4). This is like the role played by the weak isotopic and hot CO_2 15 μm bands in the atmospheres of the terrestrial planets. The strong cooling produces a negative temperature gradient, smaller than the adiabatic lapse rate, and causes the CH_4 and C_2H_2 bands to heat this region. At higher regions, C_2H_2 becomes optically thin and is the major coolant between about 400 and 500 km, while CH_4 ν_4 is still a source of heat. At even higher altitudes, CH_4 takes over the cooling and the weaker C_2H_6 ν_9 band is a source of heat. This band is the primary source of heating at the mesopause (~600 km) with a contribution larger than the solar heating in the CH_4 near-IR bands or the solar UV. Again, we have its counterpart in the terrestrial

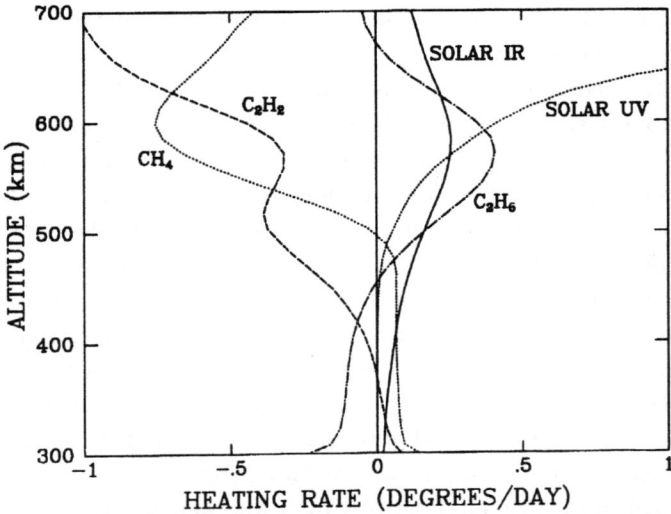

Fig. 10.34   Cooling and heating rates by the most important infrared emissions in Titan's mesosphere. Near the mesopause, at around 600 km ($0.1\,\mu$b), C_2H_6 is the dominant heat source and
CH_4 the dominant cooler.

planet atmospheres in the heating produced by the CO_2 isotopic bands for the
increased k_{vv} v_2 rate in Mars (Fig. 10.9).

Recently, the first observations of fluorescence non-LTE emission in the fundamental and hot v_3 bands of CH_4 in Titan's atmosphere by Kim and colleagues
allowed the derivation of rotational temperature and activation rates of the CH_4 v_3
and v_3+v_4 vibrational levels. The derived temperature from the observed emissions,
originated mostly around the mesopause, are consistent with the Titan's upper atmosphere thermal structure derived by Yelle and discussed above.

Similar studies for calculating the thermal structure which include the solar
heating and thermal cooling in non-LTE of the relevant bands have been carried
out for the atmospheres of Pluto, Triton and Io (see the 'References and Further
Reading' section). In the case of Pluto, the active species is CH_4, involving the same
bands as already discussed, i.e., the non-LTE solar heating at 2.3 and 3.3 μm and
the non-LTE cooling at 7.6 μm. In the case of Io, more exotic bands are involved,
including heating in the near-IR by SO_2 and non-LTE cooling by the SO_2 v_1, v_2,
and v_3 bands. Cooling in the pure rotational band of SO_2 is also important, but it
is considered to be in LTE.

10.8   Comets

The interpretation of measurements of molecular infrared emissions from comets
and the development of physical-chemical models of their structure and behaviour
require a non-LTE treatment. While low-resolution photometric measurements of
IR radiation from comets is dominated by radiation from dust grain, at high res-

olution, the IR molecular line emissions from molecules in the comet's extended atmosphere or coma can dominate.

The cometary coma contain gases such as H_2O, CO, CO_2 and CH_4 which are evaporated from a small, icy nucleus when the comet approaches the Sun. The coma is generally very cold and rarefied, and extends several tens of thousands of km from the nucleus. Collisions are infrequent, particularly in the outer parts, and solar radiation is always present, its intensity increasing as the comet's highly elliptical or hyperbolic orbit carries it near the Sun. The cometary coma offers very favourable scenario for studying rotational as well as vibrational non-LTE. Until now, when considering the atmospheres of the Earth, the terrestrial and the giant planets and their satellites, we have started with the assumption that a kinetic (translational) temperature can be defined, and hence the concept of LTE makes sense (see Sec. 3.5). In the mid and outer parts of the cometary coma (10^3–10^5 km), the density is so low and collisions are so infrequent that the translational velocities of gases can no longer be described by a Maxwellian distribution, and hence the concept of local thermodynamic equilibrium becomes meaningless. This applies even in the inner coma (first 10^3 km) for the 'daughter' species, such as H_2, H, and OH, which are produced after the photodissociation of the 'parent' molecules sublimated from the nucleus. These species can have translational velocity distributions which are very different from those of the main components even quite close to the nucleus.

It is interesting to consider the coupling between the translational and rotational energy distributions in this case. In the inner coma, both the Maxwellian (for translational) and Boltzmann (for rotational) distributions generally apply and are coupled to each other by collisions so that the kinetic and rotational temperatures for most ground or excited vibrational states are equal. Let us now move outwards from the nucleus and take the case of the most abundant molecule, water vapour, and consider the connection between the translational and rotational distributions in water-water collisions. Since H_2O is a polar molecule, very distant water-water collisions (nearly elastic) occur which can exchange rotational energy with a probability about one order of magnitude larger than those (inelastic) collisions which efficiently exchange kinetic energy. Hence, as the molecules move outwards, the kinetic distribution becomes non-isotropic and non-Maxwellian before the rotational distribution decouples from the kinetic distribution. So, there can be regions where we have a non-Maxwellian velocity distribution but the rotational levels can still be represented by a Boltzmann term at a given temperature. Essentially all of the vibrational bands are in non-LTE everywhere in the coma, even in the inner part, where they produce a high IR fluorescence emission due to solar pumping. In the outer coma, for the example of CO, even the low-J rotational levels in the ground vibrational state are in non-LTE.

Compared to planetary atmospheres, the non-LTE modelling of the vibrational level populations in cometary coma is rather simple. The processes involved are the absorption of solar radiation in the IR or near-IR, spontaneous emission, and thermal collisional losses. In these tenuous, optically thin atmospheres, radiative transfer between molecules in the coma can often be ignored. However, an accurate treatment of the cooling by, and populations of, the optically thick pure rotational and vibration-rotation lines of H_2O bands in the inner coma does require us to incorporate radiative transfer, at least approximately. Solar excitation is computed, as in planetary atmospheres, by the photoabsorption coefficient (e.g., Eq. 4.22 and Sec. 6.3.3.2), which is commonly referred to as the g-factor.

One complexity inherent to cometary atmospheres, which we do not encounter in planetary atmospheres, is the fact that the atmospheres of these small objects follows a r^{-2} density law as they expand outwards, encountering rapidly varying conditions. The temperature at the surface of the nucleus is about $170\,K$, drops down to about $15\,K$ at 50–$100\,km$ distance, and then increases rapidly beyond, reaching a steady state in the outer coma. Hence, accurate modelling of the non-LTE populations must include the time evolution as the molecules expand from the nucleus.

Many observations of comets at high spectral and spatial resolution at near-IR and radio wavelengths have shed much light on their physical and chemical characteristics over the last decade. For example, Mumma and colleagues deduced the abundances of several species in comet Hyakutake by non-LTE modelling of measured emissions of H_2O in the $(1,1,1$–$1,0,0)$ band at $1.7\,\mu m$, CO(1–0) near $4.7\,\mu m$, CH_4 in the ν_3 band at $3.3\,\mu m$, and C_2H_6 in the ν_7 band near $3.3\,\mu m$. Observations of comets Hale–Bopp and Hyakutake, both ground-based and by the Infrared Space Observatory, obtained emission spectra of H_2O (at 2.7 and $6.3\,\mu m$), CO ($4.7\,\mu m$) and CO_2 ($4.3\,\mu m$), as well as several hydrocarbons (CH_4, C_2H_2 and C_2H_6), HCN, NH_3, and OCS. Observations by Glinski and Anderson of the $NH_2(0,3,0)$ band in Hale–Bopp showed a larger contribution from high J lines when the comet was closer to the Sun but almost no difference with nucleocentric distance. Mumma *et al.* observed Comet Lee in the 2.9–$3.7\,\mu m$ range with a ground-based high resolution infrared spectrometer, revealing emission from water, carbon monoxide, methanol (CH_3OH), methane (CH_4), ethane (C_2H_6), acetylene (C_2H_2), hydrogen cyanide (HCN), and many multiplets of OH.

We have mentioned above that the rotational levels, both in the ground and excited vibrational levels, are in non-LTE in the outer coma. Thus, the interpretation of rotational line radiance of a range of species (including CO, CH_3OH, HCN, H_2O, H_2S, CS, H_2CO, and CH_4CN) observed at radio wavelengths shows that they, too, require a non-LTE formulation in many instances. This is in contrast with the Earth and planetary atmospheres where rotational LTE normally holds for radio observations of the low J ground vibrational levels, and hence a non-LTE formulation is not required (see Sec. 8.9).

Measurements of the rotational temperature by simultaneous observations of several lines of the same molecule show that collisional excitation prevails in the inner coma, and the rotational temperature reflects the kinetic temperature. In the outer coma, however, the rotational temperature is determined by radiative equilibrium (solar pumping versus spontaneous emission). Then, since the spontaneous emission of lines depends on J^3 (Sec. 8.9), we expect the lower J levels to be greatly enhanced in comparison with high J levels and hence most of the emission of a vibration-rotation band to be concentrated in the low J lines. The need to invoke non-LTE modelling of the rotational lines to derive temperatures and species abundances from radio observations of comets depends on which line is being observed.

When we talk about rotational LTE in the inner coma we refer normally to the rotational levels in the ground or low vibrational levels of the molecules. Cases have also been observed in excited electronic states, where the rotational levels are far away from LTE even in the inner coma. A good example is the $C_2(d^3\Pi_g - a^3\Pi_u)$ Swan system in the visible spectrum, where the low $(J < 15)$ rotational states in the (0–0) band show a very large departure from LTE, caused by the exothermic formation of C_2 from parent molecules.

Infrared radiative cooling is also important in comets since it has a significant effect on the temperature of the inner coma, and on the outflow velocity and the density of the mid and outer coma. Most of the cooling takes place in the IR bands of water. As for the Earth and the other planets, two regions can be distinguished and treated separately: the inner optically thick coma, in LTE, and the outer optically thin regions, which are in non-LTE and can be treated with a line-by-line escape probability formalism.

10.9   References and Further Reading

The earliest models of Venus and Mars using simplified treatments were by Goody and Belton (1967), and Ramanathan and Cess (1974). Dickinson (1972, 1976) produced the first comprehensive model for Venus. This was adapted for Mars by Bougher and Dickinson (1988) who focussed on the thermal structure and heat budget of the Martian thermosphere. The model includes a fairly complete collisional scheme, but still with some approximations for the redistribution of energy from solar excited states. Another simplified scheme, using the 'diffusion approximation' for radiative transfer, was presented by Battaner *et al.* (1982).

A model specifically to study the CO_2 $4.3\,\mu m$ emissions under daytime conditions was developed by Stepanova and Shved (1985). This included radiative transfer between atmospheric layers but only for the CO_2 $4.3\,\mu m$ and CO $4.7\,\mu m$ fundamental bands, and assumes the CO_2 v_2 levels are in LTE. They also analysed the atmospheric heating due to absorption of solar radiation in the near-infrared by CO_2 and CO bands but, again, V–V collisional coupling was studied for only the two extreme cases of: no V–V interaction at all, and instantaneous V–V exchange.

To compute the CO_2 infrared heating/cooling contribution to the energy balance of the mesosphere, Bittner and Fricke (1987) used a simple scheme for the deactivation of the high-energetic vibrational levels. They also assumed that the energy levels all have the same vibrational temperature within each of the 15 μm and 4.3 μm systems.

The most comprehensive non-LTE radiative transfer model for CO_2 and CO infrared emissions in a mainly-CO_2 atmosphere is that developed by López-Valverde and López-Puertas (1994a,b) and López-Puertas and López-Valverde (1995) which is the basis for the detailed Martian model described in this chapter. For the COSPAR Mars reference atmosphere, see Seiff (1982), Stewart and Hanson (1982), and Kaplan (1988). The neutral composition of the Martian atmosphere was calculated with a non-stationary one-dimensional model for equinox conditions, mid-latitude, moderate solar activity, and clear atmosphere (small amounts of dust) by Rodrigo *et al.* (1990).

Earth-based measurements of CO_2 emission at 10 μm and of CO at 4.7 μm have been carried out by Johnson *et al.* (1976), Mumma *et al.* (1981), Deming *et al.* (1983), Küfl *et al.* (1984), and Billebaud *et al.* (1991), and interpreted as laser emission by Mumma *et al.* (1981), and Deming and Mumma (1983). The former authors used the 'removal-of-radiation' approximation, while the latter used the radiative transfer scheme developed by Dickinson. Gordiets and Panchenko (1983) developed independent studies of the CO_2 10 μm. Stepanova and Shved (1985b) have shown that this natural laser emission is more probable to occur in Mars than in Venus and that it is favoured by the propagation of atmospheric waves.

The PMIRR remote temperature sounding experiment for Mars (which, at the time of writing, has made two unsuccessful journeys to the planet and will shortly be launched a third time) is described by McCleese *et al.* (1986), and an earlier version which studied Venus from the Pioneer orbiter by Taylor *et al.* (1980). For some basic considerations of outer planet temperature sounding, see Taylor (1972).

Wide-band and line-by-line models for calculating the CO_2 15 μm cooling in the LTE region for the Martian lower atmosphere have been developed by Crisp *et al.* (1986), Crisp (1990), and Hourdin (1992). *In situ* temperature profiles measured by the ASI/MET Mars Pathfinder instrument at night-time and by the Viking descent in daytime have been interpreted on the basis of a radiatively driven Martian mesosphere with a non-LTE radiative model by López-Valverde *et al.* (2000).

Recently, measurements of the Martian atmosphere in the 2.4–45 μm spectral range by the Infrared Space Observatory (ISO) have been reported by Lellouch *et al.* (2000). These data show the infrared bands in emissions of CO_2, H_2O and CO. Evidence of fluorescence and non-LTE emissions in the CO_2 4.3 and 2.7 μm has been shown.

The non-LTE theory for methane on Jupiter, Saturn, Uranus, and Neptune was developed by Appleby (1990), and Yelle (1991) has performed a similar study for the atmosphere of Titan (see also Coustenis and Taylor, 1999). More recently, cooling and heating rates under non-LTE have been calculated for the atmospheres of Pluto and Triton by Strobel *et al.* (1996), and for Io by Strobel *et al.* (1994) and Austin and Goldstein (2000).

A detailed discussion about the coupling between the kinetic and rotational distributions in cometary comas can be found in Combi (1996). Early work in modelling IR emission from comets was carried out by Yamamoto (1982) and Encrenaz *et al.* (1982) who considered the emission in the whole IR molecular bands (they do not consider the rotational distribution). Subsequent work in this area was carried out by Crovisier and Encrenaz (1983) and Weaver and Mumma (1984). They demonstrated that the molecular (vibrational) bands are in non-LTE everywhere in the coma, and that the rotational levels are not in LTE in the middle and outer part of the coma.

Detailed non-LTE models of the rotational structure within the vibrational bands were developed by Crovisier and Le Bourlot (1983) and Chin and Weaver (1984) for CO, and by Crovisier (1984) for water vapour. More complete non-LTE modelling, including the contribution of radiative transfer, was undertaken by Bockeleé-Morvan (1987); Crovisier (1987); and Bockeleé-Morvan and Crovisier (1987a). Observation of high resolution IR-emissions from comets have been reported by many investigators. Mumma *et al.* (1996) reported measurements of infrared emissions of the $H_2O(1,1,1-1,0,0)$ band at $1.7\,\mu m$, $CO(1-0)$ near $4.7\,\mu m$, CH_4 in the ν_3 band $(3.3\,\mu m)$, and C_2H_6 in the ν_7 band near $3.3\,\mu m$ from comet Hyakutake. Recently, Glinski and Anderson (2000) have shown observations of NH_2 in Hale–Bopp. Measurements of Hale–Bopp as observed by ISO have been reported by Crovisier *et al.* (1997). Two surveys of cometary infrared emissions are given by Crovisier (1997) and Mumma *et al.* (2000). Biver *et al.* (1997) gives a good summary of the species that can be observed in a comet at radio wavelengths. For examples of rotational non-LTE in electronically excited states see, e.g., Lambert *et al.* (1990), and Krishna Swamy (1997). For studies on radiative cooling in comets see Crovisier (1984), Bockeleé-Morvan (1987b), and the review of Combi (1996).

Appendix A

List of Symbols, Abbreviations, and Acronyms

The symbols, abbreviations, and acronyms most frequently used throughout the book are defined here. Those defined but used in short passages are not listed. See also about terminology and units in Appendix C.

A.1 List of Symbols

α_D	Doppler width
α_L	Lorentz width
α_N	Natural width
A, A_{21}	Einstein coefficient for spontaneous emission
β	Diffusivity factor
B	Rotational constant
B_{12}	Einstein coefficient for absorption
B_{21}	Einstein coefficient for induced emission
B_ν	Planck function
χ	Anharmonicity constant (Chapter 2), Solar zenith angle
c	Speed of light
c_p	Specific heat at constant pressure
c_v	Specific heat at constant volume
$\mathcal{C}$	Curtis matrix
$\mathrm{Ch}(x, \chi)$	Chapman function
$\mathcal{D}$	Diagonal matrix multiplying the heating rate vector $\mathbf{h}$
d	Mean Earth-Sun distance
D_e	Dissociation energy of a molecule
ϵ	$\simeq l_t/A_{21}$
e_ν	Molecular extinction coefficient
E	Energy
E_k	Kinetic energy
E_p	Potential energy
E_r	Rotational energy
E_v	Vibrational energy
E_n	Exponential integral of order n

$f(\nu)$	Normalized line shape
$f(v)$	Vibrational nascent distribution
f_{iso}	Isotopic ratio abundance
$F, F_\nu, F_{\Delta\nu}$	Flux of the radiant energy
$F(\nu)$	Instrumental frequency response function
g	Earth gravitational acceleration constant
g_v	Statistical degeneracy of level v
h	Planck constant
h_ν, h_{12}, h_{ij}	Divergence of the radiative flux (with sign reversed), Heating rate (see also q)
H	Scale height
$\mathcal{I}$	Identity matrix
I	Moment of inertia
j_ν	Emission coefficient
J_ν	Source function
J	Source function, Total angular momentum (Chapter 2), Rotational quantum number
k	Boltzmann constant, Force constant of a linear spring (Chapter 2)
k_ν	Absorption coefficient
$\bar{k}_\nu$	Absorption coefficient in LTE
k_t	Rate coefficient of vibrational-translational processes in the direct sense
k_t'	Rate coefficient of vibrational-translational processes in the reverse sense
k_{vv}	Rate coefficient of vibrational-vibrational processes
k_{ev}	Rate coefficient of electronic-vibrational processes
K	Angular momentum along the axis of symmetry
λ	Wavelength
$L, L_\nu, L_{\Delta\nu}$	Radiance
$\bar{L}_\nu, \bar{L}_{\Delta\nu}$	Mean radiance of the radiation field
l_t	Specific loss in V–T collisional processes
l_{nt}	Specific loss in non-thermal collisional processes
m	Absorber amount (Chapters 2 and 4), Molecular mass (Chapter 2)
M, [M]	Air molecule, Number density of molecule M
M	Mean molecular mass of the atmosphere
μ	Cosine of polar coordinate θ
n, n_v, n_r, n_i	Number density of energy level v, r, i
$\bar{n}_2, \bar{n}_1$	Number density of energy level 2, 1 in LTE
n_a	Total number density of absorbing molecules
ν	Frequency
ν_0	Central frequency of a vibration-rotation band
ν_i	Vibrational mode

$\tilde{\nu}$	Wavenumber
ω	Angular velocity (Chapter 2), Solid angle
ω_r	Frequency of rotation
ω_v	Frequency of vibration
ϕ	Polar coordinate
p	Pressure
p_t	Specific production in V–T collisional processes
p_{nt}	Specific production in non-thermal collisional processes
P_t	Production in V–T collisional processes
P_{nt}	Production in non-thermal collisional processes
P_c	Production in a chemical reaction
q	Cooling rate (see also h_ν, h_{12}, h_{ij})
$q_{r,v}$	Normalized factor for the rotational states distribution in vibrational level v
Q_i	Normal coordinate
Q_r	Rotational partition function
Q_{vib}	Vibrational partition function
ρ	Density
R	Universal gas constant
$R_\odot$	Solar radius
$R_\oplus$	Earth radius
σ	Stefan–Boltzmann constant, Surface
s	Distance
S	Band strength
θ	Polar coordinate
τ_ν	Optical depth
$\bar{\tau}_\nu$	Optical thickness
$\mathcal{T}_\nu(z',z,\mu)$	Monochromatic transmission
$\mathcal{T}(z',z,\mu)$	Mean transmission over $\Delta\nu$
$\mathcal{T}_{F,\nu}(z',z)$	Monochromatic flux transmission
$\mathcal{T}_F(z',z)$	Mean flux transmission over $\Delta\nu$
$\mathcal{T}^*$	Probability escape function
t	Time, Dummy variable of integration
T	Temperature
T_k	Kinetic temperature
T_e	Effective temperature
T_r	Rotational temperature
T_v	Vibrational temperature
(v_1,v_2,v_3)	Vibrational level excited in the vibrational modes 1, 2 and 3
W	Equivalent width
z	Altitude (vertical coordinate)

A.2 List of Abbreviations and Acronyms

ALI	Accelerated Lambda Iteration
ATMOS	Atmospheric Trace MOlecule Spectroscopy
CIRA	COSPAR International Reference Atmosphere
CIRRIS	Cryogenic InfraRed Radiance Instrumentation for Shuttle
CIRS	Composite InfraRed Spectrometer
CRISTA	CRyogenic Infrared Spectrometers and Telescopes for the Atmosphere
CVF	Circular Variable Filter
DYANA	DYnamics Adapted Network for the Atmosphere
EBC	Energy Budget Campaign
ENVISAT	ENVIronmental SATellite
FWI	Field-Widened Interferometer
GEISA	Gestion et Etude des Informations Spectroscopiques Atmosphériques
GENLN2	GENeral LiNe-by-line atmospheric transmittance and radiance model
GOMOS	Global Ozone Monitoring by Occultation of Stars
HALOE	HALogen Occultation Experiment
HIRDLS	HIgh Resolution Dynamics Limb Sounder
HIRIS	HIgh-Resolution Interferometer Spectrometer
HITRAN	HIgh-resolution TRANsmission molecular absorption
HWHM	Half Width at Half Maximum
ICECAP	Infrared Chemistry Experiments–Co-ordinated Auroral Program
ISAMS	Improved Stratospheric And Mesospheric Sounder
ISO	Infrared Space Observatory
KOPRA	Karlsruhe Optimized and Precise Radiative transfer Algorithm
LIMS	Limb Infrared Monitor of the Stratosphere
LTE	Local Thermodynamic Equilibrium
MAP/WINE	Middle Atmosphere Program/WInter in Northern Europe
MIPAS	Michelson Interferometer for Passive Atmospheric Sounding
MLS	Microwave Limb Sounder
MSISE	Mass Spectrometer and Incoherent Scatter Extended
MSX	Midcourse Space eXperiment
NESR	Noise Equivalent Spectral Radiance
NIMS	Near Infrared Mapping Spectrometer
NLC	NoctiLucent Clouds
NOAA	National Oceanic and Atmospheric Administration
PMC	Pressure Modulated Cell
PMIRR	Pressure Modulator InfraRed Radiometer
PSC	Polar Stratospheric Clouds
PV	Pioneer Venus
RFM	Reference Forward Model

RTE	Radiative Transfer Equation
SABER	Sounding of the Atmosphere using Broadband Emission Radiometry
SAMS	Stratospheric And Mesospheric Sounder
SEE	Statistical Equilibrium Equation
SISSI	Spectroscopic Infrared Structure Signatures Investigation
SME	Solar Mesosphere Explorer
SPIRE	SPectral Infrared Rocket Experiment
SPIRIT	SPectral InfraRed Interferometric Telescope
STP	Standard Temperature (273.15 K) and Pressure (1 atm)
TES	Tropospheric Emission Spectrometer
TIMED	Thermosphere, Ionosphere, Mesosphere, Energetics and Dynamics
TIME-GCM	Thermosphere-Ionosphere-Mesosphere-Electrodynamics General Circulation Model
UARS	Upper Atmospheric Research Satellite
VMR	Volume Mixing Ratio
V–T	Vibrational-Translational
V–V	Vibrational-Vibrational
WINDII	WIde-aNgle Doppler Imaging Interferometer

A.3 List of Chemical Species

C_2H_2	Acetylene
C_2H_6	Ethane
CH_3OH	Methanol
CH_4	Methane
CO	Carbon monoxide
CO_2	Carbon dioxide
$ClONO_2$	Chlorine nitrate
H	Atomic hydrogen
H_2	Hydrogen
H_2O	Water
H_2O_2	Hydrogen peroxide
HCl	Hydrogen chloride
HCN	Hydrogen cyanide
HF	Hydrogen fluoride
HNO_3	Nitric acid
N	Atomic nitrogen
N_2	Nitrogen
N_2O	Nitrous oxide
NH_2	Amino radical
NH_3	Ammonia
NO	Nitric oxide
NO_2	Nitrogen dioxide

NO_x	$NO + NO_2$
$O(^1D)$	Electronically excited atomic oxygen
$O(^3P)$	Atomic oxygen
O_2	Oxygen
O_3	Ozone
OCS	Carbonyl sulphide
OH	Hydroxyl radical
SO_2	Sulphur dioxide

Appendix B

Physical Constants and Useful Data

B.1 General and Universal Constants

General constants

Avogadro constant	$N_A = 6.02214199 \times 10^{23} \text{ mol}^{-1}$
Boltzmann constant	$k = 1.3806503 \times 10^{-23} \text{ J K}^{-1}$
Electron rest mass	$m_e = 9.10938188 \times 10^{-31} \text{ kg}$
Elementary charge	$e = 1.602176462 \times 10^{-19} \text{ C}$
Energy per unit wavenumber	$hc = 1.9864454404 \times 10^{-25} \text{ J m}$
First radiation constant	$c_1 = 2hc^2 = 1.191042722 \times 10^{-16} \text{ W m}^2 \text{ sr}^{-1}$
Gravitational constant	$G = 6.673 \times 10^{-11} \text{ m}^3 \text{ s}^{-2} \text{ kg}^{-1}$
Universal gas constant	$R = N_A k = 8.314472 \text{ J K}^{-1} \text{ mol}^{-1}$
Planck constant	$h = 6.62606876 \times 10^{-34} \text{ J s}$
Second radiation constant	$c_2 = hc/k = 1.4387752 \times 10^{-2} \text{ m K}$
Speed of light (in vacuo)	$c = 2.99792458 \times 10^8 \text{ m s}^{-1}$
Stefan–Boltzmann constant	$\sigma = 5.670400 \times 10^{-8} \text{ W m}^{-2} \text{ K}^{-4}$
Wien displacement constant	$b = \lambda_{max} T = 2.8977686 \times 10^{-3} \text{ m K}$

Sun

Average radius	$R_\odot = 6.995508 \times 10^8 \text{ m}$
Effective temperature	$T_\odot = 5777 \text{ K}$
Flux at Sun surface	$F_\odot = 6.312 \times 10^7 \text{ W m}^{-2}$
Luminosity	$L_\odot = 3.845 \times 10^{26} \text{ W}$
Mass	$M_\odot = 1.989 \times 10^{30} \text{ kg}$
Mean angular diameter from Earth	$\theta_\odot = 31.988 \text{ arc min}$
Mean solid angle from Earth	$\omega_\odot = 6.8000 \times 10^{-5} \text{ sr}$
Surface area	$S_\odot = 6.087 \times 10^{18} \text{ m}^2$

Earth

Albedo	$A = 0.29$
Distance from Sun (mean)	$d = 1.49598 \times 10^{11}$ m
Earth orbital period	$= 365.25463$ days
Eccentricity of orbit	$\epsilon = 0.016709$
Equatorial escape velocity	$v_{esc} = 1.18 \times 10^4$ m s^{-1}
Inclination of rotation axis	$i = 23.44°$
Mass	$M_\oplus = 5.9737 \times 10^{24}$ kg
Mass of atmosphere	$M_{atm} = 5.136 \times 10^{18}$ kg
Mean angular rotation rate	$\Omega_\oplus = 7.292 \times 10^{-5}$ rad s^{-1}
Radius (average)	$R_\oplus = (R_a^2 R_c)^{1/3} = 6.371 \times 10^6$ m
Radius (equatorial)	$R_a = 6.378136 \times 10^6$ m
Radius (polar)	$R_c = 6.356753 \times 10^6$ m
Solar constant	$S = 1367 \pm 2$ W m^{-2}
Standard surface gravity	$g_0 = 9.80665$ m s^{-2}
Standard surface pressure	$p_0 = 1.01325 \times 10^5$ Pa
Standard temperature	$T_0 = 273.15$ K
Surface area	$S_\oplus = 5.1007 \times 10^{14}$ m^2

Dry air

Mean molecular mass	$M = 28.964$ g mole^{-1}
Dry adiabatic lapse rate	$\Gamma = -9.75$ K km^{-1}
Mass density at 273.15 K and 1 atm (STP)	$\rho_0 = 1.2928$ kg m^{-3}
Molecular density at 273.15 K and 1 atm	$N_0^* = 2.6867775 \times 10^{25}$ m^{-3}
Specific gas constant	$R/M = 287.06$ J K^{-1} kg^{-1}
Specific heat at constant pressure	$c_p = 1005$ J K^{-1} kg^{-1}
Specific heat at constant volume	$c_v = 717.6$ J K^{-1} kg^{-1}

*Loschmidt number

B.2   Planetary Characteristics

	Mass (kg)	Equatorial radius (km)	Surface gravity ($m\,s^{-2}$)	Mean surface temp. (K)	Mean surface pressure (bar)	Rotation period (Earth days)
Mercury	3.30×10^{23}	2439.7	3.70	440		58.65
Venus	4.87×10^{24}	6051.8	8.87	730	90	-243.0
Earth	5.97×10^{24}	6378.1	9.81	288–293[*]	1	1
Mars	6.42×10^{23}	3397	3.71	183–268[*]	0.007–0.010	1.026
Jupiter	1.90×10^{27}	71492	23.12	165[†]	~0.3[‡]	0.413
Saturn	5.69×10^{26}	60268	8.96	134[†]	~0.4[‡]	0.444
Uranus	8.68×10^{25}	25559	8.69	76[†]		-0.718
Neptune	1.02×10^{26}	24764	11.00	73[†]		0.671
Pluto	1.29×10^{22}	1195	0.81		8×10^{-5}[‡]	-6.387

[*]Observed variability. [†]At the 1 bar pressure level. [‡]Pressure at visible cloud surface.

	Semi-major axis of orbit (A.U.)	Revolution period[*] (sidereal)	Inclination of rotation axis (degrees)	Orbit inclination (degrees)	Orbit eccentricity
Mercury	0.387	87.97 days	0.00	7.00	0.206
Venus	0.723	224.7 days	177.30	3.39	0.007
Earth	1.000	365.25 days	23.44	0.00	0.0167
Mars	1.524	686.94 days	25.19	1.85	0.093
Jupiter	5.203	11.86 years	3.12	1.31	0.048
Saturn	9.537	29.42 years	26.73	2.48	0.054
Uranus	19.19	83.75 years	97.86	0.77	0.047
Neptune	30.07	163.7 years	29.58	1.77	0.009
Pluto	39.48	248.0 years	119.61	17.14	0.248

[*]in Earth days or years.

	Visual geometric albedo	Atmospheric components
Mercury	0.11	Trace amounts of H_2 and He
Venus	0.65	96% CO_2, 3% N_2, 0.002% H_2O
Earth	0.37	78% N_2, 21% O_2, 1% Ar
Mars	0.15	95% CO_2, 3% N_2, 1.6% Ar
Jupiter	0.52	86% H_2, 16% He, 0.2% CH_4
Saturn	0.47	94% H_2, 6% He, 0.2% CH_4
Uranus	0.51	85% H_2, 15% He, ≤2% CH_4
Neptune	0.41	85% H_2, 15% He, ≤2% CH_4
Pluto	Variable	Perhaps CH_4 and N_2

References: National Institution of Standards and Technology, (http://physics.nist.gov), and Cox (2000).

Appendix C

Terminology and Units

An energy level, or the energy of a transition, can be specified in a number of different units. These are all equivalent to each other and all found from time to time in the literature, since experimental spectroscopists, radiative transfer specialists, and theoreticians working on the quantum mechanics of molecules, all have somewhat different traditions. Standard units tend to be cumbersome in these fields, and so it is still common to find wavelength, for example, stated in micrometers (also called microns μm), especially in the infrared, or in nanometers (nm), especially in the visible or ultraviolet.

Figure 2.1 relates energy levels in electron volts and the wavelength of the photon emitted from that level during a transition to the ground (zero energy) state. The energy E, more commonly expressed in ergs or joules, is of course related to the frequency of the photon ν by $E = h\nu$, where h is Planck's constant, and frequency is in turn related to wavelength λ by $c = \nu/\lambda$, where c is the velocity of light. In the experimental spectroscopy literature in particular, the reciprocal of wavelength, called wavenumber, written $\tilde{\nu}$ and expressed in cm^{-1}, is frequently used in preference to wavelength, because $\tilde{\nu}$ is proportional to photon energy. It also sometimes replaces ν, because values for line positions and widths can be expressed in wavenumbers without the use of large exponents, and so they are easier to remember and less cumbersome to use. More confusingly, it is not uncommon for energy levels to be expressed in cm^{-1} as well, in which case a factor of hc is implied, and needs to be included explicitly in formulae.

In this book we will normally describe photons by their frequency, ν, spectral bands by the approximate wavelength λ of the band centre, and spectral lines by their position and width in wavenumbers, $\tilde{\nu}$. By doing so, we hope to guide the reader in the use of the most common conventions, and employ the terms which, in a given context, either give simpler expressions, or are easier to visualise, than the alternatives.

In astrophysics, the transfer of energy by photons is usually described by the *specific intensity of radiation*, often shortened to *specific intensity* or just *intensity*. This has units of energy (Joules) per unit of: time (seconds), solid angle (steradians), area (square meters) and frequency interval (hertz), i.e. $\mathrm{W\,m^{-2}\,sr^{-1}\,Hz^{-1}}$. Because the pioneering work on radiative transfer, and non-LTE, was done on stellar atmospheres, this terminology is found in much of the early (and some more recent) literature on terrestrial atmospheric radiation as well. However, the modern usage has shifted gradually to the use of the term *radiance* for this quantity, and we follow this convention in the present book.

Radiance is written L_ν where ν is the (monochromatic) frequency at which it is defined. (Books which use intensity instead usually write I_ν for the same quantity). Radiance (or intensity) sometimes appears integrated over a specific spectral range ($\Delta\nu$ centred on ν), when it has units of $\mathrm{W\,m^{-2}\,sr^{-1}}$, and will be written $L_{\Delta\nu}$. There is also inconsistency over the use of 'intensity' or 'strength' to describe the integrated absorption for a unit absorber amount of a spectral line or band. Here we use 'strength', to avoid any possibility of confusion with the intensity of radiation.

Finally, for working with atmospheric heating rates, we need to define the *flux*, which is the energy passing through a unit surface area, in $\mathrm{W\,m^{-2}}$, obtained by integrating the intensity over wavelength and solid angle. If a spectral range $\Delta\nu$ is defined then the symbol $F_{\Delta\nu}$ is used, although the subscript may be dropped if the context makes it redundant, or if the flux is integrated over all frequencies from 0 to ∞. Fluxes in the atmosphere are usually defined separately in the upward and downward hemispheres; the difference between the two is the *net flux* (or total net flux if all frequencies are considered).

The terminology used about the modes of vibration, the bands arising from them, and the vibrational levels is as followed. Fundamental modes of vibration are designated ν_1 to ν_n where the ν-numbers are used consistently to designate a particular type of motion, for example, ν_1 is always a symmetric stretching mode. We also called the bands arising from these modes of vibration by ν, e.g., the ν_2 or the ν_3 band. The use of the symbol ν to designate both frequency and fundamental modes is unfortunately fairly universal. Vibrational levels are represented by v, e.g., (v_1, v_2, v_3) for the level excited in the vibrational modes ν_1, ν_2, and ν_3. When only one mode is mentioned, we assume that there is no excitation in the other vibrational modes, e.g., the v_3 level is equivalent to the $(0,0,v_3)$ level, and with 'the v_2 levels' we refer to the $(0,nv_2,0)$ states with $n = 1$, 2, 3, etc. Hot bands are represented as Δv, e.g., $\Delta v_3 = 1$ (see also Table 2.2).

Appendix D

The Planck Function

The Planck function represents the emission spectrum of a blackbody. Taking into account the following relationships,

$$\lambda = 1/\tilde{\nu} = c/\nu; \quad \text{and}$$
$$B_\lambda \, d\lambda = B_{\tilde{\nu}} \, d\tilde{\nu} = B_\nu \, d\nu,$$

it may be written in terms of wavelength λ in cm, by

$$B_\lambda = \frac{c_1}{\lambda^5 \left[\exp(c_2/\lambda T) - 1\right]}, \tag{D.1}$$

where B_λ is the radiance in $\text{W m}^{-2} \, \text{sr}^{-1} \, \text{cm}^{-1}$ of a blackbody at temperature T in K; in terms of wavenumber $\tilde{\nu}$ in cm^{-1}, by

$$B_{\tilde{\nu}} = \frac{c_1 \tilde{\nu}^3}{\exp(c_2 \tilde{\nu}/T) - 1}, \tag{D.2}$$

where $B_{\tilde{\nu}}$ is radiance in $\text{W m}^{-2} \, \text{sr}^{-1} \, (\text{cm}^{-1})^{-1}$; and in terms of frequency ν in s^{-1}, by

$$B_\nu = \frac{2h\nu^3}{c^2 \left[\exp(h\nu/kT) - 1\right]}, \tag{D.3}$$

where B_ν is radiance in $\text{W m}^{-2} \, \text{sr}^{-1} \, (\text{s}^{-1})^{-1}$.

The first and second radiation constants, c_1 and c_2, respectively, have the values[*]:

$$c_1 = 1.191042722 \times 10^{-8} \quad \text{W m}^{-2} \, \text{sr}^{-1} \, (\text{cm}^{-1})^{-4} \quad \text{and}$$
$$c_2 = 1.4387752 \quad \text{K} \, (\text{cm}^{-1})^{-1}.$$

Wien's displacement law is obtained by taking the derivative of the Planck function with respect to wavelength and setting it equal to zero. It gives the wavelength

---

[*]Note that they have different units than in Sec. B.1.

$\lambda_{\max}$ in cm at which the radiance B_λ peaks, i.e.,

$$\lambda_{\max} T = 0.28978 \quad \text{cm K}.$$

Analogously, the wavenumber $\tilde{\nu}$ in cm^{-1} of maximum radiance $B_{\tilde{\nu}}$ is given by

$$T/\tilde{\nu}_{\max} = 0.50995 \quad \text{cm K},$$

and the frequency ν in s^{-1} of maximum radiance B_ν is given by

$$c\, T/\nu_{\max} = 0.50995 \quad \text{cm K}.$$

Reference: Goody and Yung (1989).

Appendix E

Conversion Factors and Formulae

Pressure	Pa	$\mathrm{dyn\,cm^{-2}}$	atm	torr	mb
1 Pa	1	10	9.86923×10^{-6}	7.50062×10^{-3}	10^{-2}
1 $\mathrm{dyn\,cm^{-2}}$	10^{-1}	1	9.86923×10^{-7}	7.50062×10^{-4}	10^{-3}
1 atm	1.01325×10^{5}	1.01325×10^{6}	1	760	1.01325×10^{3}
1 torr	1.33322×10^{2}	1.33322×10^{3}	1.31579×10^{-3}	1	1.33322
1 mb	10^{2}	10^{3}	9.86923×10^{-4}	7.50062×10^{-1}	1

Energy	J	eV	erg	$\mathrm{cm^{-1}}$*	K*
1 J	1	6.24151×10^{18}	10^{7}	5.03412×10^{22}	7.24296×10^{22}
1 eV	1.60218×10^{-19}	1	1.60218×10^{-12}	8.06554×10^{3}	1.16045×10^{4}
1 erg	10^{-7}	6.24151×10^{11}	1	5.03412×10^{15}	7.24296×10^{15}
1 $\mathrm{cm^{-1}}$	1.98645×10^{-23}	1.23984×10^{-4}	1.98645×10^{-16}	1	1.43878
1 K	1.38065×10^{-23}	8.61734×10^{-5}	1.38065×10^{-16}	6.95036×10^{-1}	1

*Derived from $E = hc\tilde{\nu} = kT$.

Heating and cooling rate (h, q)

$$h\,[\mathrm{erg\,cm^{-3}\,s^{-1}}] = \frac{N_A}{M\,N}\,h\,[\mathrm{erg\,g^{-1}\,s^{-1}}] = 8.64\times10^4\,\frac{N_A}{c_p\,M\,N}\,h\,[\mathrm{K\,day^{-1}}],$$

where N_A is Avogadro constant in $\mathrm{mol^{-1}}$, M is the mean molecular mass in $\mathrm{g\,mole^{-1}}$, N is the total number density in $\mathrm{cm^{-3}}$, and c_p is the specific heat at constant pressure in $\mathrm{erg\,g^{-1}\,K^{-1}}$.

Reaction rate (k_t)

$$k_t\,[\mathrm{cm^3\,s^{-1}}] = 1.03558\times10^{-19}\,T\,k_t\,[\mathrm{torr^{-1}\,s^{-1}}] = 1.36261\times10^{-22}\,T\,k_t\,[\mathrm{atm^{-1}\,s^{-1}}],$$

where T is temperature in K.

Radiance (L)

$$L_\lambda \left[\frac{R}{\mu m} \right] = 2.013647 \, \pi \, \lambda \, [\mu m] \, 10^{13} \, L_\lambda \left[\frac{W}{cm^2 \, sr \, \mu m} \right],$$

where R is Rayleigh, or, equivalently,

$$1 \frac{R}{\mu m} = \frac{1.580763055 \times 10^{-14}}{\lambda \, [\mu m]} \, \frac{W}{cm^2 \, sr \, \mu m},$$

which, taking into account that

$$\Delta \lambda \, [\mu m] = 10^{-4} \, \lambda^2 \, [\mu m] \, \Delta \tilde{\nu} \, [cm^{-1}],$$

leads to

$$1 \frac{R}{\mu m} = 1.580763055 \times 10^{-18} \, \lambda \, [\mu m] \, \frac{W}{cm^2 \, sr \, cm^{-1}}$$

$$= \frac{1.580763055 \times 10^{-14}}{\tilde{\nu} \, [cm^{-1}]} \, \frac{W}{cm^2 \, sr \, cm^{-1}}.$$

References: Baker (1974), Baker and Pendleton (1976), and Chamberlain (1995).

Appendix F

CO_2 Infrared Bands

Table F.1   Principal CO_2 infrared bands.

Group	Isotope†	Upper Level*	Lower Level*	Upper Level†	Lower Level†	$\bar{\nu}_0$, cm^{-1}	Strength**
15 FB	626	01^10	00^00	01101	00001	667.380	79451.727
15 FH	626	02^00	01^10	10002	01101	618.029	1364.291
	626	02^20	01^10	02201	01101	667.752	6257.199
	626	10^00	01^10	10001	01101	720.805	1395.613
15 SH	626	03^10	02^00	11102	10002	647.062	212.882
	626	03^10	02^20	11102	02201	597.338	49.145
	626	03^10	10^00	11102	10001	544.286	2.534
	626	03^30	02^20	03301	02201	668.115	368.838
	626	11^10	02^00	11101	10002	791.447	8.663
	626	11^10	02^20	11101	02201	741.724	76.730
	626	11^10	10^00	11101	10001	688.671	144.014
15 TH	626	04^00	03^10	20003	11102	615.897	6.985
	626	04^00	11^10	20003	11101	471.511	0.010
	626	04^20	03^10	12202	11102	652.552	15.810
	626	04^20	03^30	12202	03301	581.776	1.968
	626	04^20	11^10	12202	11101	508.166	0.056
	626	12^00	03^10	20002	11102	738.673	2.445
	626	12^00	11^10	20002	11101	594.287	0.931
	626	04^40	03^30	04401	03301	668.468	18.270
	626	12^20	03^10	12201	11102	828.255	0.147
	626	12^20	03^30	12201	03301	757.479	2.337
	626	12^20	11^10	12201	11101	683.869	8.173
	626	20^00	03^10	20001	11102	864.666	0.041
	626	20^00	11^10	20001	11101	720.280	3.981

*Herzberg notation. †HITRAN notation. **Values shown are $\times 10^{22}$ (cm^{-1}/cm^{-2}) at $T = 296$ K.

Table F.1   Principal CO_2 infrared bands (continued).

Group	Isotope[†]	Upper Level[*]	Lower Level[*]	Upper Level[†]	Lower Level[†]	$\tilde{\nu}_0$, cm^{-1}	Strength[**]
15 FB	636	01^10	00^00	01101	00001	648.478	824.749
15 FH	636	02^00	01^10	10002	01101	617.350	19.480
	636	02^20	01^10	02201	01101	648.786	70.505
	636	10^00	01^10	10001	01101	721.584	17.052
15 SH	636	03^10	02^00	11102	10002	630.710	2.483
	636	03^10	02^20	11102	02201	599.274	0.662
	636	03^10	10^00	11102	10001	526.475	0.032
	636	03^30	02^20	03301	02201	649.087	4.511
	636	11^10	02^00	11101	10002	771.266	0.135
	636	11^10	02^20	11101	02201	739.829	0.679
	636	11^10	10^00	11101	10001	667.031	1.521
15 TH	636	04^00	03^10	20003	11102	610.996	0.106
	636	04^20	03^30	12202	03301	585.328	0.033
	636	04^20	03^10	12202	11102	635.140	0.199
	636	12^00	11^10	20002	11101	607.975	0.019
	636	12^00	03^10	20002	11102	748.530	0.019
	636	04^40	03^30	04401	03301	649.409	0.246
	636	12^20	11^10	12201	11101	663.171	0.095
	636	12^20	03^10	12201	11102	803.726	0.0014
	636	20^00	11^10	20001	11101	713.503	0.035
15 FB	628	01^10	00^00	01101	00001	662.374	317.399
15 FH	628	02^00	01^10	10002	01101	597.052	4.900
	628	02^20	01^10	02201	01101	662.768	25.356
	628	10^00	01^10	10001	01101	703.470	7.930
15 SH	628	03^10	02^00	11102	10002	642.311	0.747
	628	03^10	02^20	11102	02201	576.596	0.157
	628	03^10	10^00	11102	10001	535.893	0.009
	628	03^30	02^20	03301	02201	663.187	1.446
	628	11^10	02^00	11101	10002	789.913	0.054
	628	11^10	02^20	11101	02201	724.198	0.304
	628	11^10	10^00	11101	10001	683.495	0.743
15 FB	627	01^10	00^00	01101	00001	664.729	59.857
15 FH	627	02^00	01^10	10002	01101	607.558	1.108
	627	02^20	01^10	02201	01101	665.114	4.701
	627	10^00	01^10	10001	01101	711.299	1.219

[*]Herzberg notation.   [†]HITRAN notation.   [**]Values shown are $\times 10^{22}$ (cm^{-1} /cm^{-2}) at $T = 296$ K.

Table F.1 Principal CO$_2$ infrared bands (continued).

Group	Isotope[†]	Upper Level*	Lower Level*	Upper Level[†]	Lower Level[†]	$\tilde{\nu}_0$, cm^{-1}	Strength**
15 SH	627	03^10	02^00	11102	10002	644.408	0.143
	627	03^10	02^20	11102	02201	586.852	0.040
	627	03^30	02^20	03301	02201	665.509	0.274
	627	11^10	02^00	11101	10002	789.812	0.009
	627	11^10	02^20	11101	02201	732.256	0.044
	627	11^10	10^00	11101	10001	686.071	0.125
15 FB	638	01^10	00^00	01101	00001	643.329	3.393
15 FB	637	01^10	00^00	01101	00001	645.744	0.6147
10	626	00^01	02^00	00011	10002	1063.735	9.686[‡]
	626	00^01	10^00	00011	10001	960.959	6.801[‡]
10	636	00^01	02^00	00011	10002	1017.659	0.053[‡]
	636	00^01	10^00	00011	10001	913.425	0.076[‡]
10	628	00^01	02^00	00011	10002	1072.687	0.035[‡]
	628	00^01	10^00	00011	10001	966.269	0.017[‡]
10	627	00^01	02^00	00011	10002	1067.727	0.006[‡]
	627	00^01	10^00	00011	10001	963.986	0.004[‡]
4.3 FB	626	00^01	00^00	00011	00001	2349.143	955357.125
4.3 FH	626	01^11	01^10	01111	01101	2336.632	73666.258
4.3 SH	626	02^21	02^20	02211	02201	2324.141	2838.857
	626	02^01	02^00	10012	10002	2327.433	1789.326
	626	10^01	10^00	10011	10001	2326.598	1079.329
4.3 TH	626	03^11	03^10	11112	11102	2315.235	14.610[‡]
	626	03^31	03^30	03311	03301	2311.667	10.260[‡]
	626	11^11	11^10	11111	11101	2313.773	72.010[‡]
4.3SH2	626	00^02	00^01	00021	00011	2324.183	20.293
4.3 FRH	626	04^01	04^00	20013	20003	2305.256	3.630[‡]
	626	04^21	04^20	12212	12202	2302.963	6.073[‡]
	626	12^01	12^00	20012	20002	2306.692	1.961[‡]
	626	04^41	04^40	04411	04401	2299.214	3.908[‡]
	626	12^21	12^20	12211	12201	2301.053	2.550[‡]
	626	20^01	20^00	20011	20001	2302.525	1.069[‡]
2.7 FB	626	10^01	00^00	10011	00001	3714.783	15796.078
	626	02^01	00^00	10012	00001	3612.842	10397.265
2.7 FH	626	11^11	01^10	11111	01101	3723.249	1270.129
	626	03^11	01^10	11112	01101	3580.326	779.872

*Herzberg notation. [†]HITRAN notation. **Values shown are $\times 10^{22}$ (cm^{-1}/cm^{-2}) at $T = 296$ K.

Table F.1   Principal CO$_2$ infrared bands (continued).

Group	Isotope[†]	Upper Level[*]	Lower Level[*]	Upper Level[†]	Lower Level[†]	$\tilde{\nu}_0$, cm^{-1}	Strength[**]
2.7 SH	626	04^21	02^20	12212	02201	3552.854	28.710
	626	04^01	02^00	20013	10002	3568.215	31.180
	626	12^01	10^00	20012	10001	3589.651	16.210
	626	12^01	02^00	20012	10002	3692.427	36.220
	626	20^01	10^00	20011	10001	3711.476	29.360
	626	12^21	02^20	12211	02201	3726.646	47.210
2.0	626	12^01	00^00	20012	00001	4977.835	351.925
	626	20^01	00^00	20011	00001	5099.661	108.997
	626	04^01	00^00	20013	00001	4853.623	77.801
4.3 FB	636	00^01	00^00	00011	00001	2283.488	9598.150
4.3 FH	636	01^11	01^10	01111	01101	2271.760	817.848
4.3 SH	636	02^21	02^20	02211	02201	2260.049	33.330
	636	02^01	02^00	10012	10002	2261.910	19.460
	636	10^01	10^00	10011	10001	2262.848	11.620
4.3 TH	636	03^11	03^10	11112	11102	2250.694	1.801[‡]
	636	03^31	03^30	03311	03301	2248.356	1.398[‡]
	636	11^11	11^10	11111	11101	2250.605	0.894[‡]
4.3 FRH	636	04^01	04^00	20013	20003	2240.536	0.046[‡]
	636	04^21	04^20	12212	12202	2239.297	0.080[‡]
	636	12^01	12^00	20012	20002	2242.323	0.023[‡]
	636	04^41	04^40	04411	04401	2236.678	0.057[‡]
	636	12^21	12^20	12211	12201	2283.570	0.033[‡]
	636	20^01	20^00	20011	20001	2240.757	0.013[‡]
2.7 FB	636	02^01	00^00	10012	00001	3527.737	94.171
	636	10^01	00^00	10011	00001	3632.911	159.960
2.7 FH	636	03^11	01^10	11112	01101	3498.754	7.271
	636	11^11	01^10	11111	01101	3639.220	15.460
2.0	636	04^01	00^00	20013	00001	4748.065	0.333
	636	12^01	00^00	20012	00001	4887.385	2.975
	636	20^01	00^00	20011	00001	4991.350	2.119
4.3 FB	628	00^01	00^00	00011	00001	2332.113	3518.852
4.3 FB	627	00^01	00^00	00011	00001	2340.014	647.575
4.3 FB	638	00^01	00^00	00011	00001	2265.972	38.781
4.3 FB	637	00^01	00^00	00011	00001	2274.088	7.147

[*]Herzberg notation. [†]HITRAN notation. [**]Values shown are $\times 10^{22}$ (cm^{-1}/cm^{-2}) at $T = 296$ K. [‡]Radiative transfer is not included for this transition.

Appendix G

O_3 Infrared Bands

Table G.1 Principal O_3 infrared bands.

Band	Upper level	Lower level	$\tilde{\nu}_0$, cm^{-1}	A, s^{-1}	Band	Upper level	Lower level	$\tilde{\nu}_0$, cm^{-1}	A, s^{-1}
1	053	213	563.810	1.160	31	043	033	634.286	0.233
2	223	142	568.843	0.466	32	033	023	634.286	0.699
3	241	024	568.844	0.699	33	302	221	634.287	0.699
4	115	105	578.912	0.233	34	053	043	639.320	0.233
5	016	006	583.946	0.233	35	250	240	639.320	0.233
6	223	213	594.014	0.233	36	411	123	639.320	0.932
7	250	005	598.041	0.233	37	500	141	639.321	1.160
8	251	170	599.048	1.160	38	113	103	640.327	0.233
9	260	114	599.048	1.400	39	123	113	641.837	0.233
10	044	034	599.048	0.233	40	023	013	642.240	0.466
11	015	005	603.075	0.233	41	133	123	644.354	0.233
12	034	024	604.082	0.233	42	142	203	644.355	0.233
13	007	133	609.116	0.233	43	212	202	646.871	0.233
14	303	222	614.150	0.699	44	105	061	649.388	1.630
15	232	222	619.184	0.233	45	340	330	649.388	0.233
16	213	203	619.184	0.233	46	152	142	649.388	0.233
17	401	113	621.701	0.466	47	322	071	649.388	0.466
18	114	104	624.218	0.233	48	240	230	649.388	0.932
19	260	250	624.218	0.233	49	430	420	649.389	0.699
20	080	104	624.218	0.233	50	013	003	650.995	0.233
21	025	015	624.218	0.233	51	241	231	654.422	0.233
22	024	014	624.722	0.466	52	251	241	654.422	0.233
23	025	250	629.252	0.466	53	322	312	654.422	0.233
24	014	004	632.776	0.233	54	340	401	654.422	0.699
25	222	212	633.783	0.466	55	312	302	654.422	0.233
26	510	151	634.286	0.233	56	420	410	654.422	0.466
27	071	231	634.286	0.233	57	132	122	654.423	0.699
28	044	241	634.286	0.932	58	122	112	655.429	0.466
29	124	080	634.286	0.466	59	112	102	655.566	0.233
30	124	114	634.286	0.233	60	034	160	659.456	1.630

O₃ Infrared Bands

Table G.1   Principal O_3 infrared bands (continued).

Band	Upper level	Lower level	$\tilde{\nu}_0$, cm^{-1}	A, s^{-1}	Band	Upper level	Lower level	$\tilde{\nu}_0$, cm^{-1}	A, s^{-1}
61	330	320	659.456	0.233	106	211	130	683.015	1.290
62	411	401	659.456	0.233	107	011	001	684.438	0.233
63	232	052	659.456	1.400	108	070	060	684.625	1.630
64	042	032	659.456	0.932	109	331	321	684.626	0.699
65	510	500	659.456	0.233	110	080	070	684.627	0.233
66	142	132	659.456	0.233	111	210	200	685.024	0.233
67	231	221	659.457	0.233	112	050	040	689.660	1.160
68	140	201	659.579	0.268	113	161	500	689.660	1.400
69	221	211	661.067	0.466	114	180	034	689.661	1.860
70	311	301	661.470	0.233	115	060	050	689.661	1.400
71	022	012	661.787	0.466	116	120	110	690.314	0.466
72	211	201	663.002	0.233	117	110	100	693.125	0.233
73	320	310	663.484	0.466	118	040	030	694.695	0.932
74	410	400	663.987	0.233	119	030	020	694.878	0.699
75	161	151	664.490	0.233	120	020	010	698.342	0.466
76	062	052	664.490	0.233	121	090	015	699.728	2.100
77	032	022	664.490	0.699	122	081	105	699.729	1.860
78	114	240	664.490	1.160	123	010	000	700.931	0.233
79	015	070	664.491	1.860	124	213	033	709.796	0.932
80	151	141	664.491	0.233	125	170	302	734.966	0.932
81	052	042	664.491	1.160	126	152	043	750.069	1.160
82	230	220	664.491	0.699	127	062	330	755.103	0.233
83	012	002	668.216	0.233	128	044	213	775.239	44.300
84	310	300	669.020	0.233	129	204	231	790.340	33.300
85	141	131	669.524	0.932	130	007	006	795.374	77.600
86	081	071	669.525	0.233	131	043	141	810.477	22.200
87	180	170	669.525	0.233	132	115	114	815.511	11.600
88	131	121	670.380	0.699	133	115	080	815.511	55.400
89	121	111	672.275	0.466	134	016	015	825.579	11.600
90	321	311	672.545	0.466	135	016	250	830.613	66.500
91	220	210	672.872	0.466	136	115	015	835.647	0.482
92	111	101	674.455	0.233	137	115	250	840.681	0.482
93	123	320	674.558	0.699	138	006	005	844.708	66.500
94	160	150	674.558	1.400	139	007	142	845.714	44.300
95	061	051	674.559	1.400	140	006	104	845.715	66.500
96	140	130	679.592	0.932	141	501	151	850.749	11.100
97	090	080	679.592	0.233	142	241	033	850.749	44.300
98	071	061	679.592	0.233	143	044	043	850.749	11.600
99	051	041	679.592	1.160	144	240	103	854.273	28.400
100	041	031	679.592	0.932	145	105	005	859.810	0.482
101	170	160	679.592	0.233	146	223	222	860.816	33.300
102	150	140	679.593	1.160	147	105	104	860.817	11.600
103	130	120	679.821	0.699	148	232	231	865.851	11.600
104	021	011	681.413	0.466	149	053	052	870.884	33.300
105	031	021	682.952	0.699	150	124	123	870.885	11.600

Table G.1   Principal O_3 infrared bands (continued).

Band	Upper level	Lower level	$\tilde{\nu}_0$, cm^{-1}	A, s^{-1}	Band	Upper level	Lower level	$\tilde{\nu}_0$, cm^{-1}	A, s^{-1}
151	501	500	875.919	11.600	196	132	131	921.226	22.200
152	080	113	878.436	44.300	197	105	070	921.226	11.100
153	114	113	878.436	11.600	198	142	212	925.756	33.300
154	133	203	880.953	44.300	199	161	160	926.259	11.100
155	250	014	881.457	55.400	200	052	051	926.260	22.200
156	312	005	884.980	0.482	201	203	202	928.272	33.300
157	340	132	885.986	33.300	202	311	310	932.300	11.100
158	312	104	885.987	55.400	203	212	211	933.408	22.200
159	322	321	885.987	22.200	204	014	013	933.709	44.300
160	401	122	885.987	33.300	205	103	102	934.954	33.300
161	303	302	885.987	22.200	206	260	160	936.327	0.964
162	034	033	885.987	11.600	207	151	150	936.327	11.600
163	025	024	885.987	55.400	208	231	301	936.328	22.200
164	015	014	886.491	11.600	209	081	080	936.328	11.600
165	124	401	891.021	44.300	210	122	121	937.183	22.200
166	104	103	894.545	44.300	211	301	300	939.850	11.100
167	133	132	896.054	11.600	212	042	041	941.361	22.200
168	204	203	896.055	1.450	213	251	114	941.361	11.100
169	043	042	896.055	0.482	214	401	023	941.361	0.482
170	152	151	896.055	11.600	215	321	320	941.361	11.100
171	213	212	900.585	11.600	216	231	230	946.395	11.600
172	402	330	901.089	22.200	217	202	201	949.539	22.200
173	312	311	904.110	11.600	218	023	022	951.429	33.300
174	402	401	906.123	11.600	219	221	220	951.429	11.100
175	411	410	906.123	11.600	220	241	240	951.429	11.600
176	123	122	906.123	11.600	221	141	140	951.429	11.100
177	222	221	906.124	22.200	222	331	123	951.429	11.100
178	114	014	906.627	0.482	223	071	070	951.430	11.600
179	080	014	906.627	0.482	224	004	003	951.928	44.300
180	071	311	909.144	22.200	225	112	111	954.029	22.200
181	401	400	910.654	11.600	226	061	060	956.463	11.100
182	232	061	911.157	22.200	227	081	015	956.464	11.100
183	500	150	911.157	11.100	228	140	210	959.811	7.900
184	142	141	911.158	11.600	229	131	130	961.497	11.100
185	302	301	911.158	11.600	230	340	033	961.497	0.482
186	104	004	915.185	0.482	231	123	023	961.497	0.482
187	330	400	915.688	11.100	232	032	031	961.497	22.200
188	124	024	916.191	0.482	233	113	013	961.900	0.482
189	024	023	916.191	44.300	234	211	210	963.234	11.100
190	062	061	916.191	11.600	235	331	330	966.531	11.600
191	005	004	916.192	55.400	236	251	250	966.531	11.600
192	113	112	919.715	33.300	237	303	203	966.532	0.482
193	152	500	921.225	22.200	238	121	120	970.938	11.100
194	033	032	921.225	33.300	239	013	012	970.976	33.300
195	302	230	921.225	11.100	240	051	050	971.565	11.100

O₃ Infrared Bands

Table G.1   Principal O₃ infrared bands (continued).

Band	Upper level	Lower level	$\tilde{\nu}_0$, cm^{-1}	A, s^{-1}	Band	Upper level	Lower level	$\tilde{\nu}_0$, cm^{-1}	A, s^{-1}
241	213	042	971.565	33.300	286	241	141	1026.940	0.482
242	133	033	971.565	0.482	287	240	140	1026.940	0.964
243	103	003	972.568	0.482	288	170	070	1026.940	0.482
244	102	101	972.918	22.200	289	501	330	1026.940	2.410
245	223	123	976.599	0.482	290	160	060	1031.973	0.482
246	022	021	979.959	22.200	291	133	104	1031.973	0.964
247	152	052	981.633	0.482	292	501	401	1031.974	0.482
248	303	132	981.633	0.964	293	180	015	1031.974	0.482
249	041	040	981.633	11.100	294	231	202	1033.987	1.450
250	201	200	985.256	11.100	295	131	031	1037.007	0.482
251	250	150	986.667	0.482	296	322	222	1037.007	1.450
252	232	132	986.667	0.482	297	600	151	1037.008	2.890
253	500	051	986.668	0.482	298	430	123	1037.008	1.930
254	003	002	988.197	33.300	299	402	302	1042.041	0.482
255	111	110	988.977	11.100	300	251	500	1042.041	0.964
256	051	121	992.557	13.900	301	231	131	1042.042	0.482
257	223	401	996.735	0.964	302	001	000	1042.084	11.100
258	142	042	996.736	0.482	303	212	112	1043.552	0.964
259	132	032	996.736	0.482	304	411	311	1045.062	0.482
260	031	030	996.736	11.100	305	203	103	1045.565	0.964
261	012	011	999.585	22.200	306	311	211	1045.666	1.450
262	161	061	1001.769	0.482	307	204	104	1047.075	0.482
263	122	022	1006.803	0.482	308	150	050	1047.076	0.482
264	170	240	1006.803	11.100	309	401	301	1047.076	0.482
265	062	410	1006.804	11.100	310	301	201	1047.198	1.450
266	034	070	1006.804	0.482	311	121	021	1049.579	0.482
267	101	100	1007.647	11.100	312	142	113	1049.593	0.964
268	021	020	1008.662	11.100	313	330	301	1052.110	1.930
269	302	202	1008.817	0.482	314	430	330	1052.110	0.482
270	114	150	1011.837	0.964	315	202	102	1052.247	0.964
271	151	051	1011.838	0.482	316	221	121	1052.965	0.964
272	180	080	1011.838	0.482	317	140	040	1057.143	0.482
273	112	012	1013.161	0.482	318	230	130	1057.144	0.964
274	002	001	1015.807	22.200	319	321	221	1057.144	1.450
275	312	212	1016.368	0.482	320	111	011	1058.717	0.482
276	402	231	1016.871	1.930	321	211	040	1060.566	0.087
277	251	151	1016.871	0.482	322	140	111	1060.750	0.174
278	007	080	1016.871	0.964	323	420	320	1062.177	1.930
279	302	131	1016.872	0.964	324	340	240	1062.177	0.482
280	071	212	1021.402	1.450	325	330	230	1062.177	0.482
281	222	122	1021.906	0.964	326	600	500	1062.178	0.482
282	213	113	1024.422	0.482	327	510	410	1062.178	2.410
283	011	010	1025.591	11.100	328	211	111	1064.173	0.964
284	102	002	1025.811	0.482	329	500	400	1066.709	0.482
285	141	041	1026.939	0.482	330	411	240	1067.211	1.450

Table G.1   Principal O_3 infrared bands (continued).

Band	Upper level	Lower level	$\tilde{\nu}_0$, cm^{-1}	A, s^{-1}	Band	Upper level	Lower level	$\tilde{\nu}_0$, cm^{-1}	A, s^{-1}
331	320	220	1067.212	1.450	376	241	221	1313.879	0.233
332	101	001	1068.700	0.482	377	240	220	1313.879	0.233
333	410	310	1071.239	1.930	378	420	400	1318.409	0.233
334	130	030	1072.246	0.482	379	231	211	1320.524	0.233
335	220	120	1072.474	0.964	380	330	310	1322.940	0.233
336	201	101	1075.626	0.964	381	042	022	1323.946	0.233
337	400	300	1076.272	1.930	382	052	032	1323.947	0.233
338	310	210	1076.600	1.450	383	221	201	1324.069	0.233
339	123	230	1077.279	1.450	384	032	012	1326.277	0.233
340	331	231	1082.313	0.482	385	062	042	1328.981	0.233
341	170	141	1082.314	0.964	386	161	141	1328.981	0.233
342	120	020	1087.303	0.482	387	022	002	1330.003	0.466
343	210	110	1089.916	0.964	388	320	300	1332.504	0.233
344	151	400	1091.879	2.410	389	321	301	1334.015	0.233
345	300	200	1092.604	1.450	390	151	131	1334.015	0.233
346	110	010	1095.331	0.482	391	230	210	1337.363	0.233
347	200	100	1098.017	0.964	392	141	121	1339.904	0.233
348	100	000	1103.137	0.482	393	131	111	1342.655	0.233
349	331	302	1107.483	1.450	394	121	101	1346.730	0.466
350	062	311	1145.743	1.930	395	081	061	1349.117	0.233
351	044	024	1203.130	0.233	396	180	160	1349.117	0.233
352	223	203	1213.198	0.233	397	170	150	1354.150	0.233
353	025	005	1227.293	0.233	398	071	051	1354.151	0.233
354	034	014	1228.804	0.233	399	061	041	1354.151	0.233
355	232	212	1252.967	0.233	400	160	140	1354.151	0.233
356	024	004	1257.498	0.233	401	331	311	1357.171	0.233
357	124	104	1258.504	0.233	402	220	200	1357.896	0.466
358	260	240	1263.538	0.233	403	051	031	1359.184	0.233
359	043	023	1268.572	0.233	404	150	130	1359.185	0.233
360	053	033	1273.606	0.233	405	140	120	1359.413	0.233
361	033	013	1276.526	0.233	406	041	021	1362.544	0.233
362	222	202	1280.654	0.233	407	090	070	1364.219	0.233
363	123	103	1282.164	0.233	408	031	011	1364.365	0.466
364	133	113	1286.191	0.233	409	021	001	1365.851	0.699
365	250	230	1288.708	0.233	410	080	060	1369.252	0.233
366	023	003	1293.235	0.233	411	130	110	1370.135	0.466
367	430	410	1303.811	0.233	412	070	050	1374.286	0.233
368	021	100	1304.798	9.280	413	060	040	1379.321	0.233
369	152	132	1308.844	0.233	414	120	100	1383.439	0.699
370	340	320	1308.844	0.233	415	050	030	1384.355	0.233
371	251	231	1308.844	0.233	416	040	020	1389.573	0.466
372	322	302	1308.844	0.233	417	030	010	1393.220	0.699
373	132	112	1309.852	0.233	418	020	000	1399.273	0.466
374	122	102	1310.995	0.233	419	016	005	1428.654	2.700
375	142	122	1313.879	0.233	420	115	104	1439.729	2.700

O₃ Infrared Bands

Table G.1   Principal O₃ infrared bands (continued).

Band	Upper level	Lower level	$\tilde{\nu}_0$, cm^{-1}	A, s^{-1}	Band	Upper level	Lower level	$\tilde{\nu}_0$, cm^{-1}	A, s^{-1}
421	044	033	1485.035	2.700	466	131	120	1641.318	2.700
422	223	212	1494.599	2.700	467	032	021	1644.449	2.700
423	025	014	1510.709	2.700	468	061	050	1646.124	2.700
424	124	113	1512.722	2.700	469	223	014	1646.627	2.700
425	114	103	1518.763	2.700	470	211	200	1648.258	5.400
426	015	004	1519.267	2.700	471	051	040	1661.225	2.700
427	034	023	1520.273	2.700	472	121	110	1661.252	5.400
428	232	221	1525.308	2.700	473	022	011	1661.372	5.400
429	053	042	1535.375	2.700	474	041	030	1676.328	2.700
430	213	202	1547.456	2.700	475	111	100	1682.102	8.100
431	133	122	1550.477	2.700	476	012	001	1684.023	8.100
432	043	032	1555.511	2.700	477	213	004	1685.389	2.700
433	024	013	1558.431	2.700	478	260	051	1686.396	2.700
434	322	311	1558.532	2.700	479	031	020	1691.614	5.400
435	152	141	1560.546	2.700	480	115	113	1693.947	11.600
436	123	112	1561.552	2.700	481	232	023	1696.464	2.700
437	312	301	1565.580	2.700	482	021	010	1707.004	8.100
438	222	211	1567.191	2.700	483	016	014	1712.070	11.600
439	411	400	1570.110	2.700	484	222	013	1719.520	2.700
440	113	102	1575.281	2.700	485	011	000	1726.522	2.700
441	142	131	1580.682	2.700	486	121	011	1730.992	0.595
442	014	003	1584.704	2.700	487	212	003	1736.732	2.700
443	033	022	1585.715	2.700	488	250	041	1741.770	2.700
444	062	051	1590.750	2.700	489	044	042	1746.804	11.600
445	132	121	1591.606	2.700	490	105	103	1755.362	11.600
446	212	201	1596.410	2.700	491	006	004	1760.900	11.600
447	241	230	1600.817	2.700	492	251	042	1766.940	2.700
448	161	150	1600.817	2.700	493	223	221	1766.940	11.600
449	311	300	1601.320	2.700	494	241	032	1771.974	2.700
450	321	310	1604.845	2.700	495	124	122	1777.008	11.600
451	251	240	1605.851	2.700	496	312	103	1780.532	2.700
452	052	041	1605.852	2.700	497	240	031	1782.042	2.700
453	122	111	1609.458	2.700	498	231	022	1782.042	2.700
454	231	220	1610.886	2.700	499	221	012	1784.372	2.700
455	023	012	1613.216	2.700	500	211	002	1791.521	5.400
456	151	140	1615.920	2.700	501	322	113	1794.627	2.700
457	042	031	1620.953	2.700	502	053	051	1797.144	11.600
458	081	070	1620.955	2.700	503	303	301	1797.145	11.600
459	221	210	1624.301	2.700	504	114	112	1798.151	11.600
460	331	320	1625.987	2.700	505	025	023	1802.178	11.600
461	112	101	1628.484	5.400	506	411	202	1804.191	2.700
462	141	130	1631.021	2.700	507	034	032	1807.212	11.600
463	071	060	1636.055	2.700	508	340	131	1807.212	2.700
464	013	002	1639.192	5.400	509	311	102	1811.376	2.700
465	007	005	1640.082	11.600	510	232	230	1812.246	11.600

Table G.1 Principal O_3 infrared bands (continued).

Band	Upper level	Lower level	$\tilde{\nu}_0$, cm^{-1}	A, s^{-1}	Band	Upper level	Lower level	$\tilde{\nu}_0$, cm^{-1}	A, s^{-1}
511	230	021	1815.606	2.700	556	103	002	1960.765	9.150
512	402	400	1816.777	11.600	557	204	004	1962.260	0.482
513	133	131	1817.280	11.600	558	260	060	1968.300	0.482
514	015	013	1820.200	11.600	559	013	011	1970.561	23.300
515	204	202	1824.327	11.600	560	102	100	1980.565	34.900
516	322	320	1827.348	11.600	561	232	032	1983.403	0.482
517	430	221	1827.349	2.700	562	213	013	1986.322	0.482
518	330	121	1828.204	2.700	563	122	021	1986.762	6.480
519	321	112	1828.355	2.700	564	022	020	1988.621	23.300
520	104	102	1829.499	11.600	565	003	001	2004.004	0.350
521	152	150	1832.382	11.600	566	303	103	2012.097	0.482
522	220	011	1832.528	5.400	567	112	011	2012.746	0.124
523	213	211	1833.993	11.600	568	203	003	2018.133	0.482
524	312	310	1836.410	11.600	569	202	101	2025.165	13.500
525	043	041	1837.416	11.600	570	012	010	2025.176	34.900
526	420	211	1839.027	2.700	571	251	051	2028.709	0.482
527	320	111	1841.023	2.700	572	222	022	2028.709	0.482
528	123	121	1843.306	11.600	573	250	050	2033.743	0.482
529	210	001	1844.094	8.100	574	102	001	2041.618	3.800
530	410	201	1847.607	2.700	575	140	110	2049.727	7.040
531	302	300	1851.008	11.600	576	402	202	2050.858	0.482
532	310	101	1851.994	5.400	577	241	041	2053.879	0.482
533	331	122	1857.552	2.700	578	212	012	2056.713	0.482
534	222	220	1857.553	11.600	579	002	000	2057.891	0.412
535	142	140	1862.587	11.600	580	121	020	2058.241	3.600
536	510	301	1862.587	2.700	581	322	122	2058.913	0.482
537	024	022	1867.620	11.600	582	312	112	2059.920	0.482
538	005	003	1868.120	11.600	583	302	102	2061.064	0.482
539	062	060	1872.654	11.600	584	202	002	2078.058	0.964
540	113	111	1873.744	11.600	585	231	031	2079.049	0.482
541	203	201	1877.811	11.600	586	501	301	2079.050	0.482
542	033	031	1882.722	11.600	587	201	100	2083.273	3.800
543	132	130	1882.723	11.600	588	240	040	2084.083	0.482
544	212	210	1896.642	11.600	589	111	010	2084.308	0.010
545	052	050	1897.825	11.600	590	340	140	2089.117	0.482
546	014	012	1904.685	11.600	591	411	211	2090.728	0.482
547	103	101	1907.872	23.300	592	401	201	2094.274	0.482
548	122	120	1908.121	11.600	593	221	021	2102.544	0.482
549	042	040	1922.994	11.600	594	311	111	2109.839	0.482
550	023	021	1931.388	11.600	595	321	121	2110.109	0.482
551	202	200	1934.795	23.300	596	101	000	2110.784	3.810
552	223	023	1938.096	0.482	597	430	230	2114.287	0.482
553	004	002	1940.125	23.300	598	330	130	2119.321	0.482
554	112	110	1943.006	23.300	599	301	101	2122.824	0.964
555	032	030	1958.233	11.600	600	211	011	2122.890	0.964

Table G.1   Principal O_3 infrared bands (continued).

Band	Upper level	Lower level	$\bar{\nu}_0$, cm^{-1}	A, s^{-1}	Band	Upper level	Lower level	$\bar{\nu}_0$, cm^{-1}	A, s^{-1}
601	331	131	2124.355	0.482	609	410	210	2147.839	0.482
602	600	400	2128.887	0.482	610	220	020	2159.777	0.964
603	420	220	2129.389	0.482	611	310	110	2166.516	0.964
604	230	030	2129.390	0.482	612	400	200	2168.876	0.964
605	510	310	2133.417	0.482	613	210	010	2185.247	1.450
606	320	120	2139.686	0.482	614	300	100	2190.621	1.450
607	500	300	2142.981	0.482	615	200	000	2201.154	0.130
608	201	001	2144.326	1.450					

Figure Credits

- Figure 4.7 adapted from Rodgers (1976).
- Figure 6.13 reprinted from López-Puertas *et al.* (1998a), copyright by AGU[†].
- Figures 7.1, 7.2, 7.3, and 7.4 reprinted from López-Puertas *et al.* (1993), copyright by AGU.
- Figures 7.12, 7.13, 7.14, 7.15 and 7.16a,b reprinted from López-Puertas *et al.* (1995), copyright by AGU.
- Figures 7.17 and 7.27 reprinted from Martín-Torres *et al.* (1998), with kind permission from Elsevier Science.
- Figure 8.2 reprinted from Stair *et al.* (1985), copyright by AGU.
- Figure 8.3 reprinted from Wintersteiner *et al.* (1992), copyright by AGU.
- Figure 8.4 reprinted from Vollmann and Grossmann (1997), with kind permission from Elsevier Science.
- Figures 8.5, 8.6 and 8.7 reprinted from López-Puertas *et al.* (1997), copyright by AGU.
- Figures 8.8, 8.9, 8.10 and Plate 1 reprinted from Edwards *et al.* (1996), copyright by AGU.
- Figure 8.11 reprinted from Nebel *et al.* (1994), copyright by AGU.
- Figure 8.12 reprinted from Vollmann and Grossmann (1997), with kind permission from Elsevier Science.
- Figures 8.13, 8.15, 8.16 and 8.17 reprinted from López-Puertas and Taylor (1989), copyright by AGU.
- Figure 8.14 reprinted from López-Puertas *et al.* (1988), with kind permission from Kluwer Academic Publishers.
- Figures 8.18, 8.19 and 8.20, and Plate 2 reprinted from López-Puertas *et al.* (1998b), copyright by AGU.
- Figure 8.21 reprinted from Green *et al.* (1986), copyright by AGU.
- Figure 8.22 reprinted from Edwards *et al.* (1994), with kind permission from Elsevier Science.

[†]American Geophysical Union.

- Figures 8.23, 8.24 and 8.25 reprinted from Zaragoza *et al.* (1998), copyright by AGU.
- Figure 8.26 reprinted from Zhou *et al.* (1999), copyright by AGU.
- Figure 8.28 reprinted from Zachor *et al.* (1985), copyright by AGU.
- Figure 8.29 reprinted from Ballard *et al.* (1993), copyright by AGU.
- Figure 8.30 has been kindly provided by Tom Slanger (see, Cosby *et al.*, 1999).
- Figure 8.31 reprinted from López-Puertas *et al.* (1992b), copyright by AGU.
- Figures 9.1 and 9.2 reprinted from López-Puertas *et al.* (1992a), with kind permission from the Royal Meteorological Society.
- Figures 9.19, 9.20, 9.21, 9.23, 9.27 and 9.28 reprinted from López-Puertas *et al.* (1990), with kind permission from the American Meteorological Society.
- Figures 10.1, 10.2, 10.3, 10.4, 10.5 reprinted from López-Valverde and López-Puertas (1994a), copyright by AGU.
- Figures 10.6, 10.7 reprinted from López-Valverde and López-Puertas (1994b), copyright by AGU.
- Figures 10.8, 10.9, 10.11, 10.12 and 10.14 reprinted from López-Puertas and López-Valverde (1995), with kind permission from Academic Press.
- Figures 10.15, 10.16, 10.17 and 10.18 reprinted from Dickinson (1972), with kind permission from the American Meteorological Society.
- Figures 10.21, 10.22, 10.24, 10.25, 10.26 and 10.27 reprinted from Roldán *et al.* (2000), with kind permission from Academic Press.
- Figures 10.29 and 10.30 reprinted from Appleby (1990), with kind permission from Academic Press.
- Figures 10.32, 10.33 and 10.34 reprinted from Yelle (1991), with kind permission from the American Astronomical Society.
- Plate 3 has been kindly provided by Katerina Koutoulaki (see Koutoulaki, 1998).
- Plate 4 reprinted from Allen *et al.* (2000), copyright by AGU.

Bibliography

Abramowitz, M. and Stegun, I.A. (1972) *Handbook of Mathematical Functions*, Dover, New York.

Adler-Golden, S.M. and Steinfeld, J.I. (1980) "Vibrational energy transfer in ozone by infrared-ultraviolet double resonance", *Chem. Phys. Lett.* **76**, 479.

Adler-Golden, S.M., Schweitzer, E.L. and Steinfeld, J.I. (1982) "Ultraviolet continuum spectroscopy of vibrationally excited ozone", *J. Phys. Chem.* **76**, 2201.

Adler-Golden, S.M., Langhoff, S.R., Bauschlicher, C.W. and Carney, C.D. (1985) "Theoretical calculation of ozone vibrational infrared intensities", *J. Chem. Phys.* **83**, 255.

Adler-Golden, S.M. (1989) "The NO+O and NO+O_3 reactions. 1. Analysis of NO_2 vibrational chemiluminescence", *J. Phys. Chem.* **93**, 684.

Adler-Golden, S.M. and Smith, D.R. (1990) "Identification of 4 to 7 quantum bands in the atmospheric recombination spectrum of ozone", *Planet. Space Sci.* **38**, 1121.

Adler-Golden, S.M., Matthew, M.W., Smith, D.R. and Ratkowski, A.J. (1990) "The 9 to 12 μm atmospheric ozone emission observed in the SPIRIT–1 experiment", *J. Geophys. Res.* **95**, 15243.

Adler-Golden, S.M., Matthew, M.W. and Smith, D.R. (1991) "Upper atmospheric infrared radiance from CO_2 and NO observed during the SPIRIT 1 rocket experiment", *J. Geophys. Res.* **96**, 11319.

Ahmadjian, M.A., Nadile, R. M., Wise, J. O. and Bartschi B. (1990) "CIRRIS 1A space shuttle experiment", *J. Spacecr. Rockets* **27**, 669.

Akmaev, R.A. and Fomichev, V.I. (1992) "Adaptation of a matrix parameterization of the middle atmospheric radiative cooling for an arbitrary vertical coordinate grid", *J. Atmos. Terr. Phys.* **54**, 829.

Alexander, J.A.F., Houghton, J.T. and McKnight, W.B. (1968) "Collisional relaxation from the ν_3 vibration of CO_2", *J. Phys. B.*, Series 2, **1**, 1225.

Allen, D.C., Scragg, T. and Simpson, C.J.S.M. (1980) "Low temperature fluorescence studies of the deactivation of the bend-stretch manifold of CO_2", *Chem. Phys.* **51**, 279.

Allen, D.C. and Simpson, C.J.S.M. (1980) "Vibrational energy exchange between CO and the isotopes of N_2 between 300 K and 80 K", *J. Chem. Phys.* **45**, 203.

Allen, M., Yung, Y.L. and Waters, J.W. (1981) "Vertical transport and photochemistry in the terrestrial mesosphere and lower thermosphere (50–120 km)", *J. Geophys. Res.* **86**, 3617.

Allen, D.R., Stanford, J.L., Nakamura, N. *et al.* (2000) "Antarctic polar descent and planetary wave activity observed in ISAMS CO from April to July 1992", *Geophys.*

Res. Lett. **27**, 665.

Amimoto, S.T., Force, A.P., Gulotty, R.G., Jr. and Wiesenfeld, J.R. (1979) "Collisional deactivation of $O(2^1 D_2)$ by the atmospheric gases" *J. Chem. Phys.* **71**, 3640.

Anderson, G.P., Clough, S.A., Kneizys, F.X. *et al.* (1986) "AFGL Atmospheric Constituent Profiles (0–120 km)", *AFGL-TR-86-0110*, AFGL (OPI), Hanscom AFB, Ma.

Andrews, D.G., Holton, J.R. and Leovy, C.B. (1987) *Middle Atmosphere Dynamics*, Academic Press, London.

Andrews, D.G. (2000) *An Introduction to Atmospheric Physics*, Cambridge University Press, London.

Appleby, J.F. (1990) "CH_4 nonlocal thermodynamics equilibrium in the atmospheres of the giant planets", *Icarus* **85**, 355.

Apruzese, J.P. (1980) "The diffusivity factor re-examined", *J. Quant. Spectrosc. Radiat. Transfer* **89**, 4917.

Apruzese, J.P., Strobel, D.F. and Schoeberl, M.R. (1984) "Parameterization of IR cooling in a middle atmosphere dynamics model. 2. Non-LTE radiative transfer and the globally averaged temperature of the mesosphere and lower thermosphere", *J. Geophys. Res.* **89**, 4917.

Armstrong, B.H. (1968) "Theory of the diffusivity factor for atmospheric radiation", *J. Quant. Spectrosc. Radiat. Transfer* **8**, 1577.

Armstrong, P.S., Lipson, S.J., Dodd, J.A. *et al.* (1994) "Highly rotationally excited $NO(v, J)$ in the thermosphere from CIRRIS 1A limb radiance measurements", *Geophys. Res. Lett.* **21**, 2425.

Athay, R.G. (1972) *Radiation Transport in Spectral Lines*, D. Reidel, Dordrecht.

Atkins, P.W. and Friedman, R.S. (1996) *Molecular Quantum Mechanics*, Oxford University Press, Oxford.

Austin, J.V. and Goldstein, D.B. (2000) "Rarefied gas model of Io's sublimation–driven atmosphere", *Icarus* **148**, 370.

Avramides, E. and Hunter, T.F. (1983) "Vibrational-vibrational and vibrational-translational-rotational processes in methane, oxygen gas-phase mixtures: Optoacoustic measurements", *Molec. Phys.* **48**, 1331.

Baker, D.J. (1974) "Rayleigh, the unit for light radiance", *App. Opt.* **13**, 2160.

Baker, D.J. and Pendleton, W.R. (1976) "Optical radiation from the atmosphere", in *Methods for Atmospheric Radiometry, SPIE Proc.*, **91**, 50.

Ballard, J., Kerridge, B.J., Morris, P.E. and Taylor, F.W. (1993) "Observations of $v{=}1{-}0$ emission from thermospheric nitric oxide by ISAMS", *Geophys. Res. Lett.* **20**, 1311.

Banwell, C.N. and McCash, E. (1994) *Fundamentals of Molecular Spectroscopy*, McGraw-Hill Book Company, London.

Barnett, J.J. and Chandra, S. (1990) "COSPAR International Reference Atmosphere Grand Mean", *Adv. Space Res.* **10**, (12)7.

Barth, C.A. (1974) "The atmosphere of Mars", *Ann. Rev. Earth Planet. Sci.* **2**, 333.

Bass, H.E. (1973) "Vibrational relaxation in CO_2/O_2 mixtures", *J. Chem. Phys.* **58**, 4783.

Bass, H.E., Keaton, R.G. and Williams, D. (1976) "Vibrational and rotational relaxation in mixtures of water vapor and oxygen", *J. Acoust. Soc. Amer.* **60**, 74.

Bates, D.R. (1951) "The temperature of the upper atmosphere", *Proc. Phys. Soc.* **B64**, 805.

Battaner, E., Rodrigo, R. and López-Puertas, M. (1982) "A first order approximation model of CO_2 infrared bands in the Venusian lower thermosphere", *Astron. Astrophys.* **112**, 229.

Bauer, H.J., and Roesler, H. (1966) "Relaxation of vibrational degrees of freedom in binary mixtures of diatomic gases", in *Molecular Relaxation Processes*, Academic Press, New York, 245.

Bauer, S.H., Caballero, J.F., Curtis, R. and Wiesenfeld, J.R. (1987) "Vibrational relaxation rates of $CO_2(001)$ with various collision partners for <300 K", *J. Chem. Phys.* **91**, 1778.

Baulch, D.L., Cox, R.A., Hampson, R.F., Jr. *et al.* (1984) "Evaluated kinetic and photochemical data for atmospheric chemistry: Supplement. II. CODATA task group on gas phase chemical kinetics", *J. Phys. Chem. Ref. Data* **13**, 1259.

Bevan, P.L.T. and Johnson, G.R.A. (1973) "Kinetics of ozone formation in the pulse radiolysis of oxygen gas", *J. Chem. Soc. Faraday I* **69**, 216.

Billebaud, F., Crovisier, J., Lellouch, E. *et al.* (1991) "High-resolution infrared spectrum of CO on Mars: Evidence for emission lines", *Planet. Space Sci.* **39**, 213.

Bittner, H. and Fricke, K.H. (1987) "Dayside temperatures of the Martian upper atmosphere", *J. Geophys. Res.* **92**, 12045.

Biver, N., Bockelée-Morvan, D., Colom, P. *et al.* (1997) "Evolution of the outgassing of comet Hale-Bopp (C/1995 O1) from radio observations", *Science* **275**, 1915.

Bockelée-Morvan, D. (1987) "A model for the excitation of water in comets", *Astron. Astrophys.* **181**, 169.

Bockelée-Morvan, D. and Crovisier, J. (1987a) "The 2.7 μm water band of comet P/Halley: interpretation of observations by an excitation model", *Astron. Astrophys.* **187**, 425.

Bockelée-Morvan, D. and Crovisier, J. (1987b) "The role of water in the thermal balance of the coma", In *Proc. Symposium on the Diversity and Similarity of Comets*, ESA SP 277, 235.

Bougher, S.W., Dickinson, R.E., Ridley, E.C. *et al.* (1986) "Venus mesosphere and thermosphere. II. Global circulation, temperature and density variations", *Icarus* **68**, 284.

Bougher, S.W. and Dickinson, R.E. (1988) "Mars mesosphere and thermosphere, 1, Global mean heat budget and thermal structure", *J. Geophys. Res.* **93**, 7325.

Bougher, S.W. and Roble, R.G. (1991) "Comparative terrestrial planet thermospheres, 1, Solar cycle variation of global mean temperatures", *J. Geophys. Res.* **96**, 11045.

Bougher, S.W., Hunten, D.M. and Roble, R.G. (1994) "CO_2 cooling in the terrestrial planet thermospheres", *J. Geophys. Res.* **99**, 14609.

Bransden, B.H. and Joachain, C.J. (1982) *Physics of Atoms and Molecules*, Longman Pub. Group, London.

Brasseur, G.P. and Solomon, S. (1986) *Aeronomy of the Middle Atmosphere*, 2nd edition, D. Reidel, Dordrecht.

Brasseur, G.P., Orlando, J.J. and Geoffrey G.S. (1999) *Atmospheric Chemistry and Global Change*, Oxford University Press, Oxford.

Breen, J.E., Quy, R.B. and Glass, G.P. (1973) "Vibrational relaxation of O_2 in the presence of atomic oxygen", *J. Chem. Phys.* **59**, 556.

Brown, L.R., Gunson, M.R., Toth, R.A. *et al.* (1995) "1995 Atmospheric Trace Molecule Spectroscopy (ATMOS) linelist", *App. Opt.* **35**, 2828.

Brückelmann, H.G., Grossmann, K.U. and Offermann, D. (1987) "Rocket-borne measurements of atmospheric infrared emissions by spectrometric techniques", *Adv. Space Res.* **7**, (10)43.

Bucher, M.E. and Glinski, R.J. (1999) "Physical chemical control of molecular line profiles of CH^+, CH and CN in single homogeneous parcels on interstellar clouds", *Mon. Not. R. Astron. Soc.* **308**, 29.

Buchwald, M.I. and Wolga, G.J. (1975) "Vibrational relaxation of $CO_2(001)$ by atoms", *J. Chem. Phys.* **62**, 2828.

Bullitt, M.K., Bakshi, P.M., Picard, R.H. and Sharama, R.D. (1985) "Numerical and analytical study of high-resolution limb spectral radiance from nonequilibrium atmospheres", *J. Quant. Spectrosc. Radiat. Transfer* **34**, 33.

Bunker, P.R. and Jensen, P. (1998) *Molecular Symmetry and Spectroscopy*, 2nd edition, NRC Research Press, Ottawa.

Chapman, S. (1931) "The absorption and dissociative or ionozing effect of monochromatic radiation in an atmosphere on a rotating Earth. II, Grazing incidence", *Proc. Phys. Soc.* **43**, 483.

Caledonia, G.E., and Kennealy, J.P. (1982) "NO infrared radiation in the upper atmosphere", *Planet. Space Sci.* **30**, 1043.

Caledonia, G.E., Green, B.D. and Nadile, R.M. (1985) "The analysis of the SPIRE measurements of atmospheric limb $CO_2(\nu_2)$ fluorescence", *J. Geophys. Res.* **90**, 9783.

Cartwright, D.C., Brunger, M.J., Campbell, L. *et al.* (2000) "Nitric oxide excited under auroral conditions: Excited state densities and bands emissions", *J. Geophys. Res.* **105**, 20857.

Chalamala B.R. and Copeland, R.A. (1993) "Collisions dynamics of $OH(X^2\Pi, v = 9)$", *J. Chem. Phys.* **99**, 5807.

Chamberlain, J.W. and Hunten, D.M. (1987), *Theory of Planetary Atmospheres: An Introduction to their Physics and Chemistry*, 2nd edition, Academic Press, London.

Chamberlain, J.W. (1995), *Physics of the Aurora and Airglow*, 2nd edition, American Geophysical Union, Washington.

Chandrasekhar, S. (1960) *Radiative Transfer*, Dover Publications Inc., New York.

Chetwynd, J.G., Wang, J. and Anderson, G.P. (1994) "FASCODE: An update and applications in atmospheric remote sensing", *SPIE Proc.*, **2266**, *Optical Spectroscopic Techniques for Atmospheric Research.*

Chin, G. and Weaver, H.A. (1984) "Vibrational and rotational excitation of CO in comets: nonequilibrium calculations", *Astrophys. J.* **285**, 858.

Clarmann, v. T., Dudhia, A., Echle, G. *et al.* (1998) "Study on the simulation of atmospheric infrared spectra", *ESA Final Report*, ESA CR12054/96/NL/CN.

Clough, S.A., Kneizys, F.X., Shettle, E.P. and Anderson, G.P. (1986) "Atmospheric radiance and transmittance: FASCOD2", in Proceedings of the Sixth Conference on Atmospheric Radiation, Williamsburg, Va., 141.

Clough, S.A., Kneizys, F.X., Anderson, G.P. *et al.* (1989) "FASCOD3: Spectral simulation" in *IRS'88: Current Problems in Atmospheric Radiation*, J. Lenoble and J. F. Geleyin (Eds.), 372.

Combi, M.R. (1996) "Time-dependent gas kinetics in tenous planetary atmospheres: the cometary coma", *Icarus* **123**, 207.

Cosby, P.C., Slanger, T.G. and Osterbrock, D.E. (1999) "High rotational levels of OH observed in nightglow Meinel band emission", *EOS Trans. on AGU*, paper SA52B-06, Spring Meet. Suppl.

Cottrell, T.L., McCoubrey, J.C. (1961) *Molecular Energy Transfer in Gases*, Butterworths, London.

Coulson, K.L. (1975) *Solar and Terrestrial Radiation*, Academic Press, S. Diego, Calif.

Coustenis, A. and Taylor, F.W. (1999) *Titan: the Earth-like Moon*, World Scientific Publishing, Singapore.

Cox, A.N. (2000) *Allen's Astrophysical Quantities*, Springer.

Crisp, D., Fels, S.B. and Schwarzkopf (1986) "Approximate methods for finding CO_2 15 μm band transmission in planetary atmospheres", *J. Geophys. Res.* **91**, 11851.

Crisp, D. (1990) "Infrared radiative transfer in the dust-free martian atmosphere", *J. Geophys. Res.* **95**, 14577.

Crovisier, J. and Le Bourgot, J, (1983) "Infrared and microwave fluorescence of carbon monoxide in comets", *Astron. Astrophys.* **123**, 61.

Crovisier, J. and Encrenaz, Th. (1983) "Infrared fluorescence of molecules in comets: the general synthetic spectrum", *Astron. Astrophys.* **126**, 170.

Crovisier, J. (1984) "The water molecule in comets: fluorescence mechanisms and thermodynamics of the inner coma", *Astron. Astrophys.* **130**, 361.

Crovisier, J. (1987) "Rotational and vibrational synthetic spectra of the linear parent molecules in comets", *Astron. Astrophys. Suppl.* **68**, 223.

Crovisier, J., Leech, K., Bockelée-Morvan, D. *et al.* (1997) "The spectrum of comet Hale-Bopp (C/1995 O1) observed with the Infrared Space Observatory at 2.9 astronomical units from the Sun", *Science* **275**, 1904.

Crovisier, J., Leech, K., Bockelée-Morvan, D. *et al.* (1999) "The spectrum of comet Hale-Bopp as seen by ISO", in *The Universe as seen by ISO*, P. Cox and M.F. Kessler (Eds.), ESA SP–427, 137.

Crutzen, P.J. (1970) Comments on "Absorption and emission by carbon dioxide in the mesosphere", *Quart. J. Roy. Meteor. Soc.* **96**, 769.

Curtis, A.R. (1956) "The computation of radiative heating rates in the atmosphere", *Proc. R. Soc.* **A236**, 156.

Curtis, A.R. and Goody, R.M. (1956) "Thermal variation in the upper atmosphere", *Proc. Roy. Soc.* **A236**, 193.

Deming, D. and Mumma, M.J. (1983) "Modelling of the $10\,\mu$m natural laser emission from the mesospheres of Mars and Venus", *Icarus* **55**, 356.

Deming, D., Espenak, F., Jennings, D. *et al.* (1983) "Observations of the $10\,\mu$m natural laser emission from the mesospheres of Mars and Venus", *Icarus* **55**, 347.

DeMore, W.B., Margitan, J.J., Molina, M.J. *et al.* (1985) *Chemical Kinetics and Photochemical Data for use in Stratospheric Modelling*, Evaluation No. 7, JPL Pub. 85–37.

DeMore, W.B., Sander, S.P., Golden, D.M. *et al.* (1997) *Chemical Kinetics and Photochemical Data for use in Stratospheric Modelling*, Evaluation No. 12, JPL Pub. 97–4.

Dickinson, R.E. (1972) "Infrared radiative heating and cooling in the Venusian mesosphere, I, Global mean radiative equilibrium", *J. Atmos. Sci.* **29**, 1531.

Dickinson, R.E. (1973) "Method of parameterization for infrared cooling between altitudes of 30 and 70 kilometers", *J. Geophys. Res.* **78**, 4451.

Dickinson, R.E. (1975) "Meteorology of the upper atmosphere", *Rev. Geophys. Space Phys.* **13**, 771.

Dickinson, R.E. (1976) "Infrared radiative emission in the Venusian mesosphere", *J. Atmos. Sci.* **33**, 290.

Dickinson, R.E. (1984) "Infrared radiative cooling in the mesosphere and lower thermosphere", *J. Atmos. Terr. Phys.* **46**, 995.

Dickinson, R.E. and Bougher, S.W. (1986) "Venus mesosphere and thermosphere, 1, Heat budget and thermal structure", *J. Geophys. Res.* **91**, 70.

Dodd, J.A., Lipson, S.J. and Blumberg, W.A.M. (1991) "Formation and vibrational relaxation of $OH(X^2\Pi_i, v)$ by O_2 and CO_2", *J. Chem. Phys.* **95**, 5752.

Dodd, J.A., Winick, J.R., Blumberg, W.A.M. *et al.* (1993) "CIRRIS-1A observation of $C^{13}O^{16}$ and $C^{12}O^{18}$ fundamental band radiance in the upper atmosphere", *Geophys. Res. Lett.* **20**, 2683.

Dodd, J.A., Lockwood, R.B., Hwang, E.S. *et al.* (1999) "Vibrational relaxation of $NO(v=1)$ by oxygen atoms", *J. Chem. Phys.* **111**, 3498.

Donnelly, V.M. and Kaufman, F. (1977) "Fluorescence lifetime studies of NO_2. I. Excitation of the perturbed $^2B_2$ state near 600 nm", *J. Chem. Phys.* **66**, 4100.

Donnelly, V.M., Keil, D.G. and Kaufman, F. (1979) "Fluorescence lifetime studies of NO_2.

III. Mechanism of fluorescence quenching", *J. Chem. Phys.* **71**, 659.

Doyennette, L., Mastrocinque, G., Chakroun, A. *et al.* (1977) "Temperature dependence of the vibrational relaxation of $CO(v=1)$ by NO, O_2, and D_2, and of the self-relaxation of D_2", *J. Chem. Phys.* **67**, 3360.

Doyennette, L., Boursier, C., Menard, J., Menard-Bourcin, F. (1992) "v_1–v_2 Coriolis-assisted intermode transfers in O_3–M gas mixtures (M=O_2 and N_2) in the temperature range 200–300 K from IR double resonance measurements", *Chem. Phys. Lett.* **197**, 157.

Drossart, P., Rosenqvist, J., Encrenaz, Th. *et al.* (1993) "Earth global mosaic observations with NIMS-Galileo", *Planet. Space Sci.* **41**, 551.

Drossart, P., Fouchet, Th., Crovisier, J. *et al.* (1999) "Fluorescence in the 3 micron bands of methane on Jupiter and Saturn from ISO/SWS observations", in Proc. of *The Universe as seen by ISO*, ESA SP–427, 169.

Drummond, J.R., Houghton, J.T., Peskett, G.D. *et al.* (1980) "The Stratospheric and Mesospheric Sounder on Nimbus 7", *Phil. Trans. Roy. Soc. Lond.* **A296**, 219.

Drummond, J.R. and Mutlow, C.T. (1981) "Satellite measurements of H_2O fluorescence in the mesosphere", *Nature* **1294**, 431.

Dudhia, A. (2000) "Michelson Interferometer for Passive Atmospheric Sounding (MIPAS): Reference Forward Model (RFM) software user's manual, Oxford Univ., Oxford, (http://www.atm.ox.ac.uk/RFM/sum.html).

Duff, J.W., Bien, F. and Paulsen, D.E. (1994) "Classical dynamics of $N(^4S)+O_2(X^3\Sigma_g^-)$", *Geophys. Res. Lett.* **21**, 2043.

Duff, J.W. and Sharma, R.D. (1997) "Quasiclassical trajectory study of NO vibrational relaxation by collisions with atomic oxygen", *J. Chem. Soc., Faraday Trans.* **93**, 2645.

Duff, J.W. and Smith, D.R. (2000) "The $O^+(^4S)+N_2(X^1\Sigma_g^+) \rightarrow NO^+(X^1\Sigma^+)+N(^4S)$ reaction as a source of highly rotationally excited NO^+ in the thermosphere", *J. Atmos. Solar-Terr. Phys.* **62**, 1199.

Dushin, V.K., Zabelinskii, I.E. and Shatalov, O.P. (1988) "Deactivation of the molecular oxygen vibrations", *Sov. J. Chem. Phys.* **7**, 1320.

Edwards, D.P. (1987) "GENLIN2: The New Oxford Line-by-Line Atmospheric Transmission/Radiance Model", Hooke Institute for Cooperative Research, Clarendon Laboratory, Oxford.

Edwards, D.P. (1992) "GENLN2: A general line-by-line atmospheric transmittance and radiance model. Version 3.0 Description and users guide", *NCAR/TN-367+STR*, NCAR, Boulder Colo.

Edwards, D.P., López-Puertas, M. and López-Valverde, M.A. (1993) "Non-LTE studies of the 15-μm bands of CO_2 for atmospheric remote sensing", *J. Geophys. Res.* **98**, 14955.

Edwards, D.P., López-Puertas, M. and Mlynczak, M.G. (1994) "Non-Local thermodynamic equilibrium limb radiance from O_3 and CO_2 in the 9–11 μm spectral region", *J. Quant. Spectrosc. Radiat. Transfer* **52**, 389.

Edwards, D.P., Kumer, J.B., López-Puertas, M. *et al.* (1996) "Non-local thermodynamic equilibrium limb radiance near 10 μm as measured by CLAES", *J. Geophys. Res.* **101**, 26577.

Edwards, D.P., López-Puertas, M. and Gamache, R.R. (1998) "The non-LTE correction to the vibrational component of the internal partition sum for atmospheric calculations", *J. Quant. Spectrosc. Radiat. Transfer* **59**, 423.

Edwards, D.P., Halvorson, C.M. and Gille, J.C., (1999) "Radiative transfer modeling for EOS Terra satellite Measurement of Pollution in the Troposphere (MOPPIT) instrument", *J. Geophys. Res.* **104**, 16755.

Edwards, D.P. and Francis, G.L. (2000) "Improvements to the correlated-k radiative transfer method: Application to satellite infrared sounding", *J. Geophys. Res.* **105**, 18135.

Edwards, D.P., Zaragoza, G., Riese, M. and López-Puertas, M. (2000) "Evidence of H_2O non-local thermodynamic equilibrium emission near $6.4\,\mu m$ as measured by CRISTA–1", *J. Geophys. Res.* **105**, 29003.

Eisberg, R.M. and Resnick, R. (1985) *Quantum Physics of Atoms, Molecules, Solids, Nuclei, and Particles*, John Wiley and Sons, New York.

Eliasson, B., Hirth, M., Kogelschatz, U. (1987) "Ozone synthesis from oxygen in dielectric barrier discharges", *J. Phys. D: Appl. Phys.* **20**, 1421.

Elsasser, W.M. (1938) "Mean absorption and equivalent absorption coefficient of a band spectrum", *Phys. Rev.* **54**, 126.

Ellingson, R.G. and Gille, J.C. (1978) "An infrared radiative transfer model. Part 1: Model description and comparison of observations with calculations", *J. Atmos. Sci.* **35**, 523.

Encrenaz, T., Crovisier, J. and Combes, M. (1982) "A theoretical study of comet Halley's spectrum in the infrared range", *Icarus* **51**, 660.

Endemann, M., Lange, G. and Fladt, B. (1993) "Michelson interferometer for passive atmospheric sounding: A high resolution limb sounder for the european polar platform", *SPIE Proc.* **1934**, 13.

Espy, P.J., Harris, C.R., Steed, A.J. *et al.* (1988) "Rocket-borne interferometer measurement of infrared auroral spectra", *Planet. Space Sci.* **36**, 543.

Evans, W.F.J. and Shepherd, G.G. (1996) "A new airglow layer in the stratosphere", *Geophys. Res. Lett.* **23**, 3623.

Fairchild, C.E., Stone, E.J. and Lawrence, G.M. (1978) "Photofragment spectroscopy of ozone in the UV region 270–$310\,nm$ and at $600\,nm$", *J. Chem. Phys.* **69**, 3632.

Feautrier, P. (1964) "Sur la résolution numérique de l'équation de transfert", *Comptes Rendus Acad. Sci. Paris* **258**, 3189.

Fels, S.B. and Schwarzkopf, M.D. (1981) "An efficient, accurate algorithm for calculating CO_2 $15\,\mu m$ band cooling rates", *J. Geophys. Res.* **86**, 1205.

Fernando, R.P. and Smith, I.M.W. (1979) "Vibrational relaxation of NO by atomic oxygen", *Chem. Phys. Lett.* **66**, 218.

Fichet, P., Jevais, J.R., Camy-Peyret, C. and Flaud, J.M. (1992) "NLTE processes in ozone: Importance of O and O_3 densities near the mesopause", *Planet. Space Sci.* **40**, 989.

Finn, G.D. (1971) "Probabilistic radiative transfer", *J. Quant. Spectrosc. Radiat. Transfer* **11**, 203.

Finzi, J., Hovis, F.E., Panfilov, V.N. *et al.* (1977) "Vibrational relaxation of water vapor", *J. Chem. Phys.* **67**, 4053.

Flaud, J.-M., Camy-Peyret, C., Malathy-Devi, V. *et al.* (1987) "The ν_1 and ν_3 bands of $^{16}O_3$: line positions and intensities", *J. Mol. Spectrosc.* **124**, 209.

Flaud, J.-M., Camy-Peyret, C., Rinsland, C.P. *et al.* (1990) *Atlas of Ozone Line Parameters from Microwave to Medium Infrared*, Academic Press, New York.

Fleming, E.L., Chandra, S., Barnett, J.J. and Corney, M. (1990) "Zonal mean temperature, pressure, zonal winds and geopotential heights as function of latitude", *Adv. Space Res.* **10**, 1211.

Flynn, G.W., Parmenter, C.S. and Wodtke, A.M. (1996) "Vibrational energy transfer", *J. Phys. Chem.* **100**, 12817.

Fomichev, V.I., Shved, G.M. and Kutepov, A.A. (1986) "Radiative cooling of the 30–

110 km atmospheric layer", *J. Atmos. Terr. Phys.* **48**, 529.

Fomichev, V.I. and Shved, G.M. (1988) "Net radiative heating in the middle atmosphere", *J. Atmos. Terr. Phys.* **50**, 671.

Fomichev, V.I., Kutepov, A.A., Akmaev, R.A. and Shved, G.M. (1993) "Parameterization of the 15 μm CO_2 band cooling in the middle atmosphere (15–115 km)", *J. Atmos. Terr. Phys.* **55**, 7.

Fomichev, V.I. and Shved, G.M. (1994) "On the closeness of the middle atmosphere to the state of radiative equilibrium: an estimation of the net dynamical heating", *J. Atmos. Terr. Phys.* **56**, 479.

Fomichev, V.I., Blanchet, J.-P. and Turner, D.S. (1998) "Matrix parameterization of the 15 μm CO_2 band cooling in the middle and upper atmosphere for variable CO_2 concentration", *J. Geophys. Res.* **103**, 11505.

Frisch, U. and Frisch, H. (1975) "Non-LTE transfer. $\sqrt{\epsilon}$ revisited", *Mon. Not. R. Astron. Soc.* **173**, 167.

Fuller-Rowell, T.J. and Rees, D. (1996) "Numerical simulations of the distribution of atomic oxygen and nitric oxide in the thermosphere and upper mesosphere", *Adv. Space Res.* **18**, 255.

Funke, B. and López-Puertas, M. (2000) "Non-LTE vibrational, rotational, and spin state distribution for the NO(v=0,1,2) states under quiescent conditions", *J. Geophys. Res.* **105**, 4409.

Funke, B., López-Puertas, M., Stiller, G. *et al.* (2001) "A new non-LTE retrieval method for atmospheric parameters from MIPAS-Envisat emission spectra", *Adv. Space Res.*, in press.

Garcia, R.R. and Solomon, S. (1983) "A numerical model of the zonally averaged dynamical and chemical structure of the middle atmosphere", *J. Geophys. Res.* **88**, 1379.

Garcia, R.R., Stordal, F., Solomon, S. and Khiel, J.T. (1992) "A new numerical model for the middle atmosphere 1. Dynamics and transport of tropospheric source gases", *J. Geophys. Res.* **97**, 12967.

Garcia, R.R. and Solomon, S. (1994) "A new numerical model of the middle atmosphere 2. Ozone and related species", *J. Geophys. Res.* **99**, 12937.

Gérard, J.-C., Shematovich, V.I. and Bisikalo, D.V. (1991) "Non thermal nitrogen atoms in the Earth's thermosphere 2. A source of nitric oxide", *Geophys. Res. Lett.* **18**, 1695.

Gille, J.C. and Russell III, J.M. (1984) "The limb infrared monitor of the stratosphere (LIMS): Experiment description, performance and results", *J. Geophys. Res.* **89**, 5125.

Gille, J.C. and Lyjak, L.V. (1986) "Radiative heating and cooling rates in the middle atmosphere", *J. Atmos. Sci.* **43**, 2215.

Glinski, R.J. and Anderson, C.M. (2000) "Spectroscopy of NH_2 in comet Hale–Bopp: Nature of the non-LTE rotational distributions", in *Proc. 24th Meeting of IAU*, Aug. 2000, Manchester, England.

Golde, M.F. and Kaufman, F. (1974) "Vibrational emission of NO_2 from the reaction of NO with O_3", *Chem. Phys. Lett.* **29**, 480.

Goody, R.M. and Belton, M.J.S. (1967) "Radiative relaxation times for Mars", *Planet. Space Sci.* **15**, 247.

Goody, R.M. and Yung, Y.L. (1989) *Atmospheric Radiation: Theoretical Basis*, Oxford University Press, Oxford.

Goody, R.M. (1995) *Principles of Atmospheric Physics and Chemistry*, Oxford University Press, Oxford.

Gordiets, B.F., Markov, M.N. and Shelepin, L.A. (1978) "IR radiation of the upper atmosphere", *Planet. Space Sci.* **26**, 933.

Gordiets, B.F., Kulikov, Yu. N., Markov, M.N. and Marov, M. Ya. (1982) "Numerical modelling of the thermospheric heat budget", *J. Geophys. Res.* **87**, 4504.

Gordiets, B.F. and Panchenko, V. Ya. (1983) "Nonequilibrium infrared radiation and the natural laser effect in the atmospheres of Venus and Mars", *Cosmic Res.* **21**, 725.

Gordiets, B.F. and Markov, N.A. (1986) "Energetics and infrared radiation of NO in the disturbed heated thermosphere", *Cosmic Res.* **24**, 699.

Gordley, L.L., Marshall, B.T. and Chu, D.A. (1994) "LINEPAK: Algorithms for modelling spectral transmittance and radiance", *J. Quant. Spectrosc. Radiat. Transfer* **52**, 563.

Goss-Custard, M., Remedios, J.J., Lambert, A. *et al.* (1996) "Measurements of water vapour distributions by the improved stratospheric and mesospheric sounder: retrieval and validation", *J. Geophys. Res.* **101**, 9907.

Green, B.D, Rawlins, W.T. and Nadile, R.M. (1986) "Diurnal variability of vibrationally excited mesospheric ozone as observed during the SPIRE mission", *J. Geophys. Res.* **91**, 311.

Grossmann, K.U. and Offermann, D. (1978) "Atomic oxygen emission at 63 μm as a cooling mechanism in the thermosphere and ionosphere", *Nature* **276**, 594.

Grossmann, K.U., Barthol, P., Frings, W. *et al.* (1983) "A new spectroscopic measurement of atmospheric 63 μm emission", *Adv. Space Res.* **2**, 111.

Grossmann, K.U., Homann, D. and Schulz, J. (1994) "Lower thermosphere infrared emissions of minor species during high latitude twilight. Part A: Experimental results", *J. Atmos. Terr. Phys.* **56**, 1885.

Grossmann, K.U. and Vollmann, K. (1997) "Thermal infrared measurements in the middle and upper atmosphere", *Adv. Space Res.* **19**, 631.

Grossmann, K.U., Kaufmann, M. and Vollmann, K. (1997a) "Thermospheric nitric oxide infrared emissions measured by CRISTA", *Adv. Space Res.* **19**, 591.

Grossmann, K.U., Kaufmann, M. and Vollmann, K. (1997b) "The fine structure emission of thermospheric atomic oxygen", *Adv. Space Res.* **19**, 595.

Gueguen H., Yzambart, F., Chakroun, A. *et al.* (1975) "Temperature dependence of the vibration-vibration transfer rates from CO_2 and N_2O excited in the (0,0,0,1) vibrational level to $^{14}N_2$ and $^{15}N_2$ molecules", *Chem. Phys. Lett.* **35**, 198.

Hanel, R.A., Conrath B.J., Jennings, D.E., and Samuelson, R.E. (1992) *Exploration of the Solar System by Infrared Remote Sounding*, Cambridge University Press, London.

Hanson, W.B., Sanatani, S. and Zuccaro, D.R. (1977) "The martian ionosphere as observed by the Viking retarding potential analyzers", *J. Geophys. Res.* **82**, 4351.

Harris, R.D. and Adams, G.W. (1983) "Where does the $O(^1D)$ energy go?", *J. Geophys. Res.* **88**, 4918.

Haus, R. (1986) "Accurate cooling rates of the 15 μm CO_2 band: Comparison with recent parameterizations", *J. Atmos. Terr. Phys.* **48**, 559.

Hedin, A.E., Niemann, H.B., Kasprazak, W.T. and Seiff, A. (1983) "Global empirical model of the Venus thermosphere", *J. Geophys. Res.* **88**, 73.

Hedin, A.E. (1991) "Extension of the MSIS thermosphere model into the middle and lower atmosphere", *J. Geophys. Res.* **96**, 1159.

Herzberg, G. (1945) *Infrared and Raman Spectra*, D. Van Nostrand Company.

Herzberg, G. (1950) *Spectra of Diatomic Molecules*, D. Van Nostrand Company.

Herzfeld, K.T. and Litovitz, T.A. (1959) *Absorption and Dispersion of Ultrasonic Waves*, Academic Press, New York.

Hess, P., Kung, A.H. and Moore, C.B. (1980) "Vibration–vibration energy transfer in methane", *J. Chem. Phys.* **72**, 5525.

Hippler, H., Rahn, R. and Troe, J. (1990) "Temperature and pressure dependence of ozone formation rates in the range 1–1000 bar and 90–370 K", *J. Chem. Phys.* **93**, 6560.

Hitschfeld, W. and Houghton, J.T. (1961) "Radiative transfer in the lower stratosphere due to the 9.6 micron band of ozone", *Quart. J. Roy. Meteor. Soc.* **87**, 562.

Hochanadel, C.J., Ghormley, J.A. and Boyle, J.W. (1968) "Vibrationally excited ozone in the pulse radiolysis and flash photolysis of oxygen", *J. Chem. Phys.* **48**, 2416.

Holstein, T. (1947) "Imprisonment of resonance radiation in gases", *Phys. Rev.* **72**, 1212.

Höpfner, M., Stiller, G.P., Kuntz, M. *et al.* (1998) "The Karlsruhe optimized and precise radiative transfer algorithm. Part II: Interface to retrieval applications", *SPIE Proc.* **3501**, 186.

Houghton, J.T. (1969) "Absorption and emission by carbon dioxide in the mesosphere", *Quart. J. Roy. Met. Soc.* **95**, 1.

Houghton, J.T., Taylor, F.W. and Rodgers C.D. (1984) *Remote Sounding of Atmospheres*, Cambridge University Press, London.

Houghton, J.T. (1986) *The Physics of Atmospheres*, 2nd edition, Cambridge University Press, London.

Hourdin, F. (1992) "A new representation of the absorption by the CO_2 15 μm band for a martian general circulation model", *J. Geophys. Res.* **97**, 18319.

Huddleston, R.K. and Weitz, E. (1981) "A laser-induced fluorescence study of energy transfer between the symmetric stretching and bending modes of CO_2", *Chem. Phys. Lett.* **83**, 174.

Hui, K.K., Rosen, D.I. and Cool, T.A. (1975) "Intermode energy transfer in vibrationally excited O_3", *Chem. Phys. Lett.* **32**, 141.

Inoue, G. and Tsuchiya, S. (1975) "Vibration-vibration energy transfer of $CO_2(00^01)$ with N_2 and CO at low temperatures", *J. Phys. Soc. Jpn.* **39**, 479.

Ivanov, V.V. (1973) *Transfer of Radiation in Spectral Lines*, Nat. Bureau of Standards, Washington.

Iwagami, N. and Ogawa, T. (1982) "Thermospheric 63 μm emission of atomic oxygen in local thermodynamic equilibrium", *Nature* **298**, 454.

Jacquinet-Husson, N., Arié, E., Ballard, J. *et al.* (1999) "The 1997 spectroscopic GEISA databank", *J. Quant. Spectrosc. Radiat. Transfer* **62**, 205.

Joens, J.A., Burkholder, J.B. and Bair, E.J. (1982) "Vibrational relaxation in ozone recombination", *J. Chem. Phys.* **76**, 5902.

Joens, J.A. (1986) "Evidence for metastable ozone in the upper atmosphere?", *J. Geophys. Res.* **91**, 14553.

Johnson, M.A., Betz, A.L., McLaren, R.A. *et al.* (1976) "Non-thermal 10 microns CO_2 emission lines in the atmospheres of Mars and Venus", *Astrophys. J.* **208**, L145.

Jones, R.L. and Pyle, J.A. (1984) "Observations of CH_4 and N_2O by the NIMBUS 7 SAMS: A comparison with *in situ* data and two-dimensional numerical model calculations", *J. Geophys. Res.* **89**, 5263.

Kaplan, D.I. (1988) *Environment of Mars*, NASA Tech. Memo., 100470.

Käufl, H.U., Rothermel, H. and Drapatz, S. (1984) "Investigation of the Martian atmosphere by 10 micron heterodyne spectroscopy", *Astron. Astrophys.* **136**, 319.

Kaye, J.A. and Kumer, J.B. (1987) "Non-local thermodynamic equilibrium effects in the stratospheric NO and implications for infrared remote sensing", *Appl. Opt.* **26**, 4747.

Kaye, J.A. (1989) "Nonlocal thermodynamic equilibrium effects in the stratospheric HF by collisional energy transfer from electronically excited O_2 and implications for infrared remote sensing", *Appl. Opt.* **28**, 4161.

Keating, G.M. and Bougher, S.W. (1987) "Neutral upper atmospheres of Venus and Mars", *Adv. Space Res.* **7**, 57.

Keeling, C.D. and Whorf, T.P. (2000) "Atmospheric CO_2 records from sites in the SIO air sampling network", in *Trends: A Compendium of Data on Global Change*, Carbon Dioxide Information Analysis Center, Oak Ridge National Laboratory, U.S. Dept. of Energy, Oak Ridge, Tenn., U.S.A.

Kenner, R.D. and Ogryzlo, E. A. (1980) "Deactivation of $O_2(A^3\Delta_u^+)$ by O_2, O, and Ar", *Int. J. Chem. Kinet.* **12**, 501.

Kerridge, B.J. and Remsberg, E.E. (1989) "Evidence from the limb infrared monitor of the stratosphere for non-local thermodynamics equilibrium in the ν_2 mode of mesospheric water vapour and the ν_3 mode of stratospheric nitrogen dioxide", *J. Geophys. Res.* **94**, 16323.

Kiefer, J.H. and Lutz, R.W. (1967) "The effect of oxygen atoms on the vibrational relaxation of oxygen", *XIth Symposium on Combustion*, The Combustion Institute, Pittsburg, Penn., 67.

Kiehl, J.T. and Solomon, S. (1986) "On the radiative balance of the stratosphere", *J. Atmos. Sci.* **43**, 1525.

Kim, S.J., Geballe, T.R. and Noll, K.S. (2000) "Three-micrometer CH_4 line emission from Titan's high–altitude atmosphere", *Icarus* **147**, 588.

Kleindienst, T. and Bair, E.J. (1977) "Vibrational disequilibrium in bulk reaction systems", *Chem. Phys. Lett.* **49**, 338.

Kockarts, G. and Peetermans, W. (1970) "Atomic oxygen infrared emission in the Earth's upper atmosphere", *Planet. Space Sci.* **18**, 271.

Kockarts, G. (1980) "Nitric oxide cooling in the terrestrial thermosphere", *Geophys. Res. Lett.* **7**, 137.

Kondratiev, K. Ya. (1969) *Radiation in the Atmosphere*, Academic Press, New York.

Koutoulaki, K. (1998) *Study of Ozone Non-Thermal IR Emissions using ISAMS Observations*, D. Phil. Thesis, Univ. of Oxford, Oxford.

Krishna Swamy, K.S. (1997) "On the rotational population distribution of C_2 in comets", *Astrophys. J.* **481**, 1004.

Kudritzki, R.P. and Hummer D.G. (1990) "Quantitative spectroscopy of hot stars", *Ann. Rev. Astron. Astrophys.* **28**, 171.

Kuhn, W.R. and London, J. (1969) "Infrared radiative cooling in the middle atmosphere", *J. Atmos. Sci.* **26**, 189.

Kumer, J.B. and James, T.C. (1974) "$CO_2(001)$ and N_2 vibrational temperatures in the $50 < z < 130$ km altitude range", *J. Geophys. Res.* **79**, 638.

Kumer, J.B. (1975) "Summary analysis of $4.3\,\mu$m data", In *Atmospheres of the Earth and the Planets*, McCormac, B.M. (Ed.), 347, D. Reidel, Dordrecht.

Kumer, J.B. (1977a) "Atmospheric CO_2 and N_2 vibrational temperatures at 40- to 140-km altitude", *J. Geophys. Res.* **82**, 2195.

Kumer, J.B. (1977b) "Theory of the CO_2 4.3-μm aurora and related phenomena", *J. Geophys. Res.* **82**, 2203.

Kumer, J.B., Stair, A.T., Jr., Wheeler, N. *et al.* (1978) "Evidence for an $OH^* \xrightarrow{vv} N_2^* \xrightarrow{vv} CO_2(v_3) \xrightarrow{vv} CO_2 + h\nu(4.3\,\mu$m) mechanism for 4.3-$\mu$m airglow", *J. Geophys. Res.* **83**, 4743.

Kumer, J.B. and James, T.C. (1982) "Non-LTE calculation of HCl earthlimb emission and implication for detection of HCl in the atmosphere", *Geophys. Res. Lett.* **9**, 860.

Kumer, J.B., Mergenthaler, J. L., Roche, A.E. *et al.* (1989) "Prospects for retrieval of HF from high precision measurements of non-LTE solar enhanced stratospheric HF earthlimb emission near $2.5\,\mu$m" in *IRS'88: Current Problems in Atmospheric Radiation*, J. Lenoble and J. F. Geleyin (Eds.), 464.

Kung, R.T.V. (1975) "Vibrational relaxation of the N_2O v_1 mode by Ar, N_2, H_2O and NO", *J. Chem. Phys.* **63**, 5313.

Kurucz, R.L. (1993) "ATLAS9 stellar atmosphere programs and $2\,km\,s^{-1}$ grid", *Harvard-Smithsonian Center for Astrophysics*, CD-ROM No. 13.

Kutepov, A.A. (1978) "Parameterization of the radiant energy influx in the CO_2 $15\,\mu m$ band for Earth's atmosphere in the spoilage layer of local thermodynamic equilibrium", *Atmos. Ocean. Phys.* **14**, 154.

Kutepov, A.A. and Shved, G.M. (1978) "Radiative transfer in the $15\,\mu m$ CO_2 band with the breakdown of local thermodynamic equilibrium in the Earth's atmosphere", *Atmos. Ocean. Phys.* **14**, 18.

Kutepov, A.A., Hummer, D.G. and Moore, C.B. (1985) "Rotational relaxation of the $00^0 1$ level of CO_2 including radiative transfer in the $4.3\,\mu m$ band of planetary atmospheres", *J. Quant. Spectrosc. Radiat. Transfer* **34**, 101.

Kutepov, A.A., Kunze, D., Hummer, D.G., Rybicki, G.B. (1991) "The solution of radiative transfer problems in molecular bands without the LTE assumption by accelerated lambda iteration method", *J. Quant. Spectrosc. Radiat. Transfer* **46**, 347.

Kutepov, A.A. and Fomichev, V.I. (1993) "Application of the second-order escape probability approximation to the solution of the NLTE vibration-rotational band radiative transfer problem", *J. Atmos. Terr. Phys.* **55**, 1.

Kutepov, A.A., Oelhaf, H. and Fischer, H. (1997) "Non-LTE radiative transfer in the 4.7 and $2.3\,\mu m$ bands of CO: vibration-rotational non-LTE and its effects on limb radiance", *J. Quant. Spectrosc. Radiat. Transfer* **57**, 317.

Kutepov, A.A., Gusev, O.A. and Ogivalov, V.P. (1998) "Solution of the non-LTE problem for molecular gas in planetary atmospheres: superiority of accelerated lambda iteration", *J. Quant. Spectrosc. Radiat. Transfer* **60**, 199.

Lambert, J.D. (1977) *Vibrational and Rotational Relaxation in Gases*, Clarendon Press, Oxford.

Lambert, D.L., Sheffer, Y., Danks, A.C. *et al.* (1990) "High–resolution spectroscopy of the C_2 Swan 0–0 band from comet P/Halley", *Astrophys. J.* **353**, 640.

Lee, E.T.P., Picard, R.H., Winick, J.R. *et al.* (1991) "Non-LTE $4.3\,\mu m$ limb emission measured from STS-39", *EOS Trans.* **72** (44), Fall Meet. Suppl., 358.

Lellouch, E., Encrenaz, T., de Graauw, T. *et al.* (2000) "The $2.4–45\,\mu m$ spectrum of Mars observed with the Infrared Space Observatory", *Planet. Space Sci.* **48**, 1393.

Lenoble, J. (1993) *Atmospheric Radiative Transfer*, A. Deepak Publishing, Hampton, Virginia.

Leovy, C.B. (1977) "The atmosphere of Mars", *Sci. Am.* **237**, 34.

Lepoutre, F., Louis, G. and Manceau, H. (1977) "Collisional relaxation in CO_2 between $180\,K$ and $400\,K$ measured by the spectrophone method", *Chem. Phys. Lett.* **48**, 509.

Lewittes, M.E., Davis, C.C. and McFarlane, R.A. (1978) "Vibrational deactivation of $CO(v{=}1)$ by oxygen atoms", *J. Chem. Phys.* **69**, 1952.

Lindal, G.F., Wood, G.E., Hotz, H.B. *et al.* (1983) "The atmosphere of Titan: An analysis of the Voyager 1 radio occultation measurements", *Icarus* **53**, 348.

Liou, K.-N. (1980) *An Introduction to Atmospheric Radiation*, Academic Press, London.

Liou, K.-N. (1992) *Radiation and Clouds Processes in the Atmosphere*, Oxford University Press, Oxford.

Lipson, S.L., Armstrong, P.S., Dodd, J.A. *et al.* (1994) "Subthermal nitric oxide spin-orbit distributions in the thermosphere", *Geophys. Res. Lett.* **21**, 2421.

Llewellyn, E.J., McDade, I.C., Moorhouse, P. and Lockerbie, M.D. (1993) "Possible reference models for atomic oxygen in the terrestrial atmosphere", *Adv. Space Res.* **13**, (1)135.

Llewellyn, E.J. and McDade, I.C. (1996) "A reference model for atomic oxygen in the terrestrial atmosphere", *Adv. Space Res.* **18**, (9/10)209.

Locker, J.R. Joens, J.A. and Bairer, E.J. (1987) "Metastable intermediate in the formation of ozone by recombination", *J. Photochem.* **36**, 235.

López-González M.J. (1990) *Las Emisiones del Oxígeno Molecular y Atómico y del OH en la Atmósfera Media Terrestre*, Tesis Doctoral, Univ. de Granada, Granada, Spain.

López-González, M.J., López-Moreno, J.J. and Rodrigo, R. (1992a) "Altitude profiles of the atmospheric system of O_2 and of the green line emission", *Planet. Space Sci.* **40**, 783.

López-González, M.J., López-Moreno, J.J. and Rodrigo, R. (1992b) "The altitude profile of the infrared atmospheric system of O_2 in twilight and early night: Derivation of ozone abundances", *Planet. Space Sci.* **40**, 1391.

López-Moreno, J.J., Rodrigo, R., Moreno, F. *et al.* (1987) "Altitude distribution of vibrationally excited states of atmospheric hydroxyl at levels $v=2$ to $v=7$", *Planet. Space Sci.* **35**, 1029.

López-Puertas, M., Rodrigo, R., Molina, A. and Taylor, F.W. (1986a) "A non-LTE radiative transfer model for infrared bands in the middle atmosphere. I. Theoretical basis and application to CO_2 15 μm bands", *J. Atmos. Terr. Phys.* **48**, 729.

López-Puertas, M., Rodrigo, R., López-Moreno, J.J. and Taylor, F.W. (1986b) "A non-LTE radiative transfer model for infrared bands in the middle atmosphere. II. CO_2 (2.7 μm and 4.3 μm) and water vapour (6.3 μm) bands and $N_2(1)$ and $O_2(1)$ vibrational levels", *J. Atmos. Terr. Phys.* **48**, 749.

López-Puertas, M., Taylor, F.W. and López-Valverde, M.A. (1988) "Evidence for non-local thermodynamic equilibrium in the ν_3 mode of mesospheric CO_2 from Stratospheric and Mesospheric Sounder measurements", in *Progress in Atmospheric Physics*, R., López-Moreno, J.J., López-Puertas, M. and Molina, A. (Eds.), Kluwer Academic Pub., Dordrecht (Holland), 131.

López-Puertas, M. and Taylor, F.W. (1989) "Carbon dioxide 4.3-μm emission in the Earth's atmosphere. A comparison between NIMBUS 7 SAMS measurements and non-LTE radiative transfer calculations", *J. Geophys. Res.* **94**, 13045.

López-Puertas, M., López-Valverde, M.A. and Taylor, F.W. (1990) "Studies of solar heating by CO_2 in the upper atmosphere using a non-LTE model and satellite data", *J. Atmos. Sci.* **47**, 809.

López-Puertas, M., López-Valverde, M.A. and Taylor, F.W. (1992a) "Vibrational temperatures and radiative cooling of the CO_2 15 μm bands in the middle atmosphere", *Quart. J. Roy. Meteor. Soc.* **118**, 499.

López-Puertas, M., López-Valverde, M.A., Rinsland, C.P. and Gunson, M.R. (1992b) "Analysis of the upper atmosphere $CO_2(\nu_2)$ vibrational temperatures retrieved from ATMOS/Spacelab 3 observations", *J. Geophys. Res.* **97**, 20469.

López-Puertas, M., López-Valverde, M.A., Edwards, D.P. and Taylor F.W. (1993) "Non-LTE populations of the $CO(1)$ vibrational state in the middle atmosphere", *J. Geophys. Res.,* **98**, 8933.

López-Puertas, M., Wintersteiner, P.P., Picard, R.H. *et al.* (1994) "Comparison of line-by-line and Curtis matrix calculations for the vibrational temperatures and radiative cooling of the CO_2 15 μm bands in the middle and upper atmosphere", *J. Quant. Spectrosc. Radiat. Transfer* **52**, 409.

López-Puertas, M. and López-Valverde, M.A. (1995) "Radiative energy balance of CO_2 non-LTE infrared emissions in the Martian atmosphere", *Icarus* **114**, 113.

López-Puertas, M., Zaragoza, G., Kerridge, B.J. and Taylor, F.W. (1995) "Non-local thermodynamic equilibrium model for H_2O 6.3 and 2.7 μm bands in the middle atmosphere", *J. Geophys. Res.* **100**, 9131.

López-Puertas, M., Dudhia, A., Shepherd, M.G. and Edwards, D.P. (1997) "Evidence of non-LTE in the CO_2 15 μm weak bands from ISAMS and WINDII observations", *Geophys. Res. Lett.* **24**, 361.

López-Puertas, M. (1997) "Assessment of NO_2 non-LTE effects on MIPAS", *ESA Final Report*, ESA CR12054/96/NL/CN.

López-Puertas, M., Zaragoza, G., López-Valverde, M.A. and Taylor, F.W. (1998a) "Non-LTE atmospheric limb radiances at 4.6 μm as measured by UARS/ISAMS I. Analysis of the daytime radiances", *J. Geophys. Res.* **103**, 8499.

López-Puertas, M., Zaragoza, G., López-Valverde, M.A. and Taylor, F.W. (1998b) "Non-LTE atmospheric limb radiances at 4.6 μm as measured by UARS/ISAMS. II. Analysis of the daytime radiances", *J. Geophys. Res.* **103**, 8515.

López-Puertas, M., Zaragoza, G., López-Valverde, M.Á. *et al.* (1998c) "Non-local thermodynamic equilibrium limb radiances for the MIPAS instrument on Envisat–1", *J. Quant. Spectrosc. Radiat. Transfer* **59**, 377.

López-Puertas, M., López-Valverde, M.A., Garcia, R.R. and Roble, R.G. (2000) "A review of CO_2 and CO abundances in the middle atmosphere", in *Atmospheric Science Across the Stratopause*, Siskind, D.E., Eckermann, S.D. and Summers, M.E. (Eds.), Amer. Geophys. Union, Geophys. Monograph **123**, 83.

López-Valverde, M.A., López-Puertas, M., Marks, C.J. and Taylor, F.W. (1991) "Non-LTE modelling for the retrieval of CO abundances from ISAMS measurements", in *Optical Remote Sensing of the Atmosphere* **18**, 31, Washington.

López-Valverde, M.A., López-Puertas, M., Marks, C.J., and Taylor, F.W. (1993) "Global and seasonal variations in the middle atmosphere carbon monoxide from UARS/ISAMS", *Geophys. Res. Lett.* **20**, 124.

López-Valverde, M.A. and López-Puertas, M. (1994a) "A non-local thermodynamic equilibrium radiative transfer model for infrared emissions in the atmosphere of Mars, 1, Theoretical basis and nighttime populations of vibrational levels", *J. Geophys. Res.* **99**, 13093.

López-Valverde, M.A. and López-Puertas, M. (1994b) "A non-local thermodynamic equilibrium radiative transfer model for infrared emissions in the atmosphere of Mars, 2, Daytime populations of vibrational levels", *J. Geophys. Res.* **99**, 13117.

López-Valverde, M.A., López-Puertas, M., Remedios, J.J. *et al.* (1996) "Validation of measurements of carbon monoxide from the improved stratospheric and mesospheric sounder", *J. Geophys. Res.* **101**, 9929.

López-Valverde, M.Á., Edwards, D.P., López-Puertas, M. and Roldán, C. (1998) "Non-local thermodynamic equilibrium in general circulation models of the Martian atmosphere 1. Effects of the local thermodynamic equilibrium approximation on the thermal cooling and solar heating", *J. Geophys. Res.* **103**, 16799.

López-Valverde, M.A., Haberle, R.M. and López-Puertas, M. (2000) "Non-LTE radiative mesospheric study for Mars Pathfinder Entry", *Icarus* **146**, 360.

Lunt, S.L., Wickham-Jones, C.T. and Simpson, C.J.S.M. (1985) "Rate constants for the deactivation of the 15 μm band of carbon dioxide by the collisions partners CH_3F, CO_2, N_2, Ar and Kr over the temperature range 300 to 150 K", *Chem. Phys. Lett.* **115**, 60.

Makhlouf, U., Picard, R.H. and Winick, J.R. (1995) "Photochemical-dynamical modeling of the measured response of airglow to gravity waves 1. Basic model for OH airglow", *J. Geophys. Res.* **100**, 11289.

Manuilova, R.O. and Shved, G.M. (1985) "The 2.7 and 6.3 H_2O band emissions in the middle atmosphere", *J. Atmos. Terr. Phys.* **47**, 423.

Manuilova, R.O. and Shved, G.M. (1992) "The 4.8 and 9.6 μm band emissions in the middle atmosphere", *J. Atmos. Terr. Phys.* **54**, 1149.

Manuilova, R.O., Gusev, O.A., Kutepov, A.A. *et al.* (1998) "Modelling of non-LTE limb radiance spectra of IR ozone bands for the MIPAS space experiment", *J. Quant. Spectrosc. Radiat. Transfer* **59**, 405.

Maricq, M.M., Gregory, E.A. and Simpson, C.J.S.M. (1985) "Non-resonant V–V energy transfer between diatomic molecules at low temperatures", *Chem. Phys.* **95**, 43.

Martín-Torres, F.J., López-Valverde, M.A. and López-Puertas, M. (1998) "Non-LTE populations of methane and nitric acid for MIPAS/Envisat-1", *J. Atmos. Solar–Terr. Phys.* **60**, 1631.

Martín-Torres, F.J. and López-Puertas, M. (2002) "A non-LTE model for the O_3 vibrational levels in the middle atmosphere", *J. Geophys. Res.*, submitted.

McCleese, D.J., Haskins, R.D., Schofield, J.T. *et al.* (1992) "Atmosphere and climate studies of Mars using the Mars Observer Pressure Modulator Infrared Radiometer", *J. Geophys. Res.* **97**, 7735.

McDade, I.C. and Llewellyn, E.J. (1986) "The photodissociation of vibrationally excited ozone in the upper atmosphere", *J. Photochem.* **32**, 133.

McDade, I.C. (1998) "The photochemistry of the MLT oxygen airglow emissions and the expected influences of tidal perturbations", *Adv. Space Res.* **21**, 787.

McElroy, M.B., Kong, T.Y. and Yung, Y.L. (1997) "Photochemistry and evolution of Mars' atmosphere: A Viking perspective", *J. Geophys. Res.* **82**, 4379.

McNeal, R.J., Whitson, M.E., Jr. and Cook, G.R. (1974) "Temperature dependence of the quenching of vibrationally excited N_2 by atomic oxygen", *J. Geophys. Res.* **10**, 1527.

Menard J., Doyennette L., Menard-Bourcin, F. (1992) "Vibrational relaxation of ozone in O_3-O_2 and O_3-N_2 gas mixtures from infrared double-resonance measurements in the 200–300 K temperature range", *J. Chem. Phys.* **96**, 5773.

Menard-Bourcin, F., Menard, J. and Doyennette, L. (1990) "Vibrational energy transfers in ozone from infrared double-resonance measurements", *J. Chem. Phys.* **92**, 4212.

Menard-Bourcin F., Doyennette L., Menard, J. (1994) "Vibrational energy transfer in ozone excited into the (101) state from double-resonance measurements", *J. Chem. Phys.* **101**, 8636.

Mertens, C.J., Mlynczak, M.G., Garcia, R. and Portmann, R.W. (1999) "A detailed evaluation of the stratospheric heat budget 1. Radiation transfer", *J. Geophys. Res.* **104**, 6021.

Mertens, C.J., Mlynczak, M.G., López-Puertas, M. *et al.* (2001) "Retrieval of mesospheric and lower thermospheric kinetic temperatures from measurements of CO_2 15 μm Earth limb emission under non-LTE conditions", *Geophys. Res. Lett.* **28**, 1391.

Mihalas, D. (1978) *Stellar Atmospheres*, 2nd edition, Freeman and Co., San Francisco.

Milne, E.A. (1930) "Thermodynamics of stars", *Handbuch der Astrophysik*, **3**, Part I, Chap. 2, 65.

Mitzel, A.A. and Firsov, K.M. (1995) "A fast line-by-line method", *J. Quant. Spectrosc. Radiat. Transfer* **54**, 549.

Mlynczak, M.G. and Drayson, S.R. (1990a) "Calculation of infrared limb emission by ozone in the terrestrial middle atmosphere, 1. Source functions", *J. Geophys. Res.* **95**, 16497.

Mlynczak, M.G. and Drayson, S.R. (1990b) "Calculation of Infrared limb emission by ozone in the terrestrial middle atmosphere, 2. Emission calculations", *J. Geophys. Res.* **95**, 16513.

Mlynczak, M.G. and Solomon, S. (1993) "A detailed evaluation of the heating efficiency in the middle atmosphere", *J. Geophys. Res.* **98**, 10517.

Mlynczak, M.G., Zhou, D.K., López-Puertas, M. and Zaragoza, G. (1999) "Kinetic requirements for the measurement of mesospheric water vapour at $6.8\,\mu$m under non-LTE conditions", *Geophys. Res. Lett.* **26**, 63.

Mlynczak, M.G., Mertens, C.J., Garcia, R. and Portmann, R.W. (1999) "A detailed evaluation of the stratospheric heat budget 2. Global radiation balance and diabatic circulations", *J. Geophys. Res.* **104**, 6039.

Moore, C.B. (1973) "Vibration→vibration energy transfer", in *Advances in Chemical Physics*, Prigogine, I. and Rice, S.A. (Eds.) **23**, 41, John Wiley, New York.

Mumma, M.J., Buhl, D., Chin, G. *et al.* (1981) "Discovery of natural gain amplification in the $10\,\mu$m CO_2 laser bands on Mars: A natural laser", *Science* **212**, 45.

Mumma, M.J., DiSanti, M.A. Dello Russo, N. *et al.* (1996) "Detection of abundant ethane and methane, along with carbon monoxide and water, in comet C/1996 B2 Hyakutake: Evidence for interstellar origin", *Science* **272**, 1310.

Mumma, M.J., McLean, I.S., DiSanti, M.A. *et al.* (2000) "A survey of organic volatile species in comet C/1999 H1 (Lee) using NIRSPEC at the Keck observatory", *Astrophys. J.* **546**, 1183.

Murgatroyd, R.J. and Goody, R.M. (1958) "Sources and sinks of radiative energy from 30 to 90 km", *Quart. J. Roy. Meteor. Soc.* **84**, 225.

Murphy, A.K. (1985) *Satellite Measurements of Atmospheric Trace Gases*, D. Phil. Thesis, Univ. of Oxford, Oxford.

National Institution of Standards and Technology, NIST, (http://physics.nist.gov).

Nebel, H., Wintersteiner, P.P., Picard, R.H. *et al.* (1994) "CO_2 non-local thermodynamic equilibrium radiative excitation and infrared dayglow at $4.3\,\mu$m: Application to spectral infrared rocket experiment data", *J. Geophys. Res.* **99**, 10409.

Oelhaf, H., and Fischer, H. (1989) "Relevance of upper atmosphere non-LTE effects to limb emission of stratospheric constituents" in *IRS'88: Current Problems in Atmospheric Radiation*, J. Lenoble and J.-F. Geleyin (Eds.), 460.

Offermann, D. (1985) "The Energy Budget Campaign 1980: Introductory review, *J. Atmos. Terr. Phys.* **47**, 1.

Ogawa, T. (1976) "Excitation processes of infrared atmospheric emissions", *Planet. Space Sci.* **24**, 749.

Ogibalov, V.P. and Kutepov, A.A. (1989) "An approximate solution for radiative transfer in the $4.3\,\mu$m CO_2 band: Thick atmosphere with breakdown of rotational LTE", *Sov. Astron.* **33**, 260.

Ogibalov, V.P., Kutepov, A.A. and Shved, G.M. (1998) "Non-local thermodynamic equilibrium in CO_2 in the middle atmosphere. II. Populations in the ν_1–ν_2 mode manifold states", *J. Atmos. Solar–Terr. Phys.* **60**, 315.

Ogibalov, V.P. and Shved, G.M. (2001) "Non-local thermodynamic equilibrium in CO_2 in the middle atmosphere. III. Simplified models for the set of vibrational states", *J. Atmos. Solar–Terr. Phys.* **63**, in press.

Orr, B.J. and Smith, I.W.M. (1987) "Collision-induced vibrational energy transfer in small polyatomic molecules", *J. Phys. Chem.* **91**, 6106.

Parker, J.G. and Ritke, D.W. (1973) "Effect of ozone on the vibrational relaxation time of oxygen", *J. Chem. Phys.* **59**, 5725.

Patten, K.O., Jr., Burley, J.D. and Johnstone, H.S. (1990) "Radiative lifetimes of nitrogen dioxide for excitation wavelengths from 400 to 750 nm", *J. Phys. Chem.* **94**, 7960.

Pemberton, D.N.C. (1993) *Radiative Emission from O_3 and HNO_3 in the Middle Atmosphere*, D. Phil. Thesis, Univ. of Oxford, Oxford.

Pendleton, W.R., Jr., Espy, P.J. and Hammond, M.R. (1993) "Evidence for non-local thermodynamic equilibrium rotation in the OH nightglow, *J. Geophys. Res.* **98**, 11567.

Penner, S.S. (1959) *Quantitative Molecular Spectroscopy and Gas Emissivities*, Addison-

Wesley, Reading, Ma., USA.

Philbrick, C.R., Schmidlin, F.J., Grossmann, K.U. *et al.* (1985) "Density and temperature structure over northern Europe", *J. Atmos. Terr. Phys.* **47**, 159.

Picard, R.H., Winick, J.R., Sharma, R.D. *et al.* (1987) "Interpretation of infrared measurements of the high-latitude thermosphere from a rocket-borne interferometer", *Adv. Space Res.* **7**, (10)23.

Picard, R.H., Lee, E.T.P., Winick, J.R. *et al.* (1992) "STS-39 measurements of 4.3 μm Earthlimb emission from CO_2 and NO^+", *EOS Trans. on AGU* **73**, 222.

Picard, R.H., Inan, U.S., Pasko, V.P. and Winick, J.R. (1997) "Infrared glow above thunderstorms?", *Geophys. Res. Lett.* **24**, 2635.

Picard, R.H., O'Neil, R.R., Gardiner, H.A. *et al.* (1998) "Remote sensing of discrete stratospheric gravity-wave structure at 4.3-μm from the MSX satellite", *Geophys. Res. Lett.* **25**, 2809.

Planck, M. (1913) *Waermestrahlung*, 2nd edition, English translation by Morton Masius (1914): *The Theory of Heat Radiation*, Reprinted by Dover Publications, Inc., New York in 1959.

Pollock, D.S., Scott, G.B.I. and Phillips, L.F. (1993) "Rate constant for quenching of $CO_2(0,1^1,0)$ by atomic oxygen", *Geophys. Res. Lett.* **20**, 727.

Ramanathan, V. and Cess, R.D. (1974) "Radiative transfer within the mesospheres of Venus and Mars", *Astrophys. J.* **188**, 407.

Ratkowski, A.J., Picard, R.H., Winick, J.R. *et al.* (1994) "Lower–thermospheric infrared emissions from minor species during high-latitude twilight–B. Analysis of 15 μm emission and comparison with non-LTE models", *J. Atmos. Terr. Phys.* **56**, 1899.

Rawlins, W.T., Caledonia, G.E. and Kennealy, J.P. (1981) "Observation of spectrally resolved infrared chemiluminescence from vibrationally excited $O_3(v_3)$", *J. Geophys. Res.* **86**, 5247.

Rawlins, W.T. (1985) "Chemistry of vibrationally excited ozone in the upper atmosphere", *J. Geophys. Res.* **90**, 12283.

Rawlins, W.T., Caledonia, G.E. and Armstrong, R.A. (1987) "Dynamics of vibrationally excited ozone formed by three-body recombination reaction. II. Kinetics and mechanism", *J. Chem. Phys.* **87**, 5209.

Rawlins, W.T., Fraser, M.E. and Miller, S.M. (1989) "Rovibrational excitation of nitric oxide in the reaction of O_2 with metastable atomic nitrogen", *J. Phys. Chem.* **93**, 1097.

Rawlins, W.T., Woodward, A.M. and Smith, D.R. (1993) "Aeronomy of infrared ozone fluorescence measured during an aurora by the SPIRIT 1 rocket-borne interferometer", *J. Geophys. Res.* **98**, 3677.

Rees, D. and Fuller-Rowell, T.J. (1988) "Understanding the transport of atomic oxygen within the thermosphere using a numerical global thermospheric model", *Planet. Space Sci.* **36**, 935.

Rees, D. and Fuller-Rowell, T.J. (1993) "Comparison of empirical and theoretical models of species in the lower thermosphere", *Adv. Space Res.* **13**, 107.

Rees, M.H. (1989) *Physics and Chemistry of the Upper Atmosphere*, Cambridge University Press, London.

Remsberg, E. E., Bhatt, P.B., Eckman, R.S. *et al.* (1994) "Effect of the HITRAN 92 spectral data on the retrieval of NO_2 mixing ratios from Nimbus 7 LIMS", *J. Geophys. Res.* **99**, 22965.

Rinsland C.P., Gunson, M.R., Zander, R. and López-Puertas, M. (1992) "Middle and upper atmosphere pressure-temperature profiles and the abundances of CO_2 and CO in the upper atmosphere from ATMOS/Spacelab 3 observations", *J. Geophys.*

Res. **97**, 20479.

Rishbeth, H. and Roble, R.G. (1992) "Cooling of the upper atmosphere by enhanced greenhouse gases –modelling of thermospheric and ionospheric effects", *Planet. Space Sci.* **40**, 1011.

Rishbeth, H. and Clilverd, M.A. (1999) "Long-term change in the upper atmosphere", *Astron. Geophys.* **40**, 3.26.

Robertshaw J.S. and Smith I.W.M. (1980) "Vibrational energy transfer from $CO(v{=}1)$, $N_2(v{=}1)$, $CO_2(001)$, $N_2O(001)$ to O_3", *J. Chem. Soc. Faraday Trans.* **76**, 1354.

Roble, R.G. and Dickinson, R.E. (1989) "How will changes in carbon dioxide and methane modify the mean structure of the mesosphere and thermosphere?", *Geophys. Res. Lett.* **16**, 1441.

Roble, R.G. (2000) "On the feasibility of developing a global atmospheric model extending from the ground to the exosphere", in *Atmospheric Science Across the Stratopause*, Siskind, D.E., Eckermann, S.D. and Summers, M.E. (Eds.), Amer. Geophys. Union, Geophys. Monograph **123**, 53.

Rodgers, C.D. and Walshaw, C.D. (1966) "The computation of infrared cooling rates in planetary atmospheres", *Quart. J. Roy. Meteor. Soc.* **92**, 67.

Rodgers, C.D. and Williams, A.D. (1974) "Integrated absorption of a spectral line with the Voigt profile", *J. Quant. Spectrosc. Radiat. Transfer* **14**, 319.

Rodgers, C.D. (1976) "Approximate methods of calculating transmission by bands of spectral lines", NCAR Technical Note, NCAR/TN–116+IA, Boulder, Colorado.

Rodgers, C.D., Taylor, F.W., Muggeridge, A.H. *et al.* (1992) "Local thermodynamic equilibrium of carbon dioxide in the upper atmosphere", *Geophys. Res. Lett.* **19**, 589.

Rodgers, C.D. (2000) *Inverse Methods for Atmospheric Sounding: Theory and Practice*, World Scientific Publishing Co., Singapore.

Rodrigo, R., López-Puertas, M., Battaner, E. and López-Moreno, J.J. (1982) "CO_2 infrared bands in the Martian atmosphere" in *The Planet Mars*, Battrick, B. and Rolfe, E. (Eds.), ESA SP-185, 53, Noordwijk.

Rodrigo, R., López-Moreno, J. J., López-Puertas, M. *et al.* (1986) "Neutral atmospheric composition between 60 and 220 km: A theoretical model for middle latitudes", *Planet. Space Sci.* **34**, 723.

Rodrigo, R., López-Moreno, J.J., López-González, M.J. and García-Álvarez, E. (1989) "Atomic oxygen concentrations from OH and O_2 nightglow measurements", *Planet. Space Sci.* **37**, 49.

Rodrigo, R., García-Alvarez, E., López-González, M.J. *et al.* (1990) "A non-steady one-dimensional theoretical model of Mars neutral atmospheric composition between 30 and 200 km", *J. Geophys. Res.* **95**, 14795.

Rodrigo, R., López-González, M.J. and López-Moreno, J.J. (1991) "Variability of the neutral mesospheric and lower thermospheric composition in the diurnal cycle", *Planet. Space Sci.* **39**, 803.

Roldán, C., López-Valverde, M.A., López-Puertas, M. and Edwards, D.P. (2000) "Non-LTE infrared emissions of CO_2 in the atmosphere of Venus", *Icarus* **147**, 11.

Rosen, D.I. and Cool, T.A. (1973) "Vibrational deactivation of $O_3(101)$ in gas mixtures", *J. Phys. Chem.* **59**, 6097.

Rosen, D.I., and Cool, T.A. (1975) "Vibrational deactivation of O_3 in gas mixtures", *J. Phys. Chem.* **62**, 466.

Rosenberg, v. C.W., Jr. and Trainor, D.W. (1973) "Observations of vibrationally excited O_3 formed by recombination", *J. Chem. Phys.* **59**, 2142.

Rosenberg, v. C.W., Jr. and Trainor, D.W. (1974) "Vibrational excitation of ozone formed by recombination", *J. Chem. Phys.* **61**, 2442.

Rosenberg, v. C.W., Jr. and Trainor, D.W. (1975) "Excitation of ozone formed by recombination. II", *J. Chem. Phys.* **63**, 5348.

Rothman, L.S., Gamache R.R., Goldman A. *et al.* (1987) "The HITRAN database: 1986 edition", *Appl. Opt.* **26**, 4058.

Rothman, L.S., Gamache, R.R., Tipping, R.H. *et al.* (1992) "HITRAN Molecular database, Editions of 1991 and 1992", *J. Quant. Spectrosc. Radiat. Transfer* **48**, 469.

Rybicki, G.B. (1971) "A modified Feautrier method", *J. Quant. Spectrosc. Radiat. Transfer* **11**, 589.

Rybicki, G.B. and Lightmann, A.P. (1979) *Radiative Processes in Astrophysics*, John Wiley and Sons Inc., New York.

Rybicki, G.B. and Hummer, D.G. (1991) "An accelerated lambda iteration method for multilevel radiative transfer. I. Non-overlapping lines with background continuum", *Astron. Astrophys.* **262**, 171.

Salby M.L. (1996) *Fundamentals of Atmospheric Physics*, Academic Press, San Diego.

Schor, H.H.R. and Teixeira, E.L. (1994) "Fundamental rotational-vibrational band of CO and NO: teaching the theory of diatomic molecules", *J. Chem. Educ.* **71**, 771.

Schwartz, R.N., Slawsky, Z.I. and Herzfeld, K.F. (1952) "Calculation of vibrational relaxation times in gases", *J. Chem. Phys.* **20**, 1591.

Seiff, A. (1982) "Post-Viking models for the structure of the summer atmosphere of Mars", *Adv. Space Res.* **2**, 1.

Sharma, R.D. and Wintersteiner, P.P. (1985) "CO_2 component of daytime earth limb emission at 2.7 micrometers", *J. Geophys. Res.* **90**, 9789.

Sharma, R.D. and Wintersteiner, P.P. (1990) "Role of carbon dioxide in cooling planetary atmospheres", *Geophys. Res. Lett.* **17**, 2201.

Sharma, R.D., Zachor, A.S. and Yap, B.K. (1990) "Retrieval of atomic oxygen and temperature in the thermosphere. II–Feasibility of an experiment based on limb emission in the OI lines", *Planet. Space Sci.* **38**, 221.

Sharma, R.D., Sun, Y. and Dalgarno, A. (1993) "Highly rotationally excited nitric oxide in the terrestrial thermosphere", *Geophys. Res. Lett.* **19**, 2043.

Sharma, R.D., Zygelman, B., von Esse, F. and Dalgarno, A. (1994) "On the relationship between the population of the fine structure levels of the ground electronic state of atomic oxygen and the translational temperature", *Geophys. Res. Lett.* **21**, 1731.

Sharma, R.D., Kharchenko, V.A., Sun, Y. and Dalgarno, A. (1996a) "Energy distribution of fast nitrogen atoms in the nighttime terrestrial atmosphere", *J. Geophys. Res.* **101**, 275.

Sharma, R.D., Doethe, H. and von Esse, F. (1996b) "On the rotational distribution of the 5.3 μm thermal emission from nitric oxide in the nighttime terrestrial atmosphere", *J. Geophys. Res.* **101**, 17129.

Sharma, R.D., Doethe, H., von Esse, F. *et al.* (1996c) "Production of vibrationally and rotationally excited NO in the nighttime terrestrial atmosphere, *J. Geophys. Res.* **101**, 19707.

Sharma, R.D. and Duff, J.W. (1997) "Determination of the translational temperature of the high altitude terrestrial thermosphere from the rotational distribution of the 5.3 μm emission from NO(v=1)", *Geophys. Res. Lett.* **24**, 2407.

Sharma, R.D., Doethe, H. and Duff, J.W. (1998) "Model of the 5.3 μm radiance from NO during the sunlit terrestrial thermosphere", *J. Geophys. Res.* **103**, 14753.

Shi, J. and Barker, J.R. (1990) "Emission from ozone excited electronic states", *J. Chem. Phys.* **94**, 8390.

Shved, G.M. (1975) "Non-LTE radiative transfer in the vibration-rotation bands of linear molecules", *Sov. Astron.* **18**, 499.

Shved, G.M. and Bezrukova, L.L. (1976) "The diffusivity coefficient in problems of thermal radiation transfer", *Atmos. Ocean. Phys.* **12**, 545.

Shved, G.M., Stepanova, G.I. and Kutepov, A.A. (1978) "Transfer of 4.3 μm CO_2 radiation on departure from local thermodynamic equilibrium in the atmosphere of the Earth", *Atmos. Oceanic Phys.* **14**, 589.

Shved, G.M., Khvorostovskaya, L.E., Potekhin, I. Yu. *et al.* (1991) "Measurement of the quenching rate for collisions $CO_2(01^10)$–O: The importance of the rate constant magnitude for the thermal regime and radiation of the lower thermosphere", *Atmos. and Oceanic Phys.* **27**, 431.

Shved, G.M. and Gusev, O.A. (1997) "Non-local thermodynamic equilibrium in N_2O, CH_4, and HNO_3 in the middle atmosphere", *J. Atmos. Solar-Terr. Phys.* **59**, 2167.

Shved, G.M., Kutepov, A.A. and Ogibalov, V.P. (1998) "Non-local thermodynamic equilibrium in CO_2 in the middle atmosphere I. Input data and populations of the ν_3 mode manifold states", *J. Atmos. Solar-Terr. Phys.* **62**, 993.

Shved, G.M. and Ogibalov, V.P. (2000) "Natural population inversion for the CO_2 vibrational states in Earth's atmosphere", *J. Atmos. Solar-Terr. Phys.* **60**, 289.

Shved, G.M. and Semenov, A.O. (2001) "The standard problem of non-LTE radiative transfer in the rovibrational band of the planetary atmosphere", *Sol. Syst. Res.* **35**, 212.

Siddles, R.M., Wilson, G.J. and Simpson, C.J.S.M. (1994a) "The vibrational deactivation of the bending modes of CD_4 and CH_4 measured down to 90 K", *Chem. Phys.* **188**, 99.

Siddles, R.M., Wilson, G.J. and Simpson, C.J.S.M. (1994b) "The vibrational deactivation of the (0001) mode of N_2O measured down to 150 K", *Chem. Phys. Lett.* **225**, 146.

Simonneau, E. and Crivellari, L. (1993) "An implicit integral method to solve selected radiative transfer problems. I. Non-LTE line formation", *Astrophys. J.* **409**, 830.

Simons, J.W., Paur, R.J., Webster III, H.A. and Bair, E.J. (1973) "Ozone ultraviolet photolysis, VI, The ultraviolet spectrum", *J. Chem. Phys.* **59**, 1203.

Sivjee, G.G. (1992) "Airglow hydroxyl emissions", *Planet. Space Sci.* **40**, 235.

Slanger, T.G. and Black, G. (1979) "Interactions of $O_2(b^1\Sigma_g^+)$ with $O(^3P)$ and O_3", *J. Chem. Phys.* **70**, 3434.

Smith, D.R., Blumberg, W.A.M., Nadile, R.M. *et al.* (1992) "Observation of high-N hydroxyl pure rotation lines in atmospheric emission spectra by the CIRRIS-1A Space Shuttle experiment", *Geophys. Res. Lett.* **19**, 593.

Smith, D.R., and Ahmadjian, M. (1993) "Observation of nitric oxide rovibrational band head emissions in the quiescent airglow during the CIRRIS-1A space shuttle", *Geophys. Res. Lett.* **20**, 2679.

Smith, D.R., Huppi, E.R. and Wise, J.O. (2000) "Observation of highly rotationally excited NO^+ emissions in the thermosphere", *J. Atmos. Solar-Terr. Phys.* **62**, 1189.

Sobolev, V.V. (1975) *Light Scattering in Planetary Atmospheres*, Pergamon Press, Oxford.

Solomon, S., Kiehl, J.T., Kerridge, B.J. *et al.* (1986) "Evidence for non-local thermodynamic equilibrium in the ν_3 mode of mesospheric ozone", *J. Geophys. Res.* **91**, 9865.

Solomon, S., Schmeltekopf, A.L. and Sanders, R.W. (1987) "On the interpretation of zenith sky absorption measurements", *J. Geophys. Res.* **92**, 8311.

Solomon, S.C. (1991) "Optical aeronomy", in *U.S. National Report 1987-1990, Rev. of Geophys. Supp. AGU*, 1089.

Sparks, R.K., Carlson, L.R., Shobatake, J. *et al.* (1980) "Ozone photolysis: A determination of the electronic and vibrational state distributions of primary products", *J. Chem. Phys.* **72**, 1401.

Spitzer, L., Jr. (1949) "The terrestrial atmosphere above 300 km", in *The Atmospheres of the Earth and Planets*, G.P. Kuiper (Ed.), 213, Univ. Chicago Press, Chicago.

Stair, A.T., Jr., Ulwick, J.C., Baker, K.D. and Baker, D.J. (1975) "Rocketborne observations of atmospheric infrared emission in the auroral region". In *Atmospheres of the Earth and the Planets*, McCormac, B.M. (Ed.), 335, D. Reidel, Dordrecht.

Stair, A.T., Jr., Pritchard, J., Coleman, I. *et al.* (1983) "Rocketborne cryogenic (10 K) high-resolution interferometer spectrometer flight HIRIS: Auroral and atmospheric IR emission spectra", *App. Opt.* **22**, 1056.

Stair, A.T., Jr., Sharma, R.D., Nadile, R.M. *et al.* (1985) "Observations of limb radiance with cryogenic spectral infrared rocket experiment", *J. Geophys. Res.* **90**, 9763.

Starr, D.F. and Hancock, J.K. (1975) "Vibrational energy transfer in CO_2-CO mixtures from 163 to 406 K", *J. Chem. Phys.* **63**, 4730.

Steinfeld, J.I. (1985) *An Introduction to Modern Molecular Spectroscopy*, MIT Press, Ma.

Steinfeld, J.I., Adler-Golden, S.M. and Gallagher, J.W. (1987) "Critical survey of data on the spectroscopy and kinetics of ozone in the mesosphere and thermosphere", *J. Phys. Chem. Ref. Data* **16**, 911.

Stepanova, G.I. and Shved, G.M. (1985a) "Radiation transfer in the $4.3\,\mu$m CO_2 band and the $4.7\,\mu$m CO band in the atmospheres of Venus and Mars with violation of LTE: Populations of vibrational states", *Sov. Astron.* **29**, 422.

Stepanova, G.I. and Shved, G.M. (1985b) "The natural 10-μm CO_2 laser in the atmospheres of Mars and Venus", *Sov. Astron. Lett.* **11**, 162.

Stewart, A.L. and Hanson, W.B. (1982) "Mars upper atmosphere: Mean and variations", *Adv. Space Res.* **2**, 87.

Stiller, G.P., Höpfner, M., Kuntz, M. *et al.* (1998) "The Karlsruhe optimized and precise radiative transfer algorithm. Part I: requirements, justification, and model error estimation", *SPIE Proc.* **3501**, 257.

Streit, G.E. and Johnston, H.S. (1976) "Reactions and quenching of vibrationally excited hydroxyl radicals", *J. Chem. Phys.* **64**, 95.

Strickland, D.J. and Donahue, T.M. (1970) "Excitation and radiative transport of OI 1304 Å resonance radiation-I.", *Planet. Space Sci.* **18**, 661.

Strickland, D.J. and Anderson, D.E., Jr. (1977) "The OI 1304-Å nadir intensity/column production rate ratio and its implications to airglow studies", *J. Geophys. Res.* **82**, 1013.

Strobel, D.F., Zhu, X., and Summers, M.E. (1994) "On the vertical thermal structure of Io's atmosphere", *Icarus* **111**, 18.

Strobel, D.F., Zhu, X., Summers, M.E. and Stevens, M.H. (1996) "On the vertical thermal structure of Pluto's atmosphere", *Icarus* **120**, 266.

Swaminathan, P.K., Strobel, D.F., Kupperman, D.G. *et al.* (1998) "Nitric oxide abundances in the mesosphere/lower thermosphere region: Roles of solar soft X rays, suprathermal $N(^4S)$ atoms, and vertical transport", *J. Geophys. Res.* **103**, 11579.

Taine, J., Lepoutre, F. and Louis, G. (1978) "A photoacoustic study of the collisional deactivation of CO_2 by N_2, CO and O_2 between 160 and 375 K", *Chem. Phys. Lett.* **48**, 611.

Taine, J. and Lepoutre, F. (1979) "A photoacoustic study of the collisional deactivation of the first vibrational levels of CO_2 by N_2 and CO", *Chem. Phys. Lett.* **65**, 554.

Taine, J. and Lepoutre, F. (1980) "Determination of energy transferred to rotation-translation in deactivation of $CO_2(00^01)$ by N_2 and O_2 and of CO(1) by CO_2", *Chem. Phys. Lett.* **75**, 448.

Taylor, F.W. (1972) "Temperature sounding experiments for the Jovian planets", *J. Atmos. Sci.* **29**, 950.

Taylor, F.W., Beer, R., Chahine, M.T. *et al.* (1980) "Structure and meteorology of the

middle atmosphere of Venus: Infrared remote sounding from the Pioneer Orbiter",
J. Geophys. Res **85**, 7963.

Taylor, F.W. (1987) "Remote sounding of the middle atmosphere from satellites: The
Stratospheric and Mesospheric Sounder experiment on Nimbus 7", *Surv. Geophys.*
9, 123.

Taylor, F.W. and Dudhia, A. (1987) "Satellite measurements of middle atmosphere composition", *Phil. Trans. Roy. Soc. Lond.* **A323**, 567.

Taylor, F.W., Rodgers, C.D., Whitney, J.G. *et al.* (1993) "Remote sensing of atmospheric structure and composition by pressure modulator radiometry from space:
The ISAMS experiment on UARS", *J. Geophys. Res.* **98**, 10799.

Taylor, R.L. (1974) "Energy transfer processes in the stratosphere", *Can. J. Chem.* **52**,
1436.

Thekaekara, M.P. (1976) "Solar radiation measurement, techniques and instrumentation",
Solar Energy **18**, 309.

Thomas, G.E. (1963) "Lymann α scattering in the Earth's hydrogen geocorona, 1.", *J.
Geophys. Res.* **68**, 2639.

Thomas, R.N. (1965) *Some Aspects of Non-Equilibrium Thermodynamics in the Presence
of a Radiation Field*, Univ. of Colorado Press, Boulder.

Thomas, R.J. (1990) "Atomic hydrogen and atomic oxygen density in the mesopause
region: Global and seasonal variations deduced from Solar Mesosphere Explorer
near-infrared emissions", *J. Geophys. Res.* **95**, 16457.

Thorne, A.P. (1988) *Spectrophysics*, Chapman and Hall, London.

Timofeyev, Yu.M., Kostsov, V.S. and Grassl, H. (1995) "Numerical investigations of the
accuracy of the remote sensing of non-LTE atmosphere by space–borne spectral
measurements of limb IR radiation: 15 μm CO_2 bands, 9.6 μm O_3 bands and 10 μm
CO_2 laser bands", *J. Quant. Spectrosc. Radiat. Transfer* **53**, 613.

Titheridge, J.E. (1988) "An approximate form for the Chapman grazing incidence function", *J. Atmos. Terr. Phys.* **50**, 699.

Tobiska, W.K., Woods, T., Eparvier, F. *et al.* (2000) "The SOLAR 2000 empirical solar
irradiance model and forecast tool", *J. Atmos. Solar-Terr. Phys.* **62**, 1233.

Toumi, R., Kerridge, B.J. and Pyle, J.A. (1991) "Highly vibrationally excited oxygen as a
potential source of ozone in the upper stratosphere and mesosphere", *Nature* **351**,
217.

Tyuterev, V.G., Tashkun, S., Jensen, P. *et al.* (1999) "Determination of the effective
ground state potential energy function of ozone from high-resolution infrared spectra", *J. Molec. Spectros.* **198**, 57.

Ulwick, J.C., Baker, K.D., Stair, A.T., Jr. *et al.* (1985) "Rocket–borne measurements of
atmospheric infrared fluxes", *J. Atmos. Terr. Phys.* **47**, 123.

Upschulte, B.L., Green, B.D., Blumberg, W.A.M. and Lipson, S.J. (1994) "Vibrational
relaxation and radiative rates of ozone", *J. Chem. Phys.* **98**, 2328.

Van Hemelrijck, E. (1981) "Atomic oxygen determination from a nitric oxide point release
in the equatorial lower thermosphere", *J. Atmos. Terr. Phys.* **43**, 345.

Vlaskov, V.A. and Henriksen, Y.K. (1985) "Vibrational temperature and excess vibrational
energy of molecular nitrogen", *Planet. Space Sci.* **33**, 141.

Vollmann, K. and Grossmann, K.U. (1997) "Excitation of 4.3 μm CO_2 emissions by $O(^1D)$
during twilight", *Adv. Space Res.* **20**, 1185.

Wang, P.-H., Deepak, A. and Hong, S.-S. (1981) "General formulation of optical paths for
large zenith angles in the Earth's curved atmosphere", *J. Atmos. Sci.* **38**, 650.

Wayne, R.P. (1985) *Chemistry of Atmospheres: An Introduction to the Chemistry of the
Atmospheres of Earth, the Planets, and their Satellites*, Oxford University Press,

Oxford.

Weaver, H.A. and Mumma, M.J. (1984) "Infrared molecular emissions from comets", *Astrophys. J.* **276**, 782.

Wehrbein, W.M. and Leovy, C.B. (1982) "An accurate radiative heating and cooling algorithm for use in a dynamical model of the middle atmosphere", *J. Atmos. Sci.* **39**, 1532.

West, G.A., Weston, R.E., Jr. and Flynn, G.W. (1976) "Deactivation of vibrationally excited ozone by $O(^3P)$ atoms", *Chem. Phys. Lett.* **42**, 488.

West, G.A., Weston, R.E., Jr. and Flynn, G.W. (1978) "The influence of reactant vibrational excitation on the $O(^3P)+O_3^*$ bimolecular reaction rate", *Chem. Phys. Lett.* **56**, 429.

Whitson, M.E. and McNeal, R.J. (1977) "Temperature dependence of the quenching of the vibrationally excited N_2 by NO and H_2O", *J. Chem. Phys.* **66**, 2696.

Wilkes, M.V. (1954) "A table of Chapman's grazing incidence integral $Ch(x, \chi)$", *Proc. Phys. Soc.* **B67**, 304.

Williams, A.P. (1971a) *Radiative Transfer in the Mesosphere*, D. Phil. Thesis, Oxford University, Oxford.

Williams, A.P., (1971b) "Relaxation of the $2.7\,\mu m$ and $4.3\,\mu m$ bands of carbon dioxide", in *Mesospheric Models and Related Experiments*, Fiocco, G. (Ed.), 177, Reidel Publ. Com., Dordrecht, Holland.

Williams, A.P. and Rodgers, C.D. (1972) "Radiative transfer by the $15\,\mu m$ CO_2 band in the mesosphere", in *Proc. of Int. Radiation Symposium*, Sendai, Japan, 253.

Wilson, E.H. and Atreya, S.K. (2000) "Sensitivity studies of methane photolysis and its impact on hydrocarbon chemistry in the atmosphere of Titan", *J. Geophys. Res.* **105**, 20263.

Winick, J.R., Picard, R.H., Sharma, R.D. and Nadile, R.M. (1985) "Oxygen singlet delta 1.58-micrometer (0–1) limb radiance in the upper stratosphere and lower thermosphere", *J. Geophys. Res.* **90**, 9804.

Winick, J.R., Picard, R.H., Sharma, R.D. *et al.* (1987) "$4.3\,\mu m$ radiation in the aurorally dosed lower thermosphere: Modeling and analysis", in *Progress in Atmospheric Physics*, Rodrigo, R., López-Moreno, J.J., López-Puertas, M. and Molina, A. (Eds.), Kluwer Academic Pub., Dordrecht (Holland), 229.

Winick, J.R., Picard, R.H., Sharma, R.D. *et al.* (1988) "Radiative transfer effects on aurora enhanced 4.3 microns emission", *Adv. Space Res.* **7**, (10)17.

Winick, J.R., Picard, R.H., Wheeler, N.B. *et al.* (1991) "STS-39 measurement of $4.7\,\mu m$ limb emission from CO", *Eos Trans.* **72**, 375.

Winick, J.R., Picard, R.H., Makhlouf, U. *et al.* (1992) "Analysis of the $4.3\,\mu m$ limb emission observed from STS-39", *EOS Trans.* **73**, 418.

Wintersteiner, P.P., Picard, R.H., Sharma, R.D. *et al.* (1992) "Line-by-line radiative excitation model for the non-equilibrium atmosphere: Application to CO_2 $15\,\mu m$ emission", *J. Geophys. Res.* **97**, 18083.

Wintersteiner, P.P., Mertens, C.T., López-Puertas, M. *et al.* (2001) "Comparison of CO_2 15, 4.3, and $2.7\,\mu m$ models", *J. Geophys. Res.*, in preparation.

Wise, J.O., Carovillano, R.L., Carlson, H.C. *et al.* (1995) "CIRRIS 1A global observation of 15-μm CO_2 and 5.3-μm NO limb radiance in the lower thermosphere during moderate to active geomagnetic activity", *J. Geophys. Res.* **100**, 21357.

Wysong, I.J. (1994) "Vibrational relaxation of $NO(X^2\Pi, v = 3)$ by NO, O_2 and CH_4", *Chem. Phys. Lett.* **227**, 69.

Yamamoto, H. (1977) "Radiative transfer of atomic and molecular resonant emissions in the upper atmosphere II. The 9.6 micrometer emission of atmospheric ozone", *J.*

Geomag. Geoelectr. **29**, 153.

Yamamoto, T. (1982) "Evaluation of infrared line emission from constituents molecules of cometary nuclei", *Astron. Astrophys.* **109**, 326.

Yardley, J.T. (1980) *Introduction to Molecular Energy Transfer*, Academic Press, New York.

Yelle, R.V. (1991) "Non-LTE models of Titan's upper atmosphere", *Astrophys. J.* **383**, 380.

Yelle, R.V., Strobel, D.F., Lellouch, E. and Gautier, D. (1997) "Engineering models for Titan's atmosphere", *Eur. Space Agency Spec. Publ., ESA SP–1177*, 243.

Zachor, A.S. and Sharma, R.D. (1985) "Retrieval of non-LTE vertical structure from a spectrally resolved infrared limb radiance profile", *J. Geophys. Res.* **90**, 467.

Zachor, A.S., Sharma, R.D., Nadile, R.M. and Stair, A.T., Jr. (1985) "Inversion of a spectrally resolved limb radiance profile for the NO fundamental band", *J. Geophys. Res.* **90**, 9776.

Zachor, A.S. and Sharma, R.D. (1989) "Retrieval of atomic oxygen and temperature in the thermosphere. I–Feasibility of an experiment based on the spectrally resolved $147\,\mu$m limb emission", *Planet. Space Sci.* **37**, 1333.

Zaragoza, G., López-Puertas, M., Lambert, A. *et al.* (1998) "Non-local thermodynamic equilibrium in H_2O $6.9\,\mu$m emission as measured by the Improved Stratospheric and Mesospheric Sounder", *J. Geophys. Res.* **103**, 31293.

Zhou, D.K., Mlynczak, M.G., Bingham, G.E. *et al.* (1998) "CIRRIS-1A limb spectral measurements of mesospheric 9.6-μm airglow and ozone", *Geophys. Res. Lett.* **25**, 643.

Zhou, D.K., Mlynczak, M.G., López-Puertas, M. and Zaragoza, G. (1999) "Evidence of non-LTE effects in mesospheric water vapour from spectrally-resolved emissions observed by CIRRIS-1A", *Geophys. Res. Lett.* **26**, 67.

Zhu, X. (1990) "Carbon dioxide $15\,\mu$m band cooling rates in the upper middle atmosphere calculated by Curtis matrix interpolation", *J. Atmos. Sci.* **47**, 755.

Zhu, X. and Strobel, D.F. (1990) "On the role of vibration–vibration transitions in radiative cooling of the CO_2 $15\,\mu$m band around the mesopause", *J. Geophys. Res.* **95**, 3571.

Zhu, X., Summers, M.E. and Strobel, D.F. (1992) "Calculation of the CO_2 15-μm band atmospheric cooling rates by Curtis matrix interpolation of correlated-k coefficients", *J. Geophys. Res.* **97**, 12787.

Zittel, P.F. and Masturzo, D.E. (1989) "Vibrational relaxation of H_2O from 295 to 1020 K", *J. Chem. Phys.* **90**, 977.

Zuev, A.P. (1985) "Analysis of experimental data on vibrational relaxation of N_2O", *Sov. J. Chem. Phys.* **2**, 1516.

Zuev, A.P. (1989) "Shock-tube laser-schlieren measurements of V–T and V–V relaxation in mixtures of N_2O with N_2 and O_2", *Sov. J. Chem. Phys.* **4**, 2439.

Index